Designing Interaction and Interfaces for Automated Vehicles

Transportation and Human Factors

Aerospace, Aviation, Maritime, Railroad, and Roads

Series Editor:
Professor Neville A. Stanton
University of Southampton, UK

PUBLISHED TITLES

Automobile Automation
Distributed Cognition on the Road
Victoria A. Banks and Neville A. Stanton

Eco-Driving
From Strategies to Interfaces
Rich C. McIlroy and Neville A. Stanton

Driver Reactions to Automated Vehicles
A Practical Guide for Design and Evaluation
Alexander Eriksson and Neville A. Stanton

Systems Thinking in Practice
Applications of the Event Analysis of Systemic Teamwork Method
Paul Salmon, Neville A. Stanton, and Guy Walker

Individual Latent Error Detection (I-LED)
Making Systems Safer
Justin R.E. Saward and Neville A. Stanton

Driver Distraction
A Sociotechnical Systems Approach
Kate J. Parnell, Neville A. Stanton, and Katherine L. Plant

Designing Interaction and Interfaces for Automated Vehicles
User-Centred Ecological Design and Testing
Neville Stanton, Kirsten M.A. Revell, and Patrick Langdon

For more information about this series, please visit: https://www.crcpress.com/Transportation-Human-Factors/book-series/CRCTRNHUMFACAER

Designing Interaction and Interfaces for Automated Vehicles

User-Centred Ecological Design and Testing

Edited by
Neville A. Stanton, Kirsten M. A. Revell,
and Patrick Langdon

CRC Press
Taylor & Francis Group
Boca Raton London New York

CRC Press is an imprint of the
Taylor & Francis Group, an **informa** business

First edition published 2021
by CRC Press
6000 Broken Sound Parkway NW, Suite 300, Boca Raton, FL 33487-2742

and by CRC Press
2 Park Square, Milton Park, Abingdon, Oxon, OX14 4RN

Library of Congress Cataloging-in-Publication Data

Names: Stanton, Neville A. (Neville Anthony), 1960–editor. | Revell, Kirsten M. A., editor. | Langdon, Patrick, 1961–editor.
Title: Designing interaction and interfaces for automated vehicles : user-centred ecological design and testing / edited by Neville A Stanton, Kirsten M. A. Revell, and Pat Langdon.
Description: First edition. | Boca Raton, FL : CRC Press/Taylor & Francis Group, LLC, 2021. |
Series: Transportation human factors : aerospace, aviation, maritime, rail, and road | Includes bibliographical references and index. |
Summary: "Driving Automation and Autonomy is already upon us and the problems that were predicted twenty years ago are beginning to appear. These problems include shortfalls in expected benefits, equipment unreliability, driver skill fade, and error-inducing equipment designs. This book investigates the difficult problem of how to interface drivers with automated vehicles by offering an inclusive, human-centered design process that focuses on human variability and capability in interaction with interfaces. This book is for designers of systems interfaces, interactions, UX, Human Factors and Ergonomics researchers, and practitioners involved with systems engineering, and automotive academics"— Provided by publisher.
Identifiers: LCCN 2020047985 (print) | LCCN 2020047986 (ebook) |
ISBN 9780367466640 (hardback) | ISBN 9781003050841 (ebook)
Subjects: LCSH: Automated vehicles. | Human-machine systems. | Automobile driving—Human factors.
Classification: LCC TL152.8 .D475 2021 (print) | LCC TL152.8 (ebook) | DDC 629.20285/5437—dc23
LC record available at https://lccn.loc.gov/2020047985
LC ebook record available at https://lccn.loc.gov/2020047986

ISBN: 9780367466640 (hbk)
ISBN: 9781003050841 (ebk)

Typeset in Times
by CodeMantra

Contents

PART I Modelling

*Kirsten M. A. Revell, Patrick Langdon, Michael Bradley,
Ioannis Politis, James W.H. Brown, Simon Thompson,
Lee Skrypchuk, Alexandros Mouzakitis, and Neville A. Stanton*

*Michael Bradley, Nermin Caber, Patrick Langdon,
P. John Clarkson, Simon Thompson, Lee Skrypchuk,
Alexandros Mouzakitis, Ioannis Politis, Joy Richardson,
Jisun Kim, James W.H. Brown, Kirsten M. A. Revell, and
Neville A. Stanton*

*Ioannis Politis, Patrick Langdon, Michael Bradley, Lee Skrypchuk,
Alexandros Mouzakitis, P. John Clarkson, and Neville A. Stanton*

PART II Lo-Fi and Hi-Fi Simulators

*Ioannis Politis, Patrick Langdon, Damilola Adebayo,
Michael Bradley, P. John Clarkson, Lee Skrypchuk,
Alexandros Mouzakitis, Alexander Eriksson, James W.H. Brown,
Kirsten M. A. Revell, and Neville A. Stanton*

PART IV   HMI Simulator

PART V   On-Road and Design Guidelines

Preface

This project has developed a new approach for the design of handovers between the vehicle automation and the human driver. The approach brings together a range of methods from user-centred design, inclusive design, and ecological interface design. This approach has been used to design customisable human–machine interfaces for automation–driver handovers and has been formally tested in driving simulators and in a Jaguar I-PACE on a British motorway. Validation studies were undertaken, and design guidelines were produced. This project discovered that there were links between customisation and user performance. In particular, the handover times were quicker (on average, but with a greater range) with customised settings than the defaults. Drivers much preferred the customised settings to the defaults, and there was no adverse effect on post-handover driving performance (in terms of lane and speed stability). This is one of the first studies of its type in the UK and Europe, with a genuine automated vehicle driving (rather than being driven by a surrogate driver from the passenger seat or rear of the vehicle).

New insights into the customisation of handovers between vehicle automation and the human driver have been generated. A new method for the design of driver–vehicle interfaces has been developed, called 'user-centred ecological interface design' (UCEID). Modelling of the handovers has been validated by comparing the predictions made in operator event sequence diagrams with videos of the behaviour of human drivers, both in simulators and on the road. An on-road comparison study (Tesla, Mercedes, I-PACE) has revealed the differences between the strategies of manufacturers and shown the way this affects the human driver. The research is being published in a series of peer-reviewed journal articles and this book.

This project has uncovered a number of important factors associated with the development of automated vehicles. Numerous key design requirements that Jaguar Land Rover (JLR) can take forward into their development process have been established. These include operator event sequence diagrams that describe the process of handover. This project has also developed intuitive human–machine interface proposals and evaluations that could form part of future vehicle interfaces. There are also novel findings around the topic of customisation that will potentially increase customer satisfaction with handovers in certain automated vehicle scenarios. This project has also created specific intellectual property for JLR that could be used in future vehicles related to the handover from automation to manual driving. Additionally, a process that goes from the basic scientific foundation through to the finalised, and evaluated, concept following a series of design iterations, called the 'UCEID process', has been defined and shared with JLR for use in future vehicle programme design. Guidelines for the design of handovers between the vehicle automation and the human driver have also been presented in the final chapter of this book.

Prof Neville A. Stanton, PhD, DSc
University of Southampton

Acknowledgements

This research and development work was supported by Jaguar Land Rover and the UK-EPSRC Grant EP/N011899/1 as part of the £11 million jointly funded 'Towards Autonomy: Smart and Connected Control (TASCC) Programme'. This project was led by Professor Neville A. Stanton at the University of Southampton with project partners at the University of Cambridge. We are very grateful for all those drivers who gave up their time to participate in our studies, both in driving simulators and in semi-automated vehicles on the road. The authors are also grateful to our intern, Charlie Davenport, for her assistance with the production of the materials in this book. Any errors or omissions that remain are entirely due to our own performance variability.

Editors

Professor Neville A. Stanton, PhD, DSc, is a Chartered Psychologist, Chartered Ergonomist, and Chartered Engineer. He holds the Chair in Human Factors Engineering in the Faculty of Engineering and the Environment at the University of Southampton in the UK. He has degrees in Occupational Psychology, Applied Psychology, and Human Factors Engineering, and has worked at the Universities of Aston, Brunel, Cornell, and MIT. His research interests include modelling, predicting, analysing, and evaluating human performance in systems as well as designing the interfaces and interaction between humans and technology. Professor Stanton has worked on design of automobiles, aircraft, ships, and control rooms over the past 30 years, on a variety of automation projects. He has published 50 books and over 400 peer-reviewed journal papers on Ergonomics and Human Factors. In 1998, he was presented with the Institution of Electrical Engineers Divisional Premium Award for research into System Safety. The Institute of Ergonomics and Human Factors in the UK awarded him The Otto Edholm Medal in 2001, The President's Medal in 2008 and 2018, The Sir Frederic Bartlett Medal in 2012, and The William Floyd Medal in 2019 for his contributions to basic and applied ergonomics research. The Royal Aeronautical Society awarded him and his colleagues the Hodgson Prize in 2006 for research on design-induced, flight-deck error published in *The Aeronautical Journal*. The University of Southampton has awarded him a Doctor of Science in 2014 for his sustained contribution to the development and validation of human factors methods.

Dr Kirsten M. A. Revell is a Human Factors Engineering research fellow in the Transportation Research Group (TRG) in the Faculty of Engineering and the Environment at the University of Southampton. She has degrees in both Psychology and Industrial Design from Exeter and Brunel University London, respectively, a PhD in Human Factors from the University of Southampton, and experience working in industry in Microsoft Ltd. Dr Revell passionately believes that human factors can offer solutions to the critical global issues we face today. Her focus is on using HF methods to show how systems shape lives, highlighting where change is needed to promote an inclusive and sustainable world. She leads the autonomous vehicle domain in the Human Factors Engineering (HFE) team and has conducted research in military, domestic energy, rail, and aviation domains collaborating extensively with government bodies and industry. She is a member of the HF and Sustainable Development Technical Committee and has been jointly awarded the Annual Aviation Safety Prize by the Honourable Company of Air Pilots and the Air Pilots Trust.

Professor Patrick Langdon is a Professor of Engineering Design, Transportation, and Inclusion at Edinburgh Napier School of Engineering and the Built Environment (SEBE). He is an Experimental Psychologist and has worked in AI, Robotics, and Engineering Design for over 20 years. Historically, he has led research in Inclusive

Design and contributed to its literature. His current research concerns the application of cognitive science and AI to multidisciplinary areas of research such as the design, use, and provision for autonomous vehicles in current transportation planning. Multidisciplinary research centred on inclusion, transport, and design engineering are key themes in his work, and his goal is to foster new programmes of research in these areas.

Contributors

Damilola Adebayo
Engineering Design Centre
University of Cambridge
Cambridge, United Kingdom

Theocharis Amanatidis
Engineering Design Centre
University of Cambridge
Cambridge, United Kingdom

Michael Bradley
Engineering Design Centre
University of Cambridge
Cambridge, United Kingdom

James W.H. Brown
Transportation Research Group
University of Southampton
Southampton, England

Nermin Caber
Engineering Design Centre
University of Cambridge
Cambridge, United Kingdom

Jediah R. Clark
Transportation Research Group
University of Southampton
Southampton, England

P. John Clarkson
Engineering Design Centre
University of Cambridge
Cambridge, United Kingdom

Alexander Eriksson
Transportation Research Group
University of Southampton
Southampton, United Kingdom

Jisun Kim
Transportation Research Group
University of Southampton
Southampton, United Kingdom

Patrick Langdon
School of Engineering and the
    Built Environment
Edinburgh Napier University
Edinburgh, Scotland

Alexandros Mouzakitis
Vehicle Engineering, Infotainment and
    Connectivity Research Department
Jaguar Land Rover
Coventry, United Kingdom
and
University of Cambridge
Cambridge, United Kingdom

Jim O'Donoghue
Research Autonomy
Jaguar Land Rover
Coventry, United Kingdom

Ioannis Politis
User Experience
MathWorks Inc.
Cambridge, United Kingdom

Kirsten M. A. Revell
Transportation Research Group
University of Southampton
Southampton, United Kingdom

Joy Richardson
Transportation Research Group
University of Southampton
Southampton, United Kingdom

Lee Skrypchuk
Human Science Research
Jaguar Land Rover
Coventry, United Kingdom
and
Engineering Design Centre
University of Cambridge
Cambridge, United Kingdom

Neville A. Stanton
Transportation Research Group
University of Southampton
Southampton, United Kingdom

Simon Thompson
Human Science Research
Jaguar Land Rover
Coventry, United Kingdom

Abbreviations

AAI	absolute angle input
ACC	adaptive cruise control
ADAS	advanced driver assistance systems
AH	abstraction hierarchy
ANOVA	analysis of variance
AS	acceptance scale
AV	autonomous vehicle
BU	bottom-up
CAN	controller area network
CB	countdown-based
CC	counter cycle
ConTA	control task analysis
CR	correct rejection
CTP	control transfer protocols
CTS	customised takeover settings
CWA	cognitive work analysis
DiP	dialogue performance
DMG	design methodological guidelines
DTS	default takeover settings
EID	ecological interface design
EMEIA	Europe, Middle East, India, and Africa
ERGO	Ethics and Research Governance Office
FA	false alarm
HAD	highly automated driving
HazLanSEA	hazards, lane, speed, exits, and actions
HDD	head-down display
HF	human factors
HI:DAVe	Human Interaction: Designing Autonomy in Vehicles
HMI	human–machine interface
HUD	head-up display
ID	infotainment display
IDG	interface and interaction design guidelines
ISO	International Organisation for Standardisation
IVIS	in-vehicle information systems
JLR	Jaguar Land Rover
M	mean
MB	multimodality-based
MR	monitoring request
NASA-TLX	NASA task load index
OEM	original equipment manufacturer
OESD	operator event sequence diagram
OOTL	out of the loop

PA	perceived acceptance
PCM	Perceptual Cycle Model
PU	perceived usability
PW	perceived workload
R	range
RepB	repetition-based
ResB	response-based
SA	situation awareness
SAE	Society of Automotive Engineers
SART	situation awareness rating technique
SD	standard deviation
SDT	signal detection theory
SOCA	social organisation and cooperation analysis
SRK	skills, rules, and knowledge
SUDS	Southampton University driving simulator
SUS	system usability scale
TD	top-down
TFT	thin film transistor
TOC	transfer of control
ToI	type of interface
TOR	takeover request
TORlt	takeover request lead time
TORrt	takeover reaction time
TOT	takeover time
UCD	user-centred design
UCEID	user-centred ecological interface design
UTG	user trials guidelines
UX	user experience
VAA	voice autopilot assistant
VP	verbal protocols
VtD	vehicle to driver
WCA	worker competencies analysis
WDA	work domain analysis
WOZ	Wizard of Oz

Part I

Modelling

1 UCEID – The Best of Both Worlds

Combining Ecological Interface Design with User-Centred Design in a Novel Human Factors Method Applied to Automated Driving

Kirsten M. A. Revell
University of Southampton

Patrick Langdon
Edinburgh Napier University

Michael Bradley
University of Cambridge

Ioannis Politis
The MathWorks Inc.

James W.H. Brown
University of Southampton

Simon Thompson, Lee Skrypchuk, and Alexandros Mouzakitis
Jaguar Land Rover

Neville A. Stanton
University of Southampton

CONTENTS

1.1   INTRODUCTION

Human factors (HF) methods exist to tackle problems relating to the interaction between human and other elements of the system – i.e. existing methods of design, evaluation, and procurement have failed to address. These types of problems are often resistant to purely technical interventions, resulting in less effective system performance (Stanton et al. 2013). With ever-increasing rates of technological advancement, it is more difficult for companies to compete on functionality, reliability, or cost (Green and Jordan 1999). HF methods offer a means to provide a competitive edge by harnessing technology to enable people to accomplish meaningful, real-world tasks.

HF methods fall into a range of categories that are relevant for application at different parts of the design process (Stanton et al. 2013). The UCEID process includes a combination of 'data collection', 'task analysis', and 'cognitive work analysis (CWA)' techniques. Figure 1.1 shows how different methods are suitable at different stages of the design process. The UCEID method is positioned early in the design process to allow 'analytical prototyping', the means of applying HF insights to systems or designs that are yet to exist in physical form. It covers a combination of 'identify needs' and 'developing concept' stages of the design process taking the analyst to 'initial design concept' stage, not final concept. A key finding from inclusive user-centred design (UCD) advises an active process of linked iteration between technology prototypes and user trials is necessary to meet the dual needs of diverse user demographics and technology delivery requirements (Langdon et al. 2014) (see Figure 1.1).

1.1.1   WHY USE UCEID?

UCEID is a novel HF method that integrates relationships between ecological interface design (EID) (McIlroy and Stanton 2015) and inclusive human-centred design

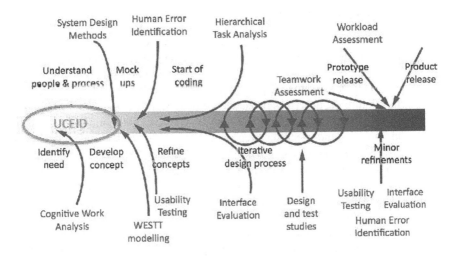

FIGURE 1.1 Diagram to show where the UCEID method fits into the design process in relation to other HF methods. (Amended from Stanton et al. 2013.)

by combining the existing methodology from the CWA framework (Rasmussen, Pejtersen and Goodstein 1994; Vicente 1999; Jenkins et al. 2008) and inclusive UCD (Czajkowski et al. 2001; Langdon and Thimbleby 2010). EID is based on Gibsonian methodology that aims to make constraints of the system and environment explicit, so that the appropriate action is apparent to the system user (McIlroy and Stanton 2015). While both EID and UCD emphasise the user at each stage of the design process, they differ in their approach. UCD has a greater focus on end user wants, needs, and limitations within single-user actions, whereas EID focuses on incorporating user wants and needs within a complex system, constrained by the values and purpose of the overall system. Some of the values relate to stakeholders that may at times be in conflict with the end user wants and needs. Both methods aim to provide the user with a visual 'mental model' to guide possible action (Norman 2013), but EID's remit within complex systems extends to providing a mental model that enables the user to troubleshooting unanticipated events (Burns and Hajdukiewicz 2004). UCDs focus on usability and can ensure solutions to meet the EID remit are easy to use and learn. EIDs focus on values and constraints based on the overall purpose of the system, and can ensure that the design solutions proposed by UCD are relevant and systematically prioritised. The UCEID approach engages with stakeholders, subject matter experts, and users to produce outputs that generate design requirements. Initial design concepts are then produced following a design workshop and concept filtering activity.

The UCEID method is best suited to complex sociotechnical systems where the user plays a critical role in the interaction. Domains that are complex exude some of the following qualities: high risk, dynamic, uncertain, with interconnected parts. Vehicle-initiated, vehicle-to-driver takeover in an SAE Level 3 (SAE International 2018) autonomous vehicle fits the criteria of a complex sociotechnical system and will be used to illustrate the application of this method.

1.2 THE UCEID METHOD

The UCEID method is compiled from a rich range of activities, starting with defining the scenario and aims of analysis, and ending with the generation of design concepts. Following the recommended criteria for depicting a method (Stanton et al. 2013), Figure 1.2 depicts a step-by-step process that can be followed 'like a recipe'. The flowchart in Figure 1.2 describes the sequence of 16 steps (rectangles), including literature review, data collection, thematic analysis, CWA, consolidation and ideas generation, and filtering and checking. Steps within boxes can be undertaken in parallel, and decision points (diamonds) and feedback loops occur at different points in the process. The type of activity is shown by the line style (Figure 1.2). Not all aspects of the method need to be done at once, or to the same level of detail. The method can be reapplied following testing cycles as the user needs and domain constraints become more clearly specified. This section will summarise the different types of activities in the UCEID flowchart to provide the reader with examples of the types of outputs at different stages. Before embarking on the specific activities, the aims of analysis, target scenario, and boundaries of analysis need to be clearly defined – e.g. analysis of *'transfer of control in autonomous vehicles'* for the scenario *'planned transfer of control from vehicle to driver on a highway in a SAE Level 3 vehicle'*.

1.2.1 LITERATURE REVIEW

Literature review occurs primarily at *Step 2. Initial Research* (Figure 1.2) but can continue throughout the process as needed. For novel systems, the domain under analysis and the parallel systems that have relevant features to the scenario should be considered. This can be through document analysis (such as user guides, reports), which will depend on what is currently available for the domain. Where possible, observation of the work in context is highly recommended. However, focus purely on current systems can risk producing design proposals that are simply modifications of current technology or practice. The objective of inclusive design step is to establish the opinions and preferences of likely users towards the target technology to be designed. Different members of a research group can undertake 'researching domain and parallel domains' (e.g. transfer of control in the aviation domain (SKYbrary 2011) and requirements of advanced driving (Roadcraft 2020)) and 'inclusive design' (e.g. reference to literature and national surveys relating to user needs for automation (Wockatz and Schartau 2015)), in parallel within the scope of the research. This initial research informs all subsequent steps of the UCEID method.

1.2.2 DATA COLLECTION

Data collection occurs at *Step 3. Initial Data Collection* and *Step 5. Focus Groups* (Figure 1.2). In Step 3, the different types of data collection can again occur in parallel. For the chosen scenario, subject matter expert, semi-structured interviews would occur with automation experts. 'Technology analysis and benchmarking' would be used to understand current commercial offerings and relevant technologies relating to autonomous vehicles. 'User interviews' would occur with target users

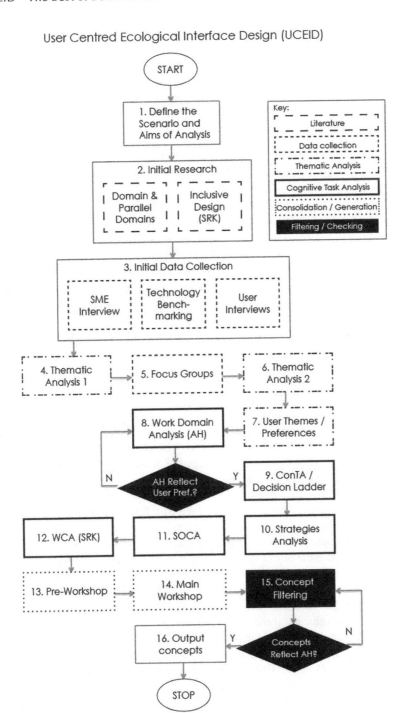

FIGURE 1.2 Flowchart showing the sequence of steps and range of activities for the UCEID method.

of autonomous vehicles following standard approaches (Glaser and Strauss 1967; Breakwell, Hammond and Fife-Schaw 2000). In Step 5, focus groups representing an inclusive range of the target users are recommended for generalisable outputs (e.g. driving population with balanced gender mix and range of legal driving age groups, from a variety of geographical regions). Two researchers (a moderator and facilitator) and a set of 5–7 participants per group are suggested.

1.2.3 THEMATIC ANALYSIS

Thematic analysis occurs at *Step 4. Thematic Analysis 1, Step 6. Thematic Analysis 2*, and *Step 7. User Themes and Preferences*. Thematic analysis is undertaken using a grounded theory approach (Glaser and Strauss 1967). The data from Step 3 is thematically analysed with the help of qualitative software such as NVivo. Generated themes should be defined and compared with independent analysts to create a final set. Themes relating to autonomous vehicles included 'trust', 'control', and 'safety'. In Step 6, data from previous steps are consolidated and triangulated according to the data source (Flick 2006). The most robust data is tabulated to capture requirements for Step 7. This takes themes from focus groups and interviews, and converts them into insights (e.g. 'auditory alert of increasing urgency' when takeover is required, as users 'anticipate being otherwise occupied' during automated mode).

1.2.4 COGNITIVE WORK ANALYSIS

The CWA elements include Steps 8–12 (Figure 1.2). Of these, Steps 8 and 9 fulfil the EID requirements (McIlroy and Stanton 2015) but other CWA steps are optional depending on focus and resources. *Step 8. Work Domain Analysis (WDA) Abstraction Hierarchy (AH)* marks the initial phase of the CWA framework. WDA describes the constraints that govern the purpose and function of the systems under analysis. The output of this step is the creation of an AH (Rasmussen, Pejtersen and Goodstein 1994; Vicente 1999). This is a diagram constructed of five levels of abstraction, from the most abstract level 'purpose' (e.g. planned transfer of control back to driver), to 'values' (e.g. safety), 'purpose-related functions' (e.g. ensure driver situation awareness), 'functional purpose' (e.g. communicated vehicle status to driver) to the most concrete form, 'physical object' (e.g. head-up displays). *Step 9. Control Task Analysis (ConTA)* looks in depth at constraints that are imposed by tasks that need to be carried out in specific situations to reveal not only the constraints, but also the level of flexibility in how activities can be achieved (Naikar, Moylan and Pearce 2006). For autonomous vehicles, three time periods were used to constrain the scenario (e.g. pre-takeover, takeover, and post-takeover). *Step 10. Strategies Analysis* is useful for exploring the flexibility within a system for different types of strategies within different contexts (Vicente 1999; Naikar 2006). The strategy adopted by an agent in a particular situation may vary based on workload, or vary between agents (e.g. different strategies for different user groups (e.g. elderly versus young drivers)) or the same user undergoing different journeys (e.g. work or leisure). *Step 11. Social Organisation and Cooperation Analysis (SOCA)* identifies how control tasks can be distributed between human and non-human agents within the system (Naikar 2006).

It determines how social (e.g. driver) and technical (e.g. automation) aspects of a system can work together to enhance performance as a whole. SOCA reveals the flexibility of the system to deal with unanticipated events. *Step 12. Worker Competencies Analysis (WCA)* is the final CWA element and involves identifying the cognitive skills required for control task performance in the system under analysis. These are classified using the skills, rules, and knowledge (SRK) framework (Rasmussen 1985).

1.2.5 Consolidation and Ideas Generation

Consolidation and ideas generation occur in Steps 13 and 14 (Figure 1.2). *Step 13. Pre-workshop* takes any new insights gained from Step 9 onwards to enable re-calibration of the scenario (Step 1), user preferences (Step 7), and AH (Step 8). This allows a clear and consistent presentation of the context and design tasks to participants in the design workshop (Step 14). The basis for concept filtering (Step 15) is also defined here. *Step 14. Design Workshop* is used to share and review the knowledge gained from previous steps about the problem and solution spaces, then to generate concept solutions to solve that problem (e.g. takeover interaction designs, alerts or alarms, multimodal variations). The inclusive design process should be adopted to explore needs and create solutions.

1.2.6 Filtering and Checking

A key element of UCEID occurs in its method of filtering and checking. There are three stages at which the outputs are sanity-checked so that they fulfil both UCD and EID requirements. After construction of the AH in Step 8, its elements are reviewed to determine if user preferences are represented. If not, additional elements are added to accommodate these preferences. This process requires reinterpreting and rewording user preferences to fit into the most appropriate AH layer. It should be made clear that only user preferences and themes applicable to the defined domain purpose are added to the AH, as this constrains the analysis. User preferences will have been generated from interviews and focus groups that fall outside the constraints of the system. The AH therefore provides a means of weighing up the relative importance of themes and preferences from user-centred activities. During *Step 15. Concept Filtering*, ideas and concepts from the workshop are transcribed, and a design criterion matrix (Pugh 1991) relating to the AH in Step 8 is constructed. Design features are rated according to inclusion and adherence to the criteria to produce candidate 'output concepts'. Based on the criteria for inclusion agreed in the pre-workshop (Step 13), the concepts need to be filtered by the agreed level of adherence to the AH (and any other factors deemed critical). This iterative process continues until only design concepts that fulfil these criteria are progressed towards Step 16 (concept development).

Following the UCEID process, final output concepts are created through an iterative process applied to the candidate concepts. Interface analysis methods are applied to assess their viability, and design methods are used to create an initial version of the concepts (Stanton et al. 2013). The output concepts should be documented with a sufficient detail to enable their implementation and evaluation in later stages of the design process, in order to arrive at the final design. The output concepts related to interactions between drivers and semi-autonomous vehicles for a planned SAE Level 3 vehicle-to-driver takeover are comprised of (1) storyboards of interactions between driver and automation, with multimodal feedback; (2) interface designs of head-up displays, head-down displays, and ambient displays to improve trust and mode awareness; and (3) verbal scripts for automated or intelligent guidance for enhancing driver situation awareness prior to taking back control of the vehicle.

1.3   METHODOLOGICAL CONSIDERATIONS

1.3.1   ADVANTAGES

The UCEID method combines a system-level view of analysis with UCD to provide design requirements that support the overall purpose of the domain, as well as the needs and expectations of users. It is a generic and highly flexible method that can be applied to a variety of different domains for different purposes and to different depths depending on requirements of the analysis. This method is based on a sound underpinning of theory by incorporating the CWA framework and inclusive design principles, and is particularly suited to complex systems. The diversity of the different approaches within UCEID promotes a comprehensive analysis. The emphasis on inclusive user groups ensures that a wide spectrum of capabilities is taken into consideration in the design phase that can provide a satisfactory user experience not only for the population with capability limitations, but for a larger proportion of the population.

1.3.2   DISADVANTAGES

The steps in the UCEID method that relate to the CWA framework can be complex, so practitioners may require considerable training in their application if the analyst lacks this experience. As some of the steps from the CWA framework are still in their infancy, there is limited proscriptive guidance available for the novice. A number of the steps can be time-consuming to apply, and some outputs (e.g. the AH) can be large, unwieldy, and difficult to present for highly complex systems. It can be challenging to recruit a user group which both complies with generalisability of the entire population and covers the range of capability limitations. It may not be reasonable for a single research or practitioner group to be skilled in the wide range of methodological approaches within UCEID. Collaborations with other research institutions or practitioners, and the use of consultants, may be required for specialist skills. The number of steps extends the application time, making this method best suited to longer delivery timeframes. As a new method, the reliability and validity have yet to be established.

1.3.3   TRAINING AND APPLICATION TIME

Due to the complexity of the steps of the process relating to the CWA framework elements, training time associated with this method is high. The training requirements for a full inclusive design approach to UCEID require practitioners that have participated in a professional development course covering inclusive or universal design. Secondarily, many of the methods used require familiarity and practice, such as interviews, focus groups, and the design workshop.

The authors recommend 6 months be allocated for the full method comprising all 16 steps. For a smaller-scale project, or where the optional steps (9–12) are omitted, the UCEID method may be applied in a few weeks.

1.3.4 TOOLS

The tools required for UCEID are modest. Most stages can be undertaken with pen and paper, though audio and visual recording equipment is strongly recommended for interviews, focus groups, and workshops to allow content to be revisited in context. The use of software tools can facilitate this method, in particular the CWA tool by the HFI-DTC, and thematic analysis and data management tools such as NVivo and MS Excel. For workshops and focus groups, rapid prototyping materials such as building blocks, toys, colouring materials, and modelling clay are beneficial to quickly understand user wants.

1.4 SUMMARY

This paper has presented a high-level overview of UCEID, incorporating the best of EID and inclusive UCD principles in a novel method to aid HF and design practitioners. The range of approaches within this method has been highlighted. Generic advice followed by examples relating to vehicle-to-driver takeover in a SAE Level 3 autonomous vehicle is provided. Practical considerations for application of this method have been described.

ACKNOWLEDGEMENTS

This work was supported by Jaguar Land Rover and the UK-EPSRC Grant EP/011899/1 as part of the jointly funded 'Towards Autonomy: Smart and Connected Control (TASCC) Programme'.

REFERENCES

Breakwell, G. M., S. Hammond, and C. Fife-Schaw. 2000. *Research Methods in Psychology* (2nd ed.). Thousand Oaks: Sage Publications Ltd.

Burns, C. M., and J. Hajdukiewicz. 2004. *Ecological Interface Design*. Boca Raton, FL: CRC Press.

Czajkowski, K., S. Fitzgerald, I. Foster, and C. Kesselman. 2001. "Grid information services for distributed resource sharing." *Proceedings 10th IEEE International Symposium on High Performance Distributed Computing*. San Francisco: IEEE. 181–194. DOI: 10.1109/HPDC.2001.945188.

Flick, U. 2006. *An Introduction to Qualitative Research* (3rd ed.). California: SAGE Publications Ltd.

Glaser, B. G., and A. L. Strauss. 1967. *The Discovery of Grounded Theory: Strategies for Qualitative Research*. Chicago: Aldine Publishing.

Green, W., and P. W. Jordan. 1999. *Human Factors in Product Design: Current Practice and Future Trends*. London: CRC Press. https://doi.org/10.1201/9781498702096.

Jenkins, D. P., N. A. Stanton, P. Salmon, and G. H. Walker. 2008. *Cognitive Work Analysis: Coping with Complexity*. Farnham: Ashgate Publishing Ltd.

Langdon, P. M., J. Lazar, A. Heylighen, and H. Dong. 2014. *Inclusive Designing: Joining Usability, Accessibility, and Inclusion*. Springer International Publishing. DOI: 10.1007/978-3-319-05095-9.

Langdon, P., and H. Thimbleby. 2010. "Inclusion and interaction: designing interaction for inclusive populations." *Interacting with Computers, 22* 439–448. https://doi.org/10.1016/j.intcom.2010.08.007.

McIlroy, R. C., and N. A. Stanton. 2015. "Ecological interface design two decades on: whatever happened to the SRK taxonomy?" *IEEE Transactions on Human-Machine Systems, 45, 2* 145–163. DOI: 10.1109/THMS.2014.2369372.

Naikar, N. 2006. "An examination of the key concepts of the five phases of cognitive work analysis with examples from a familiar system." *Proceedings of the Human Factors and Ergonomics Society Annual Meeting, 50, 3* 447–451. https://doi.org/10.1177/154193120605000350.

Naikar, N., A. Moylan, and B. Pearce. 2006. "Analysing activity in complex systems with cognitive work analysis: concepts, guidelines and case study for control task analysis." *Theoretical Issues in Ergonomics Science, 7, 4* 371–394. https://doi.org/10.1080/14639220500098821.

Norman, D. 2013. *The Design of Everyday Things: Revised and Expanded Edition*. New York: Basic Books.

Pugh, S. 1991. *Total Design: Integrated Methods for Successful Product Engineering*. Wokingham: Addison-Wesley.

Rasmussen, J. 1985. "The role of hierarchical knowledge representation in decisionmaking and system management." *IEEE Transactions on Systems, Man, and Cybernetics, SMC-15, 2* 234–243. DOI: 10.1109/TSMC.1985.6313353.

Rasmussen, J., A. M. Pejtersen, and L. P. Goodstein. 1994. *Cognitive Systems Engineering*. New York: John Wiley & Sons, Inc.

Roadcraft. 2020. *Roadcraft: The Police Driver and Rider Handbooks for Better, Safer Driving*. Accessed May 1, 2020. http://www.roadcraft.co.uk.

SAE International. 2018. *SAE International Releases Updated Visual Chart for Its "Levels of Driving Automation" Standard for Self-Driving Vehicles*. December 11. Accessed May 1, 2020. https://www.sae.org/news/press-room/2018/12/sae-international-releases-updated-visual-chart-for-its-%E2%80%9Clevels-of-driving-automation%E2%80%9D-standard-for-self-driving-vehicles.

SKYbrary. 2011. *Handover/Takeover of Operational ATC Working Positions/Responses*. Accessed May 1, 2020. https://www.skybrary.aero/index.php/Handover/Takeover_of_Operational_ATC_Working_Positions/Responses.

Stanton, N. A., P. M. Salmon, L. A. Rafferty, G. H. Walker, C. Baber, and D. P. Jenkins. 2013. *Human Factors Methods: A Practical Guide for Engineering and Design*. London: CRC Press. https://doi.org/10.1201/9781315587394.

Vicente, K. J. 1999. *Cognitive Work Analysis: Toward Safe, Productive, and Healthy Computer-Based Work*. Mahwah: Lawrence Erlbaum Associates Publishers.

Wockatz, P., and P. Schartau. 2015. *Intelligent Mobility: Traveller Needs and UK Capability Study*. Milton Keynes: Transport Systems Catapult.

2 Using UCEID to Include the Excluded

An Autonomous Vehicle HMI Inclusive Design Case Study

Michael Bradley and Nermin Caber
University of Cambridge

Patrick Langdon
Edinburgh Napier University

P. John Clarkson
University of Cambridge

Simon Thompson, Lee Skrypchuk,
and Alexandros Mouzakitis
Jaguar Land Rover

Ioannis Politis
The MathWorks, Inc.

Joy Richardson, Jisun Kim, James W.H. Brown,
Kirsten M. A. Revell, and Neville A. Stanton
University of Southampton

CONTENTS

2.1   INTRODUCTION

There are many methods in the human factors field that attempt to support designing systems where humans interact with complex technologies and within sociotechnical systems, such as those contained in Jenkins et al. (2008), Salvendy (2012), Stanton and Karwowski (2006), Stanton et al. (2013), and Stanton (2007). There are also many methods which attempt to support designing products and services for consumer needs, wants, satisfaction, and aesthetic preferences such as those contained in Etchell and Yelding (2004), Karwowski, Soares and Stanton (2011), and Stanton and Young (1999).

The user-centred ecological interface design (UCEID) method was developed to allow the rigour of extant human factors methods to reflect the safety context of the task to be carried with an interface, along with the benefits of a creative inclusive design process to reflect the wide diversity of capability, preferences, expectations, and needs typified by users of consumer products, including passenger vehicles (see Figure 2.1).

This paper describes the inclusive design approach for the design of a semi-autonomous vehicle human–machine interface (HMI), in the context of the framework of the UCEID method developed in parallel, in order to deliver a safe, and inclusive, consumer-friendly interface, which requires minimal learning and/or training for the user.

2.1.1   This Case Study: Designing an HMI for Level 3/4 Autonomous Car Takeover

Fully autonomous cars, those operating at SAE Level 5 (SAE J3016 2016), offer the tantalising opportunity of unassisted mobility for those who are unable to drive themselves, whether due to physical, sensory, or motor impairment or incapacity.

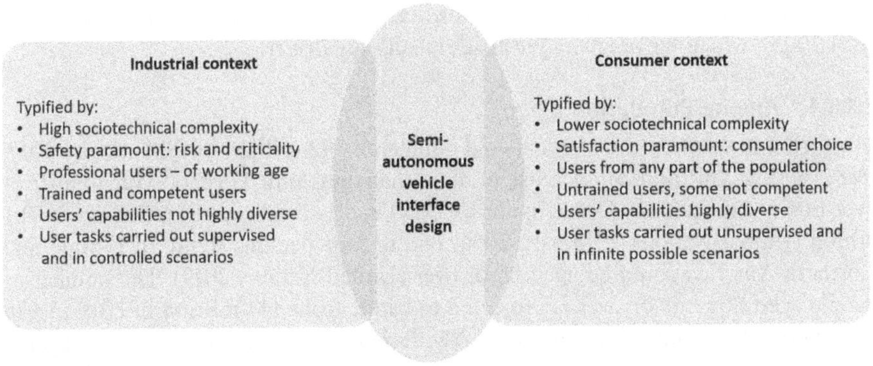

Industrial context

Typified by:
- High sociotechnical complexity
- Safety paramount: risk and criticality
- Professional users – of working age
- Trained and competent users
- Users' capabilities not highly diverse
- User tasks carried out supervised and in controlled scenarios

Semi-autonomous vehicle interface design

Consumer context

Typified by:
- Lower sociotechnical complexity
- Satisfaction paramount: consumer choice
- Users from any part of the population
- Untrained users, some not competent
- Users' capabilities highly diverse
- User tasks carried out unsupervised and in infinite possible scenarios

FIGURE 2.1 Semi-autonomous vehicle interface design in contexts for industrial and consumer interface designs.

Semi-autonomous vehicles (those operating at Level 3 or 4) offer lesser opportunities, but still highly valuable – to engage in other activities while driving. These other activities are entirely at the discretion of the driver and could include reading and responding to emails, making phone calls, sleeping, or watching a film. However, the driver is required to be able to respond to a request from the vehicle to take over the role of driving within a period of time.

Thus, the design of driver takeover is intrinsically challenging due to the need for a driver to regain situational awareness from the driver's chosen activity or activities undertaken while the car was driving. Similar situations arise in other domains such as aircraft piloting and air traffic control. However, in these scenarios, the operators are typically highly trained, competent, supervised under the terms of their employment, and are of working age.

For the driver takeover task, the user characteristics are only constrained by the need to pass a single driving test (under current UK regulations) taken usually early in life, and permission is conferred indefinitely until such time as a driver's behaviour flouts sufficient rules that the driving license is revoked through the legal system, they are deemed medically unfit, or they decide to cease driving. Accordingly, in the UK, the population of potential users of an autonomous car interface can be drivers of any age above 17.

Currently, there are no plans in the UK for drivers of passenger vehicles who intend to use semi-autonomous features to be required to undertake any training. Purchasers of new vehicles from a dealership may be offered some familiarisation and guidance, as per current practice in the industry; however for additional drivers, drivers of vehicles not bought through a dealership hire car users and any other category of user is extremely unlikely to be exposed to any familiarisation or training related to the specifics of an autonomy feature of a vehicle they may come to drive.

For the Human Interaction: Designing Autonomy in Vehicles (HI:DAVe) project, the work from which this paper describes, the challenge therefore was to design an interface that would be safe, to support the driver to take over the driving task when they may have lost awareness of their situation, and to ensure that it would be usable by 'as many people as reasonably possible' as per the definition of inclusive design (BSI 2005), which for practical purposes means any driver.

2.1.1.1 Ageing Population

The global population over 60 years old numbered 382 million in 1980, but by 2017, there were 962 million, an increase of more than double in 37 years. The number of over 60s is expected to double again by 2050 and is predicted to reach nearly 2.1 billion (United Nations 2018). By 2050, one in four persons living in Europe and Northern America could be aged 65 or over (United Nations 2019). The number of people aged 80 years or over is projected to triple, from 143 million in 2019 to 426 million in 2050 (United Nations 2019). Yet despite these huge numbers of an ageing population, the diverse needs and capabilities of this group are often not reflected in designs of products and services that reach the market (Hosking, Waller and Clarkson 2010). This is an often ignored, and therefore excluded, segment of the population.

2.1.1.2 Ageing and Capability Impairment

Ageing is one of the key drivers of capability variation in the population (Baltes and Lindenberger 1997; Grundy et al. 1999). Accordingly, severe capability losses giving rise to disabling conditions rise with age, and the co-occurrence of capability impairments increases with age. It therefore follows that products and services for an ageing population need to be designed to take into account the capabilities of the population, in particular the aged consumer (Waller et al. 2015).

2.1.1.3 Ageing and Digital Technological Interface Capability

The relationship between age and reduction in capability for use of digital interfaces has been demonstrated (Bradley et al. 2012, 2013; Goodman-Deane, Bradley and Clarkson 2020), and in combination with the other relevant factors such as motivation, self-efficacy, and attitudes towards technology, it very often can result in frustration and goal failure in product interaction for older users.

This is especially so for technological products or services, where novel digital interfaces can exclude users due to some older user's lack of familiarity and experience (Bradley et al. 2013; Czaja and Sharit 1998; Hawthorn 2000, 2006; Murad et al. 2012). Previous work has identified a disconnect between the needs and capabilities of older drivers, and modern vehicle interface designs which don't appear to take these needs into account (Bradley, Langdon and Clarkson 2016). The work this paper reports sought to minimise the exclusion wrought by ignorance of the effects on capabilities of ageing, by including the older population and those who are less technologically capable during the design and development process.

2.1.1.4 Inclusive Design

One design response to an ageing population is the movement to design inclusively (Etchell and Yelding 2004; Clarkson and Coleman 2015; Keates et al. 2004). Inclusive design is defined as

> the design of mainstream products and/or services that are accessible to, and usable by, as many people as reasonably possible, on a global basis, in a wide variety of situations

and to the greatest extent possible without the need for special adaptation or specialised design.

(BSI 2005)

Inclusive design therefore aims to expand the proportion of the population who are able to use products and services. It is conceptually very similar to 'Design for All' and 'Universal Design' approaches (BSI 2019).

The inclusive design approach therefore has application to any product, service, activity, or experience for which the general population may be reasonably expected to benefit from, including those who are currently excluded from that engagement. An inclusive design process should therefore explore the needs of those who would not normally be included but would be reasonably expected to benefit from the ability to participate. These might include people, e.g., who do not own – or do not confidently use – a smartphone, but drive long distances, or those experiencing poor near vision when driving without their reading glasses.

Although not all the population are drivers, it is true that those with capability impairments, such as age-related long-sightedness, hearing loss, reduced reach, strength, dexterity, and mobility, and those who have a limited technological capability are still capable of, and licensable to, drive a vehicle. Consequently, the solution to the HMI design challenge should accommodate their needs and capabilities.

These interactions might be experienced by those who have age-related long-sightedness so that they are struggling to focus on a steering wheel control to see its label while driving (Hosking 2020), or by those whose digital interface experience is sufficiently limited so that they are unable to use a menu system to be able to control a key feature (Goodman-Deane, Bradley and Clarkson 2020).

Designing inclusively offers benefits not only in terms of more people being able to use the products or service, but also potentially an improved user experience for all users (Waller et al. 2015). However, it can be difficult to engage participants who represent edge case users in the different phases of the design process – those with reduced capabilities, non-users, and those who don't find it easy or like to use the product or service.

In a driver takeover design and evaluation project, it was expected that those drivers who are less technologically capable would be excluded from using complex automotive interfaces such as those found on premium brand vehicles. However, typically, it is difficult to recruit these drivers for such a study, as they do not volunteer for studies involving technology advertised in the conventional way.

There is an associated problem: many HMIs in premium manufacturers do not appear to be inclusive in terms of digital interface exclusion due to complexity of the interfaces and the requirements for the user to have prior digital interface experience, a willingness to explore, some technical self-efficacy, and potentially, to have access to support to help address insurmountable difficulties during learning or use (Barnard et al. 2013; Bradley, Langdon and Clarkson 2016).

2.2   APPROACH AND ACTIVITIES

This paper reports on a number of methods used to explore and create solutions to the problem of the driver takeover after a period of the vehicle driving autonomously.

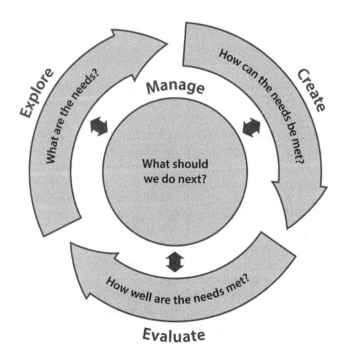

FIGURE 2.2 Overview inclusive design process diagrams from University of Cambridge (2019).

The framework followed was based on the inclusive design wheel (University of Cambridge 2019; Waller et al. 2015) (see Figure 2.2). Each of the three main activities of 'Explore', 'Create', and 'Evaluate' is normally mediated by the 'Manage' activity, which facilitates the iterative approach to optimise the design process through understanding appropriate parts of the problem, expanding the potential solution space, and ensuring that the proposed solutions do, in fact, address the identified problems.

For this project, a modified version of the design process was proposed (see Figure 2.3), where the 'Manage' activity was removed (as implicit in the project management process), and instead, 'modelling and design' were made more explicit to highlight the complex cognitive modelling and design development work that would be required for novel HMIs. Similarly, the 'Evaluate' activity was augmented to show the nature of the progression from driving simulator studies, through test track studies, to on-road trials. This highlighted the expected progression of the HMI development fidelity and safety to release participants to engage with the interface in more ecologically valid scenarios as the project unfolded.

2.2.1 OVERVIEW OF EXPLORE AND EVALUATE STAGE

In early project meetings, the team developed the UCEID approach (described in more detail by Revell et al. 2019), which integrated ecological interface design (EID) with an inclusive user-centred design (UCD) process. Figure 2.4 shows the

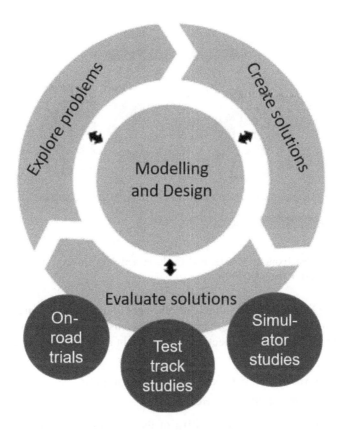

FIGURE 2.3 Inclusive design process as proposed at start of HI:DAVe project modified to incorporate three types of evaluative user studies.

relationship between the activities within UCEID and the other explore, create, and evaluate activities carried out during the remainder of the project. This chapter primarily focuses on the create steps, describing the inputs, activities, and results to show the effect of the selected methods on the design solutions.

The exploration part of the process was undertaken to understand the takeover problem in more detail, and the evaluation elements were to assess how well current solutions for similar functions in production vehicles and concept interfaces appeared to meet driver capabilities and needs.

As only a tiny proportion of the current driving population have any experience of takeovers from automated driving, where the driver is able to delegate driving to the car (i.e. at SAE Level 3 or higher – which enables the driver to attend to non-driving activities), a very wide phase of exploration was taken. This enabled the examination of some of the wider driver needs in automated driving, e.g., around confidence and trust, in addition to the more specific interface requirements. Attempts were made to ensure that participants were recruited who would not necessarily be at the front of the queue to volunteer for driving research, such as low mileage and reluctant drivers, as well as those with cognitive and sensory impairments. Participants who were

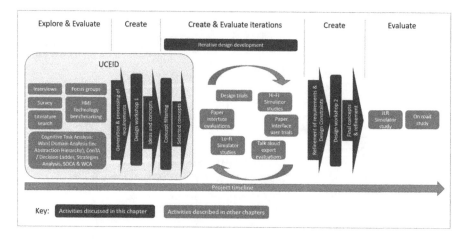

FIGURE 2.4 The Explore, Create, and Evaluate activities in context of other activities undertaken as part of the HI:DAVe project.

into their late 80s at the time of their involvement in the research were recruited, in addition to younger participants to attempt to help ensure that a balanced view of the wide diversity of people's perspectives were brought to bear on the topic.

2.2.2 EVALUATE ACTIVITY: GENERATION AND PROCESSING OF REQUIREMENTS – METHOD

From the focus groups, a combined list of themes was generated from the nodes and other sources, such as the research literature, expert input, and benchmarking information through individual researcher reviews and subsequent team discussion. These items were processed and coded to ensure the rationale, source, and any other relevant categorisation were captured. At this stage, this list contained both relevant and potentially actionable comments, as well as comments which would likely be discarded as not directly relevant to the specifics of the design of the HMI.

Those themes that were actionable and relevant to inclusivity for the design, such as 'It will be necessary to unlearn older control sequence to adapt to autonomous interfaces, seen as an onerous and dangerous requirement for ownership (of an autonomous vehicle)', were interpreted as a requirement to provide an HMI which did not require drivers to 'unlearn' well-established and highly familiar behaviours relating to current vehicular interfaces.

2.2.3 EVALUATE ACTIVITY: GENERATION AND PROCESSING OF NEEDS LISTS – RESULTS

From the survey, interviews, and focus groups, 102 themes were identified, of which four are shown in Table 2.1.

Figure 2.5 gives a couple of examples of identified needs for discussion, including 'multimodality of takeover warning' and 'system capability'. The complete list

TABLE 2.1

Example List of Themes Collated from Survey, Interviews, and Focus Groups

Name	Description	Rationale	Notes
Multimodality of takeover warning	In addition to a visual dashboard warning, there should be a verbal communication from the car, or a series of increasingly strident chimes or bleeps as the takeover point approaches	Anticipation of being otherwise occupied and needing to be alerted (e.g. facing backwards and using infotainment)	Multimodality is important for inclusion but also for the general designs
Countdown to takeover	In a complex traffic environment, the suggestion was for an auditory warning with a countdown information to the takeover, possibly accompanied by haptic stimulation through the seat	Identified as being from around 5 min down to event in increasing salience (longer times were suggested). Approximately 1–2 min	Not a complex traffic environment, not in scenario
Countdown to takeover	Seen as similar to Satnav countdown instructions to a junction		Applies to almost all scenarios for takeover
Uncertain countdown duration	Different times and distances were mentioned for planned takeover	Saw value of timing lane change for planned events and unplanned road contexts, e.g. loss of hard shoulder	A range between roughly 30 s and 5 min was expected

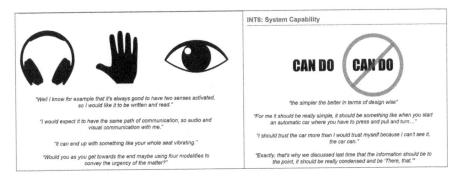

FIGURE 2.5 Shows actual presentation of identified needs for discussion in workshop.

of themes was refined through researcher discussions, and distilled into a short set of communicable, actionable requirements pertinent to the HMI design, which were presented in annotative form on a current Jaguar Land Rover (JLR) vehicle in Figure 2.6. This was visible throughout the workshop to ensure that participants were able to refer to it at any time.

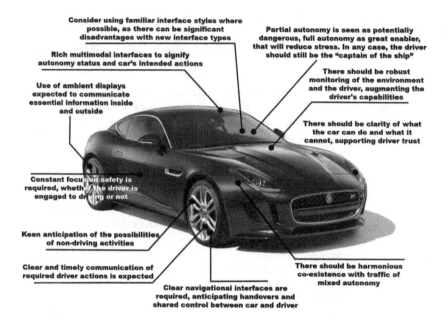

Consider using familiar interface styles where possible, as there can be significant disadvantages with new interface types

Partial autonomy is seen as potentially dangerous, full autonomy as great enabler, that will reduce stress. In any case, the driver should still be the "captain of the ship"

Rich multimodal interfaces to signify autonomy status and car's intended actions

There should be robust monitoring of the environment and the driver, augmenting the driver's capabilities

Use of ambient displays expected to communicate essential information inside and outside

There should be clarity of what the car can do and what it cannot, supporting driver trust

Constant focus on safety is required, whether the driver is engaged to driving or not

Keen anticipation of the possibilities of non-driving activities

Clear and timely communication of required driver actions is expected

There should be harmonious co-existence with traffic of mixed autonomy

Clear navigational interfaces are required, anticipating handovers and shared control between car and driver

FIGURE 2.6 Summary of requirements presented as a poster for use as continuous reference input into Design Workshop 1.

2.2.4 CREATE ACTIVITY: DESIGN WORKSHOP 1

2.2.4.1 Input

The inputs and activities for the workshop were shown in the form of PowerPoint slides which covered:

- The workshop rules (see Figure 2.9).
- The purpose of the workshop – described as 'To develop HMI concepts for Autonomous Vehicle Takeover'.
- Definitions relating to autonomous car takeovers to ensure discussions are communicated and understood effectively.
- Design approaches – including 'design council double diamond' (Design Council 2007) and the 'inclusive design wheel' (Waller et al. 2015).
- Inclusive design thinking – the relationship between capability of a user and demand of the task.
- Findings from the literature, interviews, focus groups, the requirements list generated, and benchmarking information relating to the interfaces for current vehicle advanced driver assistance systems features, and publicly available information on concept and future vehicle system features and autonomy interfaces. One competitor interface from the Autopilot Tesla Model S was shown to have significant inclusivity issues as per Figure 2.7.
- A persona-based workshop needs identification exercise – the workshop participants were allocated into three groups to identify the needs of each of

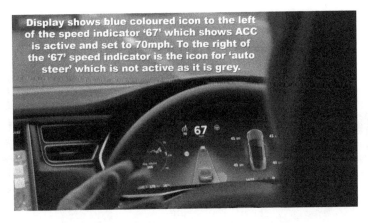

FIGURE 2.7 Showing Tesla Model S Autopilot interface (in 2016) highlighting the codified and hard to discriminate mode identification between autopilot off (steering wheel in grey as shown, but adaptive cruise control on by symbol in blue) and autopilot on (steering wheel and adaptive cruise symbols would both be in blue).

three personas representing boundary case users. Persona sheets were used to describe the relevant characteristics of the user, such as their sensory, physical, and cognitive capabilities; their self-efficacy with new technologies; and how much free time they would have to allocate to learning how to use new technologies.

- An exercise to identify stakeholders and capture their relevant needs was undertaken as a plenary activity.
- A fixation exercise to sensitise the participants to the implicit assumptions made in thinking using the 'path of least resistance', and how making the assumptions explicit can be helpful in allowing them to be challenged.

2.2.4.2 Activity

A two-day workshop programme was devised to provide the academic and industrial expert participants with sufficiently distilled information from the interviews, focus groups, and other research and guidance to enable the participants to contribute in a targeted, goal-oriented yet un-fixated way, to the idea and concept development processes. It followed a divergent, then convergent process (see Figure 2.8), to allow exploration and unfettered creativity on the first day, while on the second day, a more focused approach brought actionable concepts together.

Day 1 followed divergent exploratory activities with familiarisation of input data, such as the results from the interviews and focus groups, but also included fixation breaking instructions and activities (Crilly 2015) (see Figure 2.9 for example) to encourage workshop participants to unshackle themselves from preconceived solution ideas and to facilitate exploration of a wider solution space.

For the creative process during the workshop, an asynchronous brainstorming approach was used to allow individuals to create ideas alone and at any time through the workshop (see Figure 2.9), and then review them collectively to foster a larger

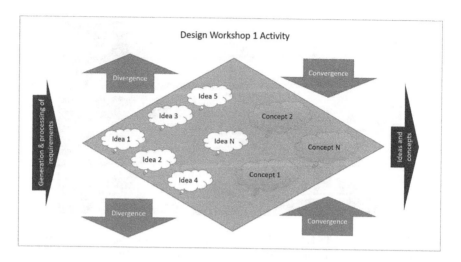

FIGURE 2.8 Diagram of Design Workshop 1 activity: divergence (day 1) followed by convergence (day 2).

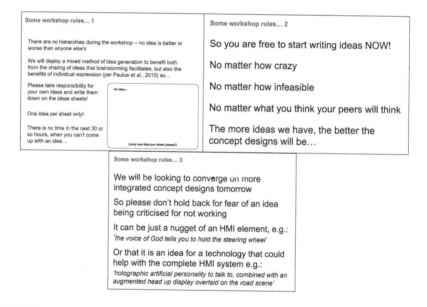

FIGURE 2.9 Instructional slides providing encouragement and permission for participants to be able to create ideas during the workshop.

number of ideas to be generated (Paulus et al. 2015). The goal was to create as many ideas as possible that may help solve the problem, and from those ideas, generate a few concepts, which would address the problem as comprehensively as possible, to take forward for evaluation.

An activity to generate a very hard to use interface for a recording device was carried out. This was intended as a novel way of participants thinking about what makes

FIGURE 2.10   Exemplar Personas for workshop needs identification.

interfaces difficult to use, and subsequently allowing for a wider discussion about the factors that cause difficulty.

There was prescription and clarification of the scenario for the HMI to be designed to address the driver taking over the driving task at the request of the vehicle, in a planned way, e.g. when the journey requires the vehicle to exit a motorway and for which autonomous driving is not possible.

Personas were presented to reflect an inclusive set of potential users, e.g. 'Amy', 75 – whose vision is only just adequate for driving, and who is highly technophobic and has relatively low self-efficacy – and 'Christina', 30 – who is short of stature and has very little spare time as a busy mum to learn how to do tasks in a new way (see Figure 2.10). These were used as prompts to generate specific user and goal requirements, which otherwise are unlikely to have been identified and considered.

Short burst ideation activities were primed to focus the participants on generating ideas for specific challenges within the scenario, such as those in Figure 2.11.

During the concept development phase, participants were invited to work alone or in self-selecting groups to combine ideas from the earlier part of the workshop, or new ideas into concepts that would address the driver takeover problem in an inclusive way. The participants could articulate the solution in any form as per instructions in Figure 2.12, but many chose simply to draw and annotate their concepts.

2.2.4.3   Results

A total of 72 ideas were generated, e.g. see Figure 2.13, during the two-day process. These ranged from specific solutions to problems that were identified in the workshop briefing, to embodiments of best practice guidance.

Eight complete concepts were generated from the workshop, including two that would progress further as the basis for development for the driver takeover HMI, 'HazLanFuSEA' (see Figure 2.14) and 'voice autopilot assistant' (see Figure 2.15),

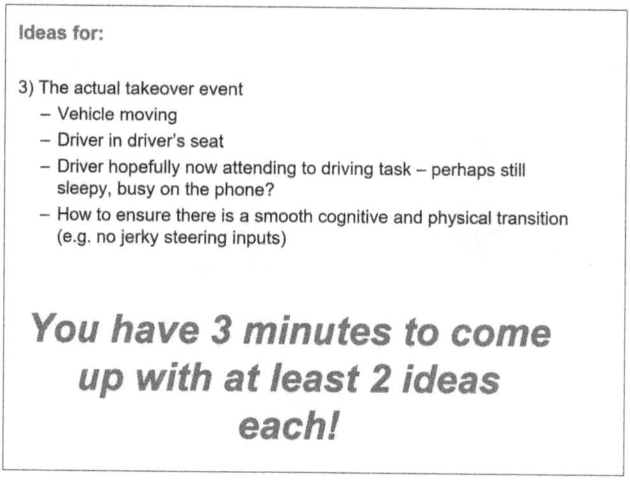

Ideas for:

3) The actual takeover event
- Vehicle moving
- Driver in driver's seat
- Driver hopefully now attending to driving task – perhaps still sleepy, busy on the phone?
- How to ensure there is a smooth cognitive and physical transition (e.g. no jerky steering inputs)

You have 3 minutes to come up with at least 2 ideas each!

FIGURE 2.11　Ideas prompt slide for generating ideas to address the driver takeover event.

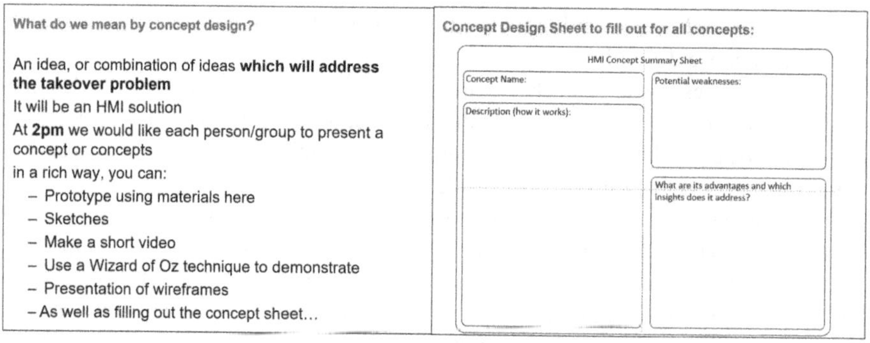

What do we mean by concept design?

An idea, or combination of ideas **which will address the takeover problem**

It will be an HMI solution

At **2pm** we would like each person/group to present a concept or concepts

in a rich way, you can:
- Prototype using materials here
- Sketches
- Make a short video
- Use a Wizard of Oz technique to demonstrate
- Presentation of wireframes
- As well as filling out the concept sheet…

Concept Design Sheet to fill out for all concepts:

HMI Concept Summary Sheet

Concept Name:

Description (how it works):

Potential weaknesses:

What are its advantages and which insights does it address?

FIGURE 2.12　Ideas prompt slide for generating concepts to address the HMI challenge.

and another that provided driver situation awareness enhancement during autonomous mode driving, 'Environment Amplification' (see Figure 2.16).

The HazLanFuSEA concept (in Figure 2.14) drew its inspiration from voice communications in the aviation sector, where in some scenarios human-to-human communication is repeated back to the originator as a method of verification. In this proposed embodiment, the vehicle provides a series of relevant information in the order of hazards, lane position, remaining fuel, speed, exit from current road, and actions required to safely take over and execute the road exit. At each stage, the driver confirms the information provided.

The voice autopilot assistant concept (in Figure 2.15) is similar to HazLanFuSEA, in that it relies on the vehicle to provide a pertinent information to the driver; however in this embodiment, it also relies heavily on a highly sophisticated conversational voice communication. This concept asks the driver to provide answers to questions

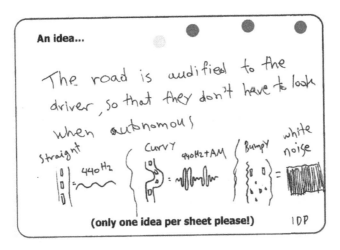

FIGURE 2.13 Exemplar idea from design workshop showing diagrammatic and text descriptions.

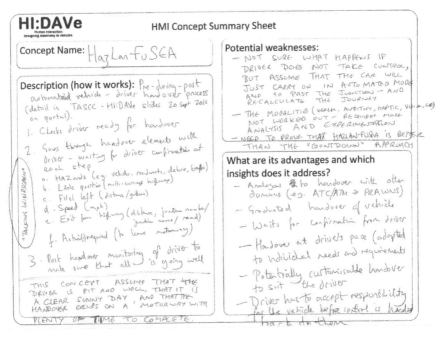

FIGURE 2.14 HazLanFueSEA concept as written up during the workshop.

which require engagement with relevant information available in and outside the vehicle to promote situation awareness.

The Environment Amplification concept in Figure 2.16 proposed increasing the salience of the normal channels providing the driver with contextual information, to improve situation awareness. The concept suggested a natural and subtle mechanism

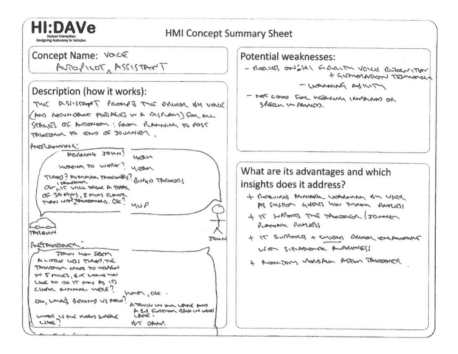

FIGURE 2.15 Voice autopilot assistant concept as written up during the workshop.

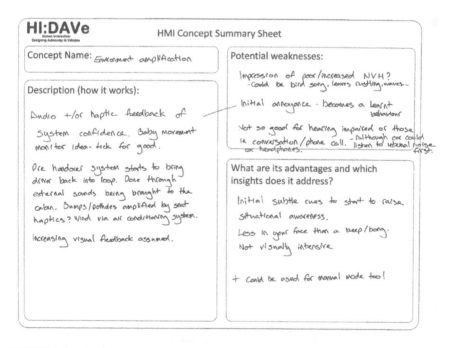

FIGURE 2.16 Environment amplification concept as written up during the workshop.

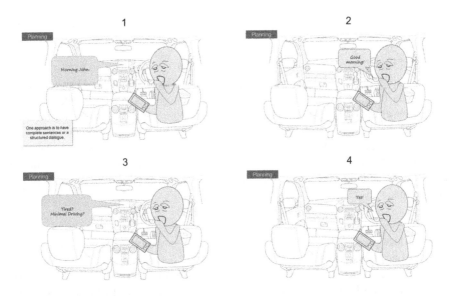

FIGURE 2.17 Part of a storyboard development for the voice autopilot assistant concept.

to bring the driver's awareness of the vehicle context through increasing or introducing sounds, vibration, and ambient visual feedback into the cockpit.

2.2.5 CREATE ACTIVITY: ITERATIVE DESIGN DEVELOPMENT

Between workshops, development took place to refine, combine, and iterate the concept designs; to ensure they could provide a complete solution; and to ensure that they were suitable and feasible for further development. These activities took many forms, including the development of storyboard prototypes, team reviews, and iteration to remove inconsistencies and feasibility issues, and to ensure the concepts could deliver a complete meaningful HMI interaction over all of the prescribed scenarios. An example storyboard section is shown in Figure 2.17 for the voice autopilot assistant concept in the planning phase of the drive, when the vehicle is interrogating the driver to understand their route priorities.

2.2.6 EVALUATE ACTIVITY: TESTING WITH EXPERTS AND USERS – OVERVIEW

The developed design concepts were user-tested and further iterated during expert evaluations, and paper interface evaluations with participants, and driving simulator studies were carried out both in Cambridge and in a higher fidelity driving simulator at Southampton. This work is reported in multiple chapters in detail elsewhere.

2.2.7 CREATE ACTIVITY: DESIGN WORKSHOP 2

A second workshop was undertaken to address specific HMI design issues using a design constraints approach. This enabled the insights gained from the 'Evaluate' activities to be condensed into a shortlist which was then used to inspire detailed design solutions.

2.2.7.1   Input

Thirty-one design constraints were identified from the previous Lo-Fi (referring to the Cambridge-based STISIM seat, steering wheel, pedals, and computer monitor screen set-up) and Hi-Fi (referring to the Southampton JLR vehicle-based) driving simulator studies and meetings (see Table 2.2). These constraints captured general guidance such as '24. Eyes out', to facilitate the driver looking out of the vehicle, and more specific such as '11. Cue driver to grab steering wheel', to provide an explicit instruction to the driver for takeover actions, as well as to the source of the constraints.

2.2.7.2   Outputs

The workshop participants focused on developing embodiment solutions from the design constraints (see Figure 2.18). These embodiments came in a wide range of forms, from abstract ideas – such as takeovers coming in different 'sizes' – to specific implementations for takeover control – such as the steering wheel which is pushed for autonomous mode and pulled for manual mode.

Six emergent themes were identified from the workshop:

1. Driver's responsibility:
   a. Clarity of driver responsibility and risk,
   b. Feedback on driver action for various situations,
   c. Displays of control and degrees of control, control salience (clear indi-cation who is in control),
   d. Ability to ignore takeover request, or speed-up takeover protocol, self-paced takeover.
2. Customisation/personalisation of takeover:
   a. Slider, checklist, dial elements to select pre-sets,
   b. Based on person (age, readiness, time out of loop, experience), situation (e.g. road), values of manufacturer,
   c. Adaptability of the above based on context.

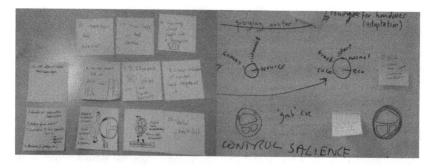

FIGURE 2.18   Exemplar design solution embodiments relating to specific design constraints.

TABLE 2.2

Design Constraints Used as Input to Design Workshop 2

Design Related Constraints	Lo-Fi	Hi-Fi
1. Allow driver to take control at any point during takeover, be sure hand on wheel/feet on pedals	X	X
2. Personalise takeover based on driver preferences (and situation)	X	X
3. Allow option to complete task (even if it means miss takeover for junction/exit	X	X
4. Allow sufficient time for takeover (big individual differences)	X	
5. Customise takeover based on duration of being outside of the control loop and frequency of takeover (and context: road, weather etc) multimodal HMI	X	
6. Querying situation awareness of driver by 'vehicle avatar'	X	
7. Make explicit who is in control of vehicle – mode awareness HMI (light up steering wheel)	X	X
8. Recommended settings based on customer profiles for customisation	X	
9. Pre-set defaults for takeover	X	
10. Graduated alert to takeover visual, audio, haptic (escalating)		X
11. Cue driver to 'grab' steering wheel		X
12. Make 'take over button' easy to access (e.g. put on gear stick)		X
13. 'Repeat' button 'OK' button?		X
14. Encourage (facilitate) visual checks in environment and controls of vehicle		X
15. Display the vehicle status and intention		X
16. Driver's HMI actions need to be clearly fed back. (Link to 1 – Volvo hands on wheel to flip both paddles)		X
Education and Training		
17. Education of drivers in rationale and technique		X
18. Training (video) before being able to use autopilot on roads	X	
General		
19. Older drivers do not like to constantly monitor automation for takeover (timer only) – *trend only*	X	
20. Differences between user preference and rankings of usefulness		X
21. Characteristics of modality – HMI (blinky tape not noticed in peripheral vision)		X
22. Synchronise multimodal cues. Combining or single modality		X
23. Longitudinal studies		X
Additional Constraints Identified at Design Meeting		
24. Eyes out		
25. Use system to aid manual driving		
26. Some level of personalisation and setting of levels		
27. Longer vehicle to driver takeover in urban environment (compared to motorway)		
28. Takeover strategy that guides visual search		
29. Feedback to every driver action (process needs adapting to driver and situation)		
30. Checklist		
31. Option to request specific information of importance to driver (if not in protocol)		

3. Car personification/avatar:
   a. Inquiring about readiness for takeover,
   b. Going through a checklist for required actions before takeover (e.g. verbally).
4. Wheel interventions:
   a. Affordances to grab wheel, cues to grab wheel, displays on wheel, wheel movement (push wheel for autonomous/pull for manual),
   b. Paddles.
5. Ambient displays:
   a. For inside the car or outside the car,
   b. Head-up display cues for traffic,
   c. Augmented reality on windscreen,
   d. To guide visual search.
6. Input and output modalities:
   a. Voice,
   b. User Interface (UI) elements to initiate, check completion, and move between takeover steps (e.g. next, previous button, dial, trackpad, toggle switch),
   c. Communicating essential information and environmental conditions before and during takeover (e.g. part of trip, time of day, road conditions, next junction, hazards, type of road).

2.2.8   CREATE ACTIVITY: FINAL CONCEPTS AND REFINEMENT

The final HMI interface tested at the JLR simulator at the HMI laboratory, attempted to embody the design constraints, and presented the emergent themes from the workshop (see Figure 2.19). This interface showed the potential for the approach and provided a testbed for carrying out experimentation to determine participant

FIGURE 2.19 Overview of JLR HMI driving simulator mocked up with Hi:DAVe autonomous HMI as used for participant trials shown in manual driving mode, and with all output modes active. In addition to the visible displays, audible instructions and tones, and seat vibrations were available for personalisation.

personalisation preferences for the output modes, including displays, and haptic and audible feedback. In addition, eye-tracking cameras were fitted to provide gaze information to help understand which displays were being employed.

Details of the studies and results carried out using the HMI simulator are available in other Hi:DAVe papers.

The HMI demonstrated in the Jaguar I-PACE for the on-road study was largely the same embodiment as for the JLR simulator (see Figures 2.20–2.22). A more complete picture of the interactions during handover to vehicle and takeover by driver can be seen in Appendix 1, which shows the high level of 'hand-holding' and mode salience that the interface provides to the user. This feature was intrinsic to the inclusive nature of the interface, so that users who were not familiar, nor confident in the use of the interface, would not need to rely on their cognitive skills to work out how the system might be activated or deactivated, nor to require to recall such knowledge. In addition, the HMI supported the modes of control that all drivers are familiar with to resume control at all times, such as grabbing the steering wheel to change

FIGURE 2.20 Jaguar I-PACE Hi:DAVe autonomous HMI as used for the on-road trials shown in manual mode ('You Are in Control' orange text, orange ambient lighting, and colour coding on displays), but when autonomous driving has become available ('Automation Available' white text).

FIGURE 2.21 Jaguar I-PACE Hi:DAVe autonomous HMI instrument cluster display detail shown just after manual mode has been activated.

FIGURE 2.22 Jaguar I-PACE Hi:DAVe autonomous HMI head-up display detail shown when in manual mode and demonstrating the explicit instructions for the driver to activate autonomous mode when automation is available.

direction, and pressing the brake to slow down. It was recognised that when all the display modalities were deployed that this represented a significant (and potentially overwhelming) degree of redundancy, and so personalisation settings were an experimental priority for both the simulator and the on-road trials. These studies enabled the participants to choose which displays were active and the level of haptic and auditory feedback provided.

2.3 DISCUSSION AND CONCLUSIONS

This inclusive autonomous HMI case study has shown how attention to user requirements throughout the explore, create, and evaluate phases can help ensure that a design is both user-centred and suitable in a safety-critical domain. However, it should be noted that in some of the simulator studies, participants preferred a simpler interface which counted down to a driver takeover event. This may highlight an intrinsic preference versus safety trade-off, where some drivers may prefer to have a simple and quick takeover, but that it would be safer for them to have a more time-consuming and effortful one. The generation of prescriptive and measurable preference and safety goals for the HMI may have provided a mechanism to determine which balance of trade-offs would be most appropriate. In this project, it was decided to allow the driver to be able to take control at all times, congruent with their expectations both in execution and in resultant action, so that the takeover protocol could be aborted at any time.

It has highlighted the difficulties of engaging with the difficult-to-reach participants and users who may have the most to give in terms of ensuring that a design is going to be inclusive; however, it points to the benefits of doing so. During this study, it became apparent that one agency used to recruit participants in user studies and market research had been using explicit screening criteria to eliminate potential participants who would not be positive towards technology. However, without testing a design with the least capable potential users, how can it be inclusive, and for this particular case study, verifiably safe? For too long and too often, user testing has been carried out with easy to access and compliant users, with the maxim seemingly being 'better to test with some users, rather than none'. However, this approach can lead to a false sense of a design's inclusivity and user friendliness when the true limitations of the participant selection are ignored.

The specifics of the embodiment of the HI:DAVe HMI shows what is achievable in terms of demystifying and decodifying the utility of a technology, in what is often a highly exclusionary and codified environment. The HI:DAVe HMI does not require a driver to be technically competent, have high self-efficacy, or exhibit the behaviours typified by younger users keen to develop new mental models of exploration, trial and error, and consequent learning. The interface succeeded in being supportive at each stage of the autonomous functionality, in a way that did not require extensive prior knowledge, training, or learning. However, it became apparent under some lighting conditions that some of the prototype test displays suffered from suboptimal legibility due to glare and reflections. In addition, the HI:DAVe HMI was highly optimised for the autonomous driving functionality, and some of the other functions typically prioritised in the HMI were omitted or de-emphasised. In a production vehicle, it is unlikely that the autonomous functionality would be permitted to have carte blanche to dominate the instrument panel real estate. The magnitude of this effect on the experience of the HI:DAVe HMI is not known.

UCEID therefore offers a way of combining inclusive design with formal human factors methods in the early conceptual design phase to address interface design problems for users who are not trained nor specifically selected for the task, but who are required to manage high-risk, critical scenarios. The outcomes suggest that offering flexibility of engagement of takeover support is good for consumer preference.

ACKNOWLEDGEMENTS

This work was supported by Jaguar Land Rover and the UK-EPSRC Grant EP/N011899/1 as part of the jointly funded 'Towards Autonomy: Smart and Connected Control (TASCC) Programme' HI:DAVe Project.

REFERENCES

Baltes, P. B., and U. Lindenberger. 1997. "Emergence of a powerful connection between sensory and cognitive functions across the adult life span: a new window to the study of cognitive aging?" *Psychology and Aging, 12, 1* 12–21. DOI: 10.1037//0882-7974.12.1.12.

Barnard, Y., M. D. Bradley, F. Hodgson, and A. D. Lloyd. 2013. "Learning to use new technologies by older adults: perceived difficulties, experimentation behaviour and usability." *Computers in Human Behavior, 29, 4* 1715–1724. https://doi.org/10.1016/j.chb.2013.02.006.

Bradley, M., J. Goodman-Deane, S. Waller, R. Tenneti, P. Langdon, and P. J. Clarkson. 2013. "Age, technology prior experience and ease of use: who's doing what?" *Contemporary Ergonomics and Human Factors: Proceedings of the international conference on Ergonomics & Human Factors 2013, Cambridge, UK, 15-18 April 2013.* 363–369.

Bradley, M., P. Langdon, and P. J. Clarkson. 2016. "An inclusive design perspective on automotive hmi trends." In *Universal Access in Human-Computer Interaction. Users and Context Diversity. UAHCI 2016. Lecture Notes in Computer Science*, vol 9739. Cham: Springer. 548–555. https://doi.org/10.1007/978-3-319-40238-3_52.

Bradley, M., S. Waller, J. Goodman-Deane, I. Hosking, R. Tenneti, P. Langdon, and P. J. Clarkson. 2012. "A population perspective on mobile phone related tasks." In *Designing Inclusive Systems*, by P. Langdon, P. J. Clarkson, P. Robinson, J. Lazar, and A. Heylighen. London: Springer. 55–64. https://doi.org/10.1007/978-1-4471-2867-0_6.

BSI. 2005. "BS 7000-6:2005, Design management systems. Managing inclusive design."

BSI. 2019. "BS EN 17161:2019 Design for all. Accessibility following a design for all approach in products, goods and services. Extending the range of users."

Clarkson, P. J., and R. Coleman. 2015. "History of inclusive design in the UK." *Applied Ergonomics, 46*, Part B 235–247. https://doi.org/10.1016/j.apergo.2013.03.002.

Crilly, N. 2015. "Fixation and creativity in concept development: the attitudes and practices of expert designers." *Design Studies, 38* 54–91. https://doi.org/10.1016/j.destud.2015.01.002.

Czaja, S. J., and J. Sharit. 1998. "Ability–performance relationships as a function of age and task experience for a data entry task." *Journal of Experimental Psychology: Applied, 4, 4* 332–351. https://doi.org/10.1037/1076-898X.4.4.332.

Design Council. 2007. Eleven *Lessons: Managing Design in Eleven Global Brands - A Study of the Design Process*. Study Report, Design Council.

Etchell, L., and D. Yelding. 2004. "Inclusive design: products for all consumers." *Consumer Policy Review, 14, 6* 186–193.

Goodman-Deane, J., M. Bradley, and P. J. Clarkson. 2020. "Digital technology competence and experience in the UK population: who can do what." In *Contemporary Ergonomics and Human Factors*, Stratford-upon-Avon.

Grundy, E., D. Ahlburg, M. Ali, E. Breeze, and A. Sloggett. 1999. *Disability in Great Britain: Results from the 1996/97 Disability Follow-Up to the Family Resources Survey*. Corporate Document Services.

Hawthorn, D. 2006. *Designing Effective Interfaces for Older Users*. PhD Thesis, The University of Waikato, Hamilton, New Zealand.

Hawthorn, D. 2000. "Possible implications of aging for interface designers." *Interacting with Computers, 12, 5* 507–528. https://doi.org/10.1016/S0953-5438(99)00021-1.

Hosking, I. 2020. *Understanding and Evaluating User Interface Visibility*. PhD Thesis, University of Cambridge, Cambridge, UK.

Hosking, I., S. Waller, and P. J. Clarkson. 2010. "It is normal to be different: applying inclusive design in industry." *Interacting with Computers, 22, 6* 496–501. https://doi.org/10.1016/j.intcom.2010.08.004.

Jenkins, D. P., N. A. Stanton, P. Salmon, and G. H. Walker. 2008. *Cognitive Work Analysis: Coping with Complexity*. Farnham: Ashgate Publishing Ltd.

Karwowski, W., M. M. Soares, and N. A. Stanton. 2011. *Human Factors and Ergonomics in Consumer Product Design: Methods and Techniques*. Boca Raton: CRC Press.

Keates, S., P. J. Clarkson, P. Langdon, and P. Robinson. 2004. *Designing a More Inclusive World*. London: Springer-Verlag. DOI: 10.1007/978-0-85729-372-5.

Murad, S., M. D. Bradley, N. Kodagoda, Y. F. Barnard, and A. D. Lloyd. 2012. "Using task analysis to explore older novice participants' experiences with a handheld touchscreen device." In *Contemporary Ergonomics and Human Factors 2012: Proceedings of the international conference on Ergonomics & Human Factors 2012*. Blackpool: CRC Press. 57–64. DOI: 10.1201/b11933-17.

Paulus, P. B., R. M. Korde, J. J. Dickson, A. Carmeli, and R. Cohen-Meitar. 2015. "Asynchronous brainstorming in an industrial setting: exploratory studies." *Human Factors: The Journal of the Human Factors and Ergonomics Society, 57, 6* 1076–1094. https://doi.org/10.1177/0018720815570374.

Revell, K., P. Langdon, M. Bradley, I. Politis, J. Brown, S. Thompson, L. Skrypchuk, A. Mouzakitis, and N. A. Stanton. 2019. "UCEID - the best of both worlds: combining ecological interface design with user centered design in a novel HF method applied to automated driving." *Advances in Human Aspects of Transportation. AHFE 2018. Advances in Intelligent Systems and Computing*, vol 786. Cham: Springer. 493–501. https://doi.org/10.1007/978-3-319-93885-1_44.

SAE J3016. 2016. *Taxonomy and Definitions for Terms Related to Driving Automation Systems for On-Road Motor Vehicles.*

Salvendy, G. 2012. *Handbook of Human Factors and Ergonomics*, Fourth Edition. Hoboken: John Wiley & Sons, Inc.

Stanton, N. A. 2007. "Human-error identification in human-computer interaction." In *The Human-Computer Interaction Handbook*, by A. Sears and J. A. Jacko. Boca Raton: CRC Press. 371–383. https://doi.org/10.1201/9781410615862.

Stanton, N. A., and M. Young. 1999. *Guide to Methodology in Ergonomics: Designing for Human Use.* Abingdon: CRC Press.

Stanton, N. A., and W. Karwowski. 2006. *International Encyclopedia of Ergonomics and Human Factors*, Volume I. London: Taylor & Francis.

Stanton, N. A., P. M. Salmon, L. A. Rafferty, G. H. Walker, C. Baber, and D. P. Jenkins. 2013. *Human Factors Methods: A Practical Guide for Engineering and Design.* London: CRC Press. https://doi.org/10.1201/9781315587394.

United Nations. 2018. *World Population Ageing 2017 Highlights.* Population and Vital Statistics Report. United Nations, New York. https://doi.org/10.18356/10e32e81-en.

United Nations. 2019. *World Population Prospects 2019: Highlights.* Population Estimates and Projections, 2019 Revision, United Nations.

University of Cambridge. 2019. *Inclusive Design Toolkit.* www.inclusivedesigntoolkit.com.

Waller, S., M. Bradley, I. Hosking, and P. J. Clarkson. 2015. "Making the case for inclusive design." *Applied Ergonomics, 46,* Part *B* 297–303. https://doi.org/10.1016/j.apergo.2013.03.012.

APPENDIX: JAGUAR I-PACE HI:DAVE HMI AUTONOMOUS STATES AND TRANSITIONS SHOWING INTERFACE CHARACTERISTICS

FIGURE 2.23 Jaguar I-PACE Hi:DAVe autonomous HMI as used for the on-road trials shown in manual mode ('You Are in Control' orange text, orange ambient lighting, and colour coding on displays), but when autonomous driving has become available ('Automation available' voice prompt).

FIGURE 2.24 Jaguar I-PACE Hi:DAVe autonomous HMI as used for the on-road trials shown in autonomous mode ('The Car is in Control' blue text, blue ambient lighting, and colour coding on displays), just after autonomous driving has been activated ('Automation activated, the car is in control' voice prompt).

FIGURE 2.25 Jaguar I-PACE Hi:DAVe autonomous HMI as used for the on-road trials shown in autonomous mode, at the beginning of the handover to driver protocol ('Get ready for handover' voice prompt).

FIGURE 2.26 Jaguar I-PACE Hi:DAVe autonomous HMI as used for the on-road trials shown in autonomous mode, during the question phase of the takeover by driver protocol ('Is the traffic busy ahead?' voice prompt). Other questions used included 'Are there any vehicles alongside?', 'What is your speed?', 'What is the weather like?', 'Is the road ahead straight?', 'Can you see any motorcycles?', and 'What traffic is behind you?'.

FIGURE 2.27 Jaguar I-PACE Hi:DAVe autonomous HMI as used for the on-road trials shown in autonomous mode, during the transition phase of the takeover by driver protocol.

FIGURE 2.28   Jaguar I-PACE Hi:DAVe autonomous HMI as used for the on-road trials shown in manual mode, just after the takeover by driver phase.

FIGURE 2.29   Jaguar I-PACE Hi:DAVe autonomous HMI as used for the on-road trials shown in manual mode, just after the takeover by driver phase.

FIGURE 2.30   Jaguar I-PACE Hi:DAVe autonomous HMI as used for the on-road trials shown in manual mode during normal driving.

3 Designing Autonomy in Cars

A Survey and Two Focus Groups on Driving Habits of an Inclusive User Group, and Group Attitudes towards Autonomous Cars

Ioannis Politis
The MathWorks Inc.

Patrick Langdon
Edinburgh Napier University

Michael Bradley, Lee Skrypchuk,
Alexandros Mouzakitis, and P. John Clarkson
University of Cambridge

Neville A. Stanton
University of Southampton

CONTENTS

3.1   INTRODUCTION

Autonomy in cars is becoming a reality, with an ever-increasing number of manufacturers predicting a widespread availability of autonomous driving solutions in the next 5 years (Sage and Lienert 2016; Abbugao 2016; Hawkins 2016). The public's expectation is that this technology will deeply affect the norms of transportation and the dynamics of car ownership (Sessa et al. 2015). In the midst of this significant transition, potential users of autonomous cars have been given the opportunity to express their views, hopes, and concerns for a future where self-driving cars will be widely available (e.g. Rödel et al. 2014; Schoettle and Sivak, 2014a, 2014b, 2015, 2016; Kyriakidis, Happee and de Winter 2015). However, two critical considerations are lacking: firstly, available work related to how current driving habits correlate with the perceived need for autonomy is sparse, especially in the UK (Clark, Parkhurst and Ricci 2016), and secondly, consideration of the population as a set of users with varying degrees of capabilities to be accommodated, as described in inclusive design (Clarkson et al. 2003), has been rare (Jeon et al. 2016).

This paper begins to address this problem by investigating driving habits and opinions of a set of users in the UK, through an online survey conducted between July and November 2016. The survey, along with general questions on demographics and driving habits, investigated self-reported identification with UK driver profiles, as identified by a UK-wide capability study conducted in 2015 by Transport Systems Catapult (Wockatz and Schartau 2015). It also provided a set of questions regarding self-reported capabilities, as used in an inclusive design methodology (Clarkson et al. 2012), never before used in the context of autonomous cars. The relation of the above factors to current driving enjoyment and acceptance of the idea of autonomous cars provided an interesting set of insights on the views of drivers about autonomy.

A subset of the survey respondents were invited to attend either of two focus groups at the University of Cambridge, to share ideas specific to the problem of takeover of control by driver from car, which is expected to be of particular importance in autonomous cars human–machine interface (HMI) design (Meschtscherjakov et al., 2015, 2016). During a takeover, vehicle control is transferred between the car and the driver, e.g. due to driver choice or limitations of automation. Focus group participants were presented with the problem and asked to share their views during an interactive session with the assistance of visual and tangible aids. In accordance with an inclusive design approach, participants of distinct characteristics regarding age, gender, and technological expertise were invited in each focus group, providing an initial but extensive set of views on autonomous cars and takeovers. Further, interesting commonalities were found, as well as unique views expressed among participants of the two focus groups.

3.2 RELATED WORK

3.2.1 USER VIEWS

As car autonomy is entering the mainstream in driving technology, recent studies have investigated user views on autonomous cars. KPMG LLP (Silberg et al. 2013) reported a focus group conducted with US drivers in 2013. Participants expressed concerns on safety, liability, and vehicle handling, but they were positive towards reduced expected commute times and avoidance of traffic. They were expecting to be able to switch automation on and off when using an autonomous car. Rödel et al. (2014) reported a survey on levels of autonomy with members of the University of Salzburg, conducted in 2014. It was found that perceived control and fun decreased as the degree of automation increased (National Highway Traffic Safety Administration 2013; SAE J3016 2016). User acceptance and perceived user experience were higher with cars similar to the ones on the roads at the time, but the pre-existing experience with advanced driver assistance systems positively affected these metrics for autonomous vehicles.

Kyriakidis, Happee, and de Winter (2015) presented a large-scale survey in 2015 on public opinion of automated driving from 109 countries. It was found that manual driving was the most enjoyable mode of transportation. Further, there were split opinions on how much respondents were willing to pay for an autonomous driving function (ranging from zero to tens of thousands of dollars). Concerns were expressed about security, liability, and safety of autonomous cars. Bazilinskyy, Kyriakidis, and de Winter (2015) processed the free text responses in a crowdsourcing study (Kyriakidis, Happee and de Winter 2015), revealing split opinions between positive and negative comments on automated driving, and further highlighting this polarisation of views.

Schoettle and Sivak (2014a, 2014b) considered the public opinion on autonomous vehicles from users in China, India, Japan, US, UK, and Australia, in two 2014 surveys. There were high expectations on benefits of autonomous cars, but high level of concerns regarding riding in these vehicles, especially in terms of security, possible system failures, and performance. The majority of respondents would not pay extra in order to have autonomous vehicles (except users in China and India). Regarding drivers' preferences of vehicle automation, Schoettle and Sivak (2015, 2016) presented the results of two surveys conducted in 2015 and 2016 with US drivers. Results were consistent in both surveys, showing that no self-driving capability was the most preferred, with partially self-driving coming next and fully self-driving being the least preferred mode. In a fully self-driving mode, the highest concerns were expressed, while a desire to have the option to take over, and to be notified multimodally on this event, was expressed.

To conclude, although excitement is high, the public worldwide has expressed reservations regarding autonomy in vehicles, especially in terms of safety and liability. Retaining the option of control seems to be of vital importance for perceived user experience. Studies with a focus in the UK are lacking, with a 2014 survey by Ipsos MORI being one of the few available (Ipsos MORI 2014). In that work, there was a limited perception of importance of driverless technology, and perceived usefulness was higher with Londoners. Older respondents (aged over 55) were less willing to use this technology, while non-driving enthusiasts were more likely to accept this technology. This study, while useful, did not explicitly address the capabilities defined in inclusive design, while interaction with users was only in the context of a survey, with no follow-up discussions. Elaborating on the views of an inclusive user group in the UK is therefore essential and would aid a design for a wider range of capabilities.

3.2.2 INCLUSIVENESS

Inclusive design (Clarkson et al. 2003) considers a diverse user group in terms of capabilities in the design cycle. Viewing the population as either being 'able-bodied' or 'disabled' can be limiting when designing for users who would not identify themselves with any of the above groups. Difficulties can be present in everyday activities, ranging from hearing or vision, memory or thinking, communication, mobility, and dexterity, without necessarily leading to a person to self-identify, or to have sufficient capability loss, to be classified as having a disability. A lack of technological prior experience can also lead to exclusion from being able to operate technologies with complex interfaces, e.g. by those who are considered digitally excluded (Jaxa-Chamiec and Fuller 2007; Milner 2007; Bradley, Langdon and Clarkson 2015). Inclusive design aims to address these varying characteristics to create complete and usable solutions, rather than designing specialist products. By detecting difficulties when using a design and considering how these affect the usability of the product, it aims to provide solutions that can satisfy a wider range of the population. In the field of autonomy in vehicles, this methodology, while promising, has not been widely applied (Jeon et al. 2016).

Körber and Bengler (2014) provide a review of potential individual factors that could influence interaction with automation. They identify dispositional factors (relating to performance), traits (relating to task engagement), driver state (relating to fatigue), attitudes (relating to trust in automation), and demographics and other factors (relating to age, experience with advanced driver assistance systems, behavioural disorders, and working memory). Although broad, this classification is not as heavily focused on the individual and less transient characteristics of the users, which under Körber and Bengler's categorisation would possibly belong to the 'demographics and other factors' category. Souders and Charness (2016) focus on age as a defining characteristic of technology acceptance for autonomous vehicles, and identify trust and familiarity in the new technology as decisive factors for adoption by older adults. To this end, along with demographics and driving habits, an assessment of self-reported performance in vision, hearing, managing tasks of daily life, mobility or physical movement, and tasks that require precise hand movements was followed, in order to create a complete picture of user capabilities in everyday activities, as suggested in inclusive design.

The present study begins to address the exciting opportunity to utilise an inclusive design thinking as early as possible in the development of autonomous car technology. Other than Kunur et al. (2015), who studied future car concepts with no explicit focus on autonomy, no study has utilised this methodology in the past in the context of autonomous vehicles. Doing so can increase focus on user needs in the resulting concepts, identifying the population as an inclusive one from early on in the design process. Combining two different tools to detect user views, namely a survey and a focus group, can elicit distinct findings, which together can inform the creation of safe and inclusive HMI concepts for autonomy. The remainder of this paper will summarise the survey used as an initial tool for identifying the characteristics of a small user population, and as a recruitment tool for a series of focus groups to follow. Two focus groups with a subset of survey participants will then be described, which helped to identify different user views on the topic of autonomous cars and takeovers of control.

3.3 SURVEY

3.3.1 DESCRIPTION

A survey comprising questions on demographics, driving habits, and a self-reported assessment of capabilities was administered, using social networks, online mailing lists, and paper advertisements distributed in the city of Cambridge. After asking whether participants had a driving licence, they reported their most regular forms of transport (car, public transport, bicycle etc.), as well as the type of journeys most frequently made with a car (rural, suburban, or urban). The frequency with which respondents drove (daily, weekly, monthly etc.), and their estimation of annual mileage were then asked. The survey continued by asking participants to self-categorise into one of the traveller profiles identified in the Traveller Needs and UK Capability study by Transport Systems Catapult (Wockatz and Schartau 2015). An effort was made to reflect the characteristics of the profiles in Wockatz and Schartau (2015) into a set a set of items that would be easy for the survey participants to respond to. The description of profiles in Wockatz and Schartau (2015) and the respective questions in the present survey can be found in Table 3.1. The survey continued with questions about how much the respondents enjoyed driving, and how much they liked the idea of autonomous cars. Both questions used 5-point Likert scale items and were used as simple measures of user acceptance for both manual and autonomous driving. Since autonomous cars are not prevalent on the roads yet, the appeal of the idea of these cars was asked instead of current enjoyment. The survey continued with a set of questions on self-reported capabilities with vision, hearing, managing tasks of daily life (assessing thinking and communication), and mobility or physical movement (assessing reach and stretch), and with tasks requiring precise hand movements (assessing dexterity). The questions were used as a short self-assessment of capabilities and were modified from Clarkson et al. (2012), using information from the Inclusive Design Toolkit (University of Cambridge 2017). All questions used a 4-point Likert scale, in line with Clarkson et al. (2012), and participants could provide optional comments on any difficulties. Finally, the respondents' age, gender, and country of residence were asked, as well as their intention to participate in the following focus group.

3.3.2 RESULTS

In total, there were 97 respondents to the survey. Out of these, 63 were living in the UK and had a driving licence, for which results are reported in this paper. There were

TABLE 3.1

The profiles identified in Traveller Needs and UK Capability study (Wockatz and Schartau 2015) (A), and their interpretation in survey items in the present study (B)

A: Traveller Types	B: For What Reasons Do You Typically Drive a Car? Choose One or More
Progressive metropolites: Living in the heart of the city, typified by the technology-savvy young professional, with significant amounts of personal and business travel. Want to reduce their transport footprint	• As part of my job • For commuting to work • For everyday small trips • For taking children to school and household needs
Default motorists: High mileage drivers, with a mix of those who enjoy driving and many for whom it is a functional choice	• For the enjoyment of driving
Dependent Passengers: Dependent on others for their mobility needs, representing a mix of students, elderly, and those with impairments	• Because I have no other way of doing my journeys • I don't typically drive, I am usually a passenger
Urban riders: City dwellers, who travel less frequently than the progressive metropolites, making use of public transport available to them	• Other
Local drivers: Mainly retirees or stay at home parents, making low mileage local journeys	

25 female respondents, 37 male, and one preferred not to say. Descriptive statistics of the responses are presented in Table 3.2.

Independent sample t-tests showed that respondents who used a car as part of their job enjoyed driving more compared to respondents who did not ($t(61) = -2.78$, $p < 0.01$). Further, respondents who used a car for everyday small trips enjoyed driving more compared to respondents who did not ($t(61) = -3.23$, $p < 0.01$). As expected, respondents who used a car for the enjoyment of driving enjoyed driving more compared to respondents who did not ($t(26.94) = -4.37$, $p < 0.001$). Finally, respondents who used a car for other reasons to the ones mentioned in the survey enjoyed driving less compared to participants who did not ($t(61) = 2.60$, $p < 0.05$). Other reasons for driving mentioned were mostly related to holidays and social visits.

A one-way analysis of variance (ANOVA) with frequency of driving as a factor revealed a significant main effect on driving enjoyment ($F(4,58) = 2.92$, $p < 0.05$). Planned contrasts revealed that respondents who drove less than once a year enjoyed driving less compared to respondents who drove once a year or more ($t(58) = -2.35$, $p < 0.05$), once a month or more ($t(58) = -2.20$, $p < 0.05$), once a week or more ($t(58) = -2.95$, $p < 0.01$), and everyday ($t(58) = -3.16$, $p < 0.01$). No other significant effects were found regarding how driving habits and user characteristics affected driving enjoyment, and no significant effects were found regarding how driving habits and user characteristics affected likeability of the idea of autonomous cars. Regarding inclusive user characteristics of the specific group of respondents, there were limited responses indicating any difficulty in everyday life (see Table 3.2).

TABLE 3.2

Descriptive Statistics of the Survey on Driving Habits

Age

16–20	21–30	31–40	41–50	51–60	61–70	71–80	Prefer not to say
1	10	20	13	9	7	2	1

Self-Reported Capabilities (N: Not at all limited, S: Somewhat limited, V: Very limited)

Vision			Hearing			Daily Life Tasks		Mobility			Dexterity	
N	S	V	N	S	V	N	S	N	S	V	N	S
59	3	1	59	3	1	61	2	55	7	1	62	1

Regularly Used Means of Transport*

Car	Public Transport	Motorcycle	Bicycle	Walking
52	33	4	30	47

Type of Journeys Primarily Driven*

Rural	Suburban	Urban
39	36	31

Typical Reasons to Drive*

As Part of Job	Commuting to Work	Everyday Small Trips	Taking Children to School/ Household Needs	Enjoyment	Because no Other Way of Doing Journeys	Not Typically Drive, Usually Passengers	Other Reason
11	24	31	10	10	18	4	20

Frequency of Driving

Less than once a Year	Once a Year or More	Once a Month or More	Once a Week or More	Everyday
2	8	8	23	22

Annual Mileage

Less than 2,000 Miles	Between 2,000 and 7,000 Miles	Between 7,000 and 12,000 Miles	More than 12,000 Miles	Not Sure
15	22	20	4	2

Driving Enjoyment ($M = 3.59$, $SD = 1.07$)

1 (Not at all)	2	3	4	5 (Very much)
3	5	21	20	14

Likeability of the Idea of Autonomous Cars ($M = 3.30$, $SD = 1.47$)

1 (Not at all)	2	3	4	5 (Very much)
10	11	11	12	19

Note that some of the above categories (indicated by *) are not mutually exclusive. In these categories, one person can have multiple responses; for example, they may both use car and public transport as means of transport. In all cases, absolute numbers are reported.

The respondents of the survey were low in number, limiting the generalisability of the results presented. Future work could distribute this survey to a wider UK population, in order to present more extensive results, using a more inclusive set of users in terms of self-reported capabilities. However, as discussed, this survey was mainly used as a tool to recruit participants for the focus groups, and as such, it achieved its goal. Further, the characteristics of the population presented were not dissimilar to the ones presented by larger-scale surveys, reviewed in Section 3.2.1. Finally, the statistically significant results observed provide confidence that a larger population could display even richer effects. A subset of the survey respondents were invited to either of two focus groups at the University of Cambridge, to be described in the next section.

3.4 FOCUS GROUPS

3.4.1 DESCRIPTION

In order to elicit more elaborate discussions on the topic of autonomous cars, two focus groups with participants of the survey were conducted at the University of Cambridge in July and August 2016, focus group 1 and focus group 2, as shown in Figure 3.1 (FG1 and FG2, respectively). The topic of interest for both was the takeover of control in autonomous cars. As discussed earlier, takeovers are situations where the car transfers control to the driver and handovers, and vice versa. These can occur for various reasons and can be voluntary or otherwise. This topic was selected as one of particular interest to the industry, as well as one that is expected to become especially relevant as cars become more autonomous (Politis, Brewster and Pollick 2015). Previous studies have explored the interactions that can occur in such a situation, e.g., by investigating appropriate warnings to deliver to drivers (Mccall et al. 2016). However, exploring participants' thoughts on takeovers in the context of a focus group had not been previously attempted. To invite the thoughts of sufficiently different user groups, the focus groups used different demographics.

FG1 consisted of five males aged between 31 and 39 years ($M=35.00, SD=4.00$), had driving experience between 11 and 19 years ($M=15.40, SD=3.05$), and practised mostly technical professions (Project Architect, Hardware Design Engineer, Motion Graphics Designer, Research Engineer, and Quality Control Assistant). FG2 consisted of three females and three males, aged between 24 and 79 years ($M=55.83, SD=18.93$), had driving experience between 5 and 62 years (one participant did not report years of driving experience, for the rest: $M=34.00, SD=21.14$), and practised a mixed set of professions (one person did not report profession, for the rest: Retired Government Officer, Designer, Retired Motor Engineer, Illustrator, and Charity Worker).

Both focus groups had an identical format. Aside from the focus group participants, two researchers were facilitating the discussions and taking notes. The focus groups were recorded and filmed with the participants' consent, and a computer presentation, corresponding to all parts of the focus group, was always projected. Further, there were schematics of different road types (a highway, a city road, and a roundabout) available as large prints, as well as miniature cars, to be used in any part

FIGURE 3.1 The set-up of the two focus groups (FG1 (a) and FG2 (b)).

of the discussion. In the beginning, participants signed consent forms, and received web store vouchers for their time, while refreshments and light food were available. They were informed that there were no right or wrong answers in the discussions, since their views were sought, and that any data received would be anonymised. After the initial introductions of participants and researchers, the topic of takeovers was introduced, using schematics and videos through the presentation. Any initial thoughts on the topic were discussed, using unstructured discussions. This led to the next part of the focus group, where two specific scenarios requiring takeovers were discussed. These scenarios were focusing on the case where a driver would need to take control from the car, either before an exit on the highway, or before a roundabout in the city. Any additional thoughts about each of the two scenarios were discussed, with a focus on what the car should be like, and how it should behave, aiming to elicit comments on the car's HMI design. For these two scenarios, aside from the unstructured discussions, a set of questions were always visible to the focus group participants in the presentation, to assist more focused discussions. These discussions were related to what would happen in these scenarios, and when and how would it happen, what would the driver need to know, who would be involved, and any other thoughts. After asking for any final thoughts or comments, the focus groups were concluded. They lasted about 2 h each. See Figure 3.1 for the set-up used.

3.4.2 RESULTS

The results of the focus groups were analysed thematically by three coders, who were members of the research team, as suggested in Flick (2009). NVivo (QSR International 2020) software was used to create thematic nodes, and each coder's analysis was iterated and revisited by the next coder. This resulted in two thematic analyses, one for each focus group. The main views discovered are summarised in Table 3.3.

As can be seen in Table 3.3, the views discovered in the two focus groups bear similarities. A need for the driver to be informed during the process of takeovers was expressed by both groups, proposing multimodal displays (FG1) or saliency in general (FG2) as solutions. Further, simplicity of the HMI in autonomous cars was desired by both groups. Improved safety when autonomous driving technology becomes mainstream was expected and required by both groups. A point made by participants of both focus groups was that autonomy is useful only if it is complete, due to potential complexities in driving partially autonomous cars. This view

TABLE 3.3

A Summary of the Views Discovered in the Focus Groups, Sorted by the Frequency of Occurrence of the Respective Themes in the Thematic Analysis

Focus Group 1	Focus Group 2
• Multimodal displays that maintain the driver's situation awareness during takeovers are required, to aid safety.	• There is concern whether autonomous cars will be able to cope with difficult road scenarios, and whether their interfaces will be robust enough.
• HMIs designed for takeovers should be simple and easy to use.	• Autonomous cars would need to address limitations in everyday life activities, such as sight problems, by simplifying the HMI, using, e.g., voice control.
• The driver should be the one who decides whether to give or take control of driving.	
• Cars should either have full autonomy or no autonomy.	• Vehicle-to-vehicle and vehicle-to-specialised infrastructure communications can create safer future roads.
• Self-driving cars can provide a better use of the driver's time, safety, and efficiency.	
• Self-driving cars can create ethical questions in case of incidents, loss of jobs, and privacy-related concerns.	• Takeovers should be communicated saliently by the HMI well before they need to happen.
• Takeovers should be safe.	• Autonomous cars are expected to increase road safety and comfort.
• Autonomous cars should enhance the driver's capabilities.	• Cars should have either full autonomy or no autonomy.
	• Autonomous cars should adjust to the driver's driving style.

received no clear consensus in either of the groups, but it was a seemingly strong one for the participants who shared it. Notable differences between the focus group views were also observed. Participants of FG1 put emphasis on retaining control of the takeover situation, having the final say on who will have control. This seemed to be less the case for FG2 participants, who were mostly concerned about whether vehicle and HMI technologies will be mature enough to accommodate an autonomous functionality, and whether it will be safe and robust enough. Conversely, FG1 participants expressed concerns related to hard ethical questions on liability in case of accidents with an autonomous car, as well as concerns related to loss of jobs and exploitation of privacy. FG1 participants, however, also saw an opportunity for a better use of the driver's time in autonomous cars. They felt that autonomous cars can assist them in becoming even better and more efficient in their activities. In contrast, FG2 participants required autonomy to address limitations of the driver's capabilities, such as sight problems, adjust to their driving style, and facilitate driver comfort through easy-to-use HMIs, such as voice-based ones. They also required a comfortable period to signify a takeover by the driver, aided by communications between cars and infrastructure.

3.5 DISCUSSION

The results of the survey, although with a small sample of respondents, showed a good distribution of group characteristics. There was representation of all age ranges and coverage of all driving profiles. Since this survey was the first, to the authors' knowledge, to combine questions on inclusive characteristics, driving habits, driving enjoyment, and likeability of the idea of autonomous cars, it is difficult to make direct comparisons with available literature in terms of responses. It is argued, however, that the combination of these questions covered a wide range of information, and can be used as tool to recruit a more inclusive participant sample for later studies, such as focus groups (as done successfully in this paper) and experiments (as planned for future work). It is also expected that a larger sample of respondents would reveal even clearer patterns on how driver characteristics correlate with driver attitudes towards manual and autonomous cars. However, even using this small sample of respondents, some clear patterns emerged. These patterns were related to driving enjoyment and showed that participants who drove more or drove due to their work enjoyed driving more. No such pattern was found for the likability of the idea of autonomous cars. This is comparable to the results of Rödel et al. (2014); Schoettle and Sivak (2015, 2016); and Kyriakidis, Happee, and de Winter (2015), where manual modes of driving were considered as the most enjoyable and accepted. There still seems to be road to cover in terms of improving acceptance of autonomous cars, as also found for the UK in Ipsos MORI (2014), and this study showed no evidence to the contrary. In the authors' view, and as also posited by Rödel et al. (2014) and Souders and Charness (2016), a more widespread exposure to the technology may mitigate concerns and improve trust. Engaging an inclusive user group during the design process is a step in this direction.

The results of the focus groups showed clear evidence on why an inclusive design process can be beneficial. This is because similar views on major topics of concern were discovered, but also distinct points were made by the groups in areas of interest for inclusive design. Participants of both focus groups were interested in a technology that will be safe and will assist their everyday needs – a popular finding also in studies like Schoettle and Sivak (2014a, 2014b); Kyriakidis, Happee, and de Winter (2015); and Silberg et al. (2013). However, FG1 participants saw the technology as augmenting their already high everyday capabilities, and allowing them to be productive in autonomous mode, with them still having the final say in things. Being a highly technical group, they were mostly worried about social implications of autonomy and less about robustness. FG2 participants, on the other hand, expected the technology to be robust, to not fail, to be comfortable, and to account for possible personal limitations in everyday life. This is a new result and highlights the different views of this technology for groups of different capabilities. It can help autonomous vehicle designers increase their inclusivity, by providing solutions that a wider spectrum of drivers could benefit from. On the topic of in-car displays, the need for clear, easy to interpret, multimodal information during takeovers was a point of consensus between the two groups. Available literature (e.g. Mccall et al. (2016)) has looked into the effectiveness of such warning mechanisms, but never before has this requirement been confirmed in the context of a focus groups with such varying demographics. Autonomous vehicle designers can benefit from this guideline, by creating usable and inclusive interfaces with salient cues that are easy to react to. Finally, the polarisation of views on whether partial autonomy is acceptable, again by both groups, raises an issue of acceptance. Some participants were concerned that partial autonomy might be more of a burden than a liberator, due to high demand from the driver. This point needs to be read carefully since the availability of autonomous cars needs to be an enabler for users, and high demand interfaces may limit perceived usefulness. Other than Bazilinskyy, Kyriakidis, and de Winter (2015), where split positive and negative views between autonomous and non-autonomous cars were discovered, the authors are not aware of any other study that highlights this split of opinions on whether the in-between step of partial automation is useful for all.

3.6   CONCLUSIONS

This paper presented a survey with UK participants as a recruitment tool for later studies, and two subsequent focus groups, with an inclusive user group of survey respondents. The survey was novel in combining self-reported inclusive characteristics, driving habits, and acceptance of manual and autonomous driving. It demonstrated how even a small number of respondents can reveal clear attitude patterns towards manual driving and achieve an acceptable spread of demographics. It also enabled later recruitment of users with distinct characteristics as part of an inclusive design methodology. The two focus groups, using survey participants of different inclusive characteristics, showed consensus in the topics of safety and usability of autonomous cars and HMIs, and the use of multimodal displays as a warning mechanism. They also revealed a split of opinions on whether partial autonomy is acceptable. However, different concerns between the focus groups were also discovered, with the younger participants seeing autonomy as an augmenter of already high capabilities, and the more inclusive group as an enabler for possible limitations. The implications of these findings for autonomous vehicle and HMI designers were discussed.

ACKNOWLEDGEMENTS

This work was funded by EPSRC and Jaguar Land Rover, as part of the project Human Interaction: Designing Autonomy in Vehicles (HI:DAVe), Project Grant Number: EP/N011899/1.

REFERENCES

Abbugao, M. 2016. "Driverless taxi firm eyes operations in 10 cities by 2020." Agence France-Presse, August 29.
Bazilinskyy, P., M. Kyriakidis, and J. de Winter. 2015. "An International Crowdsourcing Study into People's Statements on Fully Automated Driving." *Procedia Manufacturing* 2534–2542. https://doi.org/10.1016/j.promfg.2015.07.540.
Bradley, M., P. Langdon, and P. J. Clarkson. 2015. "Assessing the inclusivity of digital interfaces - a proposed method." Antona M., Stephanidis C. (eds) *Universal Access in Human-Computer Interaction. Access to Today's Technologies. UAHCI 2015. Lecture Notes in Computer Science.* Springer International Publishing. 25–33. https://doi.org/10.1007/978-3-319-20678-3_3.
Clark, B., G. Parkhurst, and M. Ricci. 2016. *Understanding the Socioeconomic Adoption Scenarios for Autonomous Vehicles: A Literature Review.* Project Report, Bristol: University of the West of England.

Clarkson, P. J., R. Coleman, S. Keates, and C. Lebbon. 2003. *Inclusive Design: Design for the Whole Population*. London: Springer-Verlag.

Clarkson, P. J., F. A. Huppert, S. Waller, J. Goodman-Deane, P. Langdon, J. Myerson, and C. Nicolle. 2012. "Towards Better Design 2010." Colchester: UK Data Archive [distributor], May 2012. SN: 6997, http://dx.doi.org/10.5255/UKDA-SN-6997-1

Flick, U. 2009. *An Introduction to Qualitative Research - Fourth Edition*. California: SAGE Publications Ltd.

Hawkins, A. J. 2016. "Delphin and Mobileye are teaming up to build a self-driving system by 2019." *The Verge*, August 23.

Ipsos MORI. 2014. "Only 18 per cent of Britons believe driverless cars to be an important development for the car industry to focus on." July 30. https://www.ipsos.com/ipsos-mori/en-uk/only-18-cent-britons-believe-driverless-cars-be-important-development-car-industry-focus (accessed 4 November 2020)

Jaxa-Chamiec, M., and R. Fuller. 2007. "Understanding digital inclusion - a research summary." UK Online Centres, UK. http://www. ukonlinecentres.co.uk (accessed 3 November 2020).

Jeon, M., I. Politis, S. Shladover, C. Sutter, J. Terken, and B. Poppinga. 2016. "Towards life-long mobility: accessible transportation with automation." *AutomotiveUI 2016-8th International Conference on Automotive User Interfaces and Interactive Vehicular Applications*. ACM. 203–208. https://doi.org/10.1145/3004323.3004348.

Körber, M., and K. Bengler. 2014. "Potential individual differences regarding automation effects in automated driving." *Proceedings of the XV International Conference on Human Computer Interaction*. Spain: Association for Computing Machinery. 1–7. https://doi.org/10.1145/2662253.2662275.

Kunur, M., P. Langdon, M. Bradley, Bichard J. A., E. Glazer, F. Doran, P. J. Clarkson, and J. J. Loeillet. 2015. "Creating inclusive HMI concepts for future cars using visual scenario storyboards through design ethnography." *International Conference on Universal Access in Human-Computer Interaction*. Springer. 139–149. https://doi.org/10.1007/978-3-319-20687-5_14.

Kyriakidis, M., R. Happee, and J. C. F. de Winter. 2015. "Public opinion on automated driving: Results of an international questionnaire among 5000 respondents." *Transportation Research Part F: Traffic Psychology and Behaviour* 127–140. https://doi.org/10.1016/j.trf.2015.04.014.

Mccall, R., F. McGee, A. Meschtscherjakov, N. Louveton, and T. Engel. 2016. "Towards a taxonomy of autonomous vehicle handover situations." *AutomotiveUI - Proceedings of the 8th International Conference on Automotive User Interfaces and Interactive Vehicular Applications*. Ann Arbor: Association for Computing Machinery. 193–200. https://doi.org/10.1145/3003715.3005456.

Meschtscherjakov, A., M. Tscheligi, D. Szostak, R. Ratan, R. McCall, I. Politis, and S. Krome. 2015. "Experiencing autonomous vehicles: crossing the boundaries between a drive and a ride." *CHI EA - Proceedings of the 33rd Annual ACM Conference Extended Abstracts on Human Factors in Computing Systems*. Seoul, Republic of Korea: Association for Computing Machinery. 2413–2416. https://doi.org/10.1145/2702613.2702661.

Meschtscherjakov, A., M. Tscheligi, D. Szostak, S. Krome, B. Pfleging, R. Ratan, I. Politis, S. Baltodano, D. Miller, and W. Ju. 2016. "HCI and autonomous vehicles: contextual experience informs design." *CHI EA - Proceedings of the 2016 CHI Conference Extended Abstracts on Human Factors in Computing Systems*. San Jose, USA: Association for Computing Machinery. 3542–3549. https://doi.org/10.1145/2851581.2856489.

Milner, H. 2007. *Digital Inclusion: A discussion of the Evidence Base*. London: Freshmiinds.

National Highway Traffic Safety Administration. 2013. *Preliminary Statement of Policy Concerning Automated Vehicles*. Washington, DC: U.S. Department of Transportation.

Politis, I., S. Brewster, and F. Pollick. 2015. "Language-based multimodal displays for the handover of control in autonomous cars." *AutomotiveUI - Proceedings of the 7th International Conference on Automotive User Interfaces and Interactive Vehicular Applications.* Nottingham: Association for Computing Machinery. 3–10. https://doi.org/10.1145/2799250.2799262.

QSR International. 2020. *Technology that Empowers Real World Change.* https://www.qsrinternational.com/.

Rödel, C., S. Stadler, A. Meschtscherjakov, and M. Tscheligi. 2014. "Towards autonomous cars: the effect of autonomy levels on acceptance and user experience." *Proceedings of the 6th International Conference on Automotive User Interfaces and Interactive Vehicular Applications.* Seattle: AutomotiveUI. 1–8. https://doi.org/10.1145/2667317.2667330.

SAE J3016. 2016. *Taxonomy and Definitions for Terms Related to Driving Automation Systems for On-Road Motor Vehicles.* https://www.sae.org/standards/content/j3016_201806/ (accessed 3 November 2020).

Sage, A., and P. Lienert. 2016. "Ford plans self-driving car for ride share fleets in 2021." *Reuters*, August 16.

Schoettle, B., and M. Sivak. 2014a. *A Survey of Public Opinion about Autonomous and Self-Driving Vehicles in the U.S., the U.K. and Australia.* Ann Arbor: University of Michigan.

Schoettle, B., and M. Sivak. 2014b. "Public opinion about self-driving vehicles in China, India, Japan, the U.S., the U.K., and Australia." University of Michigan, Ann Arbor, Transportation Research Institute.

Schoettle, B., and M. Sivak. 2015. "Motorists' preferences for different levels of vehicle automation." University of Michigan, Ann Arbor, Transportation Research Institute.

Schoettle, B., and M. Sivak. 2016. "Motorists' preferences for different levels of vehicle automation. University of Michigan, Ann Arbor, Transportation Research Institute.

Sessa, C., F. Pietroni, A. Alessandrini, D. Stam, P. Delle Site, C. Holguin, M. Flament, and S. Hoadley. 2015. "Results on the on-line DELPHI survey." *CityMobil2.*

Silberg, G., M. Manassa, K. Everhart, D. Subramanian, M. Corley, H. Fraser, and V. Sinha. 2013. *Self-Driving Cars: Are We Ready?* KPMG LLP.

Souders, D., and N. Charness. 2016. "Challenges of older drivers' adoption of advanced driver assistance systems and autonomous vehicles." *International Conference on Human Aspects of IT for the Aged Population.* Springer. 428–440. https://doi.org/10.1007/978-3-319-39949-2.

University of Cambridge. 2017. *Inclusive Design Toolkit.* www.inclusivedesigntoolkit.com.

Wockatz, P., and P. Schartau. 2015. *Intelligent Mobility: Traveller Needs and UK Capability Study.* Transport Systems Catapult.

Part II

Lo-Fi and Hi-Fi Simulators

4 An Evaluation of Inclusive Dialogue-Based Interfaces for the Takeover of Control in Autonomous Cars

Ioannis Politis
The MathWorks Inc.

Patrick Langdon
Edinburgh Napier University

*Damilola Adebayo, Michael Bradley,
P. John Clarkson, Lee Skrypchuk,
and Alexandros Mouzakitis*
University of Cambridge

Alexander Eriksson
Swedish National Road and Transport Research Institute

*James W.H. Brown, Kirsten M. A. Revell,
and Neville A. Stanton*
University of Southampton

CONTENTS

4.1   INTRODUCTION

Autonomous cars are reaching the roads, and expected to become more widespread in the near future (Hawkins 2017; Phys.org 2017), although not without concerns from the public over their impact (Rödel et al. 2014). This has motivated a significant body of research in public opinion about these vehicles (Kyriakidis, Happee and de Winter 2015; Rödel et al. 2014; Schoettle and Sivak, 2014a, 2014b, 2015, 2016), anticipated behaviours when interacting with the technology within these cars (Naujoks et al. 2017), and user interface interventions to make this interaction more safe and usable (Forster, Naujoks and Neukum 2016; Large et al. 2017; van Veen and Terken 2017; Walch et al. 2016). As described in SAE J3016 (2016), car autonomy is a spectrum rather than a binary state. This standard defines levels of autonomy, ranging from 0 (no driving automation) to 5 (full driving automation), with decreasing expected driving engagement as levels increase. Both ends of this spectrum have relatively straightforward driver engagement expectations; level 0 expects full engagement, and level 5 expects no engagement. The problem arguably becomes harder in the intermediate levels, where *some* driving engagement is expected. This has triggered extensive research on how to design for *some* driving engagement, and better support this transition between car and driver control, the takeover. Available approaches have investigated, among others, new interfaces utilising rich displays, expected to keep the driver alert (Borojeni et al. 2016; Eriksson et al. 2019; Naujoks, Mai and Neukum 2014; Politis, Brewster and Pollick, 2015, 2017), as well as made efforts in modelling the tasks that take place during takeovers, so as to better design for them (Banks, Stanton and Harvey 2014; Mccall et al. 2016; Walch, Lange et al. 2015).

The approach of this paper is one of inclusive designs (Clarkson et al. 2003; Jeon et al. 2016), considering a diverse user group in terms of capabilities in the design cycle. We are interested to investigate the opinions and difficulties of such a group regarding the issue of takeovers, and aim to design solutions that will satisfy users of varying demographics and degrees of functional capability. There are studies investigating opinions of such groups on the topic (e.g. Langdon et al. 2017; Politis, Langdon et al. 2017). An emerging requirement is simple, safe, and easy-to-use interface, enhancing the driver's capabilities. Voice control has explicitly been suggested as a means of achieving this (Politis, Langdon et al. 2017). Further, existing studies in automation have outlined the merits of using dialogue interaction as a means to reduce workload in other modalities, but also as a facilitator of situation awareness (SA) (Forster, Naujoks and Neukum 2017; Lee, Joo and Nass 2014; Naujoks et al. 2016). We therefore decided to take the approach of designing a set of dialogue systems as a start. We explored different types of potential dialogue interactions, in terms of how users perceive them and how they behave when exposed to them. We designed an experiment evaluating the dialogue interfaces, and results confirmed that simplicity of the interfaces is paramount, even at the cost of available information regarding the road. The rest of this paper will describe the experiment and conclude with views on how to better design simple takeover interfaces that facilitate SA.

4.1.1 DIALOGUE-BASED INTERFACES DESIGNED

A set of dialogue-based interfaces were designed, addressing the takeover process in a step-based approach and aiming to be facilitators of varying degrees of SA (Endsley 1995). Using input from prior studies (Forster, Naujoks and Neukum 2017; Politis et al. 2017), the goal was to achieve a natural verbal interaction with the vehicle, which is likely to put less strain on the driver's attentional resources when resuming control (Horrey and Wickens 2004; Wickens 1980, 2008). Although there are available takeover strategies involving speech output from the car (e.g. Forster, Naujoks and Neukum 2017; Politis, Brewster and Pollick 2015), there are no approaches, to the authors' knowledge, utilising a natural dialogue for the takeover. This difference is important as it involves a significant added effort from the driver in order to achieve the takeover. To reflect some fundamentally different dialogue structures, four systems were designed. They envisaged a case of an SAE Level 3 autonomous vehicle system, and a takeover due to sensor degradation. All systems were adaptations of emerging ideas from a design workshop with members of the automotive industry and academia, which had taken place in the authors' home institution. All systems required an initial confirmation of readiness to assume control of the vehicle, which was a basic form of dialogue. After the confirmation was given, there were the following alternatives:

- A **countdown-based (CB)** system, where there was no further dialogue required, only a button press from the driver to assume control, and a countdown of 60 s in blocks of 10 s. This concept was simulating common available takeover concepts from the industry (e.g. Volvo 2017), with the addition of an initial dialogue-based confirmation of readiness.
- A **repetition-based (RepB)** system, prompting the driver to repeat information, uttered by the system, about essential driving features when resuming control, including hazards, current lanes, current fuel level, current speed, intended exit, and immediate actions to be taken. This concept was based on aviation readbacks (Civil Aviation Authority 2015), a standard practice of pilots repeating important information to air traffic controllers. Although readbacks aim to reduce misunderstandings between air traffic controllers and pilots, the rationale of using such a technique in the automotive domain was that it might benefit a similar interaction during a takeover, and possibly increase driver SA.
- A **response-based (ResB)** system, asking the driver to provide the same information as the RepB system by answering to related questions, posed by the system, instead of asking them to repeat it. This provided an analogous interaction to RepB, expecting higher engagement from the driver's side, as they would need to find and verbalise the required information rather than solely repeating it, similarly to answering SA-related questions in the SA global assessment technique (Endsley 1988).
- A **multimodality-based (MB)** system, identical to the RepB system, with added multimodal information, i.e. LED strips in amber colour, animating

on either side of the driver and simulating the position of surrounding traffic, as well as ambient vibrations, augmenting the utterances of the dialogue system. The rationale was that multimodal cues have repeatedly shown benefits, informing the driver about events on the road (Politis, Brewster and Pollick, 2015; 2017), and multimodal information has emerged as a requirement in qualitative studies in this domain (Politis et al. 2017).

An experiment evaluating the above dialogue systems was designed, aiming to evaluate users' performance and views when exposed to the dialogue-based takeover interfaces.

4.2 EXPERIMENT

The designed experiment used a within-subject design, with type of interface (ToI – CB, RepB, ResB, MB) as the single independent variable. The dependent variables were perceived workload (using the NASA task-load index (NASA-TLX), raw TLX adaptation (Hart 2006; Hart and Staveland 1988)), perceived acceptance (using the acceptance scale (AS) (Van Der Laan, Heino and De Waard 1997)), perceived usability (using the system usability scale (SUS) (Brooke 1996)), perceived SA (using the situation awareness rating technique (SART) (Taylor 1989)), dialogue performance (performance during the verbal interaction with the different dialogue systems), takeover time (TOT – the time from the start of the dialogue interaction until participants resumed control of the car), and driving performance (speed, longitudinal acceleration, and absolute angle input (AAI)). It was hypothesised that ToI would influence these measures.

4.2.1 PARTICIPANTS

The experiment had 49 participants (24 females and 25 males), aged between 17 and 86 years ($M = 45.51$, $SD = 17.36$). They all held a valid driving license and had between 1 and 70 years of driving experience ($M = 25.68$, $SD = 17.58$). In order to adhere to the principles of inclusive design (Keates and Clarkson 2002), effort was made to recruit representative population samples in terms of age and gender. Thus, there were 7 males and 5 females in the age band of 17–34, 11 males and 13 females in the age band of 35–56, and 7 males and 6 females in the age band of 57–86. The recruitment was opportunistic; however, the resulting sample displayed a wide variety of ages, distributed between genders.

4.2.2 EQUIPMENT

A low-fidelity driving simulator was used for the experiment (STISIM 2017). It consisted of three computer screens, depicting the front and sides of the road scene; a gaming car seat; a gaming steering wheel; and centre stack software, running on a separate computer, depicting current speed, state of automation, upcoming exits, desired lane to be in, and a pictorial representation of surrounding traffic. The simulator was depicting a circa ten-mile-long route, with straight rural roads and a motorway with no major turns. When in autonomous mode, participants were using a 10-inch tablet to perform a secondary task. Their movements were filmed with a wide-angle camera. Only used in MB interface, there were also two LED strips mounted on the wall on either side of the car seat, and a tactile-acoustic device (Karam, Russo and Fels 2009), translating audio information into ambient vibrotactile stimuli.

4.2.3 PROCEDURE

Participants were welcomed and had the purpose of the experiment explained to them. After responding to demographics questions, they were given a £20 web voucher for their time. They were then introduced to the driving simulator, completing a short drive. Other than a general description of the dialogue systems, provided by the experimenter, there was no prior training with the systems before the main part of the experiment. Their interaction with the different types of interfaces followed, counterbalanced through a Latin square. Participants were instructed to play a tablet memory game, also used in Politis, Brewster, and Pollick (2015, 2017), and to resume driving whenever requested. In each interaction, there were three takeover requests, approximately when 10%, 40%, and 70% of the route was completed. Requests were spoken by a female text-to-speech voice (Microsoft Anna (Microsoft 2012)). Participants were expected to interact with the dialogue interfaces and press a button on the steering wheel to take over control once the dialogues were completed. Recognition of their responses was based on Wizard of Oz, meaning that there was no actual dialogue system, but the experimenter was manipulating software on another computer, choosing the appropriate utterances based on the participants' responses, in line with studies like Dhillon et al. (2011) and Large et al. (2016). Any wrong or missed answers by participants regarding road-related information were counted as errors, affecting their dialogue performance. The latter was scored as 0 if no mistakes were made, or 1 if any number of mistakes were made, including failing to acknowledge readiness to takeover. In the case of CB interface, failing to acknowledge readiness of takeover was the only possible mistake, as there was no further dialogue. TOT was calculated from the time any dialogue interface required verbal confirmation of readiness, until participants had completed all steps of the verbal interaction and took over by pressing the button when ready (in the cases of RepB, ResB, and MB), or by the time participants pressed the button to take over control, as no further confirmation was required (in the case of CB). When participants assumed control in each of the three takeovers, they were asked to drive as normal for 60 s, during which time their driving performance was logged, and they were then reminded to give control back to the car. After the interaction with each interface, lasting about 15 min, the NASA-TLX, AS, SUS, and SART questionnaires were administered, in this order. After all interfaces had been tested, a ranking of concepts was performed, concluding with a short interview (not to be discussed in detail in this paper). The experiment lasted a total of 90 min.

4.2.4 RESULTS

Data for perceived workload, perceived acceptance, perceived usability, and perceived SA were analysed using a one-way analysis of variance (ANOVA), with ToI as a factor, and planned contrasts between conditions. A significant effect of ToI on perceived workload ($F(3,191) = 2.95$, $p < 0.05$, $\omega = 0.17$) revealed that ResB was perceived as inducing significantly higher workload than CB and RepB. A significant effect of ToI on perceived acceptance, in the satisfaction score ($F(3,191) = 8.71$, $p < 0.001$, $\omega = 0.33$), revealed that CB had significantly higher perceived satisfaction than RepB, ResB, and MB. A significant effect of ToI on perceived usability ($F(3,190) = 5.97$, $p < 0.01$, $\omega = 0.27$) revealed that CB had significantly higher perceived usability than RepB, ResB, and MB. There were no significant effects observed in terms of perceived SA. Further, a Friedman test revealed that participants' rankings of the ToI were significantly different across interfaces ($\chi^2(3) = 47.12$, $p < 0.001$). Wilcoxon tests with Bonferroni correction revealed that CB was significantly preferred to RepB, ResB, and MB, and ResB was significantly preferred to MB. Data for dialogue performance was treated as dichotomous and analysed with Cochran's Q tests. These revealed that ResB induced more errors than CB ($Q(1) = 22.00$, $p < 0.001$), RepB ($Q(1) = 19.88$, $p < 0.001$), and MB ($Q(1) = 15.38$, $p < 0.001$). In terms of TOT, a one-way ANOVA, with ToI as a factor, and planned contrasts between conditions, found a significant effect of ToI on perceived workload ($F(3,187) = 156.16$, $p < 0.001$, $\omega = 0.84$), revealing that TOT in CB was significantly shorter than in RepB, ResB, and MB. Concluding, a one-way ANOVA, analysing driving performance, specifically AAI, with ToI as a factor, and planned contrasts between conditions, found a significant effect of ToI on AAI ($F(3,190) = 4.62$, $p < 0.01$, $\omega = 0.23$), revealing that AAI in CB was significantly lower than in RepB, ResB, and MB. No other driving metrics were found to elicit any significant effects. See Table 4.1 for all results. Finally, two-way mixed ANOVAs revealed no significant effect of age on the results.

4.3 DISCUSSION

The subjective results of the experiment favoured simplicity, regardless of age. The CB interface, which required the shortest TOT, and allowed the participants to choose the moment of takeover, induced higher perceived usability and acceptance, and was the highest ranked interface. This suggests that the desired interaction during takeover needs to be succinct and under participant's timing control. This is in line with Langdon et al. (2017) and Politis et al. (2017), where a simplicity of interaction was found to be desirable. The longer interactions of the rest of the concepts, although arguably doing more to increase SA, were less preferred. Interview results supported this argument since participants acknowledged that CB was doing the least to support their knowledge of what was happening on the road but appreciated its short duration and sense of control it elicited. We note that the shorter interaction time and less errors of CB were a result of the design and were not an indicator of higher performance or SA, since no driver information was provided in CB. Arguably, participants may have established SA to their satisfaction through the simulation. As a guideline, intelligent dialogue systems for takeovers should enable a straightforward interaction, providing only essential information. The only quantitative result in terms of driving performance that differed across types of interface

TABLE 4.1

Results of Perceived Workload (PW – Using NASA-TLX), Perceived Usability (PU Using SUS), Satisfaction Score of Perceived Acceptance (PA – Using AS), Ranking of Preferences, Dialogue Performance (DiP), Takeover Time (TOT), and Absolute Angle Input (AAI)

	CB	RepB	ResB	MB
PW (*M, SD, SE*)	22.38, 14.35, 2.05	22.91, 11.80, 1.69	29.72, 15.56, 2.22	26.32, 13.52, 1.95
PU (*M, SD, SE*)	85.15, 11.55, 1.65	77.35, 12.63, 1.80	75.26, 13.62, 1.97	75.89, 14.42, 2.08
PA (*M, SD, SE*)	1.05, 0.76, 0.11	0.14, 1.05, 0.15	0.37, 1.01, 0.14	0.20, 1.08, 0.16
Ranking (*Median*)	1	3	2	3
DiP (*% Correct*)	86.36	85.71	32.65	75.00
TOT in *sec* (*M, SD, SE*)	19.79, 9.30, 1.37	48.61, 4.67, 0.67	47.79, 10.04, 1.43	49.65, 6.39, 0.93
AAI in *deg* (*M, SD, SE*)	0.36, 0.34, 0.05	1.12, 1.26, 0.17	1.45, 2.53, 0.36	1.21, 1.19, 0.17

Mean (M), standard deviation (SD), standard error (SE), median, and percentage values are reported as appropriate.

was a lower AAI for CB after the takeover, indicating a smoother steering behaviour. This finding could arguably be attributed to lower disruption to the driving task (as was observed in Politis, Brewster, and Pollick (2015).

The next concept in order of preference was voice autopilot assistance, which interestingly was the concept inducing the highest perceived workload and resulting in the most dialogue mistakes. This is an expected result, as ResB required the longest interaction, and required answers rather than repetitions (like RepB and MB) or no action (like CB). This supports the argument that making the driver think about what is happening on the road before taking over was not disliked as an idea. Interview responses mentioned that if there were some shortcuts in the interaction of ResB, allowing participants to choose the moment of takeover, it would be preferable as ResB was less laborious than simply repeating information like in RepB and MB. This will be attempted in future studies, using the findings of this experiment to design interfaces that will still facilitate SA without requiring long utterances. Finally, the use of multimodal displays in MB made little difference in subjective results, possibly because the cues were subtle, so as not to annoy, but as a result, they were often not noticed. Future studies will evaluate more prominent cues during takeover interactions.

We argue that the observed results reflected the full embodiment of the tested techniques; CB simulated a short interaction with the least engagement; RepB and MB required repetition of all phrases, offering no shortcuts; and ResB required answers to all questions, with no shortcuts either. Future studies will attempt to amalgamate the tested approaches, offering an augmented CB interface, which will enable succinct interaction (see also Eriksson and Stanton (2017), but also inform the driver to facilitate SA. Increasing SA is imperative in a takeover interface (Forster, Naujoks and Neukum 2017; Naujoks et al. 2016), and a desirable quality recognised in user views. More objective measures of SA will be used in future studies, requiring specific responses to more critical road events, as opposed to an uneventful scenario simulated in this study, since richer road scenes may have an effect on observed driver behaviours.

4.4 CONCLUSIONS

This study evaluated a set of dialogue-based interfaces for the takeover of control in an autonomous car. The simplicity and personal choice about when to takeover of the CB interface was appreciated the most, and the interface requiring response to questions about the road, thus engaging the driver the most, came next. The interfaces requiring repetition of phrases were the least preferred, even when multimodally augmented. In order to better facilitate SA, an amalgamation of the best-performing techniques is proposed.

ACKNOWLEDGEMENTS

This work was funded by EPSRC and Jaguar Land Rover, as part of the project Human Interaction: Designing Autonomy in Vehicles (HI:DAVe), Project Grant Number: EP/N011899/1.

REFERENCES

Banks, V. A., N. A. Stanton, and C. Harvey. 2014. "Sub-systems on the road to vehicle automation: hands and feet free but not "mind" free driving." *Safety Science 62* 505–514. http://doi.org/10.1016/j.ssci.2013.10.014.

Borojeni, S. S., L. Chuang, W. Heuten, and S. Boll. 2016. "Assisting drivers with ambient take-over requests in highly automated driving." *AutomotiveUI - Proceedings of the 8th International Conference on Automotive User Interfaces and Interactive Vehicular Applications.* Ann Arbor: Association for Computing Machinery. 237–244. https://doi.org/10.1145/3003715.3005409.

Brooke, J. 1996. "SUS: a 'Quick and Dirty' usability scale." In *Usability Evaluation In Industry,* by P. W. Jordan, B. Thomas, I. L. McClelland, and B. Weerdmeester, 189–194. London: Taylor & Francis Group. https://doi.org/10.1201/9781498710411.

Civil Aviation Authority. 2015. "Radiotelephony manual." https://publicapps.caa.co.uk/modalapplication.aspx?catid=1&pagetype=65&appid=11&mode=detail&id=6973 (accessed 4 November 2020).

Clarkson, P. J., R. Coleman, S. Keates, and C. Lebbon. 2003. *Inclusive Design: Design for the Whole Population.* London: Springer-Verlag.

Dhillon, B., R. Kocielnik, I. Politis, M. Swerts, and D. Szostak. 2011. "Culture and facial expressions: a case study with a speech interface." *IFIP Conference on Human-Computer Interaction – INTERACT 2011.* Berlin: Springer. 392–404. https://doi.org/10.1007/978-3-642-23771-3_29.

Endsley, M. R. 1988. "Situation awareness global assessment technique (SAGAT)." *Proceedings of the IEEE 1988 National Aerospace and Electronics Conference.* Dayton: IEEE. 789–795. doi: 10.1109/NAECON.1988.195097.

Endsley, M. R. 1995. "Toward a theory of situation awareness in dynamic systems." *Human Factors: The Journal of the Human Factors and Ergonomics Society 37, 1* 32–64. http://doi.org/10.1518/001872095779049543.

Eriksson, A., and N. A. Stanton. 2017. "The chatty co-driver: A linguistics approach applying lessons learnt from aviation incidents." *Safety Science 99* 94–101. http://doi.org/10.1016/j.ssci.2017.05.005.

Eriksson, A., S. M. Petermeijer, M. Zimmermann, J. C. F. de Winter, K. J. Bengler, and N. A. Stanton. 2019. "Rolling out the red (and green) carpet: supporting driver decision making in automation-to-manual transitions." *IEEE Transactions on Human-Machine Systems 49, 1* 20–31. doi: 10.1109/THMS.2018.2883862.

Forster, Y., F. Naujoks and A. Neukum. 2016. "Your turn or my turn?: design of a human-machine interface for conditional automation." *AutomotiveUI - Proceedings of the 8th International Conference on Automotive User Interfaces and Interactive Vehicular Applications.* Ann Arbor: Association for Computing Machinery. 253–260. https://doi.org/10.1145/3003715.3005463.

Forster, Y., F. Naujoks, and A. Neukum. 2017. "Increasing anthropomorphism and trust in automated driving functions by adding speech output." *IEEE Intelligent Vehicles Symposium (IV)* 365–372. doi: 10.1109/IVS.2017.7995746.

Hart, S. G. 2006. "Nasa-task load index (NASA-TLX); 20 years later." *Proceedings of the Human Factors and Ergonomics Society Annual Meeting* 904–908. https://doi.org/10.1177/154193120605000909.

Hart, S. G., and L. E. Staveland. 1988. "Development of NASA-TLX (task load index): results of empirical and theoretical research." *Advances in Psychology 52* 139–183. http://doi.org/10.1016/S0166-4115(08)62386-9.

Hawkins, A. 2017. *Ford and Lyft will Work Together to Deploy Autonomous Cars.* September 27. https://www.theverge.com/2017/9/27/16373574/ford-lyft-self-driving-car-partnership-gm.

Horrey, W. J., and C. D. Wickens. 2004. "Driving and side task performance: the effects of display clutter, separation, and modality." *Human Factors 46, 4* 611–624. http://doi.org/10.1518/hfes.46.4.611.56805.

Jeon, M., I. Politis, S. Shladover, C. Sutter, J. Terken, and B. Poppinga. 2016. "Towards life-long mobility: Accessible transportation with automation." *AutomotiveUI 2016–8th International Conference on Automotive User Interfaces and Interactive Vehicular Applications.* ACM. 203–208. https://doi.org/10.1145/3004323.3004348.

Karam, M., F. A. Russo, and D. I. Fels. 2009. "Designing the model human cochlea: an ambient crossmodal audio-tactile display." *IEEE Transactions on Haptics 2, 3* 160–169. http://doi.org/10.1109/TOH.2009.32.

Keates, S., and P. J. Clarkson. 2002. "Countering design exclusion through inclusive design." *ACM SIGCAPH Computers and the Physically Handicapped* 69–76. https://doi.org/10.1145/960201.957218.

Kyriakidis, M., R. Happee, and J. C. F. de Winter. 2015. "Public opinion on automated driving: Results of an international questionnaire among 5000 respondents." *Transportation Research Part F: Traffic Psychology and Behaviour* 127–140. https://doi.org/10.1016/j.trf.2015.04.014.

Langdon, P., I. Politis, M. Bradley, L. Skrypchuk, A. Mouzakitis, and P. J. Clarkson. 2017. "Obtaining design requirements from the public understanding of driverless technology." *International Conference on Applied Human Factors and Ergonomics - AHFE 2017: Advances in Human Aspects of Transportation.* Springer. 749–759. http://doi.org/10.1007/978-3-319-60441-1_72.

Large, D. R., G. Burnett, B. Anyasodo, and L. Skrypchuk. 2016. "Assessing cognitive demand during natural language interactions with a digital driving assistant." *AutomotiveUI - Proceedings of the 8th International Conference on Automotive User Interfaces and Interactive Vehicular Applications.* Ann Arbor: Association for Computing Machinery. 67–74. http://doi.org/10.1145/3003715.3005408.

Large, D. R., V. Banks, G. Burnett, and N. Margaritis. 2017. "Putting the joy in driving: investigating the use of a joystick as an alternative to traditional controls within future autonomous vehicles." *AutomotiveUI - Proceedings of the 9th International Conference on Automotive User Interfaces and Interactive Vehicular Applications*. Oldenburg: Association for Computing Machinery. 31–39. https://doi.org/10.1145/3122986.3122996.

Lee, K. J., Y. K. Joo, and C. Nass. 2014. "Partially intelligent automobiles and driving experience at the moment of system transition." *CHI 2014: Proceedings of the SIGCHI Conference on Human Factors in Computing Systems*. Toronto: Association for Computing Machinery. 3631–3634. https://doi.org/10.1145/2556288.2557370.

Mccall, R., F. McGee, A. Meschtscherjakov, N. Louveton, and T. Engel. 2016. "Towards a taxonomy of autonomous vehicle handover situations." *AutomotiveUI - Proceedings of the 8th International Conference on Automotive User Interfaces and Interactive Vehicular Applications*. Ann Arbor: Association for Computing Machinery. 193–200. https://doi.org/10.1145/3003715.3005456.

Microsoft. 2012. *Microsoft Speech Platform*. http://msdn.microsoft.com/en-us/library/hh361572.aspx.

Naujoks, F., C. Mai, and A. Neukum. 2014. "The effect of urgency of take-over requests during highly automated driving under distraction conditions." *5th International Conference on Applied Human Factors and Ergonomics*. Krakow: AHFE. 431–438.

Naujoks, F., D. Befelein, K. Wiedemann, and A. Neukum. 2017. "A review of non-driving-related tasks used in studies on automated driving." *International Conference on Applied Human Factors and Ergonomics - Advances in Human Aspects of Transportation*, by N. A. Stanton. Springer. 525–537. https://doi.org/10.1007/978-3-319-60441-1_52.

Naujoks, F., Y. Forster, K. Wiedemann, and A. Neukum. 2016. "Speech improves human-automation cooperation in automated driving." *Mensch und Computer 2016 – Workshopband* http://doi.org/10.18420/muc2016-ws08-0007.

Phys.org. 2017. *GM Unit Says It Has 'Mass Producible' Autonomous Cars*. September 11. https://phys.org/news/2017-09-gm-mass-autonomous-cars.html.

Politis, I., S. Brewster, and F. Pollick. 2015. "Language-based multimodal displays for the handover of control in autonomous cars." *AutomotiveUI - Proceedings of the 7th International Conference on Automotive User Interfaces and Interactive Vehicular Applications*. Nottingham: Association for Computing Machinery. 3–10. https://doi.org/10.1145/2799250.2799262.

Politis, I., S. Brewster, and F. Pollick. 2017. "Using multimodal displays to signify critical handovers of control to distracted autonomous car drivers." *International Journal of Mobile Human-Computer Interaction 9, 3* 1–16. https://doi.org/10.4018/ijmhci.2017070101.

Politis, I., P. Langdon, M. Bradley, L. Skrypchuk, A. Mouzakitis, and P. J. Clarkson. 2017. "Designing autonomy in cars: a survey and two focus groups on driving habits of an inclusive user group, and group attitudes towards autonomous cars." *International Conference on Applied Human Factors and Ergonomics - Advances in Design for Inclusion*. Springer. 161–173. https://doi.org/10.1007/978-3-319-60597-5_15.

Rödel, C., S. Stadler, A. Meschtscherjakov, and M. Tscheligi. 2014. "Towards autonomous cars: the effect of autonomy levels on acceptance and user experience." *Proceedings of the 6th International Conference on Automotive User Interfaces and Interactive Vehicular Applications*. Seattle: AutomotiveUI. 1–8. https://doi.org/10.1145/2667317.2667330.

SAE J3016. 2016. *Taxonomy and Definitions for Terms Related to Driving Automation Systems for On-Road Motor Vehicles*.

Schoettle, B., and M. Sivak. 2014a. *A Survey of Public Opinion about Autonomous and Self-Driving Vehicles in the U.S., the U.K. and Australia*. Ann Arbor: University of Michigan.

Schoettle, B., and M. Sivak. 2014b. "Public opinion about self-driving vehicles in China, India, Japan, the U.S., the U.K., and Australia." University of Michigan, Ann Arbor, Transportation Research Institute.

Schoettle, B., and M. Sivak. 2015. "Motorists' preferences for different levels of vehicle automation." University of Michigan, Ann Arbor, Transportation Research Institute; 2015.

Schoettle, B., and M. Sivak. 2016. "Motorists' preferences for different levels of vehicle auto-mation. University of Michigan, Ann Arbor, Transportation Research Institute; 2016."

STISIM. 2017. http://stisimdrive.com/.

Taylor, R. M. 1989. "Situational awareness rating technique (Sart): the development of a tool for aircrew systems design." *Proceedings of the AGARD AMP Symposium on Situational Awareness in Aerospace Operations*. CP478.

Van Der Laan, J. D., A. Heino, and D. De Waard. 1997. "A simple procedure for the assess-ment of acceptance of advanced transport telematics." *Transportation Research Part C: Emerging Technologies 5, 1* 1–10. http://doi.org/10.1016/S0968-090X(96)00025-3.

van Veen, T., J. Karjanto, and J. Terken. 2017. "Situation awareness in automated vehicles through proximal peripheral light signals." *AutomotiveUI - Proceedings of the 9th International Conference on Automotive User Interfaces and Interactive Vehicular Applications*. Oldenburg: Association for Computing Machinery. 287–292. https://doi.org/10.1145/3122986.3122993.

Volvo. 2017. *Autopilot - Self-driving Volvos by 2017*. https://www.volvocars.com/au/about/innovations/intellisafe/autopilot.

Walch, M., K. Lange, M. Baumann, and M. Weber. 2015. "Autonomous driving: investigating the feasibility of car-driver handover assistance." *AutomotiveUI - Proceedings of the 7th International Conference on Automotive User Interfaces and Interactive Vehicular Applications*. Nottingham: Association for Computing Machinery. 11–18. https://doi.org/10.1145/2799250.2799268.

Walch, M., T. Sieber, P. Hock, M. Baumann, and M. Weber. 2016. "Towards cooperative driv-ing: involving the driver in an autonomous vehicle's decision making." *AutomotiveUI - Proceedings of the 8th International Conference on Automotive User Interfaces and Interactive Vehicular Applications*. Ann Arbor: Association for Computing Machinery. 261–268. https://doi.org/10.1145/3003715.3005458.

Wickens, C. D. 1980. "The structure of attentional resources." In *Attention and Performance VIII, Volume 8*, by R. S. Nickerson, 239. Cambridge: Lawrence Erlbaum Associates.

Wickens, C. D. 2008. "Multiple resources and mental workload." *Human Factors: The Journal of the Human Factors and Ergonomics Society 50, 3* 449–455. http://doi.org/10.1518/001872008X288394.

5 The Design of Takeover Requests in Autonomous Vehicles
Low-Fidelity Studies

Patrick Langdon
Edinburgh Napier University

Nermin Caber, Michael Bradley, and Theocharis Amanatidis
University of Cambridge

James W.H. Brown
University of Southampton

Simon Thompson
Jaguar Land Rover

Joy Richardson
University of Southampton

Lee Skrypchuk
Jaguar Land Rover

Kirsten M. A. Revell and Jediah R. Clark
University of Southampton

Ioannis Politis
The MathWorks Inc.

P. John Clarkson
University of Cambridge

Neville A. Stanton
University of Southampton

CONTENTS

5.1   INTRODUCTION

The introduction of autonomous driving technology in cars is rapidly becoming a reality (Walker, Stanton and Salmon 2015). Public discussion responding to advertising and original equipment manufacturer (OEM) media focus has included debates over issues of technological achievement; the specifications of different levels of automation, infrastructure, and legal requirements; the usage cases for transportation; and assessing the need for driver intervention during the ride.

This last topic is not novel for researchers in the human factors field who have studied the effects of automation and associated interfaces on the capabilities of the driver to safely control the vehicle on public roads (Young and Stanton 2007; Gold et al. 2013). Before 2017, all automated driving features required the driver to constantly monitor the automated driving system throughout its operation. With increased degrees of automation, the ability to safely take control has been found to decrease even more (Strand et al. 2014; Young and Stanton 2007). All these studies focused on automation that required the operator to monitor the dynamic system's environment. It is also likely that there will be a negative impact on human driving performance for a takeover in SAE (Society of Automotive Engineers) Level 3 cars as the task of monitoring is greater. The takeover process has been identified as a major concern in vehicle automation (Shaikh and Krishnan 2012), and it is clear that the driver needs to receive clear alerts in order to safely resume control of the car (Cranor 2008).

We describe research carried out to investigate the transfer of control from vehicle to driver in autonomous vehicles as a result of a takeover request (TOR). The target scenario focuses on the planned transfer of control from car to driver, while on a highway, in an SAE Level 3 vehicle under benign conditions and among moderate traffic. According to the SAE, Level 3 requires the human driver to act as a fall-back driver who performs the dynamic driving as necessary, or as the automated system reaches its limits (SAE J3016 2016).

5.1.1 INCLUSIVE DESIGN

The present study also brings to bear inclusive design thinking as early as possible in the development of autonomous car technology (Keates and Clarkson 2004). Doing so increases the focus on user needs in the resulting concepts, identifying the population as an inclusive one from early on in the design process.

A key element of this approach was the synthesis of human-centred design for inclusion with advanced human factors methodology. Hence, the work described aimed to develop a new method for user-centred ecological interface design (UCEID). UCEID is a novel human factors method that integrates the relationship between ecological interface design (EID) and inclusive user-centred design (Vicente and Rasmussen 1992; Burns and Hajdukiewicz 2004; Langdon, Persad and Clarkson 2010) by combining the existing methodology from the cognitive work analysis framework (Vicente 1999). The approach engages with stakeholders, from inclusive populations, subject matter experts, and users to produce outputs that generate human–machine interface (HMI) design requirements. Initial design concepts were then produced as storyboards based on usage scenarios, following a design workshop and concept filtering activity. The resulting concepts were implemented as factorial conditions in a quantitative study of human driving performance.

5.1.2 BACKGROUND AND MOTIVATION

The aims of the project included the development of new methodology to tackle the challenges of combining user-centred design for population inclusion with ergonomics and human factors. To this end, the development of the HMI for TOR in autonomous conditions was embedded as part of a design process that was, in itself, under development. We next outline the structure of the UCEID process.

5.1.3　The UCEID: Project Design Context

Human factors methods fall into a range of categories that are relevant for application at different parts of the design process (Langdon 2015). The UCEID process traverses the 'design stage', 'requirements phase', and 'early prototype', where each stage is iterative in its own right (Figure 5.1).

Figure 5.2 shows a schematic of the process. The UCEID method was positioned early in the design process to allow 'analytical prototyping', i.e. the means of applying human factor insights to systems or designs that are yet to exist in physical form. It involves a combination of 'identify needs' and 'developing concepts' in the design process. The UCEID process described in this document was intended to generate initial design concepts for experimental testing. It follows the key finding from inclusive user-centred design that an active process of linked iteration between technology prototypes and user trials is necessary to meet the dual needs of diverse driver demographics and technology delivery requirements (Langdon 2015). Many further iterations are needed throughout the design process to reach the final concept, although for UCEID a flowchart of the main elements is straightforward (Figure 5.2).

5.1.4　Theoretical Background

EID (Vicente and Rasmussen 1992; Burns and Hajdukiewicz 2004) is based on Gibson's ecologically valid perception methodology (Gibson 1979). That aims

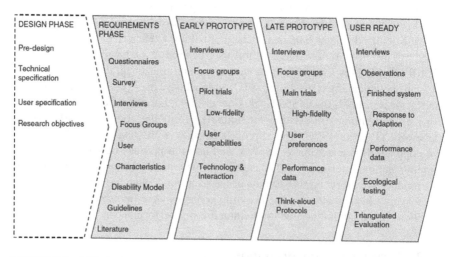

FIGURE 5.1　UCEID fits into proposed appropriate approaches at differing iterative design stages of the design process, in the first three phases up to 'Early Prototype'. (After Van Velsen et al. 2008.)

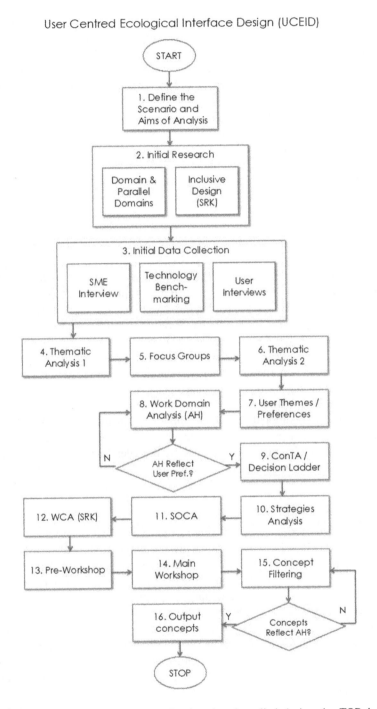

FIGURE 5.2 The UCEID process as developed and applied during the TOR interface design project.

to make constraints of the system and environment explicit, so that the appropriate action is apparent to the system user (McIlroy and Stanton 2015). UCEID is a novel human factors method that integrates relationships between EID and inclusive human-centred design (Keates and Clarkson 2004; Langdon and Thimbleby 2010) by combining the existing methodology from the cognitive work analysis framework (Rasmussen, Pejtersen and Goodstein 1994; Vicente 1999; Jenkins et al. 2008) and inclusive design (Clarkson et al. 2003; Langdon and Thimbleby 2010; Langdon, Persad and Clarkson 2010). The essential approach engages with stakeholders, experts, and inclusive samples of users to produce outputs that generate design requirements. Initial design concepts are then produced and refined following a design workshop and further concept filtering activity.

5.1.5　Definition of the Scenario, Aims, and Boundaries of Analysis

This initial research informs all subsequent steps of the UCEID method. In the design of novel systems, information regarding activity in similar, existing systems is required. This can be through document analysis (such as user guides or industry reports), which will depend on what is currently available for the domain. In the present method, initial research included the development of an HMI for an autonomous car takeover scenario. The relevant literature therefore concerned automotive research papers, trade journals, and press releases. The objective of this method was to establish the opinions and preferences of likely users towards the target technology to be designed. This therefore included public opinion in the form of surveys. These sources then act as input to the subsequent stages of the UCEID design process, converging on the design stage, with final outputs in the form of initial design concepts ready for operationalising and testing before further iterations. In this case, an important source of information was accurate, and recent, national survey data. This took the form of the Intelligent Mobility, Traveller Needs, and UK Capability Study (Wockatz and Schartau 2015), an extensive survey of UK journeys and transportation usage.

5.1.6　Initial Data Collection: Experts' Semi-structured Interview

The information gained from document analysis and observation of current and parallel domains provided the basis for semi-structured interviews with domain experts. The process for this stage was to acquire the data using standard social science techniques to analyse the data using qualitative research approaches, such as variants of grounded theory (Glaser and Strauss 1967). Use was made of general qualitative design tools such as NVivo (QSR International 2016), yielding a thematic analysis with an accompanying interpretation by the researcher, and the merging of themes using triangulated approach with other data sources to generate a definitive interpretation and representative examples of the theme (Flick 2009; Goodman-Deane, Langdon and Clarkson 2010).

An example of how distinct triangulated methods can be used for acquiring potential users' views about the target technologies and their use, are semi-structured interviews, which were carried out. This qualitative technique involved a one-to-one meeting between the researcher and a potential user of the target technology. Methods are allowed to vary in the UCEID, but in this case, a set of general questions were prepared in advance in an interview schedule and preparatory information was given to the interviewee before commencing. Following initial introductions, consent, briefing material, and ethical statements, the interviewer showed some short video material to the respondents.

- Video 1: Volvo concept autonomous interior video – essentially a countdown concept;
- Video 2: Google video of fully autonomous car – an SAE Level 5 demonstration with blind users.

These preparatory materials were chosen to reflect only the topic of direct interest to avoid priming the respondent with other information. The aim was to extract the respondents' own comments without specific prompting from the interviewer. In particular, the interviewer avoided asking leading or prompting questions related to the outcomes, and the respondents were being encouraged to develop their own trains of thought. A number of simple techniques were used to maintain the flow of the interaction following standard methodology for interviewing, such as that in Breakwell et al. (2006). Current ethical practice, BPS code (The British Psychological Society 2018), was followed, and after informed consent was given, interviewees were instructed that they could stop at any time. With consent, interviews were recorded in audio format, and these were transcribed as a single corpus, exactly as they took place.

Interview participants were sampled to cover an appropriate range of the population that it was intended to generalise the results to. Some criteria for choice were established consistent with an inclusive design requirement. Age, gender, demographic group, and geographical region were considered in sampling, with the additions of issue-related sampling, such as technology experience, where necessary. No respondents with impairments of vision, physical, or hearing capability were used, and the respondents were expected to rely on PC accessibility features. The sampling was chosen to include young, middle-aged males and females from metropolitan and rural areas. Two interviewers carried out four interviews each with the selection of respondent based on convenience and the main categories of user, identified by the IM UK Traveller Needs survey. There were three females and five males. The sampling is summarised in Table 5.1. The emergent themes from both coders were combined and submitted to the next stage of thematic analysis.

5.1.6.1 Technology Analysis and Benchmarking

To ensure that the appropriate technological landscape and competitive interaction pattern context for the interface under development was considered, a benchmarking process was carried out. Its purpose was to provide insights into unmet user needs which could inform the UCEID requirements and help define the solution space. It was carried out in a number of ways, through expert appraisal of OEM competitor interfaces and literature, and web research to establish the state of the art for the feasibility of current technologies, the expectations of users of the UCEID interface, and the identification of potential weaknesses of current solutions. The domain sampling

TABLE 5.1

Sampling for Experts Semi-structured Interviews

Name/Identifier	Descriptive
P1 transcript	Older male, musician, major UK city (urban rider, dependant passenger)
P2 interview	Middle-aged female, creative media, major UK city (progressive metropolite)
P3 interview	Middle-aged male, operations manager, major UK town (default motorist, petrol head)
P4 interview	Older female, therapist, major UK town/rural (default motorist)
Transcript JM	A middle-aged male, university professor living in a major UK town (default motorist)
Transcript NB	A middle-aged, UK male, living in a suburban area, university educated and researcher (local driver)
Transcript PH	A male FE lecturer in engineering in the north of England (default motorist, petrol head)
Transcript SA	A female university professor in London (urban rider)

FIGURE 5.3 The Tesla P90S instrument cluster display showing active, adaptive cruise control functionality (indicated through a blue-coloured speedometer icon), and inactive auto-steer functionality (indicated through a grey steering wheel icon) (Tesla 2020).

analysis in this example consisted of two representations of extant technology. These included a benchmarking analysis of current commercial offerings from OEMs, analysed by features (see Figure 5.3).

5.1.6.2 Thematic Analysis 1

The process for this step was to acquire the data using standard social science techniques, as described previously, and to analyse the data using qualitative research approaches, such as that of grounded theory (Glaser and Strauss 1967). A general qualitative design tool NVivo (QSR International 2016) yields a thematic analysis with an accompanying interpretation by the researcher. Importantly, a continued process of coding during the theme development ensured that themes were recoded and monitored iteratively. In this case, two coders (individuals) should compare their

TABLE 5.2

Example of the Top-Level Thematic Nodes Organised by Number of References in the Eight Sources

Name	Sources	References
HMI issues	8	137
Autonomous cars negative	6	68
Activities during automated driving	8	47
Trust	6	40
Autonomous cars positive	7	39
Road governance restrictions	5	19
Control	6	18
Nature of journey	6	12
Ownership of vehicle	4	11
Safety	4	11
Attitudes towards driverless cars	5	11
Negative experiences of driving	5	10
The joy of driving	4	7
Car characteristics	3	7
Driver's competence	3	5
Autonomy for disability	2	5
Future market for driverless cars	2	4
Cost	2	3

Data is from User Interviews in the HI:DAVe method prototype activity.

coding and converge on a final set. It was also necessary to merge themes using a triangulated approach to generate a definitive interpretation, representative examples of the theme, and a rationale (see Table 5.2). Descriptive analysis was undertaken with the main themes described as a hierarchical list. Each top-level item theme subsumes subthemes, as described here. A detailed interpretation was made in text format, breaking down each theme after providing an overview of the entire theme.

5.1.6.3 Focus Groups

A series of two focus groups on the topic of takeover of control in autonomous cars were conducted (see Table 5.1). Initial insights were cross-referenced and extended with a more inclusive and gender-balanced user group that followed. All participants were recruited either through an online questionnaire or opportunistically.

5.1.6.4 Thematic Analysis 2

Initial interviews, focus groups, and research were followed by an interpretation and consolidation stage where the participating researchers agree on a triangulated interpretation taking into account all the data sources (Flick 2009; Goodman-Deane, Langdon and Clarkson 2010). Triangulation here refers to the consideration of all sources of data and the extent of their agreement and disagreement, moderated by the

TABLE 5.3

An Example of a Concept and Requirements Table from Interpreted Qualitative Thematic Analysis from the HI:DAVe Method Prototype Activity

ID	Source/Theme	Name	Description	Rationale	Notes for Scenario
INT1	Interview – HMI issues	Multimodality of takeover warning	In addition to a visual dashboard warning, there should be a verbal communication from the car, or a series of increasingly strident chimes or bleeps as the takeover point approaches	Anticipation of being otherwise occupied and needing to be alerted (e.g. facing backwards and using infotainment)	Multimodality is important for inclusion but also for the general designs
INT2	Interview – HMI issues	Countdown to takeover	In a complex traffic environment, the suggestion was for an auditory warning with countdown information to the takeover, possibly accompanied by haptic stimulation through the seat	Identified as being from around 5 min down to event in increasing salience (longer times were suggested). Approximately 1–2 min	Not a complex traffic environment, not in scenario
FG50	Focus group 2 – part II	Countdown to takeover	Seen as similar to Satnav countdown instructions to a junction		Applies to almost all scenarios for takeover

methodological strength of their methods. It was necessary to use a compressed format for the outcomes of the triangulated interpretation stage, especially as the subject concerned technical or engineering designs. At a minimum, this gave each captured entity a unique identifier for later referencing as a name, a description, and a rationale. The chosen representation of this sort was deemed to be a concept or requirements table (see Table 5.3). The former is generally useful, the latter is orientated towards engineering design processes that are useful as input to a design process.

5.1.6.5 Preferences and User Themes Interpreted

Once the thematic analysis for the interviews and focus groups was completed, interpretation of the resulting themes provided a set of insights. These contained a high-level description of what the design needed to cover, with reference to the user input that led to each insight.

For the insight set to be constructed, the design team collaboratively processed the output of the triangulated thematic analyses and identified dominant themes found in the initial data collection steps (interviews and focus groups). These themes, and the accompanying insights, were enumerated in a document, linking back to each insight's original source or sources. The insights intentionally differed from requirements in that they were not considered binding features of the final design (Langdon 2015; Spath et al. 2012). Rather, they were a set of properties the final design should encompass, open-ended in nature, and subject to modifications by the design team members, based on their automotive expertise.

Once initial research had been undertaken, the design team engaged in a session where all the themes were discussed, categorised, and sorted into enumerated insights. A subset of the whole corpus was chosen to eliminate repetition resulting from triangulation of several data sources. In this way, the design tools for the main design workshop were identified and used as building blocks for the design exercises conducted, leading to the output concepts.

5.1.6.6   Work Domain Analysis (WDA) Abstraction Hierarchy

In parallel with the stages of UCEID described so far, a series of analyses from EID were carried out and their results fed into the common flow of data. The WDA provided a description of the constraints that govern the purpose and function of the systems under analysis.

This revealed the potential for unexpected behaviours that could occur in future systems, allowing human factors focus to address any issues before the design process was progressed too far. The output of this step was the abstraction hierarchy (AH) (Rasmussen 1985; Vicente 1999). This diagram was constructed of five levels of abstraction, ranging from the most abstract level (purpose) to the most concrete level of form (Vicente 1999). The top three levels consider the overall objectives of the domain, and what it can achieve, whereas the bottom two levels concentrate on the physical components and their affordances. In UCEID, the AH was evaluated against, and informed by, the user themes and preferences identified as described. The elements in the AH were reviewed to determine if user preferences were represented. If not, additional elements were added to accommodate these preferences.

5.1.6.7   Control Task Analysis

Where the AH developed considered constraints imposed by the overall purpose of the domain, regardless of activity, control task analysis looked in depth at constraints that were imposed by tasks that needed to be carried out in specific contexts. The contextual activity template by Naikar, Moylan, and Pearce (2006) supports control task analysis, used in this case. It took the form of a grid with the left-hand column listing key purpose-related functions identified from the AH and the top row breaking down the chosen situation by either time or location (or a combination of the two). This analysis helped reveal not only the constraints, but also the level of flexibility in how activities could be achieved. It may be appropriate to conduct multiple control task analyses to meet a range of goals. In the prototype, three timer periods were used to define the scenario: (1) pre-takeover, (2) takeover, and (3) post-takeover. Vicente (1999) recommends using a decision ladder when undertaking this analysis (see Figure 5.4). The decision ladder is a tool commonly used within cognitive work analysis to describe decision-making activity. It focuses on the entire decision-making

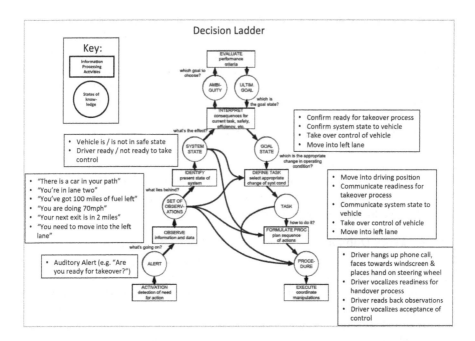

FIGURE 5.4 Example decision ladder taken from HI:DAVe method prototype for UCEID.

activity for a specific scenario, rather than a single selection between options. Rather than focusing on a single actor, it represents decision-making from the combined work system, which may be collaborative and distributed between a range of human and technical agents. The ladder contains two different types of node: data processing activities are represented by rectangles and the resulting states of knowledge are represented by circles (see Figure 5.4). The left-hand side of the ladder represents observations of the current system state, whereas the right-hand side represents the planning and execution of tasks and procedures to achieve a target system state.

5.1.6.8   Social Organisation and Cooperation Analysis

The purpose of this analysis is to determine how social and technical aspects of a system can work together in a way that enhances performance as a whole. Using the existing outputs from the WDA and AH, it was possible to map where specific functions could be undertaken by specific agents, either agent, or both agents working in collaboration. This followed Naikar (2006) who argued that social organisation and cooperation analysis reveals the flexibility of system to deal with unanticipated events. The greater the flexibility, the more likely the system will be able to adapt to varying dynamic demands of sociotechnical systems.

5.1.6.9   Design Workshop

The goal of the workshop was to share and review the knowledge gained in the previous steps about the problem and solution spaces, and then to generate concept solutions to solve that problem. With reference to Figure 5.2, a pre-workshop allowed re-calibration of the scenario (Step 1), user preferences (Step 7), and AH (Step 8), enabling a clear and consistent presentation of the context and design tasks to participants in the main design workshop. This workshop followed a structured inclusive design process (Clarkson et al. 2003) where participants were asked to explore needs ('Explore') and solutions ('Create'). Participants were chosen to be experienced in key disciplines such as human factors, inclusive design, automotive design, research and development, and also included members of the UCEID research team. The heterogeneity of the group was intended to add design scope. The concepts arising from the design workshop were extended in scope beyond the system constraints identified in the AH (Step 8) and subsequent cognitive work analysis steps. Based on the criteria for inclusion agreed in the pre-workshop, the concepts were filtered by the agreed level of adherence to the AH and any other factors deemed critical. The flowchart in Figure 5.2 shows an iterative loop linked back to concept filtering (Step 15). This iterative process repeated until the design concepts that fulfilled the criteria were selected for refinement.

5.1.6.10   Concept Refinement and Filtering

Following the main workshop, the ideas and concepts created were transcribed and a design criterion matrix, based on Pugh (1991) containing the AH Level 2 and Level 3 nodes, was constructed. These were the candidate *output concepts*. Any necessary modifications to the concepts based on the expertise of the design team were performed in this step. They were also made in the next one where the set of concepts and ideas related to autonomous takeovers created in the main workshop were transcribed into an electronic repository in the form of a single document.

From these concepts, a set to be taken forward and tested during Low-fidelity simulator trials were selected on the basis of elimination of duplicates or similar alternatives. Eliminations were also made using a Pugh matrix comparison of concept features (Pugh 1991) and a workshop verifying the concepts' accuracy with respect to the originating ideas and their operational status in a realistic engineering development process. The concepts were thus reduced to a smaller set that were allocated to conditions in a single-factor experimental design with four levels of interface type. This process is described further in this book.

These final *output concepts* were described with enough detail to enable their implementation and evaluation in later stages of the design process, in order to arrive at the final design. All the concepts were pictured in a slide deck, where their general functionality was described with storyboards. All the candidate concepts selected were elaborated and described in concept sheets. This included their title, summary, description, storyboards of one scenario involving the concepts, design rationale, inclusion rationale, and a summary of the modalities and technologies involved (Figure 5.5).

From these concepts, a set to be taken forward and tested during Low-fidelity simulator trials were selected on the basis of elimination of duplicates or similar alternatives. Eliminations were also made using a Pugh matrix comparison of concept features (Pugh 1991) and a workshop verifying the concepts' accuracy with respect to the originating ideas and their operational status in a realistic engineering development process. The concepts were thus reduced to a smaller set that were allocated to conditions in a single-factor experimental design with four levels of interface type. This process is described further in this book.

5.2   THE DESIGN AND FORMATIVE DEVELOPMENT PROCESS

5.2.1   Automotive Takeover Requests (TORs)

With the launch of the first autonomous system by OEM Audi (McNamara 2017), automated driving features no longer required the driver to monitor the automated

Voice, Autopilot Assistant

Dialogue-based system for all stages of autonomy

Concept Summary

The Voice Autopilot Assistant (VAA) concept facilitates multimodal dialogue. VAA prompts the driver by natural language interaction and redundant messages in a standard HMI format for all stages of autonomy: from planning to post-Vehicle-to-Driver transfer of control (post-VtD) to end of journey. The distinguishing feature of this dialogue is that the car requests information from the driver, and checks the veracity of the response, thereby prompting improvement of the driver's Situational Awareness (SA) (see Figure 1).

Assumptions

The concept assumes that the driver is fit and well, and that it is a clear sunny day. The concept also assumes that VtD is planned, so that it is known beforehand to the car systems and the driver that the VtD will happen. The concept can function in two scenarios:

- Junction to junction highway (see https://drive.google.com/file/d/0B3I1ZnJBWcB6TzVPbU84b1IONEU/view?usp=sharing and Table 1)
- Full urban drive (see https://drive.google.com/file/d/0B3I1ZnJBWcB6bE9ZbFIKMXNIRFE/view?usp=sharing)

The main difference of full urban drive scenario is that timings of events and contextual cues can potentially be adjusted for a shorter travelling time.

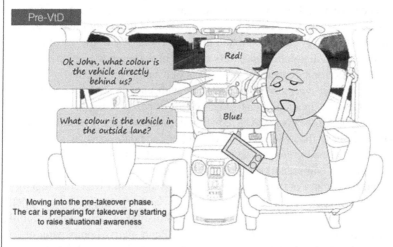

Figure 1: An ongoing dialogue takes place between car and driver to facilitate VtD.

Concept Description

The car engages in dialogue with the user at appropriate phases of the journey (planning, pre-VtD, VtD and post-VtD). In this way there is an ongoing engagement between car and driver, facilitating driver's Situational Awareness (SA) when needed.

During planning, the car greets the driver, using their name or alias (see Table 1.1), and the driver can address the car in the same way (see Table 1.2). The car can detect potential limitations in the driver's performance (e.g. fatigue), and verbalise them. It can also propose alternative driving styles to address these limitations (see Table 1.3). After any modifications to the planned trip due to the driver's limitations

FIGURE 5.5   An example of a selected design concept sheet after the refinement and filtering process.

driving system throughout its operation. The American Society of Automotive Engineers (SAE) issued widely accepted standard levels for automation ranging from 0 (no automation) to 5 (full automation). In this scheme, SAE Level 3 is the first level allowing the driver to take hands and feet off the driving controls, steering wheel and pedals (SAE J3016 2016). The driver may engage with other tasks but cannot completely disengage as the automated driving system may request the driver to take back control of the car. The SAE Level 3 requires the human driver to perform the dynamic driving task in the event that the automated system reaches its limits (SAE J3016 2016). Level 3 systems only work in predefined driving environments. For example, in the case of congestion on the highway, all conditions must be met prior to possible activation (McNamara 2017). Consequently, when the system reaches its operational limit, the driver is expected to resume driving. Since this intermediate level of automation has been considered to present a significant risk, some research argues for avoiding SAE Level 3 altogether (Seppelt and Victor 2016).

5.2.1.1 TOR Timing

Parasuraman and Riley (1997) define the range of possible interactions between humans and automation. Several factors are associated with this spectrum and influence the human operator's performance (Endsley and Kaber 1999; Parasuraman and Riley 1997).

- **Cognitive workload**: Cognitive overload, as well as underload, is known to influence human performance negatively (Desmond and Hoyes 1996; Hancock and Parasuraman 1992; Hancock and Verwey 1997; Recarte and Nunes 2003; Young and Stanton 2002, 2007).
- **Situation awareness (SA)** (Endsley and Kiris 1995): It has been identified as a critical factor for road safety (Salmon, Stanton and Young 2012) where drivers are required to permanently monitor their dynamic driving environment, to predict future changes, and to react appropriately (Stanton, Dunoyer and Leatherland 2011).
- **Trust**: It is an important precondition for automation's usage. Over-trust can lead to over-reliance on an automated system resulting in a failure to monitor the automation technology (Dzindolet et al. 2003; Lee and Moray 1992, 1994; Muir 1987).

All these elements and the resulting performance issues have been associated with operator out-of-the-loop performance problems and the limited capability of the user to get back into the system control loop by resuming manual control and performing necessary actions (Endsley and Kiris 1995; Kaber and Endsley 2004; Kessel and Wickens 1982; Young 1969). This is particularly important for SAE Level 3 automated systems, as the driver, while not being required to monitor the automation, is required to resume driving when requested.

The majority of studies conducted on transfers of control in automotive applications due to system failures show poorer human driving performance in comparison with permanent manual control (Stanton, Chambers and Piggott 2001; Vollrath, Schleicher and Gelau 2011). Furthermore, with increased levels of automation, the ability to safely take control is reduced further (Strand et al. 2014; Young and Stanton 2007). TORs have been studied widely in order to determine the optimal takeover request lead time (TORlt), i.e. the time from the issued TOR to an important event such as an obstacle. The takeover reaction time (TORrt) is therefore the time the driver needs to take control of the car after the TOR was issued. The takeover process has been identified as a major concern in this respect (Shaikh and Krishnan 2012) as the driver needs to be alerted effectively in order to safely resume control of the car (Cranor 2008). For a safe transition characterised by similar performance as for manual control, a TORlt of 7–10 s has been established (Damböck et al. 2012; Gold et al. 2013; Melcher et al. 2015). For non-critical TORs, a median TORrts of 4.5 s (with secondary task) and 6 s (without secondary task) have been found. However, there is a paucity of research into non-critical transitions and the corresponding TORlt (Eriksson and Stanton 2017). Since TORs with high urgency may induce a different human response, we therefore addressed non-critical TORs.

5.2.1.2 TOR Interfaces

Visual HMIs are useful to assist drivers' decision-making process concerning the required action after takeover (Eriksson and Stanton 2017). Multimodal alerts result in lower response times and better driving performance than the visual mode alone (Naujoks, Mai and Neukum 2014; Politis, Brewster and Pollick 2017). Furthermore, 'ambient' displays give rise to shorter reaction times without increasing workload (Borojeni et al. 2016). Auditory cues in combination with visual ones reduce drivers' hands-on times, i.e. the time needed to put hands on the steering wheel and improve lateral lane stability (Naujoks, Mai and Neukum 2014). While additional interventions, such as brake jerks, do not show any effect (Melcher et al. 2015), haptic or vibrotactile warnings do reduce drivers' reaction times (Petermeijer et al. 2017; Petermeijer, Cieler and de Winter 2017; Scott and Gray 2008). In general, there is a clear potential for multimodal ambient interfaces which have the advantage of being potentially perceivable irrespective of the drivers' current position or behaviour. These were used in the concept designs.

5.2.2 The Design Concepts

Taking into account findings on TORs and HMIs, and the outcomes of the UCEID design process, the following four concepts were chosen for the summative trials and are illustrated in Figure 5.6.

- A **countdown-based (CB)** system, only requiring a button press from the driver for them to assume control, and a countdown of 60 s in blocks of 10 s until this occurred. This concept was simulating common offerings of takeover concepts from the industry.
- A **repetition-based (RepB)** system that prompted the driver to read back information spoken in the form of oral dialogue from the car. Statements used included hazards, lanes, fuel level, speed, and intended exit.
- A **response-based (ResB)** system that used oral dialogues from the car to request the driver to answer prompts for similar information as the RepB

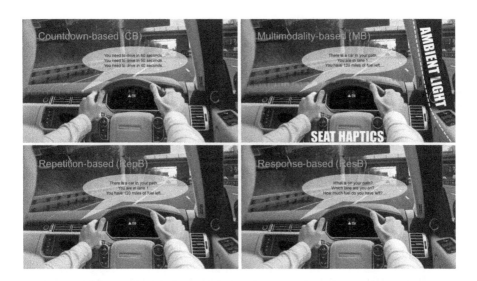

FIGURE 5.6 The four concepts operationalised into driving simulator tasks.

system, instead of asking them to repeat it. This provided a dialogue inter-action requiring greater engagement from the driver as they would need to cognitively engage and verbalise a response. This resembled the SA global assessment technique (Endsley 1988).

- A **multimodality-based (MB)** or **ambient assist** system added additional display information presented in a number of modalities: visual, sound based, haptic, and dialogue. The condition was effective when compared with the RepB system, with added ambient multimodal information, i.e. LED strips in amber colour, animating on either side of the driver and simu-lating the position of surrounding traffic, as well as ambient vibrations, aug-menting the utterances of the dialogue system. Ambient multimodal cues have shown benefits in the design of user interfaces (Politis, Brewster and Pollick 2015, 2017).

5.3 THE SUMMATIVE TRIALS

5.3.1 EXPERIMENT 1

A repeated-measures experimental design was conducted to evaluate the four TOR interfaces developed. The sequence of the interfaces was counterbalanced over the participants. For each interface, one trial with three TORs was performed. However, in an autonomous mode, participants performed a secondary task, which kept them out of the loop. There were 49 participants (24 females and 25 males) in the age range of 17 to 86 years ($M = 45.51$, $SD = 17.36$) recruited from the general public. Participants represented a wide range of ages and were gender-balanced: (1) age band 17–34 (12 participants, 5 females), (2) age band 35–56 (24 participants, 13 females), and (3) age band 57–86 (13 participants, 6 females). Participants had an average

FIGURE 5.7　The simulator set-up showing the tablet in use and the cluster simulation.

driving experience of 25 years with a maximum of 70 years and a minimum of less than a year.

A low-fidelity driving simulator, pictured in Figure 5.7, presented a ten-mile-long route on rural roads and the motorway with different curve radii using the 'STISIM Drive®' (STISIM 2017). Displays were used to depict the road scene, the dashboard, and the instrument cluster, which showed driving and car-related information, such as state of automation, current speed, and surrounding traffic. A secondary task was performed on a 10-inch tablet. On both sides of the driving seat, an LED strip was mounted on the wall to indicate traffic movement and position. A haptic device (Karam, Russo and Fels 2009) was implemented in the driving seat to deliver vibro-tactile stimuli based on the audio information presented. After a short welcoming, participants gave consent and were briefed with safety procedures and familiarised with the experiment. A memory game served as a secondary task (Politis et al. 2017).

5.3.1.1　Trials

A short acclimatisation drive was completed, and trials started under participants' control, then switching to hand over control to the vehicle. In an autonomous mode, participants interacted with the secondary task while a TOR was issued, timed at points approximately 10%, 40%, and 70% into the route. The automation was toggled by pressing two steering wheel buttons. A Wizard of Oz protocol was implemented with vocal utterances from the vehicle and pseudo-responses by the experimenter using a dialogue generator. This required the participant to interact in dialogue with the interface as the participant deemed appropriate, but instructions did not require the completion of the protocol. After completion of the condition, participants took control of the car using the wheel button and drove manually for 1 min before handing back using the button control. At the end of each trial, participants completed a range of standardised questionnaires: NASA-Task Load Index (NASA-TLX), acceptance scale, system usability scale (SUS), and situation awareness rating technique. After completing all four trials, participants were asked to rank the experienced concepts from their most to their least preferred.

5.3.1.2 Results

The experimental design was a one-factor repeated measures, and the qualitative and quantitative measures reported were as follows:

- **Takeover time (TOT)**: Total duration in seconds for completing the dialogue protocol to the point the participant pressed the takeover button on the steering wheel.
- **Driving performance**: The absolute angle input measuring the lateral driving stability. After each takeover, the data was logged for 60 s.
- **Workload**: NASA-TLX (Hart and Staveland 1988).
- **Acceptance**: System acceptance scale (Van Der Laan, Heino and De Waard 1997).
- **Usability**: System usability scale (SUS) (Brooke 1996).
- **Post-experiment ranking**: Interfaces were ranked by the participants from 1, most preferred, to 4, least preferred, at the end of the trials.

The average TOTs in seconds for the CB protocol were significantly lower than the remaining three protocols that were essentially undifferentiated (see Figure 5.8): 'CB countdown' ($M = 19.79$, $SD = 9.3$), 'RepB readback' ($M = 48.61$, $SD = 4.67$), 'ResB

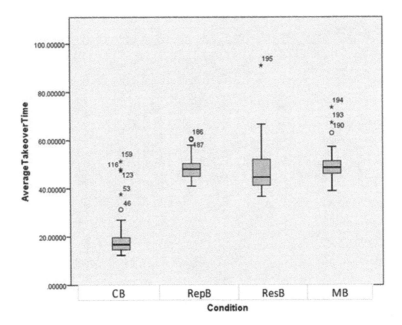

FIGURE 5.8 A box-and-whisker plot showing medians, and interquartile ranges for time to takeover.

answering questions' ($M = 47.79$, $SD = 10.04$), and 'MB multimodality' ($M = 49.65$, $SD = 6.39$).

Using one-way repeated-measures analysis of variance (ANOVA) determined that the interface type significantly affected TOT ($F(2.563,112.785) = 128.053$, $p < 0.001$), Greenhouse–Geisser ($\varepsilon = 0.854$). Post hoc comparison indicated that TOT was significantly lower for 'countdown' than for all other interfaces ($p < 0.001$).

The system acceptance scale consisted of two subscores: usefulness and satisfaction. The 'countdown' interface gave the highest usefulness as well as satisfaction ratings from the participants. However, only the satisfaction subscale reached statistical significance (see Figure 5.9): 'CB countdown' ($M = 1.05$, $SD = 0.76$), 'RepB readback' ($M = 0.14$, $SD = 0.76$), 'ResB answering questions' ($M = 0.37$, $SD = 1.01$), and 'MB multimodality' ($M = 0.2$, $SD = 1.08$).

The type of interface showed a significant effect on the satisfaction subscale ($F(3,141) = 11.979$, $p < 0.001$) of the acceptance scale. The concept 'CB countdown' obtained a higher satisfaction score than all other concepts ($p < 0.001$). The 'countdown' concept was also the highest rated for the SUS, with the other concepts again performing on similar levels ($F(3,138) = 6.826$, $p < 0.001$). 'CB countdown' gained a higher perceived usability score than 'RepB readback' ($p = 0.005$), 'ResB answering questions' ($p < 0.001$), and 'MB multimodality' ($p < 0.001$).

5.3.1.2.1   Workload TLX

There was a significant effect of interface on perceived workload ($F(3,141) = 6.826$, $p < 0.001$). Pairwise comparisons revealed that 'answering questions' was perceived to induce higher workload than the 'countdown' ($p = 0.008$) and 'readback' ($p = 0.003$).

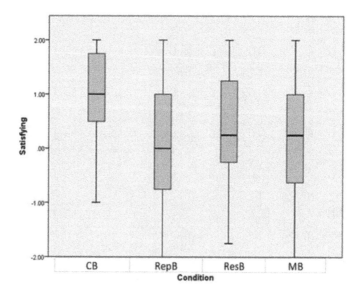

FIGURE 5.9   A box-and-whisker plot showing medians, and interquartile ranges for the system acceptance scale, satisfaction subscale.

5.3.1.2.2   Post-Experiment Rankings

For the post-experiment interview rankings, a Wilcoxon test with Bonferroni correction showed that 'CB countdown' was significantly higher ranked than 'ResB answering questions' ($T = -0.896$, $p = 0.004$), 'RepB readback' ($T = -1.521$, $p < 0.001$), and 'MB multimodality' ($T = -1.583$, $p < 0.001$). There was no significant difference in average completion time of the secondary task between the TOR interfaces.

5.3.1.2.3   Driving Performance

Driving performance was taken from the values and standard deviations for the absolute steering angle input calculated based on the logged data of 60s after the takeover: 'CB countdown' ($M = 0.36$, $SD = 0.34$), 'RepB readback' ($M = 1.12$, $SD = 1.22$), 'ResB answering questions' ($M = 1.45$, $SD = 2.53$), and 'MB multimodality' ($M = 121$, $SD = 1.19$). The driving performance was not significantly affected by the interface concept.

5.3.1.3   Discussion: Experiment 1

To summarise the main findings:

- There was an overall preference for the countdown concept. Time to takeover was significantly lower as were disturbances in driving post-takeover. The system acceptance scale scores were significantly higher, especially for satisfaction, and system usability was seen as significantly higher.
- Concepts that involved longer and more scripted dialogues with the vehicle were less preferred.
- The workload for the ResB 'answering questions' condition was significantly higher than the others, and more errors were made in dialogues.

It was apparent from the data and the post-experiment interviews that the dialogue-based protocols were largely seen as repetitive and tedious. It should be noted that the participants were instructed to complete the protocols resulting in more protracted interactions during the non-countdown interactions. No shortcuts to regaining control were offered although it was notable that some participants took control early anyway. The 'countdown' condition was shortest, most preferred, and was used by participants to take control when they wanted. Although 'answering questions' increased workload, it did not improve SA, from data based on participants' reports. This was contrary to expectations and may be attributed to the fact that *perceived* SA is a subjective measure that is influenced by participants' impressions. A more objective measure would be their response to a simulated critical road event. This may reflect more accurately the degree of in-depth SA attained.

The most dialogue mistakes were made for 'answering questions', which was expected since this interface required thinking and not purely repetition. Participants mentioned in the interviews that they would prefer 'answering questions' over 'readback' since it was less laborious. However, they had a clear preference for a shortcut to independently decide the moment of takeover, irrespective of the protocol.

Interviews revealed that participants perceived 'countdown' as the least supportive to their SA but appreciated the short duration and sense of control it elicited. The 'RepB readback' and 'MB multimodality' conditions were based on the same principle and were ranked the lowest. While such an interface may prove beneficial in aviation, it was not appreciated by the participants in this context. A possibility is that one of the main functions of readbacks in aviation is to ensure accurate comprehension of utterances made in radio telegraphy between the pilot and the air traffic control (Civil Aviation Authority 1999). Any SA improvements were not apparently directly perceived by drivers. The benefits of information perceived in the MB ambient, multimodal interface were not apparent in the objective and subjective results. The design of these cues may have been too subtle as some participants mentioned that they did not notice them.

5.3.1.4   Conclusion

The pattern of results suggest that the predominant effect was elicited by the repetitive and tedious nature of the dialogue protocols, especially when repeated multiple times. This apparently led to a preference for the clean interaction of the countdown, which required little cognition and allowed the drivers a stronger perception of control. Interviews suggested that some drivers were clearly prepared to trade SA against speed of takeover in the simplified simulation task to achieve a satisfactory result.

5.3.2   EXPERIMENT 2

Based on the insights and interfaces from experiment 1, and a review of other evidence from the UCEID process (Revell et al. 2018), a new revised set of four takeover interfaces was created. Three of these concepts represented interfaces for non-critical takeovers, and one concept constituted a takeover interface for an urgent, or critical, event. This design also enabled a comparison of driving performance and subjective measures between interfaces for critical and non-critical events. The non-critical interfaces required drivers to vocally confirm their readiness before the actual takeover dialogue started, as in experiment 1.

Since many participants did not notice the multimodal cues in the first experiment, the brightness of the LED light strip and the amplitude of the haptic vibration motor were increased. The potential beneficial effects of multimodality were also supported by recent findings (Naujoks, Mai and Neukum 2014; Petermeijer et al. 2017; Petermeijer, Cieler and de Winter 2017; Politis, Brewster and Pollick 2015, 2017).

Apart from the baseline interface, 'baseline countdown', all interfaces were augmented with the multimodal approach that consisted of LED light strips indicating the movement of surrounding traffic and ambient seat vibration reinforcing the system's voice, as before. The new interfaces listed below are illustrated in Figure 5.10.

- **'Baseline countdown'**: A simple CB system in accordance with experiment 1 was used as a baseline. However, since the mean TOT in experiment 1 was 19.79 s, it was decided to reduce the allocated time for takeover to 20 s.

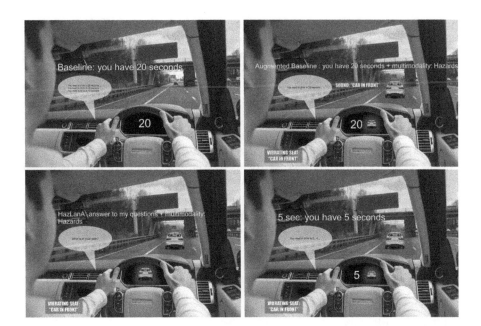

FIGURE 5.10 The four revised concepts operationalised into driving simulator tasks.

A reminder auditory message was issued every 5 s with increasing urgency in the wording, starting with *'Please takeover driving as soon as you are ready.'* and ending with *'Takeover driving now!'*.

- **'Augmented baseline'**: This interface applied the baseline 'countdown' interface but augmented it with additional information about the driving environment following the 5-s reminder message by the auditory dialogue system. The driver was not required to repeat or read back the information provided and did not have to complete the whole dialogue before taking control.

- **'Answering questions – HazLanA'**: The driver was allowed to take over driving when desired and did not have to complete the whole dialogue. This enabled the driver to take control at an earlier point during the protocol. The questions asked covered the same information as 'augmented baseline'.

- **'5-sec'**: It is based on the average TORrt found in the literature (Eriksson and Stanton 2017). An interface for critical takeovers was designed giving the driver 5-s reaction time to take control of the car. The system activated multimodal cues and the voice message *'Takeover driving now!'* and repeated this message if the driver did not take control within 5 s.

5.3.2.1 Trials

The experiment design was adapted from the first experiment; a repeated-measured design was applied, the sequence of the concepts was counterbalanced over the participants, and three TORs were issued per trial. To assess the driving performance and SA, a predefined simulator event, a traffic-cone on the carriageway, was triggered at the moment the driver took control of the car in some trials of the third takeover. This gave the driver a lead time of 7 s to attempt to avoid a collision.

There were 28 participants divided into three different age bands: (1) age band 17–34 (15 participants, 4 females), (2) age band 35–56 (7 participants, 4 females), and (3) age band 57–86 (6 participants, 3 females). Participants' driving experience ranged from 1 up to 68 years ($M = 19.61$, $SD = 18.64$). The same driving simulator and equipment was used as described for experiment 1.

Participants gave consent and received a financial incentive. After entering the simulator, an acclimatisation run helped participants to familiarise themselves with the interface and secondary task. Participants were told that, following the initial readiness confirmation, they could take over whenever they felt ready. As in experiment 1, TORs were issued at specific points through the route, at 10%, 40%, and 70%. Apart from the 'critical 5-sec' interface concept, participants had to vocally confirm their readiness to take over prior to the start of the protocol. Participants had the freedom to decide on the eventual moment of takeover by pressing the steering wheel button and were not required to finish the whole dialogue protocol. Once in manual mode, participants were required to drive manually for 30 s before handing back control to the car.

5.3.2.2   Results

5.3.2.2.1   Time to Takeover

The means and standard deviations of TOT for all four conditions are shown for the three age bands in Figure 5.11. The critical response concept '5-sec' showed

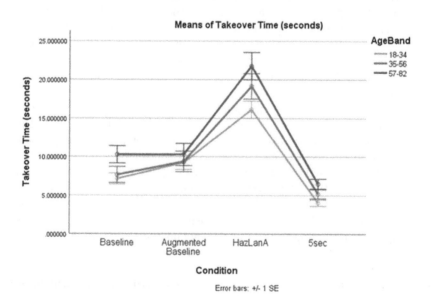

FIGURE 5.11   Interaction diagram for 2F ANOVA mean time to takeover by age for conditions in experiment 2.

the shortest TOT followed by 'baseline countdown', 'augmented baseline', and 'answering questions – HazLanA', in increasing order. A two-way, repeated-measures ANOVA revealed a significant difference in TOT among the interface concepts $(F(3,81) = 123.81, p < 0.001)$. Pairwise comparisons using Bonferroni adjustment indicated that '5-sec' was significantly lower than all other interface concepts $(p < 0.001)$.

While 'baseline countdown' and 'augmented baseline' did not differ significantly in TOT, 'answering questions – HazLanA' showed a higher TOT than both $(p < 0.001)$. Participants required different amounts of time to complete the takeover depending on their age $(F(2,25) = 4.07, p < 0.05)$. Bonferroni-adjusted pairwise comparisons revealed that participants in younger age band (17–34) took over significantly quicker than participants in the older age band (57–86) $(p < 0.05)$. There were no significant interactions between concept and age for TOT.

5.3.2.2.2 Workload

The TLX perceived workloads are shown in a simple box-and-whisker plot in Figure 5.12. 'Answering questions – HazLanA' had the lowest mean workload, followed by 'augmented baseline', 'baseline countdown', and '5-sec'. The TOR concepts induced different levels of workload $(F(2.41,65.17) = 4.74, p < 0.05)$. Degrees of freedom were adjusted using a Greenhouse–Geisser $(\varepsilon = 0.805)$ correction. The 'answering questions – HazLanA' showed a significantly lower level of perceived workload than '5-sec' $(p < 0.05)$ in Bonferroni post hoc comparisons.

5.3.2.2.3 Preference Rankings

The concepts differed significantly for preference rankings (see Table 5.4) as revealed by a Friedman test $(\chi^2(3) = 13.29, p < 0.005)$. Wilcoxon tests with Bonferroni

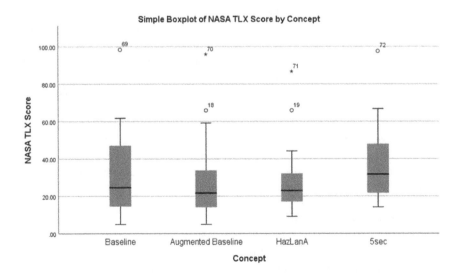

FIGURE 5.12 Box-and-whisker plots of TLX overall workload for TOR concepts.

TABLE 5.4

Preference Rankings for the Takeover Conditions

Mean	Rank
Baseline	2.04
Augmented baseline	2.18
HazLanA	2.61
5-sec	3.18

correction indicated that '5-sec' was significantly lower ranked than 'augmented baseline' ($T = -1$, $p = 0.023$) and 'baseline countdown' ($T = -1.14$, $p = 0.006$).

5.3.2.2.4   System Acceptance and Usability

Both system scales indicated that the concepts were above average and there was very little differentiation between concepts in their performance. For the satisfaction score of the SUS, all concepts performed in the positive range and there were no significant differences between TOR concepts. The SUS also indicated above-average scores (>68) with no differentiation between concepts.

5.3.2.2.5   Driving Performance, Situation Awareness, and Secondary Task

The driving performance data was examined for a time window of 10 s after the participant had taken control of the car. The following measures of degrees of absolute angle input were recorded for the different concepts: 'baseline countdown' ($M = 2.28$, $SD = 0.84$), 'augmented baseline' ($M = 2.33$, $SD = 0.94$), '5-sec' ($M = 2.48$, $SD = 0.59$), and 'answering questions – HazLanA' ($M = 2.55$, $SD = 0.99$). There were no significant differences between the TOR concepts overall in the ANOVA.

The predefined STISIM event intended to assess the driver's SA was coded as '1', indicating the driver hit the cone or caused an accident when trying to avoid the cone, and '0', indicating the driver managed to avoid the cone by either circumventing or braking. The concept '5-sec' showed the most driving errors ($M = 0.29$, $SD = 0.46$), followed by 'augmented baseline' ($M = 0.14$, $SD = 0.36$), 'baseline countdown' ($M = 0.11$, $SD = 0.32$), and 'answering questions – HazLanA' ($M = 0.11$, $SD = 0.32$). Cochran's Q test did not show any significant difference between the concepts for the predefined STISIM event assessing driver's SA ($\chi^2(3) = 4.44$, $p = 0.22$). The secondary task performance showed no significant differences between concepts for all completed memory games. There was no significant difference in average completion time of the secondary task between the TOR interfaces. Hence, it does not appear that the secondary task interacted with the primary driving task.

5.3.2.3   Discussion: Experiment 2

It is unclear why the concepts 'critical 5-sec' and 'answering questions – HazLanA' were equally well accepted although the 'critical 5-sec' showed a significantly higher workload than 'answering questions'. The 'answering questions' protocol generated the lowest perceived workload and is the only concept which gave rise to a significantly lower level of workload than the '5-sec'. Participants may have been able to use the supporting information from the dialog questions to assist their SA. However, as a dialogue, it also generated the highest TOTs, indicating that a number of participants decided to engage with the questions and take the necessary time to feel confident for the takeover. It is clear that these findings should be formative in developing the interface design for further trials. 'Answering questions' may have somehow put less pressure on participants than the '5-sec'. The critical '5-sec' protocol exhibited a lower TOT, and this demonstrated that the aim of provoking an emergency response to the critical TOR had been achieved. In contrast, the 'answering questions' protocol gave rise to the highest TOTs, and again, this was unsurprising as it was likely due to the concurrent dialogue interaction that was required by the protocol without any indication of urgency. The concepts did not differ in their effects on drivers' responses to the sudden SA event, suggesting that it is possible that even the 5 s allowed by the critical condition was enough to generate sufficient SA to adequately cope with a hazard.

All four concepts in experiment 2 were perceived as effective, and all gave participants the power to take over at any time. Despite the clear cognitive and perceptual demand of the task, the secondary task performance was unaffected by the protocol, and it was clear that the drivers were able to deploy a satisfactory degree of attention to safely completing all the driving protocols based on their judgement. The 'answering questions' protocol was clearly differentiated from the remaining conditions, and this is shown by times to takeover and perceived workload. It could be argued that this protocol afforded less opportunities to takeover due to its conversational structure. Participants perceived all concepts as being useful but not as satisfying to the same level. The 'countdown baseline' and 'augmented baseline' may have been higher ranked for preference because they required the least interaction from the participant and a clearer salient period for takeover. Participants may have accepted modifying their SA by adopting the minimal effort for the perceived situation, taking into account the demands of the secondary task.

Surprisingly, although increasing age effected TOTs, perceived workload was not affected by age for the different protocols. The older drivers required longer times to complete the takeover on request by the vehicle. This suggests that this driver target group could be given a longer lead time prior to the actual takeover. Allowing drivers to choose their TORlt in a customisable interface would also accommodate some proportion of both older and younger drivers. Based on the cognitive capability ranges found in inclusive design data, it was anticipated that older drivers would take longer than younger drivers to take control of the car. This general reduction in a wide range of cognitive functionalities with age is well established in the literature (Langdon and Thimbleby 2010), and therefore, these results suggest that it should be taken into consideration when designing a TOR interface, ideally in a process of personalisation. This finding was further supported by the longer completion times obtained for the secondary task by the older participants, in comparison with younger participants. However, there was no difference in accuracy between the groups, indicating that although older participants took longer to complete the task, they nevertheless managed to perform at a similar level of accuracy and with similar levels of workload, unaffected by age.

5.4 CONCLUSIONS

The aim was the formative development of a TOR design for non-critical, benign driving situations with a secondary task (Politis et al. 2017). These should apply a range of modalities sufficient to accommodate a wide scope of user capabilities (inclusive design) and to ensure an effective SA for a safe takeover process (Politis et al. 2018; Langdon and Thimbleby 2010). Four concepts applying different interaction methods were tested to determine the most effective HMI for further investigation in a high-fidelity simulation, and as formative precursors to a final professional interface design for road trials. The results suggest that the concepts used in the conditions were all capable of eliciting an adequate TOR response, generating high levels of satisfaction, and appreciated as usable. The concepts were not differentiated by their ability to generate SA but the 'answering questions' protocol combined with multimodal ambient warnings also gave rise to low workload and high satisfaction ratings. The CB protocols also produced high satisfaction and usability but were higher ranked in the drivers' preferences, likely as they allowed more control of the point and time of takeover. Although our result was fast compared to Eriksson and Stanton's (2017), with a mean of around 6s with a secondary task, the critical incident '5-sec' interface gave rise to a higher workload and a low preference, which is consistent with it being perceived as a critical rather than a non-critical takeover. Lower workload was associated with the dialogue-based 'answering questions' protocol, which was intended to raise driver's SA. The results are also consistent with the identification of a wide range of TORrts due to population variation, and this is in agreement with the general finding of the present study, suggesting that design principle allowing personalisation of settings by each driver may give the greatest satisfaction and usability.

ACKNOWLEDGEMENTS

This work was supported by Jaguar Land Rover and the UK-EPSRC Grant EP/N011899/1 as part of the jointly funded 'Towards Autonomy: Smart and Connected Control (TASCC) Programme' HI:DAVe Project.

REFERENCES

Borojeni, S. S., L. Chuang, W. Heuten, and S. Boll. 2016. "Assisting drivers with ambient take-over requests in highly automated driving." *AutomotiveUI - Proceedings of the 8th International Conference on Automotive User Interfaces and Interactive Vehicular Applications*. Ann Arbor: Association for Computing Machinery. 237–244. https://doi.org/10.1145/3003715.3005409.

Breakwell, G. M., S. Hammond, C. Fife-Schaw, and J. A. Smith. 2006. *Research Methods in Psychology* (3rd ed.). London: Sage Publications, Inc.

Brooke, J. 1996. "SUS: a 'quick and dirty' usability scale." In *Usability Evaluation In Industry*, by P. W. Jordan, B. Thomas, I. L. McClelland, and B. Weerdmeester, 189–194. London: Taylor & Francis Group. https://doi.org/10.1201/9781498710411.

Burns, C. M., and J. Hajdukiewicz. 2004. *Ecological Interface Design*. Boca Raton: CRC Press.

Civil Aviation Authority. 1999. *RT Discipline (For Pilots & ATC).* Accessed May 25, 2020. https://publicapps.caa.co.uk/docs/33/SRG-NATS_RTDISCIP.PDF.

Clarkson, P. J., R. Coleman, S. Keates, and C. Lebbon. 2003. *Inclusive Design: Design for the Whole Population.* London: Springer-Verlag.

Cranor, L. F. 2008. "A framework for reasoning about the human in the loop." *Proceedings of the 1st Conference on Usability, Psychology, and Security.* San Francisco: USENIX Association. 1–15.

Damböck, D., M. Farid, L. Tönert, and K. Bengler. 2012. "Übernahmezeiten beim hoch-automatisierten Fahren [Takeover times for highly automated driving]." *Tagung Fahrerassistenz [Driver Assistance Conference].* 16–28.

Desmond, P. A., and T. W. Hoyes. 1996. "Workload variation, intrinsic risk and utility in a simulated air traffic control task: evidence for compensatory effects." *Safety Science, 22, 1* 87–101. https://doi.org/10.1016/0925-7535(96)00008-2.

Dzindolet, M. T., S. A. Peterson, R. A. Pomranky, L. G. Pierce, and H. P. Beck. 2003. "The role of trust in automation reliance." *International Journal of Human-Computer Studies, 58, 6* 697–718. https://doi.org/10.1016/S1071-5819(03)00038-7.

Endsley, M. R. 1988. "Situation awareness global assessment technique (SAGAT)." *Proceedings of the IEEE 1988 National Aerospace and Electronics Conference.* Dayton: IEEE. 789–795. DOI: 10.1109/NAECON.1988.195097.

Endsley, M. R., and D. B. Kaber. 1999. "Level of automation effects on performance, situation awareness and workload in a dynamic control task." *Ergonomics, 42, 3* 462–492. https://doi.org/10.1080/001401399185595.

Endsley, M. R., and E. O. Kiris. 1995. "The out-of-the-loop performance problem and level of control in automation." *Human Factors, 37, 2* 381–394. https://doi.org/10.1518/001872095779064555.

Eriksson, A., and N. A. Stanton. 2017. "Takeover time in highly automated vehicles: non-critical transitions to and from manual control." *Human Factors: The Journal of the Human Factors and Ergonomics Society, 59, 4* 689–705. https://doi.org/10.1177/0018720816685832.

Flick, U. 2009. *An Introduction to Qualitative Research* (4th ed.). California: SAGE Publications Ltd.

Gibson, J. J. 1979. *The Ecological Approach to Visual Perception.* Boston: Houghton Mifflin.

Glaser, B. G., and A. L. Strauss. 1967. *The Discovery of Grounded Theory : Strategies for Qualitative Research.* Chicago: Aldine Publishing.

Gold, C., D. Damböck, L. Lorenz, and K. Bengler. 2013. ""Take over!" How long does it take to get the driver back into the loop?" *Proceedings of the Human Factors and Ergonomics Society Annual Meeting, 57, 1* 1938–1942. https://doi.org/10.1177/1541931213571433.

Goodman-Deane, J., P. Langdon, and P. J. Clarkson. 2010. "Key influences on the user-centred design process." *Journal of Engineering Design, 21, 2* 345–373. https://doi.org/10.1080/09544820903364912.

Hancock, P. A., and R. Parasuraman. 1992. "Human factors and safety in the design of intelligent vehicle-highway systems (IVHS)." *Journal of Safety Research, 23, 4* 181–198. https://doi.org/10.1016/0022-4375(92)90001-P.

Hancock, P. A., and W. B. Verwey. 1997. "Fatigue, workload and adaptive driver systems." *Accident Analysis & Prevention, 29, 4* 495–506. https://doi.org/10.1016/S0001-4575(97)00029-8.

Hart, S. G., and L. E. Staveland. 1988. "Development of NASA-TLX (task load index): results of empirical and theoretical research." *Advances in Psychology, 52* 139–183. http://doi.org/10.1016/S0166-4115(08)62386-9.

Jenkins, D. P., N. A. Stanton, P. Salmon, and G. H. Walker. 2008. *Cognitive Work Analysis: Coping with Complexity.* Farnham: Ashgate Publishing Ltd.

Kaber, D. B., and M. R. Endsley. 2004. "Out-of-the-loop performance problems and the use of intermediate levels of automation for improved control system functioning and safety." *Process Safety Progress, 16, 3* 126–131. https://doi.org/10.1002/prs.680160304.

Karam, M., F. A. Russo, and D. I. Fels. 2009. "Designing the model human cochlea: An ambient crossmodal audio-tactile display." *IEEE Transactions on Haptics, 2, 3* 160–169. http://doi.org/10.1109/TOH.2009.32.

Keates, S. L., and P. J. Clarkson. 2004. *Countering Design Exclusion – An Introduction to Inclusive Design*. London: Springer-Verlag.

Kessel, C. J., and C. D. Wickens. 1982. "The transfer of failure-detection skills between monitoring and controlling dynamic systems." *Human Factors: The Journal of the Human Factors and Ergonomics Society, 24, 1* 49–60. https://doi.org/10.1177/001872088202400106.

Langdon, P. 2015. "Developing an interactive TV for the elderly and impaired: an inclusive design strategy." In *A Multimodal End-2-End Approach to Accessible Computing. Human–Computer Interaction Series*, by P. Biswas, C. Duarte, P. Langdon, and L. Almeida, 43–67. London: Springer. https://doi.org/10.1007/978-1-4471-6708-2_3.

Langdon, P., and H. Thimbleby. 2010. "Inclusion and interaction: designing interaction for inclusive populations." *Interacting with Computers, 22* 439–448. https://doi.org/10.1016/j.intcom.2010.08.007.

Langdon, P., U. Persad, and P. J. Clarkson. 2010. "Developing a model of cognitive interaction for analytical inclusive design evaluation." *Interacting with Computers, 22, 6* 510–529. https://doi.org/10.1016/j.intcom.2010.08.008.

Lee, J., and N. Moray. 1992. "Trust, control strategies and allocation of function in human-machine systems." *Ergonomics, 35, 10* 1243–1270. DOI: 10.1080/00140139208967392.

Lee, J. D., and N. Moray. 1994. "Trust, self-confidence, and operators' adaptation to automation." *International Journal of Human-Computer Studies, 40, 1* 153–184. https://doi.org/10.1006/ijhc.1994.1007.

McIlroy, R. C., and N. A. Stanton. 2015. "Ecological interface design two decades on: whatever happened to the SRK taxonomy?" *IEEE Transactions on Human-Machine Systems, 45, 2* 145–163. DOI: 10.1109/THMS.2014.2369372.

McNamara, P. 2017. *How did Audi Make the First Car with Level 3 Autonomy?* 12 July. Accessed May 2, 2018. https://www.carmagazine.co.uk/car-news/tech/audi-a3-level-3-autonomy-how-did-they-get-it-to-market/.

Melcher, V., S. Rauh, F. Diederichs, H. Widlroither, and W. Bauer. 2015. "Take-over requests for automated driving." *Procedia Manufacturing, 3* 2867–2873. https://doi.org/10.1016/j.promfg.2015.07.788.

Muir, B. M. 1987. "Trust between humans and machines, and the design of decision aids." *International Journal of Man-Machine Studies, 27, 5* 527–539. https://doi.org/10.1016/S0020-7373(87)80013-5.

Naikar, N. 2006. "An examination of the key concepts of the five phases of cognitive work analysis with examples from a familiar system." *Proceedings of the Human Factors and Ergonomics Society Annual Meeting, 50, 3* 447–451. https://doi.org/10.1177/154193120605000350.

Naikar, N., A. Moylan, and B. Pearce. 2006. "Analysing activity in complex systems with cognitive work analysis: concepts, guidelines and case study for control task analysis." *Theoretical Issues in Ergonomics Science, 7, 4* 371–394. https://doi.org/10.1080/14639220500098821.

Naujoks, F., C. Mai, and A. Neukum. 2014. "The effect of urgency of take-over requests during highly automated driving under distraction conditions." *5th International Conference on Applied Human Factors and Ergonomics*. Krakow: AHFE. 431–438.

Parasuraman, R., and V. Riley. 1997. "Humans and automation: use, misuse, disuse, abuse." *Human Factors: The Journal of the Human Factors and Ergonomics Society, 39, 2* 230–253. https://doi.org/10.1518/001872097778543886.

Petermeijer, S. M., P. Bazilinskyy, K. Bengler, and J. de Winter. 2017. "Take-over again: investigating multimodal and directional TORs to get the driver back into the loop." *Applied Ergonomics, 62* 204–215. https://doi.org/10.1016/j.apergo.2017.02.023.

Petermeijer, S. M., S. Cieler, and J. C. F. de Winter. 2017. "Comparing spatially static and dynamic vibrotactile take-over requests in the driver seat." *Accident Analysis & Prevention, 99* 218–227. https://doi.org/10.1016/j.aap.2016.12.001.

Politis, I., S. Brewster, and F. Pollick. 2015. "Language-based multimodal displays for the handover of control in autonomous cars." *AutomotiveUI - Proceedings of the 7th International Conference on Automotive User Interfaces and Interactive Vehicular Applications.* Nottingham: Association for Computing Machinery. 3–10. https://doi. org/10.1145/2799250.2799262.

Politis, I., S. Brewster, and F. Pollick. 2017. "Using multimodal displays to signify critical handovers of control to distracted autonomous car drivers." *International Journal of Mobile Human-Computer Interaction, 9, 3* 1–16. https://doi.org/10.4018/ijmhci.2017070101.

Politis, I., P. Langdon, D. Adebayo, M. Bradley, P. J. Clarkson, L. Skrypchuk, A. Mouzakitis, A. Eriksson, J. W. Brown, K. M. A. Revell, and N. A. Stanton. 2018. "An evaluation of inclusive dialogue-based interfaces for the takeover of control in autonomous cars." *23rd International Conference on Intelligent User Interfaces.* Tokyo: Association for Computing Machinery. 601–606. https://doi.org/10.1145/3172944.3172990.

Politis, I., P. Langdon, M. Bradley, L. Skrypchuk, A. Mouzakitis, and P. J. Clarkson. 2017. "Designing autonomy in cars: a survey and two focus groups on driving habits of an inclusive user group, and group attitudes towards autonomous cars." *International Conference on Applied Human Factors and Ergonomics - Advances in Design for Inclusion.* Springer. 161–173. https://doi.org/10.1007/978-3-319-60597-5_15.

Pugh, S. 1991. *Total Design: Integrated Methods for Successful Product Engineering.* Wokingham: Addison-Wesley.

QSR International. 2016. *What is NVIVO?* Accessed November 4, 2020. http://www.qsrinter national.com/what-is-nvivo.

Rasmussen, J. 1985. "The role of hierarchical knowledge representation in decisionmaking and system management." *IEEE Transactions on Systems, Man, and Cybernetics, SMC-15, 2* 234–243. DOI: 10.1109/TSMC.1985.6313353.

Rasmussen, J., A. M. Pejtersen, and L. P. Goodstein. 1994. *Cognitive Systems Engineering.* New York: John Wiley & Sons, Inc.

Recarte, M. A., and L. M. Nunes. 2003. "Mental workload while driving: effects on visual search, discrimination, and decision making." *Journal of Experimental Psychology: Applied, 9, 2* 119–137. https://doi.org/10.1037/1076-898X.9.2.119.

Revell, K., P. Langdon, M. Bradley, I. Politis, J. Brown, and N. Stanton. 2018. "User centered ecological interface design (UCEID): a novel method applied to the problem of safe and user-friendly interaction between drivers and autonomous vehicles." *International Conference on Intelligent Human Systems Integration.* Cham: Springer. 495–501. https://doi.org/10.1007/978-3-319-73888-8_77.

SAE J3016. 2016. *Taxonomy and Definitions for Terms Related to Driving Automation Systems for On-Road Motor Vehicles.* Accessed November 3, 2020. https://www.sae. org/standards/content/j3016_201806/

Salmon, P. M., N. A. Stanton, and K. L. Young. 2012. "Situation awareness on the road: review, theoretical and methodological issues, and future directions." *Theoretical Issues in Ergonomics Science, 13, 4* 472–492. https://doi.org/10.1080/1463922X.2010. 539289.

Scott, J. J., and R. Gray. 2008. "A comparison of tactile, visual, and auditory warnings for rear-end collision prevention in simulated driving." *Human Factors: The Journal of the Human Factors and Ergonomics Society, 50, 2* 264–275. https://doi.org/10.1518/ 001872008X250674.

Seppelt, B. D., and T. W. Victor. 2016. "Potential solutions to human factors challenges in road vehicle automation." In *Road Vehicle Automation 3. Lecture Notes in Mobility*, by G. Meyer, and S. Beiker, 131–148. https://doi.org/10.1007/978-3-319-40503-2_11. Cham: Springer.

Shaikh, S., and P. Krishnan. 2012. "A framework for analysing driver interactions with semi-autonomous vehicles." *Electronic Proceedings in Theoretical Computer Science, 105* 85–99. https://doi.org/10.4204/EPTCS.105.7.

Spath, D., F. Hermann, M. Peissner, and S. Sproll. 2012. "User requirements collection and analysis." In *Handbook of Human Factors and Ergonomics* (4th ed.), by G. Salvendy, 1313–1322. Hoboken: John Wiley & Sons, Inc. https://doi.org/10.1002/9781118131350.ch47.

Stanton, N. A., P. R. G. Chambers, and J. Piggott. 2001. "Situational awareness and safety." *Safety Science, 39, 3* 189–204. https://doi.org/10.1016/S0925-7535(01)00010-8.

Stanton, N. A., A. Dunoyer, and A. Leatherland. 2011. "Detection of new in-path targets by drivers using stop & go adaptive cruise control." *Applied Ergonomics, 42, 4* 592–601. https://doi.org/10.1016/j.apergo.2010.08.016.

STISIM. 2017. *STISIM Drive.* Accessed May 25, 2020. http://stisimdrive.com/.

Strand, N., J. Nilsson, I. C. M. Karlsson, and L. Nilsson. 2014. "Semi-automated versus highly automated driving in critical situations caused by automation failures." *Transportation Research Part F: Traffic Psychology and Behaviour, 27* 218–228. https://doi.org/10.1016/j.trf.2014.04.005.

Tesla. 2020. *Tesla P90S Instrument Cluster Display.* Accessed May 25, 2020. https://www.tesla.com/en_gb/models.

The British Psychological Society. 2018. *Code of Ethics and Conduct (2018).* 18 April. Accessed May 25, 2020. https://www.bps.org.uk/news-and-policy/bps-code-ethics-and-conduct.

Van Der Laan, J. D., A. Heino, and D. De Waard. 1997. "A simple procedure for the assessment of acceptance of advanced transport telematics." *Transportation Research Part C: Emerging Technologies, 5, 1* 1–10. http://doi.org/10.1016/S0968-090X(96)00025-3.

Van Velsen, L., T. Van der geest, R. Klaassen, and M. Steehouder. 2008. "User-centered evaluation of adaptive and adaptable systems: a literature review." *The Knowledge Engineering Review, 23, 3* 261–281. https://doi.org/10.1017/S0269888908001379.

Vicente, K. J. 1999. *Cognitive Work Analysis: Toward Safe, Productive, and Healthy Computer-Based Work.* Mahwah: Lawrence Erlbaum Associates Publishers.

Vicente, K. J., and J. Rasmussen. 1992. "Ecological interface design: theoretical foundations." *IEEE Transactions on Systems, Man, and Cybernetics, 22, 4* 589–606. http://dx.doi.org/10.1109/21.156574.

Vollrath, M., S. Schleicher, and C. Gelau. 2011. "The influence of cruise control and adaptive cruise control on driving behaviour – a driving simulator study." *Accident Analysis & Prevention, 43, 3* 1134–1139. https://doi.org/10.1016/j.aap.2010.12.023.

Walker, G. H., N. A. Stanton, and P. M. Salmon. 2015. *Human Factors in Automotive Engineering and Technology - Human Factors in Road and Rail Transport.* Ashgate: CRC Press.

Wockatz, P., and P. Schartau. 2015. *Intelligent Mobility: Traveller Needs and UK Capability Study.* Transport Systems Catapult.

Young, L. R. 1969. "On adaptive manual control." *Ergonomics, 12, 4* 635–674. https://doi.org/10.1080/00140136908931083.

Young, M. S., and N. A. Stanton. 2002. "Malleable attentional resources theory: a new explanation for the effects of mental underload on performance." *Human Factors: The Journal of the Human Factors and Ergonomics Society,* 365–375. https://doi.org/10.1518/0018720024497709.

Young, M. S., and N. A. Stanton. 2007. "What's skill got to do with it? Vehicle automation and driver mental workload." *Ergonomics, 50, 8* 1324–1339. DOI: 10.1080/00140130701318855.

6 How Was It for You? Comparing How Different Levels of Multimodal Situation Awareness Feedback Are Experienced by Human Agents during Transfer of Control of the Driving Task in a Semi-Autonomous Vehicle

Kirsten M. A. Revell, James W.H. Brown,
Joy Richardson, Jisun Kim, and Neville A. Stanton
University of Southampton

CONTENTS

6.1   INTRODUCTION

Semi-autonomous cars are already on the road, and highly autonomous cars will soon be with us. Different levels of automation vary in the need for users and automation to work together in a human-autonomy team. SAE Level 3 allows the user to be hands, feet, and mind free of the driving task when automation is operating, but they must take back manual control of the vehicle when requested. SAE Level 4 differs in that a request by the automation to take back control may be dismissed by the user as the automation has the capability to complete a safe stop (SAE International 2018). Requests by the automation for the user to take control occur when the operational boundaries for functioning automation are breached. This can happen due to changes in weather conditions and light levels preventing sufficient camera data to be captured, or if sections of the journey include road conditions where the layout, gradient, or lane lines to not meet the requirements for automated driving (Banks and Stanton 2016; Eriksson and Stanton 2017; Stanton and Marsden 1996). SAE Levels 3 and 4 require greatest synergy between automation and users during the takeover task, and this need for synergy creates a vulnerability in the system that warrants focus (Clark, Stanton and Revell 2019). Current road vehicles operate at lower levels of automation, so have not had to contend with this need for synergy. As a result, traditional in-vehicle interfaces can offer little in terms of guidance to achieve this end.

During automated driving, the user is at liberty to undertake secondary tasks, taking them cognitively 'out of the loop' (OOTL) of the main driving task. Roadcraft (2013) emphasises the need to prepare drivers for manoeuvres by ensuring they have situation awareness (SA) of the vehicle state, driving task, road conditions, and other road users. The takeover task is, in effect, a 'manoeuvre' from automated control to human control and as such, the same requirements apply. However, the nature of being OOTL has the effect of reducing SA of the driving environment and vehicle status (Endsley 1995; de Winter et al. 2014). A reduction in SA has also been shown to have safety consequences resulting from an increase in errors and incidents (Stanton, Young and McCaulder 1997), and this is particularly relevant in semi-autonomous driving when receiving control from automation (Banks et al. 2018; Brandenburg and Skottke 2014; Merat and Jamson 2009; Li et al. 2019; Yoon, Kim and Ji 2019). To effectively take over control from the automation, it is necessary for the user to become 'in the loop' of the main driving task to allow appropriate SA to be gained.

In addition to a reduction in SA, the feature of SAE Levels 3 and 4 automation offering two different drive modes ('automated drive' and 'manual drive') creates the condition for safety risks relating to mode awareness (Bainbridge 1983; Norman 1990; Leveson 2004; Revell et al. 2020). Risks can occur where users intentionally allow themselves to become OOTL because they believe the vehicle is being controlled by automation, or where they intervene unnecessarily because they mistakenly believe that the vehicle is no longer being controlled by automation (Revell et al. 2020). One of the ways to support mode awareness and SA for safe transfer of control (TOC) within the human–machine interface is the use of multiple modes of feedback (Eriksson et al. 2019). When considering interaction design for TOC, it is therefore essential to ensure effective mode awareness. In this paper, mode awareness is incorporated as part of SA to ensure in the loop readiness for TOC.

While safety is a key driver for providing SA-promoting feedback for SAE Level 3 vehicles, the importance of the user experience (UX) should not be underestimated. If this is diminished by the modality and format of feedback provided, perceived usability and adoption of semi-autonomous vehicles by consumers could be threatened (Petersson, Fletcher and Zelinsky 2005). Current SAE Level 2 vehicles (e.g. Tesla, Mercedes, JLR), which can offer a hands and feet-free semi-autonomous experience, do not prioritise SA feedback linked to mode awareness of TOC compared to the primary driving feedback (e.g. speed). However, this may be partly due to SAE Level 2 requiring the human to constantly monitor the driving task, ultimately demanding that they are in the loop at all times. As there is currently no standardisation or regulation of the type of feedback that must be provided in semi-autonomous vehicles for safe takeover of control, should SA feedback be unpopular with consumers, manufacturers may feel under commercial pressure to offer interfaces that prioritise UX over preparation for the driving task.

Workload and usability are both key measures of driver–autonomous vehicle interaction (de Winter et al. 2014). UX, in terms of workload and usability of the system, has an impact on whether users will accept and ultimately make a decision to use (Jordan 1998). Workload is the perceived relationship between the amount of mental processing capability or resources and the amount required by the task (Hart and Staveland 1988). Usability is defined as the 'extent to which a product can be used by specified users to achieve specified goals with effectiveness, efficiency and satisfaction in a specified context of use' (ISO 2018).

There is little understanding in the literature of how drivers will respond to multimodal, SA-promoting feedback, compared to non-SA-promoting feedback and the impact on their UX in terms of workload and usability. In response to this gap, this paper examines vehicle-initiated planned takeover, from (automated) vehicle to (human) driver, in a simulated SAE Level 3 vehicle, using the Southampton University driving simulator (SUDS). This study compares an industry-style non-SA-promoting interface, with three multimodal interfaces (increasing in their levels of SA-promoting feedback). UX is measured through subjective questionnaires focusing on driver workload (NASA-Task Load Index (NASA-TLX)) (Hart and Staveland 1988) and system usability (system usability scale (SUS)) (Brooke 1996) with a view to exploring if drivers respond positively to multimodal SA guidance designed to support safe TOC. It will also assess if there is a clear preference for a particular SA-promoting interface design that manufacturers should emulate.

6.2 METHOD

6.2.1 Participants and Study Design

A repeated-measures study in the SUDS exposed 60 human agents (drivers) in three gender-balanced age groups to multiple TOCs in four experimental conditions: (1) no SA guidance (baseline), (2) basic SA guidance, (3) enhanced SA guidance, and (4) enhanced plus SA guidance (see Table 6.1). These varied in the mode and type of feedback which were visually presented either on a head-down display (HDD) or on a head-up display (HUD), provided through audio feedback or through different types of ambient feedback and alerts (see Table 6.1).

TABLE 6.1

Experimental Conditions for Study Showing Variations in Feedback Provided to Participants

Feedback Type	HDD		Audio		HUD		Ambient	
Condition	Countdown Timer	SA Guidance	Alerts	SA Guidance	Mode Icon	SA Guidance	Haptic Seat Mode Alert	LED SA Guidance
C1 – Control: No SA Guidance	X							
C2 – Exp.: Basic SA Guidance		X	X	X	X			
C3 – Exp.: Enhanced SA Guidance		X	X	X	X	X		
C4 – Exp.: Enhanced Plus SA Guidance		X	X	X	X	X	X	X

The 60 participants were recruited to ensure a gender-balanced, inclusive age range of drivers comprising (1) young drivers 18–34 (mean = 24.4, SD = 2.6), (2) mid-range drivers 35–56 (mean = 48.6, SD = 6.9), and (3) older drivers 57–82 (mean = 62.9, SD = 3.8). Each group was gender-balanced (ten males, ten females) and all participants held current UK driving licenses. The research was ethically approved by the University of Southampton Ethics and Research Governance Office (ERGO number: 29615).

UX, in terms of workload and usability scores, was considered a dependent variable and the experimental condition an independent variable. The hypotheses tested were as follows:

- **Hypothesis 1**: Is UX improved when SA-promoting feedback is provided?
- **Hypothesis 2**: Is a single level of SA-promoting feedback preferred overall by participants in terms of UX?

6.2.2 EQUIPMENT

The SUDS comprised a Land Rover Discovery Sport with a 140° view provided by three frontal screens, and rear views provided by LED screens on side mirrors and a projection reflected in the rear-view mirror. STISIM Drive Version 3 simulator software was used to provide the virtual environment, which displayed a custom route with two lengths of highway and two lengths of urban road. Each length was separated by a roundabout, which triggered the necessity for a vehicle-to-driver planned takeover. Highway and urban lengths took 5 and 2.5 min each to navigate respectively, with the whole route taking around 20–25 min. The road conditions were dry and well-lit with traffic density of approximately one vehicle for 500 ft in all lanes. To activate a takeover of control, two buttons were attached to the steering wheel, positioned in line with the driver's thumbs. A Wizard of Oz application on a Microsoft Surface tablet was employed to allow experimenters to configure the elements in the automation system in response to the route, condition, and drivers' responses. Variations in the feedback to participants were provided using the equipment listed in Table 6.2.

6.2.3 PROCEDURE

The participants were briefed about the experiment, and the concept of SAE Level 3 automation, and signed a consent form. Training was given on the route and position of planned takeovers before practising the entire route (with minimal feedback) to experience TOC via the steering wheel buttons and operation of the memory game during automated periods. For the actual experiment, the route was started manually, and the participants were informed via audio that automation was available. They then activated the steering wheel buttons and played the memory game. Available feedback varied depending on the condition presented. Experimental conditions and presentation of route sections were counterbalanced to minimise learning effects. For the control condition (C1), the key feedback was a countdown clock on the HDD during automated drive, based on the Volvo future concept design (Volvo

TABLE 6.2

Equipment Used for Different Feedback Types

Feedback Type	Equipment	Type of feedback
HDD	Microsoft Surface tablet	Cluster feedback
		SA icons/text (C2–C4)
		Countdown clockface graphic (C1)
Audio	Vehicle's internal speakers	Alerts (e.g. verbalised prompts for takeover status)
		Navigation guidance (all conditions)
		SA guidance (e.g. questions asking about outside road environment and vehicle state)
HUD	STISIM's open module, placing objects in drivers' field of view	2D objects (mode icon), 3D objects (range of radar sensing and other road users graphically highlighted)
Ambient SA Guidance	LED strips	Position of vehicles adjacent to the driver
Ambient Alert	Haptic seat (Arduino system)	Takeover mode change alert
Secondary Task	Mini-tablet	Memory game

Cars 2015). For conditions C2–C4, the audio SA feedback consisted of a protocol with five elements of information regarding hazards, lane, speed, exit, and action. The protocol was activated at a set point in the route, approximately one minute prior to the required takeover point. When asked about each element, the participant had to answer correctly before moving to the next question (checked and operated by the experimenter using the Wizard of Oz tablet). The HUD, HDD, or ambient feedback varied according to the condition (see Table 6.1). At the end of each condition, the participant was asked to fill in UX questionnaires via a laptop, and they were provided with space to comment on their opinion of the interface presented and the features present.

6.2.4 METHOD OF ANALYSIS

A one-way, repeated-measures, analysis of variance (ANOVA) was conducted on SUS and NASA-TLX scores, comparing the four experimental conditions (C1–C4). Post hoc tests were applied for significant results to understand where the differences occurred.

Raw TLX scores (Hart and Staveland 1988; Hart 2006) were used as these are considered more valid (Bustamante and Spain 2008). SUS scores were calculated according to guidelines from Brooke (1996). Boxplots overall, and split by age group, were constructed for NASA-TLX raw scores and SUS scores to show the median, interquartile range, and outliers. SUS scores were judged according to the criteria in Table 6.3.

6.3 RESULTS AND DISCUSSION

The first hypothesis sought to explore if UX improved when SA-promoting feedback was provided, in terms of workload and usability. To accept this hypothesis, a statistically

TABLE 6.3
Table of SUS Scores Split into Grades and Adjective Rating (UsabiliTEST 2011)

SUS Score	Letter Grade	Adjective Rating
Above 80.3	A	Excellent
Between 68 and 80.3	B	Good
68	C	OK
Between 51 and 67	D	Poor
Below 51	F	Awful

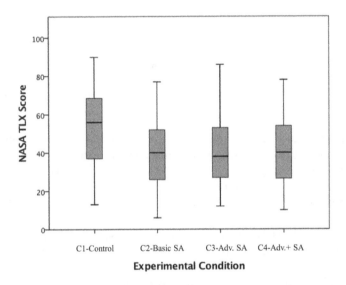

FIGURE 6.1 Boxplot showing NASA-TLX scores for each condition overall.

significant difference is required between the control conditions and experimental conditions, whereby NASA-TLX raw scores are lower and SUS scores are higher for the experimental conditions than for the control condition. The second hypothesis sought to explore if a single level of SA-promoting feedback provided a better UX to participants. To accept this hypothesis, a statistically significant higher usability score and lower workload score need to be found for a *single* experimental condition.

6.3.1 WORKLOAD

Figures 6.1 and 6.2 show boxplots for workload (NASA-TLX raw score) overall, and split by age, respectively.

While the range of scores is similar for all conditions, the interquartile range is markedly higher for the control condition (C1) in comparison with the experimental conditions (C2–C4) (see Figure 6.1). This trend is consistent when examining the trend

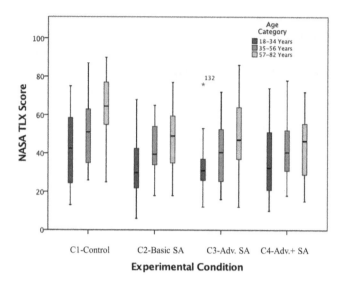

FIGURE 6.2　Boxplot showing NASA-TLX scores for each condition by age group.

split by age group each age (see Figure 6.2). The median score for each condition shows the lowest workload for the youngest age group, with workload scores increasing with age such that the oldest age group have the highest median workload in each condition.

The results from the ANOVA show that self-reported workload was significantly affected by the type of interaction design presented (F (2.45, 130.03) = 15.51, $p < 0.001$).

Post hoc tests showed a significant difference was seen between the control group (C1) and the experimental groups (C2–C4), but not between the experimental groups themselves.

The trend of higher workload in each condition for the older age range was not unexpected, as older drivers experience greater workload in general when driving (Cantin et al. 2009). This can be partly explained by age-related degradation in cognitive capacity (Rogers and Fisk 2001), as well as a decline in speed of information processing and decision-making (Shaheen and Niemeier 2001).

The reduction in workload found in the experimental conditions suggests that, for the TOC task, the SA guidance, mode awareness, and alerts presented in the interface better supported the needs of users during the TOC. This included guidance regarding where to pay attention, through verbalisations and graphical highlighting on the HUD, and what to do at different stages of the takeover, through icons and verbal instructions. In addition, the multimodal alerts signalling different stages of the task allow different user preferences to be accommodated.

6.3.2　Usability

Figures 6.3 and 6.4 show boxplots for usability (SUS score) overall, and split by age, respectively.

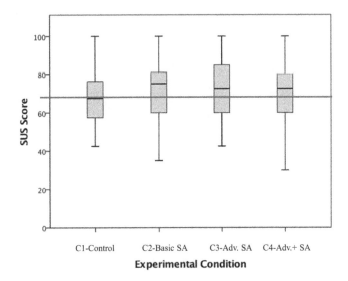

FIGURE 6.3 Box plots showing results for SUS scores for each condition overall. The red line represents 'average' usability with a SUS score of 68.

While the range of scores are comparable for all conditions, the range for condition 4 is slightly larger. The median is lowest for the control condition (C1) in comparison with the experimental conditions (C2–C4) (see Figure 6.3). There is less consistency in trends for usability when split by age group, although the median values are lowest for the oldest age group in each condition (see Figure 6.4). The results from the ANOVA show that system usability was significantly affected by the type of interaction design presented (F (3, 162) = 3.64, $p < 0.05$). Post hoc tests showed the significant difference was seen between the control group (C1) and the experimental groups (C2–C4), but not between experimental groups themselves.

Comparing the scores to the usability grades in Table 6.3, it can be seen that the median overall SUS scores (Figure 6.3) is graded at 'C' (average, Table 6.3) for the control condition (C1), and graded at 'B' (good, Table 6.3) for all the experimental conditions (C2–C4). When broken down by age group, there appears a greater variation in experience between conditions. In two conditions, C1 and C3, the older age group showed a median SUS score graded at 'D' (poor, Table 6.3).

An explanation of why the lowest SUS score for the older age group was found in the control condition (C1) may be linked to the value of guidance for safety compliance when driving. Van Der Laan, Heino, and De Waard (1997) found, when comparing usefulness and satisfaction between younger and older drivers, that alerts and tutoring for safe and compliant manual driving were far better received by the older age group (60–75 year olds) than by the younger (30–45 year olds – comparable to our mid-range age group). The relative lack of guidance and alerts compared in C1 compared to the experimental conditions (C2–C4) may contribute to this effect. The 'poor' usability rating by the older age group for the C3 experimental condition requires further investigation; however, this may be due to the balance between

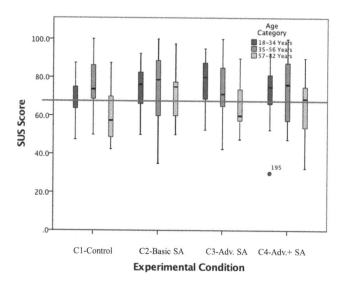

FIGURE 6.4 Box plots showing results for SUS scores for each condition by age group. The red line represents 'average' usability with a SUS score of 68.

driving SA feedback and mode feedback/alerts. The key difference between C2 and C3 was the introduction of SA guidance on the HUD showing reach of automation sensors and highlighting other road users. The key difference between C3 and C4 was the introduction of a haptic seat alert highlighting a change of mode, LED lighting showing current mode, and an LED blind spot indicator (both contain the HUD SA guidance). As the median score by the older age group for C4 was rated as 'good', the HUD SA guidance is not sufficient to explain the 'poor' rating for C3. However, it may be that the older age group rely on alerts more for signalling the start and end of the TOC process, and value more mode awareness feedback than younger age groups to help guide safe and compliant behaviour.

The results of workload and usability UX support hypothesis 1, as both workload and usability scores improved when using an interface with SA feedback to support TOC compared to non-SA feedback. The results did not support hypothesis 2 as there was no significant difference found between the different levels of SA feedback provided to support TOC. Further work to explore the post-experiment comments, in terms of variations in feedback preferences and trends by age, is recommended, to feed into the iterative design process.

6.4 CONCLUSION

UX improved from baseline with experimental designs in terms of both workload and usability. The introduction of SA guidance and alerts in the interface design of semi-autonomous vehicles should therefore be embraced by manufacturers to ensure a safe TOC without fear of diminishing UX and uptake. Despite three levels of SA feedback being presented to participants (basic, enhanced, and enhanced plus), there was no clear winner in terms of the amount of feedback offered. Further work should investigate the variations in user preferences to better understand the modes and types of feedback that best support users. It was concluded that feedback that offers SA guidance should be provided to ensure human agents are in the loop when transferring control.

ACKNOWLEDGEMENTS

This work was supported by Jaguar Land Rover and the UK-EPSRC Grant EP/011899/1 as part of the jointly funded 'Towards Autonomy: Smart and Connected Control (TASCC) Programme'.

REFERENCES

Bainbridge, L. 1983. "Ironies of Automation." *Analysis, Design and Evaluation of Man–Machine Systems. Proceedings of the IFAC/IFIP/IFORS/IEA Conference.* Baden-Baden, Germany: Elsevier Ltd. 129–135. https://doi.org/10.1016/B978-0-08-029348-6.50026-9.

Banks, V. A., and N. A. Stanton. 2016. "Keep the driver in control: Automating automobiles of the future." *Applied Ergonomics, 53,* Part B 389–395. DOI: 10.1016/j.apergo.2015.06.020.

Banks, V. A., A. Eriksson, J. O'Donoghue, and N. A. Stanton. 2018. "Is partially automated driving a bad idea? Observations from an on-road study." *Applied Ergonomics, 68* 138–145. DOI: 10.1016/j.apergo.2017.11.010.

Brandenburg, S., and E.M. Skottke. 2014. "Switching from manual to automated driving and reverse: Are drivers behaving more risky after highly automated driving?" *17th International IEEE Conference on Intelligent Transportation Systems (ITSC).* Qingdao, China: IEEE. 2978–2983. DOI: 10.1109/ITSC.2014.6958168.

Brooke, J. 1996. "SUS: A 'Quick and Dirty' usability scale." In *Usability Evaluation In Industry,* by P. W. Jordan, B. Thomas, I. L. McClelland and B. Weerdmeester, 189–194. https://doi.org/10.1201/9781498710411. London: Taylor & Francis Group.

Bustamante, E. A., and R. D. Spain. 2008. "Measurement Invariance of the Nasa TLX." *Proceedings of the Human Factors and Ergonomics Society Annual Meeting, 52,* 19 1522–1526. DOI: 10.1177/154193120805201946.

Cantin, V., M. Lavallière, M. Simoneau, and N. Teasdale. 2009. "Mental workload when driving in a simulator: Effects of age and driving complexity." *Accident Analysis & Prevention, 41,* 4 763–771. DOI: 10.1016/j.aap.2009.03.019.

Clark, J. R., N. A. Stanton, and K. M. Revell. 2019. "Conditionally and highly automated vehicle handover: A study exploring vocal communication between two drivers." *Transportation Research Part F: Traffic Psychology and Behaviour, 65* 699–715. DOI: 10.1016/j.trf.2018.06.008.

de Winter, J. C.F., R. Happee, M. H. Martens, and N. A. Stanton. 2014. "Effects of adaptive cruise control and highly automated driving on workload and situation awareness: A review of the empirical evidence." *Transportation Research Part F: Traffic Psychology and Behaviour, 27,* Part B 196–217. DOI: 10.1016/j.trf.2014.06.016.

Endsley, M. R. 1995. "Toward a theory of situation awareness in dynamic systems." *Human Factors: The Journal of the Human Factors and Ergonomics Society 37,* 1 32–64. DOI: 10.1518/001872095779049543.

Eriksson, A., and N. A. Stanton. 2017. "Takeover time in highly automated vehicles: Noncritical transitions to and from manual control." *Human Factors: The Journal of the Human Factors and Ergonomics Society, 59,* 4 689–705. DOI: 10.1177/0018720816685832.

Eriksson, A., S. M. Petermeijer, M. Zimmermann, J.C.F. de Winter, K. J. Bengler, and N. A. Stanton. 2019. "Rolling out the red (and green) carpet: Supporting driver decision making in automation-to-manual transitions." *IEEE Transactions on Human-Machine Systems, 49,* 1 20–31. DOI: 10.1109/THMS.2018.2883862.

Hart, S. G. 2006. "Nasa-Task Load Index (NASA-TLX); 20 Years Later." *Proceedings of the Human Factors and Ergonomics Society Annual Meeting,* 904–908. DOI: 10.1177/154193120605000909.

Hart, S. G., and L. E. Staveland. 1988. "Development of NASA-TLX (Task Load Index): Results of empirical and theoretical research." *Advances in Psychology, 52* 139–183. DOI: 10.1016/S0166-4115(08)62386-9.

ISO. 2018. *ISO 9241-11:2018 Ergonomics of human-system interaction — Part 11: Usability: Definitions and concepts.* March. Accessed November 4, 2020. https://www.iso.org/obp/ui/#iso:std:iso:9241:-11:ed-2:v1:en.

Jordan, P. W. 1998. *An Introduction To Usability.* Boca Raton, FL: CRC Press.

Leveson, N. 2004. "A new accident model for engineering safer systems." *Safety Science, 42,* 4 237–270. DOI: 10.1016/S0925-7535(03)00047-X.

Li, S., P. Blythe, W. Guo, A. Namdeo, S. Edwards, P. Goodman, and G. Hill. 2019. "Evaluation of the effects of age-friendly human-machine interfaces on the driver's takeover performance in highly automated vehicles." *Transportation Research Part F: Traffic Psychology and Behaviour,* 67 78–100. DOI: 10.1016/j.trf.2019.10.009.

Merat, N., and A. H. Jamson. 2009. "Is Drivers' situation awareness influenced by a fully automated driving scenario?" *Human Factors, Security and Safety. Human Factors and Ergonomics Society Europe Chapter Conference.* Soesterberg, the Netherlands: Shaker Publishing.

Norman, D. A. 1990. "The "problem" with automation: inappropriate feedback and interaction, not "overautomation". *Philosophical Transaction of the Royal Society of London, 327* 585–593.

Petersson, L., L. Fletcher, and A. Zelinsky. 2005. "A framework for driver-in-the-loop driver assistance systems." *Proceedings. 2005 IEEE Intelligent Transportation Systems.* Vienna, Austria: IEEE. 771–776. DOI: 10.1109/ITSC.2005.1520146.

Revell, K., J. Richardson, P. Langdon, M. Bradley, I. Politis, S. Thompson, L. Skrypchuk, and et al. 2020. "Breaking the cycle of frustration: Applying Neisser's perceptual cycle model to drivers of semi-autonomous vehicles." *Applied Ergonomics, 85* 103037. DOI: 10.1016/j.apergo.2019.103037.

Roadcraft. 2013. *Roadcraft: The Police Driver's Handbook.* London, UK: The Stationery Office.

Rogers, W. A., and A. D. Fisk. 2001. "Understanding the role of attention in cognitive aging research." In *Handbook of the Psychology of Aging,* by J. E. Birren and K. W. Schaie, 267–287. Cambridge, MA: Academic Press.

SAE International. 2018. *SAE International Releases Updated Visual Chart for Its "Levels of Driving Automation" Standard for Self-Driving Vehicles.* 11 December. Accessed May 1, 2020. https://www.sae.org/news/press-room/2018/12/sae-international-releases-updated-visual-chart-for-its-%E2%80%9Clevels-of-driving-automation%E2%80%9D-standard-for-self-driving-vehicles.

Shaheen, S. A., and D. A. Niemeier. 2001. "Integrating vehicle design and human factors: minimizing elderly driving constraints." *Transportation Research Part C: Emerging Technologies, 9,* 3 155–174. DOI: 10.1016/S0968-090X(99)00027-3.

Stanton, N. A., and P. Marsden. 1996. "From fly-by-wire to drive-by-wire: Safety implications of automation in vehicles." *Safety Science, 24* 35–49. DOI: 10.1016/S0925-7535(96)00067-7.

Stanton, N. A., M. Young, and B. McCaulder. 1997. "Drive-by-wire: The case of driver workload and reclaiming control with adaptive cruise control." *Safety Science, 27,* 2 149–159. DOI: 10.1016/S0925-7535(97)00054-4.

UsabiliTEST. 2011. *System Usability Scale (SUS) Plus*. Accessed July 14, 2020. https://www.usabilitest.com/system-usability-scale.

Van Der Laan, J. D., A. Heino, and D. De Waard. 1997. "A simple procedure for the assessment of acceptance of advanced transport telematics." *Transportation Research Part C: Emerging Technologies, 5*, 1 1–10. DOI: 10.1016/S0968-090X(96)00025-3.

Volvo Cars. 2015. "Volvo Cars: Explore the User-Interface of Tomorrow [Video]." https://www.youtube.com/watch?v=xYqtu39d3CU.

Yoon, S. H., Y. W. Kim, and Y. G. Ji. 2019. "The effects of takeover request modalities on highly automated car control transitions." *Accident Analysis & Prevention, 123* 150–158. DOI: 10.1016/j.aap.2018.11.018.

7 Human Driver Post-Takeover Driving Performance in Highly Automated Vehicles

James W.H. Brown, Kirsten M. A. Revell, and Joy Richardson
University of Southampton

Ioannis Politis
The MathWorks Inc.

Patrick Langdon
Edinburgh Napier University

Michael Bradley
University of Cambridge

Simon Thompson and Lee Skrypchuk
Jaguar Land Rover

Neville A. Stanton
University of Southampton

CONTENTS

7.1 INTRODUCTION

One of the major challenges in designing automated vehicles is the manner by which the driver might ideally take back manual control of the vehicle, specifically the form of the takeover from the automated vehicle to the human driver (Gold et al. 2013; Merat et al. 2019). Much research has been carried out into the efficacy of different feedback modalities (Naujoks, Mai and Neukum 2014; van den Beukel, van der Voort and Eger 2016; Yoon, Kim and Ji 2019). Safety is clearly the primary concern with regard to takeover performance, both in terms of the driver's situation awareness (SA) and in terms of their immediate behaviour post-takeover. This study examines four different takeover protocols, three of which were specifically designed to raise SA. The question is therefore, 'What effect does the amount and type of feed-forward information have on driver performance immediately post-takeover?'

Driving is considered to be a complex activity, constituting multiple tasks that must be carried out in parallel (Regan, Lee and Young 2008). Indeed, Walker, Stanton, and Salmon (2018) identified over 1600 unique tasks that may form part of the driving role. Despite the complexity, in 2017, over 6 trillion road miles were covered worldwide (Marn 2017). Within the first world countries such as the UK and the US, over 70% of the population have driving licenses (Hedges & Company 2020; Department for Transport 2018). The complex nature of driving combined with its popularity means that accidents inevitably occur; the WHO estimated that worldwide, 1.35 million people were killed in 2018 (World Health Organization 2018). One of the major rationales for the development and introduction of autonomous vehicles is the potential to dramatically increase safety on the roads. The National Highway Traffic Safety Administration in the US states that of recorded serious accidents, 94% are due to human error, and by removing humans from the driving task, these accidents could be avoided (National Highway Traffic Safety Administration 2019). The development of fully autonomous cars is highly complex, and it is therefore being addressed via multiple levels as new technology becomes available and capable. The Society of Automotive Engineers (SAE) outline these six distinct levels of automation, from zero, which represents a vehicle with no driving aids, through to level five where a vehicle is fully autonomous in all conditions, and driving controls would not be necessary (Shuttleworth 2019). Several Level 2 semi-autonomous vehicles are currently available to consumers, such as those developed by Tesla (2019), BMW (2019), and Mercedes Benz (2019). These vehicles provide lateral and longitudinal control, rendering them ostensibly automated; however, they have been designed both technically and legally to require drivers to maintain full awareness of the driving scenario and be ready to assume control immediately (Tesla 2019; BMW 2019; Mercedes Benz 2019). Level 3 cars are distinguished from those in Level 2 in that they allow the driver to attend to secondary tasks such as reading or using a mobile phone. The SAE specifically state that a driver is no longer considered to be driving in a Level 3 car when automation is enabled. However, Level 3 cars retain the requirement for the driver to assume manual control if necessary (Shuttleworth 2019).

de Winter et al. (2014) carried out a meta-analysis into the effects of advanced driver assistance systems and highly automated driving on SA. Their research revealed that highly automated driving in combination with a non-driving-related task could reduce SA. Vogelpohl et al. (2018) state that in complex road traffic scenarios, drivers responding to takeover requests (TORs) can take considerably longer to gain a sufficient level of SA. They also advocate additional support for drivers as they transition to manual, and the introduction of some method by which SA can be raised. The unique challenge of Level 3 is therefore how to raise the SA of the driver sufficiently after a period of non-driving to ensure that they can safely resume manual control of a moving vehicle in all conditions. Considerable research has been carried out into how to bring the driver back into the loop via multiple modalities (Politis, Brewster and Pollick 2015; Petermeijer et al. 2017; van den Beukel, van der Voort and Eger 2016). Indeed, the use of multimodal TORs in place of unimodal TORs is advocated (Yoon, Kim and Ji 2019; Politis, Brewster and Pollick 2015). For example, Yoon, Kim, and Ji (2019) found that employing multiple TOR modalities resulted in a faster hands-on time, and visual-only TORs contributed towards longer takeover times. Politis, Brewster, and Pollick (2015) and Melcher et al. (2015) also specifically advised against the use of solely unimodal visual TORs. Biondi et al. (2017), however, found that while multimodal alerts work well in low driver demand scenarios, the advantage over unimodal alerts is reduced as driver demand increases. In a crowdsourcing survey, people have expressed a preference for multimodal warnings in high-urgency scenarios, whereas in low-urgency scenarios, visual TORs were preferred to haptic TORs (Bazilinskyy et al. 2018). This apparent contradiction in research findings suggests what people prefer may not always reflect what they perform best with. Also, it may be the case that an effective and desired set of TOR modalities for some drivers may be ineffective and undesirable for others (Parasuraman and Riley 1997).

In a review of the automated vehicle takeover literature, McDonald et al. (2019) report that the safety of a takeover process is mainly influenced by two factors, namely the time budget available (urgent or non-urgent takeover) and the effectiveness of the takeover action. The quality of the takeover can be measured using multiple metrics. McDonald et al. (2019) identified 17 separate methods from the literature; the primary areas for analysis appear to be lateral and longitudinal control; these can be measured as velocities, accelerations, control inputs, distances, or times. Eriksson and Stanton (2017a) also reviewed the literature, specifically on non-critical takeover times, in addition to conducting their own study. Their findings revealed that the time to transition to manual control from automated mode could take between 1.9 and 25.7 s after TOR presentation. This represents a large variation, with a clear influence on the safety of the takeover process. Dogan et al. (2019) examined the effects of non-driving-related tasks on takeover performance using a range of post-takeover metrics. They state that the definition of a safe takeover is likely to be situationally dependent. It is therefore advantageous to consider a set of post-takeover performance metrics to quantify takeover quality. Eriksson and Stanton (2017b) experimented with driver-paced non-critical control transitions to explore their effect on driving performance when resuming control. Their findings revealed that driver-paced transitions result in improved performance in the 20 s post-takeover, compared with system-paced systems. This therefore is an additional influence on takeover safety in addition to those proposed by McDonald et al. (2019).

A key factor affecting the quality of takeover performance is the existence and modalities of the TORs (McDonald et al. 2019; Brandenburg and Chuang 2019). There have been multiple studies into the post-takeover performance of human–machine interface (HMI) modalities, both individually and in combination. In terms of advocating multimodality, Naujoks, Mai, and Neukum (2014) found that lateral control performance is improved when both visual and audio modalities are employed to raise SA. Toffetti et al. (2009) also discovered that a vocal modality resulted in lower average speeds post-takeover when compared to a purely auditory modality. Brandenburg and Chuang (2019) found that presenting participants with skeuomorphic visual TORs was both considered preferable and resulted in more accurate and faster post-takeover performance when compared with abstract visual TORs (both being multimodal due to the addition of an auditory signal). However, in contrast, Petermeijer et al. (2017) found mixed benefits when they examined the performance of unimodal auditory and vibrotactile TORs, and then combined them into a multimodal TOR. Their experiments showed multimodal TORs resulted in significantly faster steering-touch times and were the preferred method of request. However, lane changing time and braking time showed no statistically significant differences between TORs. This suggests that for some aspects of performance, multimodality does not affect post-takeover performance. Mirnig et al. (2017) carried out a literature analysis into the control transition interface in autonomous cars. They found a wide variance of output modalities – visual being the most prevalent – and additional modalities such as audio and haptic tend to be introduced to cope with scenarios where distraction may be an issue, or the visual TOR is not within the field of view. This is suggestive that the visual TOR should perhaps be treated as of primary importance. Eriksson et al. (2019) looked at the effectiveness of four HMIs designed to support driver decisions in transitions from automated to manual driving when presented with lane changing or braking requirements. Their findings indicated that while reaction times were not significantly affected, decision-making was improved by two of the HMI designs. They advocate the use of visual HMIs to assist driver decisions, and further recommend additional modalities to render the HMI more authoritative, again reinforcing the case for multimodality and the use of visuals as the primary mode.

The literature clearly suggests that in most cases, performance improvements are possible through the combination of modalities; there, however, remains the question as to whether the level of feedback that a multimodal HMI provides has a significant effect on takeover performance.

It should be noted that in addition to the presentation of a TOR in any modality, there often exist accompanying parameters, such as magnitude and position. Timing the presentation of a TOR based upon the driver's task can also affect performance. Wintersberger et al. (2018) examined the effects of presenting TORs at low demand task boundaries prior to takeover; they found that presenting visual TORs within the in-vehicle information system resulted in longer takeover times and worse takeover performance. Presenting TORs in the drivers' non-driving-related task had the advantageous effect of preventing them from continuing to use them.

The aim of this study was to examine the post-takeover performance effects of an HMI featuring four differing levels of feedback, three of which were designed specifically to raise SA. It was anticipated that HMI variations with higher levels of feedback would produce better performance post-takeover (Naujoks, Mai and Neukum 2014; Toffetti et al. 2009). Three metrics of longitudinal and lateral behaviour were selected to analyse takeover performance in the initial 10 seconds post-takeover: speed, lane deviation, and steering angle.

7.2 METHOD

7.2.1 PARTICIPANTS

All participants were required to hold full UK driving licenses and provide a signed consent in order to take part. The University of Southampton's Ethics and Research Governance Office approved the experiment (ERGO number: 29615). A group of 25 participants were recruited that exhibited a gender split of 11f/14m and an age range split of 5 (18–34, mean=24.4, SD=2.6), 8 (35–56, mean=48.6, SD=6.9), and 12 (57–82, mean=62.9, SD=3.8). The age ranges were derived from personal communications (Langdon and Stanton, Personal communication, 2017).

7.2.2 EXPERIMENTAL DESIGN

The experiment featured a repeated-measures design. Four differing vehicle-to-driver control transfer protocols (CTPs) were tested on both urban and highway road types; the CTP and road type therefore represented the independent variables. There were four trials carried out per participant, one for each CTP, with two road types experienced twice per trial (highway, highway, urban, urban). Two transfers of control from automation to manual took place within each road type.

The first CTP was based on a design by Volvo. This featured a clock-style timer that appeared on the cluster tablet with a text-based message requesting the driver to take control by pressing a button. The 'clock' counted down from 60 s in 2-s intervals.

The following three CTPs were designed specifically to increase SA, by requiring the driver to repeat out loud the presented information, in order, regarding hazards, lane, speed, exits, and actions (HazLanSEA). On urban road types, the lane aspect of SA was removed. These CTPs were differentiated based upon the modalities of information that were presented as described in Table 7.1. Figure 7.1 pictures the driver's view of the final stage of a TOR for the HazLanSEA 2 protocol. Drivers' repetitions were assessed by the experimenter; if correct, they would be presented with the next protocol element or prompted to take control.

Dependent variables were represented by vehicle performance data generated by the simulation software during the experiments; these variables included speed, lane position, steering, throttle, braking, automation modes, and timestamps.

The order of CTPs was counter-balanced for the HazLanSEA protocols; however, participants always experienced the timer-based CTP first.

TABLE 7.1

Information Modalities Employed by Each of the HazLanSEA Protocols

Protocol	Text	Icons	Vocals	HUD Highlighting	Haptics	Blind spot Indicator
HazLanSEA 1	Cluster+HUD	Cluster+HUD	Yes	No	No	No
HazLanSEA 2	Cluster+HUD	Cluster+HUD	Yes	Yes	No	No
HazLanSEA 3	Cluster+HUD	Cluster+HUD	Yes	Yes	Yes	Yes

FIGURE 7.1 Driver view of the last stage of a HazLanSEA 2 TOR.

7.2.3 EQUIPMENT

The Southampton University Driving Simulator laboratory was utilised for the experiment. This constituted a Land Rover Discovery Sport fixed-base vehicle, coupled with STISIM Drive V3.0 software and a 135° projected field of view. The cluster in the car was replaced with a Microsoft Surface tablet, allowing a custom interface to be displayed featuring icons and text. Wing mirrors were simulated using small monitors, and the vehicle's rear-view mirror reflected a rear-facing projector screen (see Figure 7.2 for a diagram of the equipment used). An Arduino-based haptic seat cover was fitted to the driver's seat to provide haptic-based alerts, and ambient LED strips were fitted to the A-pillars to act as blind spot warning indicators. Software was written for STISIM's Open Module that provided the ability to display head-up display (HUD)-based information directly on the main frontal screens. Additional software was also written within the open module that enabled automation when activated by

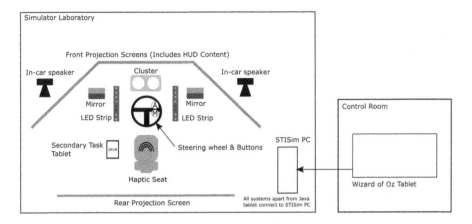

FIGURE 7.2 A block diagram illustrating the systems employed within the Southampton University Driving Simulator. The A and M buttons on the steering wheel correspond to the automation and manual modes, respectively.

a steering wheel button. A tablet PC in the control room provided the ability to control settings for automation availability, cluster display, HUD, haptics, and ambient warnings. This tablet also controlled the generation of vocal instructions/questions as part of the CTPs. A small tablet with a Java-based memory game was employed as a secondary task for participants when the vehicle was in automated mode. A test route was developed for STISIM with two sections of motorway followed by two sections of urban road, separated by roundabouts. The motorway sections took approximately 5 min to complete and the urban sections, 2.5 min, when complying with UK traffic law. The driving scenario took place in dry conditions during simulated daytime, and the density of traffic was set to approximately one vehicle every 500 ft for all lanes.

The HUD highlighting displayed a green portion of a circle projected on the road in front of the car that represented the area within which the automation system can detect vehicles. Detected vehicles were also highlighted using 3D boxes drawn around them. This combination of highlighting was designed to simulate the Mobileye automation technology (Mobileye 2019). If the CTP included it, the HUD highlighting was visible in both automated and manual modes. The haptics system presented a haptic cue via the seat when the takeover process started. The blind spot indicator lit a series of LEDs that provided a visual representation of vehicles as they passed by the side of the participant's vehicle. For example, as a car moved past the participants' car, the LEDs would light from the top of the A-pillar, slightly behind and above the driver, forward to the bottom, and in doing so vehicles in the blind spot were revealed. The blind spot indicator was functional in both automated and manual modes for the relevant CTP.

7.2.4 PROCEDURE

Upon arrival, participants were greeted and provided with the relevant health and safety information. An introduction to the experiment and its objectives was then given, and they were asked to read the participant information sheet. Following an opportunity to ask any questions, they were then asked to sign a consent form to confirm that they agree to take part in the study and a biographical sheet to collect additional demographical information. A short video was provided to train participants in the use of verbal protocol, then the BAST 3 levels of automation were explained, together with an overview of the technology used in each of the CTPs. They were shown a route map, with locations of slip roads and roundabouts, that required a transition to manual driving from automated, highlighted. They were also informed that navigational instructions would be provided when necessary. Participants were taught how to operate the secondary task memory game tablet and requested to play it every time that automation was enabled. Once complete, they were introduced to the vehicle and shown the relevant systems in situ. All participants were given a short (<5 min) test run on the highway road type to familiarise them with the controls and the operation of the systems and interfaces.

Each trial started with the participant being requested via a synthesised vocal command initiated by the experimenter via the Wizard of Oz (WOZ) software to drive from a motorway slip road and merge onto the motorway. Once on the motorway, the experimenter issued a second vocalisation via the WOZ interface that automation was available, this was also indicated by an icon and text on the custom cluster. Participants then activated the automation by pressing a button marked 'A' on the steering wheel. The automation mode was then activated and confirmed with an icon on the cluster and a synthesised vocalisation. Once automated, participants picked up the secondary task tablet and played the memory game. At the mid-point of each road type, and prior to each roundabout, the experimenter used the WOZ interface to initiate a takeover. For each CTP, the participant was expected to notice and acknowledge the request to take control, put down the secondary task tablet, and follow any given instructions. This involved, depending upon the CTP, simply taking control, or repeating SA information as it was presented, and then pressing a button marked 'M' on the steering wheel when requested by the HMI. The HMI then provided a vocalisation and displayed text and an icon to inform that manual mode was now active. After a short period (approximately 70 s) of manual control, either mid-road type or post-roundabout, the experimenter then reissued the notification that automation was available again and the process was repeated until the full route was completed. A total of eight takeovers therefore took place per CTP, two for each occurrence of road type.

7.2.5 ANALYSIS

Data was recorded in STISIM at 10 Hz; however due to the processing demand on the simulation PC, this frequency was not consistent. To account for this, data was down-sampled into bins of 0.2 s, resulting in a consistent data at 5 Hz. For each participant, the data was then processed to identify the changes in automation mode from automated to manual; 10-s sections from the data were then extracted and saved as individual Comma Separated Values (CSVs) for analysis. Speed and lane position data were normalised, as the start values differed between participants. Values for steering and lane deviation were converted to absolute values. Four forms of analysis were employed; mean values and standard deviations per bin across all participants were calculated. To indicate effect sizes, Cohen's D values were calculated per bin for all dependent variables. Significance was tested using per-bin *t*-tests for all dependent variables, and then graphed as $-\log10(p)$; therefore, higher values indicate more significance. For all data, time bins of 0.2 s periods were analysed independently with Bonferroni alpha-level corrections in place. A Bonferroni correction was calculated, and the resultant threshold was displayed on each graph. It should be noted that the effect size and significance test graphs do not start at zero as they were calculated using non-absolute and non-normalised values.

7.3 RESULTS

7.3.1 SPEED

Figure 7.3 displays the mean speed over the 10 s following takeover, grouped into trials (timer and HazLanSEA 1, 2, and 3). The HazLanSEA protocols appear to elicit very similar trends despite their modality variations. Mean speeds appear to reduce at a similar rate and by approximately 6 mph, 10 s post-takeover for the HazLanSEA protocols (trials 2, 3, and 4). This is suggestive that despite the differences between modalities and associated feedback, speed post-takeover is not greatly impacted. Mean speeds on the urban road type showed similar trends of an initial deceleration of approximately 2 mph over the initial 7.5 s post-takeover. This again suggests that the differences in modalities and feedback are not affecting post-takeover speed. The post-takeover speed for the Timer protocol appeared to show a considerably different trend to the HazLanSEA protocols, with deceleration ending after approximately 4 s, followed by gentle acceleration. A speed differential of approximately 6 mph

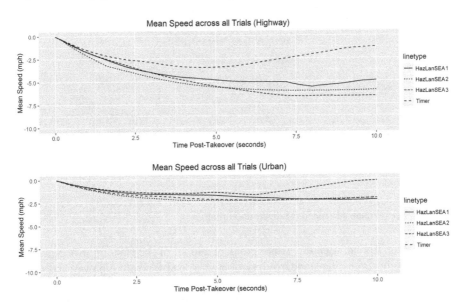

FIGURE 7.3 Mean absolute speed (highway and urban).

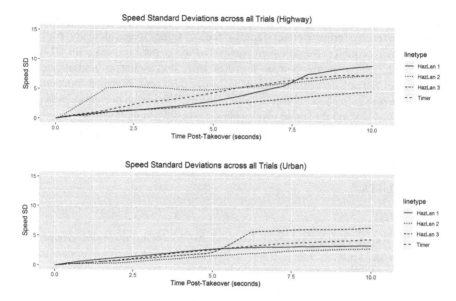

FIGURE 7.4   Speed standard deviations (highway and urban).

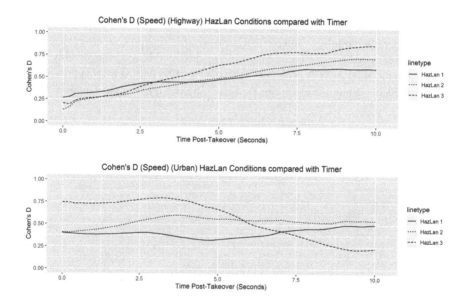

FIGURE 7.5   Cohen's D speed (highway and urban).

existed after 10 s between the Timer protocol and all of the HazLanSEA protocols. Figure 7.4 illustrates the standard deviations of speed for both highway and urban road types.

The graphs in Figure 7.5 illustrate the effect size for speed, which mirror the results for the significance graphs in Figure 7.6. These are initially broadly similar

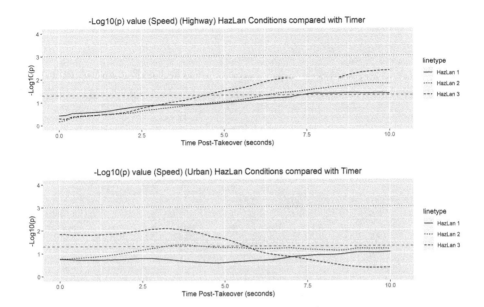

FIGURE 7.6 -Log10(p) speed (highway and urban).

for the HazLanSEA protocols on the highway road type with some divergence from 3 s where the effect size increases in Trial 4. There is more variation in Cohen's D values on the urban road type with the largest differences occurring between trials at around 3.5 s. T-tests carried out on a per-bin basis showed that there were no significant differences between trials for all time bins [$p < 3.0086$] when compared with trial 1 data (Figure 7.6).

7.3.2 STEERING

Figure 7.7 displays the mean absolute steering over the 10 s following takeover, grouped into trials (Timer and HazLanSEA 1, 2, and 3). T-tests carried out on a per-bin basis (Figure 7.8) indicated that there was no consistent sustained statistical significance for steering angle bins on the highway road type [$p < 3.0086$]. This again suggests that the different levels of feedback and modality of the HazLanSEA protocols are not significantly affecting steering post-takeover on highway road types. The associated standard deviations and Cohen's D effect sizes are found in Figures 7.9 and 7.10, respectively. The urban road type did, however, indicate significance for bins between 5 and 10 s post-takeover for all the HazLanSEA protocols [$p > 3.0086$]. There also appeared to be a trend in these significance levels on both road types from which it may be inferred that the HazLanSEA protocols, despite their differing modalities, do not have a significant effect on steering angle post-takeover. This is reinforced through examination of the mean absolute values for steering (Figure 7.7) where after an initial steering input over approximately the first 1 s post-takeover, the levels are sustained with little variance. Steering inputs when utilising the Timer

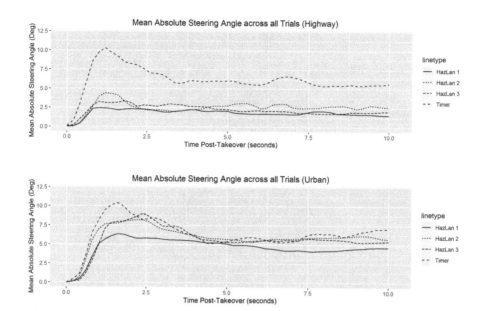

FIGURE 7.7   Mean absolute steering angles (highway and urban).

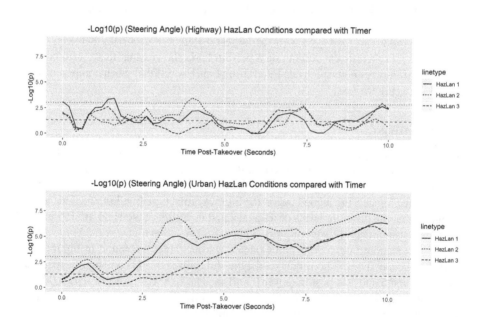

FIGURE 7.8   -Log10(p) steering angle (highway and urban).

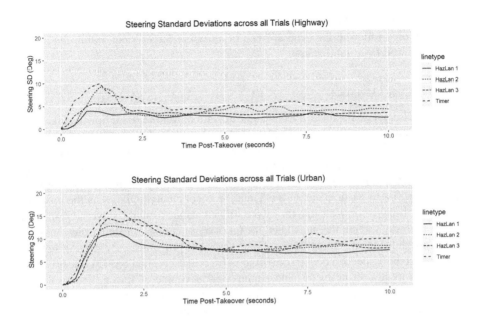

FIGURE 7.9 Steering standard deviations (highway and urban).

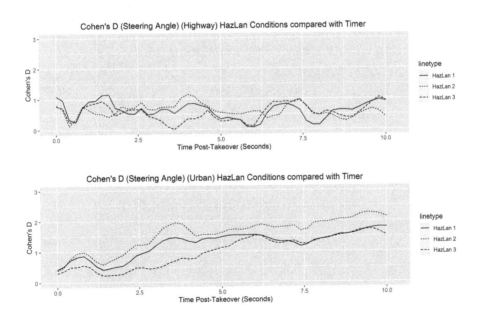

FIGURE 7.10 Cohen's D steering angle (highway and urban).

protocol appeared to be of larger magnitudes than those of the HazLanSEA protocol, although they followed the general trend of an initial larger input within the initial 2.5 s, followed by a stabilisation.

7.3.3 LANE DEVIATION

Figure 7.11 displays the mean absolute lane deviation over the 10 s following takeover, grouped into trials (Timer and HazLanSEA 1, 2, and 3). Their associated standard deviations and Cohen's D effect sizes are found in Figures 7.12 and 7.13, respectively. Lane deviation values per bin showed significance [$p > 3.0086$] on the highway road type for initial values post-takeover up to between 2 and 4.8 s (Figure 7.14). This may have been due to participants in the Timer protocol changing lanes as the protocol took little time to complete and they were still a distance from the slip road at the end of the highway section that mandated the takeover of control. Indeed, examining the mean absolute lane deviation values (Figure 7.11), the probable lane change effect can be seen on the Timer protocol trace (3 ft in 2 s). Each of the HazLanSEA traces exhibits similar trends, deviating by a maximum of approximately 2 ft over the 10-s period post-takeover. This indicates that the differences between the HazLanSEA protocols result in minimal variations in performance. Lane deviations on the urban road type were only significant for less than 1 s post-takeover [$p > 3.0086$] (Figure 7.14), and there is a clear trend visible in the mean absolute values graph (Figure 7.11), with initial deviations occurring in the first 2 s post-takeover and stabilising between 1 and 2 ft to the 10-s mark. These results tend to suggest that varying the information modalities does not appear to significantly affect lane deviation performance.

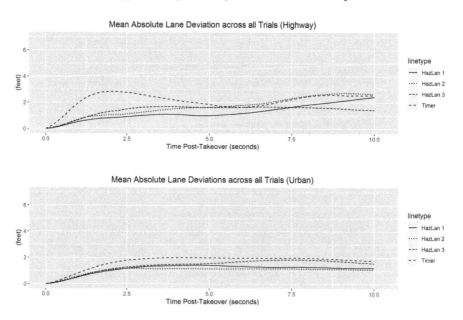

FIGURE 7.11   Mean absolute lane deviations (highway and urban).

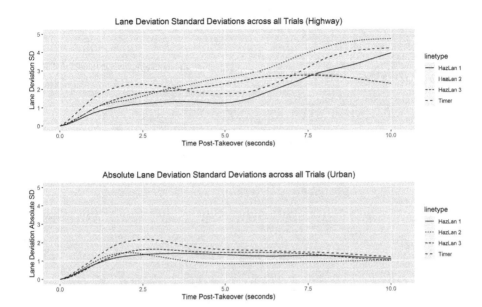

FIGURE 7.12 Lane deviation standard deviations (highway and urban).

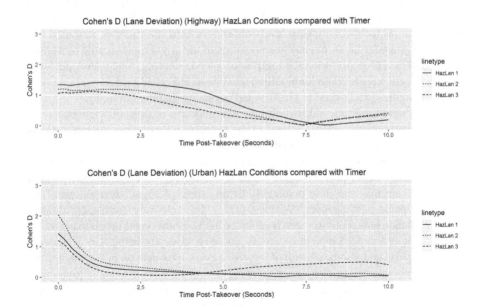

FIGURE 7.13 Cohen's D lane deviation (highway and urban).

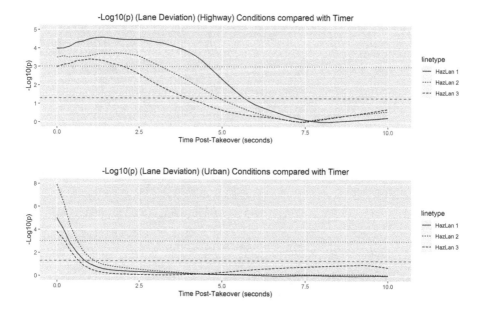

FIGURE 7.14   -Log10(p) lane deviation (highway and urban).

7.4 DISCUSSION

This study examined the post-takeover performance effects of varying levels of TOR feed-forward information to drivers in automated vehicles. Four conditions were examined: a unimodal Timer protocol, similar to that used in Volvo cars, and three multimodal protocols of varying levels of feed-forward information, designed specifically to transfer SA from vehicle automation to the driver as advocated by Stanton et al. (2017). Three performance metrics of speed, steering, and lateral deviation were selected to reflect both longitudinal and lateral effects as carried out in multiple previous studies (McDonald et al. 2019; Eriksson and Stanton 2017a, b).

Speeds were lower post-takeover for the multimodal protocols; this may be a result of the vocal modality included as part of the HazLanSEA protocols, and this may go some way to confirm the findings of Toffetti et al. (2009) that the vocalised modality can result in lower speeds compared to a purely auditory modality. It can be argued that a vocal modality provides more scope for conveying information than an auditory one; there may therefore potentially be an aspect of cognitive workload affecting participants post-takeover (Wickens 2002). The relatively consistent, although non-significant, reductions in speed both on the highway and on the urban road types of approximately 6 and 2.5 mph, respectively, could be viewed as being positive in terms of safety and might also be a result of increased caution brought on by augmented SA. This is likely due to participants not knowing how much throttle to apply to maintain speed and erring on the side of caution by applying little or no throttle and then applying it to match. It is of note that after experiencing the Timer protocol, participants appeared to allow their speed to drop less and after 5 s were already accelerating back to the speed at which the takeover took place. This might indicate a level of haste that does not occur after experiencing the longer HazLanSEA protocols. The HazLanSEA protocols are designed specifically to raise SA, and therefore, they may be instilling participants with a potentially beneficial sense of caution. This lends further weight to the findings of Politis, Brewster, and Pollick (2015) and Melcher et al. (2015), in advocating the use of multimodal TORs over unimodal, due to the potential performance and safety benefits that may be bestowed. The unimodal Timer protocol required less time to complete than the multimodal HazLanSEA protocols, and this may have had an effect on takeover performance. Dogan et al. (2019) asserted that a faster takeover may not result in improved SA; indeed, it may have the opposite effect if drivers are simply asked to resume control, as they may attempt to gain SA after the control transition and control inputs and performance may reflect this in terms of larger control inputs. The International Transport Forum (2018) describe speed as having a direct influence on both the seriousness and occurrence of accidents, and they cite this relationship as being confirmed through multiple models, including Nilsson's power model. Minimal reductions in speed, provided there is a safe separation to vehicles behind, could therefore be considered to correlate with safety. The International Transport Forum (2018) also cite additional advantages to reductions in speed of decreased noise and emissions. The main finding with regard to speed in the context of the paper is that the level of feedback does not appear to have a significant effect.

The similarity in mean steering angle trends for the HazLanSEA protocols on the highway road type, coupled with the transient and minimal levels of significance, indicates that level of feedback is not affecting steering angle inputs. On the urban road type, the least feedback dense HazLanSEA protocol (trial 2) appeared to elicit slightly, but nevertheless significantly, lower steering angle inputs. However, the unimodal Timer protocol (trial 1) with the lowest level of feedback generated a similar trend of steering angle input as both of the HazLanSEA protocols featuring higher levels of feedback. This coupled with the low level of steering angle reduction suggests that overall, the effect of level of feedback is minimal. The larger initial steering angle inputs on the urban road types might be due to participants steering away from the oncoming traffic (Triggs 1980). The same effect seen on the highway road type may be reflecting a similar desire for participants to create more distance from traffic in the faster lane(s). Both graphs for mean absolute steering angle show sustained values over time, which will result in a turning radius. The difference between the Timer protocol and the HazLanSEA protocols in terms of steering angle magnitude on the highway road type is likely due to the Timer protocol taking less time to complete than the other protocols; as a result of this, the takeover has occurred on a section of highway with a smaller radius. The unimodal Timer protocol appeared to show a non-significant trend for higher initial steering input on both road types at approximately 1.25 s post-takeover, which is in agreement with the findings of Naujoks, Mai, and Neukum (2014), whose study revealed improved lateral control performance with multimodal TORs. The HazLanSEA protocols appear to result in improved steering inputs; they are of lower magnitude and vary less over the 10 s post-takeover. This may be due to the improved levels of SA resulting in raised caution levels and smoother, potentially safer, and more measured control inputs.

Lane deviation data on the urban road type showed significance only over the initial 1 s post-takeover; examination of the mean absolute values reveals trends for all protocols to be very similar; the deviations are most likely the result of participants steering away from the oncoming vehicles (Triggs 1980). It is therefore apparent that lane deviations on the urban road type appear to be unaffected by the level of TOR feedback. The unimodal Timer protocol (which represents a minimal level of feedback) showed a similar trend to the HazLanSEA protocols. This lends further weight to the finding that lane deviations in urban settings are unaffected by TOR feedback. This agrees with the findings of Petermeijer et al. (2017), who found no statistically significant differences in lane changing time or braking time between multimodal and unimodal TORs. On the highway road type, the HazLanSEA protocols again appeared to result in similar minor lane deviation trends. The lane deviations were significant for all HazLanSEA protocols used on the highway road type for the initial 2.5 s post-takeover; this is potentially due to the higher levels of lane deviation in the Timer protocol. The increased lane deviations when using the timer protocol are potentially a result of lane changes that occurred due to its short duration. Participants experiencing the Timer protocol sometimes considered that they had enough time prior to the next slip road to overtake traffic. There may also therefore be an element of multimodality reducing lane deviations (Naujoks, Mai and Neukum 2014), although it is interesting that this is predominantly on the highway road type. This could potentially be due to the nature of the highway environment through the provision of wider lanes. It could also be the case that it may not be appropriate to compare lane deviations for protocols of considerably differing durations.

There appear to be some performance benefits of multimodality, as revealed by this study which is in agreement with the findings of Biondi et al. (2017), particularly in terms of speed post-takeover. Clearly, multimodal TORs will provide more feedback than those that are unimodal, Politis Brewster and Pollick (2015) found that participants considered multimodal TORs to be both more effective and communicative of urgency. Overall, examining the mean values for all variables, the three HazLanSEA condition results showed similar trends compared to the unimodal Timer protocol. This may be indicative that multimodal HMIs generate more consistent behaviour post-takeover. This finding is in agreement with those of previous studies, such as Naujoks, Mai, and Neukum (2014).

7.5 CONCLUSION

This study examined post-takeover driver performance after experiencing four TOR HMI variations, each with progressively increasing levels of feedback. They were designed to represent a set of interfaces that might result from a fully customisable HMI. Participants experienced four trials, one for each HMI variation. During each trial, they experienced a TOR four times each on both highway and on urban-style roads.

Examination of data from the 10-s period post-takeover indicated that the different levels of feedback afforded by the HazLanSEA protocol variations had no significant effects on speed. Trends indicated that speeds tended to reduce in all cases, but at a low rate, this behaviour is generally considered to be safe (International Transport Forum 2018). The overall trends for the multimodal HazLanSEA protocol lane deviation remained similar however, with minimal and benign movements. Steering angles also showed some significance on the urban road type; however, trends between the HazLanSEA protocols again remained similar. The findings represent a positive outcome, indicating that the level of information presented to drivers may be increased, thus improving SA without impacting post-takeover performance in such a way that safety is compromised. There is scope for future work to explore the performance of participants utilising a fully customisable takeover protocol to examine the potential benefits. Attention should, however, be paid to the nature and fidelity of the customisations, ensuring that it is not overly complex or too simple to match the user's requirements. Examination of the unimodal Timer protocol's performance revealed increased steering and lane deviation post-takeover compared with the HazLanSEA protocols. Speed was also seen to increase after approximately 5 s post-takeover. Both of these behaviours potentially indicate a less smooth, considered, and therefore safe transition of control has taken place. This is broadly in line with the literature stating unimodal TORs should be avoided if possible (Politis, Brewster and Pollick 2015); therefore, the minimum level of feedback specified through customisation should also be carefully considered. It may also be beneficial for future work to be carried out to identify the most appropriate default TOR matched to target users' requirements, and implement this as a customisation starting point for individuals wishing to improve the usability of their protocol.

ACKNOWLEDGEMENTS

This work was supported by Jaguar Land Rover and the UK-EPSRC Grant EP/N011899/1 as part of the jointly funded Towards Autonomy: Smart and Connected Control (TASCC) Programme.

REFERENCES

Bazilinskyy, P., S. M. Petermeijer, V. Petrovych, D. Dodou, and J. C.F. de Winter. 2018. "Take-Over Requests in Highly Automated Driving: A Crowdsourcing Survey on Auditory, Vibrotactile, and Visual Displays." *Transportation Research Part F: Traffic Psychology and Behaviour, 56* 82–98. DOI: 10.1016/j.trf.2018.04.001.

Biondi, F., M. Leo, M. Gastaldi, R. Rossi, and C. Mulatti. 2017. "How to Drive Drivers Nuts: Effect of Auditory, Vibrotactile, and Multimodal Warnings on Perceived Urgency, Annoyance, and Acceptability." *Transportation Research Record: Journal of the Transportation Research Board, 2663* 34–39. DOI: 10.3141/2663-05.

BMW. 2019. *The Path to Autonomous Driving.* Accessed October 16, 2019. https://www.bmw.com/en/automotive-life/autonomous-driving.html.

Brandenburg, S., and L. Chuang. 2019. "Take-Over Requests During Highly Automated Driving: How Should They Be Presented and Under What Conditions?" *Transportation Research Part F: Traffic Psychology and Behaviour, 66* 214–225. DOI: 10.1016/j.trf.2019.08.023.

de Winter, J. C.F., R. Happee, M. H. Martens, and N. A. Stanton. 2014. "Effects of Adaptive Cruise Control and Highly Automated Driving on Workload and Situation Awareness: A Review of the Empirical Evidence." *Transportation Research Part F: Traffic Psychology and Behaviour, 27,* Part B 196–217. DOI: 10.1016/j.trf.2014.06.016.

Department for Transport. 2018. *Reported Road Casualties in Great Britain: Quarterly Provisional Estimates Year Ending June 2018*. 8 November. Accessed October 1, 2019. https://assets.publishing.service.gov.uk/government/uploads/system/uploads/attachment_data/file/754685/quarterly-estimates-april-to-june-2018.pdf.

Dogan, E., V. Honnêt, S. Masfrand, and A. Guillaume. 2019. "Effects of Non-Driving-Related Tasks on Takeover Performance in Different Takeover Situations in Conditionally Automated Driving." *Transportation Research Part F: Traffic Psychology and Behaviour, 62* 494–504. DOI: 10.1016/j.trf.2019.02.010.

Eriksson, A., and N. A. Stanton. 2017a. "Takeover Time in Highly Automated Vehicles: Noncritical Transitions to and From Manual Control." *Human Factors: The Journal of the Human Factors and Ergonomics Society, 59,* 4 689–705. DOI: 10.1177/0018720816685832.

Eriksson, A., and N. A. Stanton. 2017b. "Driving Performance After Self-Regulated Control Transitions in Highly Automated Vehicles." *Human Factors, 59,* 8 1233–1248. DOI: 10.1177/0018720817728774.

Eriksson, A., S. M. Petermeijer, M. Zimmermann, J.C.F. de Winter, K. J. Bengler, and N. A. Stanton. 2019. "Rolling Out the Red (and Green) Carpet: Supporting Driver Decision Making in Automation-to-Manual Transitions." *IEEE Transactions on Human-Machine Systems, 49,* 1 20–31. DOI: 10.1109/THMS.2018.2883862.

Gold, C., D. Damböck, L. Lorenz, and K. Bengler. 2013. ""Take over!" How Long Does It Take to Get the Driver Back into the Loop?" *Proceedings of the Human Factors and Ergonomics Society Annual Meeting, 57,* 1 1938–1942. DOI: 10.1177/1541931213571433.

Hedges & Company. 2020. *How Many Licensed Drivers Are There in the US?* Accessed February 26, 2020. https://hedgescompany.com/blog/2018/10/number-of-licensed-drivers-usa/.

International Transport Forum. 2018. *Speed and Crash Risk.* Research Report, OECD.

Marn, J. 2017. *The Future of Cars 2040: Miles Traveled Will Soar While Sales of New Vehicles Will Slow, New IHS Markit Study Says.* Accessed February 26, 2020. https://www.business-wire.com/news/home/20171113006466/en/Future-Cars-2040-Miles-Traveled-Soar-Sales.

McDonald, A. D., H. Alambeigi, J. Engström, G. Markkula, T. Vogelpohl, J. Dunne, and N. Yuma. 2019. "Toward Computational Simulations of Behavior During Automated Driving Takeovers: A Review of the Empirical and Modeling Literatures." *Human Factors: The Journal of the Human Factors and Ergonomics Society, 61,* 4 642–688. DOI: 10.1177/0018720819829572.

Melcher, V., S. Rauh, F. Diederichs, H. Widlroither, and W. Bauer. 2015. "Take-Over Requests for Automated Driving." *Procedia Manufacturing, 3* 2867–2873. DOI: 10.1016/j.promfg.2015.07.788.

Merat, N., B. Seppelt, T. Louw, J. Engström, J. D. Lee, E. Johansson, C. A. Green, and et al. 2019. "The "Out-of-the-Loop" Concept in Automated Driving: Proposed Definition, Measures and Implications." *Cognition, Technology & Work, 21* 87–98. DOI: 10.1007/s10111-018-0525-8.

Mercedes Benz. 2019. *Driving Assistance Systems in S-Class: Intelligent Drive Next Level.* Accessed October 16, 2019. https://www.mercedes-benz.com/en/innovation/the-new-s-class-intelligent-drive-next-level/.

Mirnig, A. G., M. Gärtner, A. Laminger, A. Meschtscherjakov, S. Trösterer, M. Tscheligi, R. McCall, and F. McGee. 2017. "Control Transition Interfaces in Semiautonomous Vehicles: A Categorization Framework and Literature Analysis." *Automotive UI: Proceedings of the 9th International Conference on Automotive User Interfaces and Interactive Vehicular Applications.* Oldenburg, Germany: Association for Computing Machinery. 209–220. https://doi.org/10.1145/3122986.3123014.

Mobileye. 2019. *Sensing the Future.* Accessed October 28, 2019. https://www.mobileye.com/.

National Highway Traffic Safety Administration. 2019. *Automated Vehicles for Safety.* Accessed October 1, 2019. https://www.nhtsa.gov/technology-innovation/automated-vehicles-safety.

Naujoks, F., C. Mai, and A. Neukum. 2014. "The Effect of Urgency of Take-Over Requests During Highly Automated Driving Under Distraction Conditions." *5th International Conference on Applied Human Factors and Ergonomics.* Krakow, Poland: AHFE. 431–438.

Parasuraman, R., and V. Riley. 1997. "Humans and Automation: Use, Misuse, Disuse, Abuse." *Human Factors: The Journal of the Human Factors and Ergonomics Society,* 39, 2 230–253. DOI: 10.1518/001872097778543886.

Petermeijer, S. M., P. Bazilinskyy, K. Bengler, and J. de Winter. 2017. "Take-Over Again: Investigating Multimodal and Directional TORs to Get the Driver Back into the Loop." *Applied Ergonomics,* 62 204–215. DOI: 10.1016/j.apergo.2017.02.023.

Politis, I., S. Brewster, and F. Pollick. 2015. "Language-Based Multimodal Displays for the Handover of Control in Autonomous Cars." *AutomotiveUI - Proceedings of the 7th International Conference on Automotive User Interfaces and Interactive Vehicular Applications.* Nottingham, UK: Association for Computing Machinery. 3–10. https://doi.org/10.1145/2799250.2799262.

Regan, M. A., J. D. Lee, and K. Young. 2008. *Driver Distraction.* Boca Raton, FL: CRC Press.

Shuttleworth, J. 2019. *SAE J3016- Levels of Driving.* 7 January. Accessed October 1, 2019. https://www.sae.org/news/2019/01/sae-updates-j3016-automated-driving-graphic.

Stanton, N. A., P. M. Salmon, G. H. Walker, E. Salas, and P. A. Hancock. 2017. "State-of-science: Situation Awareness in Individuals, Teams and Systems." *Ergonomics,* 60, 4 449–466. DOI: 10.1080/00140139.2017.1278796.

Tesla. 2019. *Autopilot.* Accessed October 16, 2019. https://www.tesla.com/en_GB/model3/design#autopilot.

Toffetti, A., E. S. Wilschut, M. H. Martens, A. Schieben, A. Rambaldini, N. Merat, and F. Flemisch. 2009. "CityMobil: Human Factor Issues Regarding Highly Automated Vehicles on eLane." *Transportation Research Record: Journal of the Transportation Research Board, 2110,* 1 1–8. DOI: 10.3141/2110-01.

Triggs, T. J. 1980. "The Influence of Oncoming Vehicles on Automobile Lateral Position." *Human Factors: The Journal of the Human Factors and Ergonomics Society, 22,* 4 427–433. DOI: 10.1177/001872088002200404.

van den Beukel, A. P., M. C. van der Voort, and A. O. Eger. 2016. "Supporting the Changing Driver's Task: Exploration of Interface Designs for Supervision and Intervention in Automated Driving." *Transportation Research Part F: Traffic Psychology and Behaviour,* 43 279–301. DOI: 10.1016/j.trf.2016.09.009.

Vogelpohl, T., M. Kühn, T. Hummel, T. Gehlert, and M. Vollrath. 2018. "Transitioning to Manual Driving Requires Additional Time After Automation Deactivation." *Transportation Research Part F: Traffic Psychology and Behaviour, 55* 464–482. DOI: 10.1016/j.trf.2018.03.019.

Walker, G. H., N. A. Stanton, and P. M. Salmon. 2018. *Vehicle Feedback and Driver Situation Awareness.* Boca Raton, FL: CRC Press.

Wickens, C. D. 2002. "Multiple Resources and Performance Prediction." *Theoretical Issues in Ergonomics Science, 3,* 2 159–177. DOI: 10.1080/14639220210123806.

Wintersberger, P., A. Riener, C. Schartmüller, A.K. Frison, and K. Weigl. 2018. "Let Me Finish before I Take Over: Towards Attention Aware Device Integration in Highly Automated Vehicles." *AutomotiveUI: Proceedings of the 10th International Conference on Automotive User Interfaces and Interactive Vehicular Applications.* Toronto, Canada: Association for Computing Machinery. 53–65. https://doi.org/10.1145/3239060.3239085.

World Health Organization. 2018. *Global Status Report on Road Safety 2018*. Accessed February 26, 2020. https://www.who.int/violence_injury_prevention/road_safety_status/2018/en/.

Yoon, S. H., Y. W. Kim, and Y. G. Ji. 2019. "The Effects of Takeover Request Modalities on Highly Automated Car Control Transitions." *Accident Analysis & Prevention, 123* 150–158. DOI: 10.1016/j.aap.2018.11.018.

8 Validating Operator Event Sequence Diagrams
The Case of Automated Vehicle-to-Human Driver Takeovers

Neville A. Stanton, James W.H. Brown, and Kirsten M. A. Revell
University of Southampton

Patrick Langdon
Edinburgh Napier University

Michael Bradley
University of Cambridge

Ioannis Politis
The MathWorks Inc.

Lee Skrypchuk, Simon Thompson, and Alexandros Mouzakitis
Jaguar Land Rover

CONTENTS

8.1 INTRODUCTION

With the impending arrival of highly automated vehicles, the takeover of control, from computing technology to human drivers, becomes an important issue (Eriksson and Stanton 2017). This issue may be exacerbated if the driver has not been monitoring the vehicle and the road environment, as defined in SAE Level 3 automation (SAE J3016 2016). It is anticipated that Level 3 automation will enable drivers to undertake non-driving activities, while the computers drive the vehicle for them. Such non-driving activities could include reading, checking emails, surfing the Internet, watching movies, making phone or video calls, eating and drinking, or even playing video games (Stanton 2015). For the purposes of our research, we envision driving automation on motorways (freeways) only, and that the driver would remain the fall-back for planned takeovers (i.e. non-emergency conditions) when the automated system reaches operational limits (such as approaching a motorway/freeway exit). Bringing the driver back into control of the vehicle after some time of non-engagement is not a trivial undertaking. This will require drivers to become aware of a range of situational and contextual features, including (but not limited to) the road environment, weather conditions, behaviour of other vehicles, hazards on the road, status of their own vehicle, as well as immediate actions required, such as vehicle navigation and guidance (Walker, Stanton and Salmon 2015; Stanton et al. 2017).

The design of the takeover interface requires some form of representation of the human and machine activities that are anticipated, so that the strategy can be implemented in the vehicle. There are a range of human factors and ergonomics methods that could be employed to represent these activities (Stanton et al. 2013; Stanton, Young and Harvey 2014). Modelling of the interaction between people and technology is becoming increasingly popular in human factors and ergonomics research and practice (Moray, Groeger and Stanton 2017). These methods enable the modelling of tasks, processes, timings, and potential errors. Choosing the most appropriate methods really comes down to the intended purpose to which the modelling is to be put (Stanton et al. 2013). Modelling the structure of tasks can be useful for understanding the nature of work design (Stanton 2006). Modelling processes helps in the design of interaction between humans and machines (Banks and Stanton 2017). Modelling timings can be used to understand time-critical interventions (Stanton and Baber 2008; Stanton and Walker 2011). Modelling errors can be helpful for anticipating non-normative behaviour (Stanton et al. 2009). For the purposes of modelling the process of takeover, from an automated vehicle to a human driver, operator event sequence diagrams (OESDs) were selected as they have been successfully used to model the interactions between drivers and vehicles previously (Banks, Stanton and Harvey 2014; Banks and Stanton 2017).

OESDs have been used to represent the interaction between humans and technology in a graphical manner (Stanton et al. 2013). They are based on the engineering technique for describing technical processes in the form of a flowchart, with the use of standard symbols for each defined process (see Table 8.1). Each separate system has its own column, colloquially called 'swim-lanes'. OESDs have

TABLE 8.1
OESD Task Elements with Description

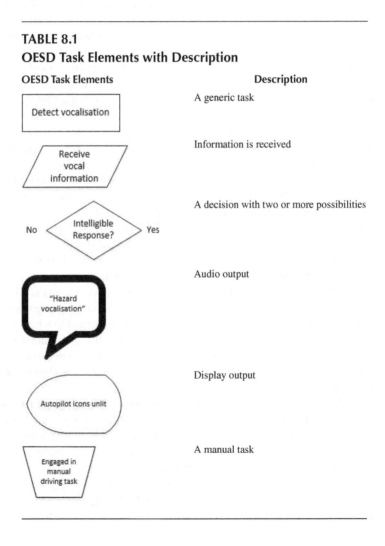

OESD Task Elements	Description
Detect vocalisation	A generic task
Receive vocal information	Information is received
No — Intelligible Response? — Yes	A decision with two or more possibilities
"Hazard vocalisation"	Audio output
Autopilot icons unlit	Display output
Engaged in manual driving task	A manual task

additional 'swim-lanes' for the human operator(s). The output of an OESD graphically depicts activities, including the tasks performed and the interaction between human and machines over time, using standardised symbols. There are numerous forms of OESDs, ranging from a simple flow diagram representing task order, to a more complex OESD, which accounts for team interaction and communication, and often includes a timeline of the scenario under analysis and potential sources of error.

Previous applications of OESDs include modelling single pilot operations in commercial aviation (Harris, Stanton and Starr 2015), aircraft landing (Sorensen, Stanton and Banks 2011), air traffic control (Walker et al. 2006), electrical energy distribution (Salmon et al. 2008), and automatic emergency braking systems in automobiles (Banks, Stanton and Harvey 2014). The latter of these applications used OESDs to compare four different levels of automation for pedestrian detection and avoidance from manual control, through decision support and automated decision-making to full automation. This analysis showed that both the decision support and automated decision-making systems actually involved the driver in more work than driving manually. All of the applications of OESDs were able to delineate between the processes undertaken by different human actors and machine agents in their respective systems. For example, Harris, Stanton, and Starr (2015) showed the effects of reducing the crew of two pilots to a single pilot. Single pilot operations reduced many of the crew communication activities but did result in more work overall for the remaining pilot, as might be expected.

In the other applications of OESDs, researchers have been able to show how work is distributed across multiple actors and agents. In particular, Sorensen, Stanton, and Banks (2011) used OESDs to illustrate how distributed cognition (Hutchins 1995a, b) worked in their analysis of an aircraft cockpit and crew for a landing task. Inspired by the original work of Hutchins (1995b), they used OESDs to show how the cognition of the cockpit is distributed among the artefacts and two human pilots as the descent of the aircraft is managed. Similarly, Walker et al. (2006) and Salmon et al. (2008) show how cognition is distributed among the artefacts and people (who are themselves in different physical locations) for aviation and energy distribution industries, respectively. In a recent review of distributed cognition and situation awareness, Stanton et al. (2017) argued that automated driving systems provide an excellent case study for distributed cognition. Clearly, the vehicle automation is performing some of the cognitive functions on behalf of the human driver, the extent to which depends upon the level of automation involved. OESDs can make the distribution of these cognitive functions more explicit, as well as identifying the interactions between the human driver and automated systems when conducting the takeover (Banks, Stanton and Harvey 2014).

While modelling of the interaction may be useful for designing the strategy for the takeover, it is only a prediction of the behaviour of the system. As such, it requires validation. Unfortunately, validation evidence is rarely reported in the literature (Stanton 2014, 2016; Stanton and Young 1998, 1999a, b). There are a few notable exceptions however, such as the development of the human error prediction methods (Stanton and Stevenage 1998; Stanton and Baber 2005; Stanton et al. 2009). Stanton et al. (2013) report the state of evidence of reliability and validity for 109 methods, which changed little since their first review in 2005. Stanton (2016) has argued strongly for reporting of this evidence when using human factors and ergonomics methods. In addition, even when there is such evidence for use of a method in one domain, validity generalisation to other domains cannot be assumed and requires testing.

Previous validation studies of human factors and ergonomics methods have used the signal detection paradigm (Stanton and Young 1999a, b, 2003). This approach has enabled analysts to distinguish between predictions of behaviours that are observed (the hit rate) and those that are not (the false alarm rate). It is also possible to calculate the overall sensitivity of the methods, taking both the hit rate and the false alarm rate into account. In summary, the purpose of this paper is to develop OESDs for the vehicle-to-driver (VtD) takeover process and then to validate them in studies using driving simulators. It was anticipated that the models of takeover would offer good predictions of actual takeover behaviour, based on modelling evidence from other methods and domains (e.g. Stanton et al. 2013; Stanton, Young and Harvey 2014).

8.1.1 OESD Development

A set of interaction concepts to facilitate VtD takeover were derived following two design workshops. These were refined into storyboards that revealed the processes, tasks, and agents involved. OESDs were selected for their ability to illustrate the tasks and processes with respect to time (Sorensen, Stanton and Banks 2011; Banks, Stanton and Harvey 2014; Harris, Stanton and Starr 2015). Events that occur within the system are modelled using a set of task unit elements, and the type of element depends on the event type. Table 8.1 illustrates the set of task unit elements that correspond to different events. Time is represented on the vertical axis of OESDs; therefore, two events occurring simultaneously would be at the same level, whereas sequential events would be vertically offset.

Four interaction designs were selected from the set of concepts, and each was represented using an OESD. Each OESD modelled the interactions between the agents throughout a complete cycle of the automation system. The cycle started with the driver in manual mode undertaking a driver-to-vehicle handover. During the automated period (when the vehicle is being driven by automation), the OESD modelled the driver interacting with a secondary task (which was engaging in a tablet-based memory game) followed by the VtD takeover that returned the driver to manual control.

Figure 8.1 illustrates a part of an OESD – the four agents in this case are driver, vehicle, takeover interaction, and environment. Phase A represents the start of the takeover of control, where the driver is informed of a forthcoming mode change and they begin to prepare. The vehicle is currently automated; however in processes not shown in the diagram, the vehicle's sensors have detected the need for the driver to take control, and this has resulted in the vehicle making the takeover interaction vocalisation. The six driver processes of phase A then follow. The driver immediately receives the information, then performs a sequence of manual actions, ceasing interaction with a secondary task device, putting down the secondary task device, resuming a driving position, and paying attention to the road. The driver then vocalises that they are ready to take control of the vehicle, followed by the vehicle detecting the vocalisation. In the event that the driver makes an unintelligible or negative response, the initial vocalisation is repeated.

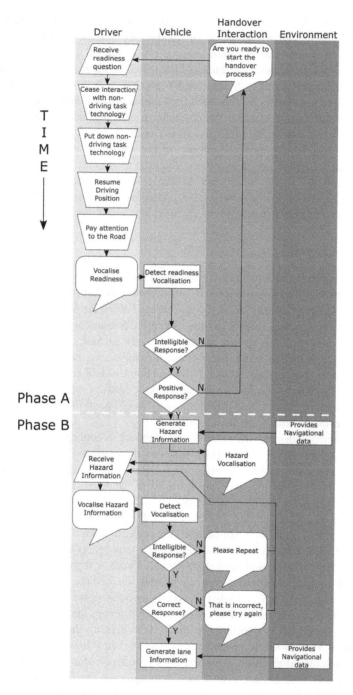

FIGURE 8.1 Example OESD section showing all six driver processes for phase A (start of VtD) and the first two driver processes of phase B (takeover protocol phase). This OESD is an example from study 2 (higher-fidelity simulator study). See supplementary online resources for all of the OESDs.

Phase B, which contains the takeover protocol processes, then begins with the vehicle generating hazard information from navigational data created by on-board sensors (see Politis et al. 2018). The vehicle vocalises the hazard information, which results in the first two driver processes of phase B – the driver receives the hazard information and then vocalises it. In the event that the vocalisation is unintelligible, the driver is asked to repeat. If the driver answers incorrectly, they are informed of their incorrect answer and asked to try again. Once the driver has provided the correct response, the next part of the protocol starts, in this case, lane information.

To determine how well takeover behaviour was modelled, simulator studies were conducted in both lower-fidelity (study 1: Politis et al. 2018) and higher-fidelity (study 2) simulators. After the completion of the lower-fidelity experiments, changes were made to the interaction designs to provide a better driver experience. The OESDs were amended to reflect these changes allowing the method to be further tested in the higher-fidelity experiment.

8.2 STUDY 1 – VALIDATION OF OESD-MODELLED DRIVER BEHAVIOUR IN A LOWER-FIDELITY DRIVING SIMULATOR

8.2.1 METHOD

8.2.1.1 Participants
A total of 49 participants took part, 24 of whom were female, and ages ranged from 17 to 86 years (mean = 45.51, SD = 17.36). All participants held a full driving licence for between 1 and 70 years (mean = 25.68, SD = 17.58). An approximate gender split between age ranges was achieved: 7 males, 5 females, aged 17–34; 11 males, 13 females, aged 35–56; 7 males, 6 females, aged 58–86. This was in line with the principles of inclusive design (Keates and Clarkson 2002). Ethical approval was granted by the Ethics Review Committee, University of Cambridge, Department of Engineering.

8.2.1.2 Experimental Design
The experiment employed a repeated-measures design with four VtD control transfer conditions, representing the single independent variable. Conditions were counterbalanced using the Latin square design. All conditions featured an initial stage whereby the system would verbally ask if the driver were ready to resume control; following a verbal confirmation, the protocol would start. The first condition 'Timer' was based on a simple timer that appeared on the dashboard display when the automation detected a need to hand control back to the driver. On confirmation that the driver was ready, it counted down in 10-s intervals from 60 s, with the driver being required to take control by pressing a button on the steering wheel before the countdown reached zero. This was based on an existing design currently undergoing testing by Volvo Cars (2015), with an auditory rather than visual countdown.

The second condition 'HazLan' uses a 'readback' principle to raise situation awareness; the system would vocalise five elements of situation awareness: potential hazards, current lane, current speed, the next required exit, and the next required action. Following each of the system's vocalisations, the participant was required to repeat them back. Incorrect or missing readbacks resulted in the system repeating the original vocalisation. Once all of the readbacks were complete, the participant was able to resume manual control.

The third condition 'VAA' (voice autopilot assistant) was response-based using the same element sequence as the second condition; however, the system provided the participants with a question regarding each element. If the participant answered the question correctly, the next question was presented. Upon completion of the sequence, the participant was able to resume manual control.

The fourth condition was an augmented version of the HazLan condition. In addition to the HazLan situation awareness aspect sequence, it incorporated multiple modalities in the form of audio-driven seat-based haptics (whereby audio signals, including vocalisations, were transmitted to the driver via a pad on the driver's seat) and two LED strips mounted on each side of the driving position. This presented a constant information on the longitudinal positions of cars in neighbouring lanes, thus providing a dynamic blind spot warning system.

Data on multiple aspects was collected. However, for the validation experiment presented in this paper, the dependent variable was driver behaviour, in terms of the processes carried out by participants during VtD control transitions. This driver behaviour data was collected using four webcams, generating footage from multiple angles within the vehicle. The processes consisted of actions, inactions, and vocalisations. Actions included those expected from the driver as predicted by the associated OESD, as well as unexpected actions, such as placing a finger on a button early or making an exaggerated glance. Inactions consisted of the failure to carry out a process predicted by the associated OESD. Vocalisations included those expected by the system as specified in the associated OESD.

8.2.1.3   Equipment

Experiments were carried out using a lower-fidelity driving simulator consisting of a gaming seat, a Logitech G25 steering wheel, and pedal set, and three screens to provide a wide field of view (as shown in Figure 8.2). An additional tablet was employed

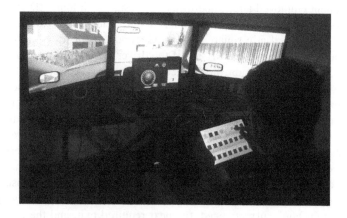

FIGURE 8.2   Study 1 (lower-fidelity) experimental configuration (Politis et al. 2018).

to act as a pseudo-dashboard, to illustrate speed, lane positioning relative to other cars, fuel level, automation mode, ideal lane, and the next required exit. The driving scenario featured a route approximately 10 miles long, consisting of a combination of highways with gentle bends and urban roads without corners. A Java-based memory game application was installed on a tablet-based PC to provide the participant with a secondary task when automation was enabled. Participant behaviour was recorded using a camera with a wide-angle lens. Two Arduino-based LED strips were fitted to the wall each side of the driving position, and C code was written to enable them to perform as blind spot indicators. A tactile acoustic device was placed on the driving seat to provide sound-based haptic information.

8.2.1.4 Procedure

After welcoming, the participants were briefed on the experiment and presented with a demographics questionnaire. The simulator was then presented to the participants, and they embarked on a short introductory test drive. No other information was provided to the participants, other than a brief overview of the vocalisation system, in order to avoid training effects. The driving scenarios were then run, using a counterbalanced design to mitigate order effects. During each automation phase, participants were requested to play the tablet-based memory game; a total of three VtD transitions were performed per scenario, at approximately 10%, 40%, and 70% progress through the route. VtD transition dialogues used synthetic vocalisations, and the participants were expected to respond vocally before switching to manual control using a button on the steering wheel. A Wizard of Oz-based system was employed to manage the synthetic vocalisations in response to the participant. At the end of the study, participants were provided with remuneration in the form of a £20 web voucher for their time.

8.2.1.5 Analysis

Signal detection theory's (SDT) primary use is to discern between 'signals' and noise (Abdi 2007). Four stimulus-response events exist: hits, misses, false alarms, and correct rejections (Nevin 1969). In the context of this experiment, it provided a method by which to compare participant behaviour observed during experiments, with predicted driver behaviour illustrated on OESDs (see Figure 8.3).

For the lower-fidelity experiments, analysis was limited to the second control transition for each condition. Three conditions were analysed: Timer, HazLan, and VAA.

Labelled template forms consisting of SDT matrices were created to allow the paper-based recording of the analysis. Footage was displayed on a large LCD screen, and printed sequential lists of VtD driver processes for each condition, drawn from the OESDs, were used for reference. The footage of the selected VtD transfers was viewed, and participants' behaviours were noted on the SDT matrices using the OESD-derived driver processes as a template of expected behaviour. To aid the analysis, the driver processes were split into three phases, namely A, B, and C, representing the participant preparing to take control, proceeding through the protocol, and taking control back from the automation, respectively. The 'Timer' condition was particularly short and therefore had no requirement for phase B. Figure 8.1 shows the six driver-based processes of phase A in the driver column.

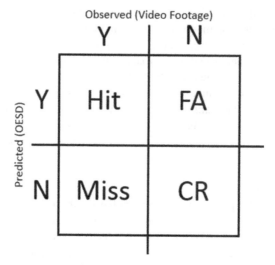

FIGURE 8.3 Signal detection theory (SDT) matrix.

A 'perfect' SDT score was attained when the participant only carried out every predicted process as part of the VtD control transfer; in this case, the SDT matrix would have an equal number of 'hits' to the number of predicted driver processes. For each predicted process that a participant failed to carry out, a 'false alarm' was recorded. In the event that a driver exhibited a behaviour in addition to that which was predicted, a 'miss' was recorded. Correct rejections were calculated at the end of the SDT analysis by subtracting the number of false alarms from the total pool of all false alarms for all participants.

The results from all the SDT matrices were collated in a spreadsheet, allowing Phi to be calculated. The Matthews correlation coefficient (Phi) was applied to the data generated by the SDT analysis; this quantified the correlation between the expected and observed behaviour, as a means to validate the OESDs. The Matthews correlation coefficient formula is given by the following equation:

$$\varphi = \frac{\text{Hit} \times \text{CR} - \text{FA} \times \text{Miss}}{\sqrt{(\text{Hit} + \text{FA})(\text{Hit} + \text{Miss})(\text{CR} + \text{FA})(\text{CR} + \text{Miss})}}. \tag{8.1}$$

8.2.1.6 Inter-Rater Reliability Method

Inter-rater reliability testing was carried out due to the subjective nature of analysing, interpreting, and categorising driver behaviour. An analyst was provided with approximately 10% of the video footage files, together with associated SDT analysis forms, a list of exceptions, and a list of driver processes split across the three phases. The analyst watched the footage and compared the driver behaviour to that which was expected as specified in the list of driver processes. SDT results were recorded on the SDT analysis forms, together with any exceptions that occurred. This was identical to the method used by the original analysts. Equal-weighted Cohen's kappa values were calculated and are reported in Sections 8.2.2 and 8.3.2.

8.2.2 Results

Equal-weighted Cohen's kappa values were calculated, resulting in a value of 0.773 for the lower-fidelity simulator. This represents a moderate agreement between the analysis in their classification of hits, misses, false alarms, and correct rejections (Landis and Koch 1977).

As shown in Figure 8.4, all three experimental conditions exhibited a relatively high number of hits per condition, and all shared identical interquartile ranges of 1. The Timer condition did not require verbal interaction with the driver, resulting in fewer possible hits than the HazLan and VAA conditions. All experimental conditions contained outliers.

In terms of misses by condition, both the HazLan and VAA conditions had identical median values (3) and interquartile ranges (4). The shorter interaction steps in the Timer condition may have contributed to the lower median value of 1 and the interquartile range of 3.

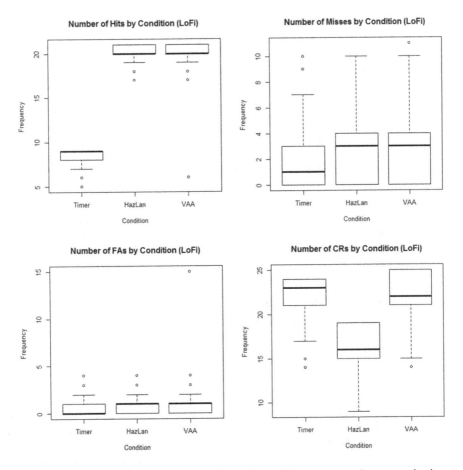

FIGURE 8.4 Lower-fidelity experiment hits, misses, false alarms, and correct rejections, by condition.

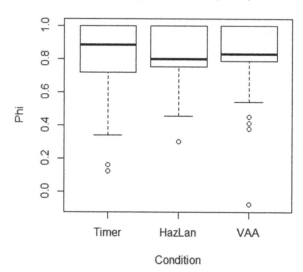

FIGURE 8.5   Study 1, lower-fidelity experiment Phi values, by condition.

In terms of false alarms, all conditions showed equal interquartile ranges of 1 and shared identical outlier values. Median values for the HazLan and VAA conditions were also equal at 1. The VAA condition exhibited a slightly higher number of correct rejections (16) than the HazLan (12) and Timer (13) conditions.

As shown in Figure 8.5, all median values for Matthews correlation coefficient (Phi) by condition were greater than 0.8 (the minimum acceptable criterion), indicating a strong positive relationship. The Timer condition scored particularly high, with a median value around 0.9. The Timer condition also exhibited a large interquartile range, varying from slightly above 0.7–1.0. The HazLan condition interquartile range was slightly lower, between around 0.75 and 1.0, whereas the smallest interquartile range was found in the VAA condition, between around 0.8 and 1.0.

8.3   STUDY 2 – VALIDATION OF OESD-MODELLED DRIVER BEHAVIOUR IN A HIGHER-FIDELITY DRIVING SIMULATOR

8.3.1   Method

8.3.1.1   Participants

A total of 60 participants were recruited (30 males and 30 females) with equal gender split across three age ranges: 18–34 (mean = 25.8, SD = 4.2), 35–56 (mean = 44.2, SD = 7.8), and 57–82 (mean = 63.2, SD = 4.8). All participants were holders of current UK driving licenses and signed the informed consent form. The research was approved by the University of Southampton Ethics and Research Governance Office (ERGO number: 29615).

8.3.1.2 Experimental Design

A repeated-measures experiment was selected; four different VtD control transfer conditions were tested. The type of takeover condition formed the independent variable, each featuring a unique combination of head-down display (HDD) format, head-up display (HUD) format, and visual or haptic ambient indicators. Participants always experienced the control condition (Timer) first, followed by the remaining three conditions, counterbalanced to mitigate learning and fatigue effects.

Timer was the control condition and consisted of a simulated version of Volvo's timer-based interaction design, which depicts a visual countdown (Volvo Cars 2015). On initiation, the HDD displayed an instruction to resume control, a 60-s countdown timer, and an associated graphic, updating at 2-s intervals. When the participant took manual control by pressing a button marked 'M' on the steering wheel, the HDD icons and automation icon would disappear.

HazLan condition 1 presented the driver with instructions via synthesised vocalisations, and icons on the HDD and HUD. The driver was first asked verbally by the system if they were ready to start the takeover process. The protocol then consisted of five elements of information for highway sections: hazards, lane, speed, exit, and action (HazLanSEA). For urban sections, the lane information was removed as the route featured a single carriageway. These elements were presented sequentially, as icons on the HDD and HUD and as corresponding synthesised vocalisations. The driver was expected to acknowledge that they had received each element of information by repeating it back, at which point the next was presented. After the last element of the protocol was acknowledged, the driver was presented with an icon and verbalisation asking them to resume manual control by pressing the 'M' button on the steering wheel.

HazLan condition 2 consisted of an augmented version of HazLan condition 1, with the addition of vehicle highlighting and radar range indicators on the HUD. The radar range was indicated by means of a large green circle at ground level, centralised on the participant's car. The radius of the circle represented the range at which other vehicles were detected. Vehicles were highlighted by placing them within a red framed box once they were within range.

HazLan condition 3 consisted of HazLan condition 2, further augmented with two ambient systems. A haptic seat base provided a short, strong vibration alert to the participant at the condition initiation. LED strips mounted on the A-pillars provided the participant with a continual feedback of the longitudinal position of vehicles in the immediately opposite lanes.

At the end of every road section of conditions 2, 3, and 4, the participant was presented with an icon, accompanied by a synthesised vocalisation, asking them to take manual control of the car by pressing the 'M' button on the steering wheel. This completed the condition by transferring control from the vehicle to the driver.

The dependent variable for the higher-fidelity study was driver behaviour, captured via video footage, in an identical way to that of the lower-fidelity study. A wide range of additional data relating to performance and subjective experience was also collected as part of this study, but only data relevant to the validation focus of this paper will be discussed.

8.3.1.3   Equipment

The driving simulator consisted of a Land Rover Discovery Sport, in combination with three frontal screens, providing a 140-° field of view. Door mirrors employed LCD screens, and the rear-view mirror reflected a rear-mounted projection. Vehicle control data was extracted from the controller area network using a hardware interface and software in C#. STISIM Drive® version 3 simulator software provided the virtual environment. The vehicle's cluster was replaced by a Microsoft Surface tablet running a custom dashboard in C#; this displayed speed, vehicle position, and icons/graphics (as shown in Figure 8.6). A HUD was generated by displaying 2D icons and 3D objects in the driver's field of view, using STISIM's Open Module. An Arduino-based haptic system was fitted to the driver's seat. Arduino-based LED strips were fitted to the A-pillars and coded to act as blind spot indicators. A Microsoft Surface tablet was employed to run a C# Wizard of Oz application, allowing control over HUD and HDD icons graphics, audio vocalisations, and haptics, as part of the take-over protocol. Four webcams were fitted within the vehicle's interior, the footage from which was saved using a stand-alone PC. A mini tablet was employed as a secondary task, running a Java-based memory game. All audio outputs were fed into the vehicle's line-in, providing the driver with stereo sound via the internal speakers.

Figure 8.6a illustrates the view from the driving seat of the Discovery. The tablet-based dashboard can be seen in place of the cluster, and a HUD icon is visible. The LED strip is visible at the base of the right A-pillar. Figure 8.6b illustrates just the HUD and HDD portions; a protocol icon, radar range, and vehicle highlighting are visible on the HUD. The HDD shows the same icon, speed, vehicle position, and automation status.

8.3.1.4   Procedure

Prior to the start of the experiment, participants were welcomed, briefed on health and safety issues, and provided with a basic overview of the experiment and hardware, before being presented with a consent form, patient information sheet, and biographical sheet. The principles of SAE Level 3 automation were explained, together with the purpose of the study and the forms of technology in each of the four conditions. Participants were shown a map of the route, indicating the placement of

FIGURE 8.6   Study 2 (higher-fidelity) experimental configuration.

roundabouts necessitating transition of control back to manual, and they were also informed that GPS-style verbal directions would be supplied. The tablet-based secondary task was explained, and they were instructed to use it when the automation mode was enabled. The participants were then asked to enter the vehicle before being introduced to the systems.

A custom route was created for the higher-fidelity study, which comprised two lengths of highway, separated by slip roads to roundabouts, followed by two urban stretches, separated by roundabouts. The route took approximately 20–25 min to negotiate when obeying UK traffic laws. Each highway section took approximately 5 min to complete, and each urban section two and a half minutes. Road conditions were consistently dry and well-lit with a traffic density of approximately one vehicle per 500 ft in all lanes.

At the start of each condition, participants were instructed via a synthesised audio voice to drive manually onto the route, and they were then informed that automation was available via a synthesised vocalisation and icon. The participant activated automated mode by pressing a button marked 'A', transferring control from the driver to vehicle. Participants then picked up and interacted with a secondary task in the form of a tablet-based memory game. At a prescribed point on the route, the VtD takeover interaction design, specific to the experimental condition, was activated by the experimenter using a Wizard of Oz interface. This process occurred twice on the highway and twice on the urban section of the route.

For HazLan conditions 1 to 3, the participant was expected to verbally acknowledge and prepare to start the protocol. From this point, the driver was presented with a sequence of situation awareness-based information, in sync with the dynamic context of the road and car status. Information was displayed using a combination of HDD, HUD, acoustic, and ambient systems, with the participant acknowledging through verbal readbacks. When all elements of the protocol were acknowledged, the experimenter used the Wizard of Oz interface to ask the participant to regain manual control of the vehicle. For the Timer condition, information was presented to the driver via the HDD, which presented a simple, circular dial-based, 60-s countdown and automation mode icon on the HDD. Neither audio cues were provided, nor readback confirmations required. The participant was expected to respond to the request to take control before the 60 s had passed. On completion of each condition, participants were asked to fill out the questionnaires. At the end of the fourth condition, they completed a freeform questionnaire and received remuneration in the form of £10 to cover their travel costs.

8.3.1.5 Analysis

Analysis of the higher-fidelity experimental data was identical to that described for the lower-fidelity in study 1.

8.3.2 RESULTS

Equal-weighted Cohen's kappa values were calculated, resulting in a value of 0.819 for the higher-fidelity simulator. This indicates good agreement between analysts in the classification of hits, misses, false alarms, and correct rejections (Landis and Koch 1977).

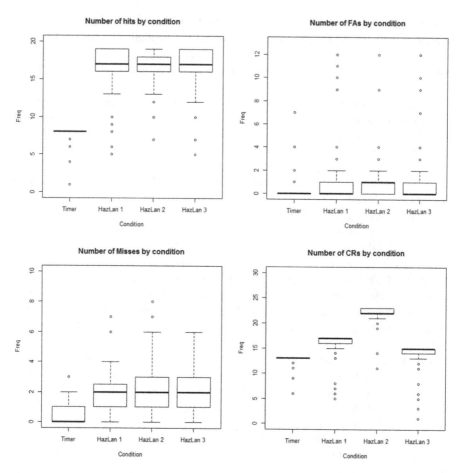

FIGURE 8.7 Higher-fidelity experiment hits, misses, false alarms, and correct rejections, by condition.

Results from the SDT analysis were used to generate box plot graphs in R. Figure 8.7 illustrates the hit/miss/false alarm and correct rejection outputs of the SDT matrices, ordered by condition.

All HazLan conditions exhibited similar median values for the number of hits by condition (17). The interquartile ranges for both HazLan1 and HazLan2 conditions were also identical. The Timer condition exhibited a much lower median value of 8 (because there were fewer steps in the transition of control), with an interquartile range of 0.

Median values for misses were similar for all HazLan conditions at interquartile value 2 and HazLan 2 and 3 conditions held the same interquartile values at 2. The median value for the Timer condition was 0 with an upper quartile range of 1.

False alarm values by condition revealed a median value of 1 for the HazLan 2 condition. HazLan 1 and 3 and Timer conditions exhibited median values of 0. Interquartile ranges for all HazLan conditions were equal at 1.

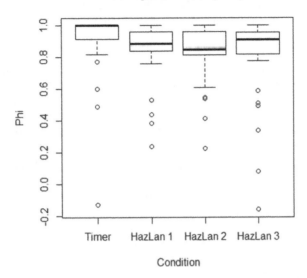

FIGURE 8.8 Study 2, higher-fidelity experiment Phi values, by condition.

The number of correct rejections showed a variation between all four conditions. The Timer condition exhibited the lowest median, with a value of 13 and an interquartile range of 0. All HazLan conditions exhibited identical interquartile ranges of 1 (lower quartile for 1 and3, upper quartile for 2), the HazLan 2 condition had the highest median score (22), followed by the HazLan 1 condition (16) and HazLan 3 (15).

Figure 8.8 illustrates a box plot displaying the Matthews correlation coefficient (Phi) for each of the four conditions. Median Phi levels above 0.8 suggest a strong positive relationship.

8.4 DISCUSSION

This chapter sought to validate a method for modelling the takeover from vehicle automation to the human driver. OESDs are offered as a way of predicting how drivers are most likely to behave during the process of takeover of control in driving simulators. The two studies show that in both lower- and higher-fidelity driving simulators, the median Phi statistic exceeded 0.8, which is the standard criterion for acceptable validation (Landis and Koch 1977). This means that OESDs can be used to make predictions about VtD takeover behaviours with some confidence. Over 100 drivers took part in the studies, with an age range of 17–86. For both studies, there were a relatively low frequency of false alarms (i.e. predicting behaviours that did not occur) and a high frequency of hits (i.e. predicting behaviours that did actually occur). This is coupled with a relative low frequency of misses (i.e. failing to predict behaviours that actually occurred) and a high frequency of correct rejections (i.e. not predicting behaviours that do not occur). In addition, the reliability of the classification of behaviour into the four categories – hits, misses, false alarms, and correct rejections – from the video data was acceptable for both studies. There is often a disappointing lack of evidence on the reliability and validity of human factors and ergonomics methods (Stanton and Young 1998; Stanton 2016), but this does not have to be the case. This and other studies have shown the way in which the evidence can be collected, analysed, and presented (Stanton et al. 2009).

The findings from the current studies are promising for the continued use of OESDs, and it is certainly comparable with the better-performing methods in the discipline of ergonomics and human factors (Stanton and Young 1998, 2003; Stanton et al. 2013; Stanton, Young and Harvey 2014). Other studies on the prediction of task performance time (Stanton and Young 1999a; Harvey and Stanton 2013a) and human error (Stanton and Stevenage 1998; Stanton and Baber 2005; Stanton et al. 2009) have all produced good levels of validity. This is not to say that all methods perform as well. In one of the first studies of its type, Stanton and Young (1999a, b) compared a range of methods, and some performed quite poorly. Stanton (2016) notes that the better-performing methods are generally quite focused in terms of their predictions, such as time and error (and now activities). Nevertheless, it is important to understand the limitations of any method before using it (Stanton and Young 1998, 2003).

The misses are probably the most interesting category of behaviours in this study as they represent behaviours conducted by the driver in the takeover of vehicle control that were not anticipated in the OESD modelling. These were unexpected behaviours that may have occurred due to the drivers being impatient or over-eager to take manual control of the vehicle. With hindsight, the early takeover behaviours could have been modelled in the OESDs. Vocalising readiness in the phase prior to carrying out processes, such as ceasing using a secondary task or resuming driving position, was not modelled as it was assumed that the driver would only vocalise readiness when they were actually ready. However, the incidence of this behaviour suggests impatience with the duration of the verbal feedback interaction, or lack of perceived value in completing the full procedure. Equally, behaviour such as removing hands from the wheel during the protocol and having to replace them during or after the protocol was not modelled as this was not anticipated in the design of the interaction. It was assumed (wrongly in some cases) that once the driving position was adopted, that it would be maintained. The duration of the takeover protocol may have encouraged drivers to take their hands off the wheel and feet of the pedals (after first placing them there) as they did not need to manually steer until the takeover procedure was complete. A design solution to overcome this could include the use of steering wheel sensors to detect correct driving position and provide feedback if hands are removed after the commencement of the takeover procedure. This demonstrates the utility of the OESD method to test design assumptions and highlight shortcuts or process failures as a means for generating mitigation strategies through design or training.

Although these studies focused on the use of OESDs to predict takeover of vehicle control, the approach could be extended to other aspects of driver behaviour (Banks, Stanton and Harvey 2014). It has shown utility in other domains, such as aviation (Sorensen, Stanton and Banks 2011; Harris, Stanton and Starr 2015; Walker et al. 2006). As a cautionary note however, validity generalisation cannot be assumed and must be tested (Stanton 2016). OESDs are good at modelling discrete events such as the stages in the takeover of vehicle control, but modelling continuous events is more challenging (such as manual control of a vehicle). This may require new notation and procedures for OESDs to handle the continuity of non-discrete activities, such as steering, maintaining speed, maintaining lane position, searching for hazards, anticipating behaviour of other road users, reading the road ahead, and so on. The nomenclature would also require some notation for the interruption of the continuous activities (such as making an emergency stop) as well as some way of representing the duality of activity in driving (such as operating the climate system while also driving the vehicle). Harvey and Stanton (2013b) presented the multimodal critical path analysis method to show how the driver could engage in activities simultaneously (spearing cognitive and physical activities by modality). This would mean that OESDs need more than one 'swim-lane' for the representation of driver behaviour. Huddlestone and Stanton (2016) borrowed notation from computer science to represent continuous activity. So, potentially at least, there are methods for extending OESDs to cope with both continuous and multiple driver activities. Future OESD modelling research could consider the incorporation of time and error data within the analysis (Stanton and Salmon 2009), to predict VtD takeover times as well as the exceptions. Data on human performance time is available in the literature and could be extended to this domain (Stanton and Baber 2008; Harvey and Stanton 2013b; Stanton, Young and Harvey 2014). This has the advantage of helping vehicle designers budget time allocation for VtD takeovers, a topic of much debate (Eriksson and Stanton 2017).

8.5 CONCLUSIONS

OESDs appear to be able to make good predictions about driver activity during simulated takeover of vehicle control from automation, in both lower- and higher-fidelity simulators. Over 100 drivers were tested in both studies with different interaction designs, and the median validity statistics were all above the criterion value. Consequently, the OESD method may be used with some confidence in modelling driver interaction for discrete events, although validity generalisation requires testing. Further development work is needed to incorporate continuous activities into OESD modelling.

ACKNOWLEDGEMENTS

This work was supported by Jaguar Land Rover and the UK-EPSRC Grant EP/N011899/1 as part of the jointly funded 'Towards Autonomy: Smart and Connected Control (TASCC) Programme'.

REFERENCES

Abdi, H. 2007. "Signal detection theory (SDT)." In *Encyclopedia of Measurement and Statistics*, by N. Salkind, 886–889. Thousand Oaks, CA: Sage.
Banks, V. A., and N. A. Stanton. 2017. *Automobile Automation: Distributed Cognition on the Road*. Boca Raton, FL: CRC Press. https://doi.org/10.1201/9781315295657.

Banks, V. A., N. A. Stanton, and C. Harvey. 2014. "Sub-systems on the road to vehicle automation: Hands and feet free but not "mind" free driving." *Safety Science 62* 505–514. DOI: 10.1016/j.ssci.2013.10.014.

Eriksson, A., and N. A. Stanton. 2017. "Takeover time in highly automated vehicles: Noncritical transitions to and from manual control." *Human Factors: The Journal of the Human Factors and Ergonomics Society, 59,* 4 689–705. DOI: 10.1177/0018720816685832.

Harris, D., N. A. Stanton, and A. Starr. 2015. "Spot the difference: Operational event sequence diagrams as a formal method for work allocation in the development of single-pilot operations for commercial aircraft." *Ergonomics, 58,* 11 1773–1791. DOI: 10.1080/00140139.2015.1044574.

Harvey, C., and N. A. Stanton. 2013a. "Modelling the hare and the tortoise: Predicting the range of in-vehicle task times using critical path analysis." *Ergonomics, 56,* 1 16–33. DOI: 10.1080/00140139.2012.733031.

Harvey, C., and N.A. Stanton. 2013b. *Usability Evaluation for In-Vehicle Systems.* London, UK: CRC Press.

Huddlestone, J. A., and N. A. Stanton. 2016. "New graphical and text-based notations for representing task decomposition hierarchies: Towards improving the usability of an Ergonomics method." *Theoretical Issues in Ergonomics Science, 17,* 5 588–606. DOI: 10.1080/1463922X.2016.1201168.

Hutchins, E. 1995a. *Cognition in the Wild.* Cambridge, MA: MIT Press.

Hutchins, E. 1995b. "How a cockpit remembers its speeds." *Cognitive Science, 19,* 3 265–288. DOI: 10.1016/0364-0213(95)90020-9.

Keates, S., and P.J. Clarkson. 2002. "Countering design exclusion through inclusive design." *ACM SIGCAPH Computers and the Physically Handicapped, 73–74* 69–76. DOI: 10.1145/960201.957218.

Landis, J. R., and G. G. Koch. 1977. "The measurement of observer agreement for categorical data." *Biometrics, 33,* 1 159–174. DOI: 10.2307/2529310.

Moray, N., J. Groeger, and N. A. Stanton. 2017. "Quantitative modelling in cognitive ergonomics: Predicting signals passed at danger." *Ergonomics, 60,* 2 206–220. DOI: 10.1080/00140139.2016.1159735.

Nevin, J. A. 1969. "Signal detection theory and operant behavior: A review of David M. Green and John A. Swets' Signal detection theory and psychophysics." *Journal of the Experimental Analysis of Behavior, 12,* 3 475–480. DOI: 10.1901/jeab.1969.12-475.

Politis, I., P. Langdon, M. Adebayo, M. Bradley, P.J. Clarkson, L. Skrypchuk, A. Mouzakitis, and et al. 2018. "An evaluation of inclusive dialogue-based interfaces for the takeover of control in autonomous cars." *Proceedings of the 2018 Conference on Human Information Interaction & Retrieval - IUI'18: 23rd International Conference on Intelligent User Interfaces.* Tokyo, Japan: ACM Press. 601–606. https://doi.org/10.1145/3172944.3172990.

SAE J3016. 2016. *Taxonomy and Definitions for Terms Related to Driving Automation Systems for On-Road Motor Vehicles.* https://www.sae.org/standards/content/j3016_201806/ (accessed 3 November 2020).

Salmon, P. M., N. A. Stanton, G. H. Walker, D. Jenkins, C. Baber, and R. McMaster. 2008. "Representing situation awareness in collaborative systems: A case study in the energy distribution domain." *Ergonomics, 51,* 3 367–384. DOI: 10.1080/00140130701636512.

Sorensen, L. J., N. A. Stanton, and A. P. Banks. 2011. "Back to SA school: Contrasting three approaches to situation awareness in the cockpit." *Theoretical Issues in Ergonomics Science, 12,* 6 451–471. DOI: 10.1080/1463922X.2010.491874.

Stanton, N. A. 2006. "Hierarchical task analysis: Developments, applications, and extensions." *Applied Ergonomics, 37,* 1 55–79. DOI: 10.1016/j.apergo.2005.06.003.

Stanton, N. A. 2014. "Commentary on the paper by Heimrich Kanis entitled 'Reliability and validity of findings in ergonomics research': Where is the methodology in ergonomics methods?" *Theoretical Issues in Ergonomics Science*, *15*, 1 55–61. DOI: 10.1080/1463922X.2013.778355.

Stanton, N. A. 2015. "Response to: Autonomous vehicles." *Ingenia*, *62*, 10.

Stanton, N. A. 2016. "On the reliability and validity of, and training in, ergonomics methods: a challenge revisited." *Theoretical Issues in Ergonomics Science*, *17*, 4 *[Methodological Issues in Ergonomics Science I]* 345–353. DOI: 10.1080/1463922X.2015.1117688.

Stanton, N. A., and C. Baber. 2005. "Validating task analysis for error identification: Reliability and validity of a human error prediction technique." *Ergonomics*, *48*, 9 1097–1113. DOI: 10.1080/00140130500219726.

Stanton, N. A., and C. Baber. 2008. "Modelling of human alarm handling responses times: A case of the Ladbroke Grove rail accident in the UK." *Ergonomics*, *51*, 4 423–440. DOI: 10.1080/00140130701695419.

Stanton, N. A., and P. M. Salmon. 2009. "Human error taxonomies applied to driving: A generic driver error taxonomy and its implications for intelligent transport systems." *Safety Science*, *47*, 2 227–237. DOI: 10.1016/j.ssci.2008.03.006.

Stanton, N. A., and S. V. Stevenage. 1998. "Learning to predict human error: Issues of acceptability, reliability and validity." *Ergonomics*, 41, 11 1737–1756. DOI: 10.1080/001401398186162.

Stanton, N. A., and G. H. Walker. 2011. "Exploring the psychological factors involved in the Ladbroke Grove rail accident." *Accident Analysis & Prevention*, *43*, 3 1117–1127. DOI: 10.1016/j.aap.2010.12.020.

Stanton, N. A., and M. S. Young. 1998. "Is utility in the mind of the beholder? A study of ergonomics methods." *Applied Ergonomics*, *29*, 1 41–54. DOI: 10.1016/S0003-6870(97)00024-0.

Stanton, N. A., and M. S. Young. 1999a. "What price ergonomics?" *Nature*, *399* 197–198. DOI: 10.1038/20298.

Stanton, N. A., and M. Young. 1999b. *Guide to Methodology in Ergonomics: Designing for Human Use*. Abingdon, UK: CRC Press.

Stanton, N. A., and M. S. Young. 2003. "Giving ergonomics away? The application of ergonomics methods by novices." *Applied Ergonomics*, *34*, 5 479–490. DOI: 10.1016/S0003-6870(03)00067-X.

Stanton, N. A., M. S. Young, and C. Harvey. 2014. *Guide to Methodology in Ergonomics: Designing for Human Use, Second Edition*. Boca Raton, FL: CRC Press.

Stanton, N. A., P. M. Salmon, D. Harris, A. Marshall, J. Demagalski, M. S. Young, T. Waldmann, and S. Dekker. 2009. "Predicting pilot error: Testing a new methodology and a multi-methods and analysts approach." *Applied Ergonomics*, *40*, 3 464–471. DOI: 10.1016/j.apergo.2008.10.005.

Stanton, N. A., P. M. Salmon, L. A. Rafferty, G. H. Walker, C. Baber, and D. P. Jenkins. 2013. *Human Factors Methods: A Practical Guide for Engineering and Design*. London, UK: CRC Press. https://doi.org/10.1201/9781315587394.

Stanton, N. A., P. M. Salmon, G. H. Walker, E. Salas, and P. A. Hancock. 2017. "State-of-science: Situation Awareness in Individuals, Teams and Systems." *Ergonomics*, *60*, 4 449–466. DOI: 10.1080/00140139.2017.1278796.

Volvo Cars. 2015. "*Volvo Cars: Explore the User-Interface of Tomorrow [Video]*." https://www.youtube.com/watch?v=xYqtu39d3CU.

Walker, G. H., N. A. Stanton, and P. M. Salmon. 2015. *Human Factors in Automotive Engineering and Technology - Human Factors in Road and Rail Transport*. Ashgate, UK: CRC Press.

Walker, G. H., H. Gibson, N. A. Stanton, C. Baber, P. M. Salmon, and D. Green. 2006. "Event Analysis of Systemic Teamwork (EAST): A novel integration of ergonomics methods to analyse C4i activity." *Ergonomics*, *49*, 12 1345–1369. DOI: 10.1080/00140130600612846.

Part III

Benchmarking

9 Breaking the Cycle of Frustration

Applying Neisser's Perceptual Cycle Model to Drivers of Semi-Autonomous Vehicles

Kirsten M. A. Revell and Joy Richardson
University of Southampton

Patrick Langdon
Edinburgh Napier University

Michael Bradley
University of Cambridge

Ioannis Politis
The MathWorks Inc.

Simon Thompson, Lee Skrypchuk,
Jim O'Donoghue, and Alexandros Mouzakitis
Jaguar Land Rover

Neville A. Stanton
University of Southampton

CONTENTS

9.1   INTRODUCTION

Self-driving cars have been predicted for some time (Stanton and Marsden 1996) and are nearly with us. Semi-automated cars, SAE Level 2 (SAE J3016 2016), are already on public roads, and within 10 years, highly automated, SAE Level 3+ cars (SAE J3016 2016) will be a reality. The largest gap in our understanding of vehicle automation is how drivers will react to this new technology and how best to design the driver–automation interaction. With high profile accidents such as the Tesla Model S fatality (National Highway Traffic Safety Administration 2013), it is clear that more work is needed to enhance safety and improve user acceptance. To this end, the authors believe it is highly relevant to adopt the approach by Stanton and Walker (2011) to collect and apply human factors analysis data from routine journeys in order to identify and mitigate for potential issues 'before the event'.

While the roles of driver and automation for each of the different SAE levels of automation continue to evolve, there is clarity that for SAE Levels 0–2, the driver is firmly in command of the vehicle and retains the need to monitor the vehicle status, the road ahead, and other road vehicles at all times. For SAE Level 0, the driver must perform all driving tasks without assistance. When vehicle assistance such as power steering, collision warning, lane detection warnings, or cruise control is present, this is classed as SAE Level 1 automation. SAE Level 2 automation is represented when more than one of the SAE Level 1 features are combined to allow both lateral control (e.g. through steering assist) and longitudinal control (e.g. adaptive cruise control (ACC)) through automation. There is similar clarity that for Levels 3–5, the automation is firmly in command of the vehicle when activated. In SAE Levels 3 and 4, the automation can only operate effectively when certain environmental conditions are met, allowing full functioning of road and vehicle sensors. The key distinction between these levels is that drivers are required to intervene when requested for SAE

Level 3 vehicles, whereas SAE Level 4 vehicles can function without driver intervention if there is an incident of system failure, or the automation reaches its boundaries of operation (e.g. through a safe stop). For SAE Level 5 vehicles, the need for human intervention is eliminated and manual driving controls, such as steering wheel and pedals, are not required (SAE J3016 2016).

On-road studies focusing on combined SAE Level 2 systems such as that found in the Tesla Model S are beginning to gain prominence in the literature (Banks and Stanton 2015, 2016; Naujoks, Purucker and Neukum 2016; Endsley 2017; Stapel, Mullakkal-Babu and Happee 2017; Heikoop et al. 2017) with an increasingly naturalistic approach demonstrating new insights into the challenges of semi-autonomous vehicles (Banks et al. 2018). On-road studies provide a 'real-life' context of use for usability testing. The presence of other road users, varying weather conditions, and changes in lighting, with genuine consequences for the driver and vehicle that cannot easily be replicated in a simulator setting. While test track studies can offer some of these conditions, the reduced complexity ultimately threatens the ecological validity of the results (Carsten, Kircher and Jamson 2013; de Winter, van Leeuwen and Happee 2012).

The safety measures required to be in place to ensure ethical on-road studies with prototype SAE Level 3 vehicles include additional safety cars and restricted road access, increasing the cost and amount of logistical arrangements necessary, as well as reducing ecological validity. Banks et al. (2018) highlight how, with the exception of the need for the driver to monitor the environment, SAE Level 2 vehicles reflect the driver experience proposed for SAE Level 3. For example, a number of existing on-road vehicles have a combination of driver assist features that allow drivers to surrender longitudinal and lateral control, through ACC combined with steering assist, lane centring technology (e.g. BMW 7 series, Honda Acura, Jaguar F-Pace, Land Rover Discovery, Mazda 6, Mercedes Benz S-Class, Infinity Q60, Tesla Model S, Lexus GS, Audi A4, Volvo S90). While the ability to perform a secondary task (e.g. eyes and mind off the road, and neglecting the monitoring task) is not possible in these cars, key interactions such as activating and deactivating automation, monitoring of mode state, and embracing a partnership with the automation features are transferrable to SAE Level 3 requirements (Banks et al. 2018). In addition, design changes in automobiles are typically iterative since the existing real estate is utilised to provide driver input and displays for new features. The authors propose that by evaluating the current interaction design for SAE Level 2 vehicles, insights for transferable interactions for SAE Level 3+ vehicles will help improve the user experience and safety in the design of vehicles with higher levels of automation.

Previous on-road studies using SAE Level 2 interaction have been focused on understanding differences between novices and expert drivers (Banks and Stanton 2015; Stapel, Mullakkal-Babu and Happee 2017), particularly in relation to workload and stress. While these studies provide insights in terms of the impact of automation on the driver and the appropriateness of their interaction, they do not provide specific design direction to aid manufacturers. Other road studies have considered driver behaviour, in terms of level of monitoring or secondary task engagement (Banks and Stanton 2016; Naujoks, Purucker and Neukum 2016). Banks et al. (2018) highlight design issues linked to four areas of non-optimal driver behaviour (system warnings, mode confusion, testing the limits, and engaging in non-driving secondary tasks) using thematic analysis of video observations. This work provided clear evidence of the risks of relying on drivers to provide a 'monitoring role' during a semi-automated drive.

However, despite these risks, it is clear that industry is forging ahead with greater levels of automation, evidenced by the first commercial offering of SAE Level 3 functionality with the Audi A7. The authors believe it is essential that manufacturers prioritise key design principles to minimise safety concerns and improve the user experience for vehicles that share control between driver and automation. This paper describes a study to examine, in a realistic 'context of use', how users interact with current semi-autonomous interaction designs available from a leading manufacturer, to inform best practice for the design of higher levels of automation. To gain a deep understanding of how the driver interacts with semi-autonomous vehicles in a routine on-road setting, the authors propose framing the analysis within the structure of Neisser's (1976) Perceptual Cycle Model (PCM). PCM has been used to account for accidents in safety-critical domains, including rail (Stanton and Walker 2011; Salmon et al. 2013), road (Salmon et al. 2014; Banks and Stanton 2016), and aviation (Plant and Stanton 2012). It reinforces the 'systems view' of human failure, with strong emphasis on the context and an evolving situation to understand behaviour.

9.1.1 THE PERCEPTUAL CYCLE MODEL

Neisser (1976) presented the view that human thought is closely coupled with a person's interaction in the world, both informing each other in a reciprocal, cyclical relationship (Plant and Stanton 2012, 2016). By considering the operator and environment together, the interaction 'in context' can be better understood. The PCM is depicted in Figure 9.1 within the context of semi-autonomous vehicles, as a relationship between world, schema, and actions. Schemata, as a concept, were first popularised in Psychology by Bartlett (1995). They can be thought of as mental 'templates' in long-term memory based on common features of similar experiences. These templates are used to interpret information in the world, predict events, and focus attention and behaviour. According to Neisser (1976), the relationships between world, schema, and actions are interrelated through a series of top-down (TD) and bottom-up (BU) processing. TD processing occurs when a schema is triggered, and particular types of information are then anticipated. BU processing often follows, whereby actions are directed to seek particular information and are interpreted within the framework of the existing schema. When what is perceived in the world contradicts expectations driven by an existing schema, modifications to schemata or selection of an alternative can occur. The actions undertaken, and the types of information sought from the world, are then directed by the new, or amended, schemata. Banks, Plant, and Stanton (2018) applied the PCM to the Tesla collision to show how the drivers' schema of reliable and trustworthy autopilot was a contributory factor. Plant and Stanton (2015) have validated the PCM with the real-world data. They found compelling evidence through post-incident interviews and verbal protocols (VPs) of significant events in aviation (studies of both fixed and rotary wing aircraft), of the existence of a 'counter cycle (CC)'. Particularly for highly skilled pilots,

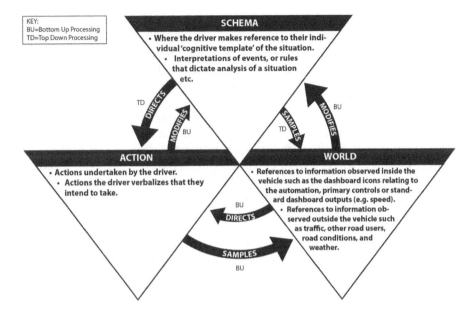

FIGURE 9.1 Simplified representation of the Perceptual Cycle Model, based on Neisser (1976) and further amended by Plant and Stanton (2015), now in the context of semi-autonomous vehicles.

information in the world informed action directly, without articulation of the schema (CC, BU processing). Similarly, examples of schema sampling the world (CC, TD processing) and actions modifying the schema (CC, BU processing) were also found. This CC is represented as the inner ring of Figure 9.1. In their study, 42% of the data were examples of CC activity, with 27% reflecting world influencing the actions, and 8% and 7%, respectively, accounting for action influencing the schema and schema sampling world. Plant and Stanton (2015) described how the prevalence of the CC varied depending on which stage of an incident they were discussing. More examples of CC activity were evident during discussions covering pre-incident, the onset of the problem, and subsequent actions. This was considered to be due to pilots focusing on describing the problem rather than reflecting on prior experience, so omitting reference to the schema. When describing immediate actions or decision-making, the prevalence of CC examples reduced, as pilots reflected more on past experience, so completing the traditional cycle between world and action data. As the majority of counter examples reflected 'mini-cycles' between world–action–world–action, Plant and Stanton (2015) focused on this relationship to understand the cognitive processes involved in the CC. They hypothesised that the phenomenon of the world–action CC could be understood through Rasmussen's (1983) skill-, rule-, and knowledge-based behaviour taxonomy. They believed what they elicited was evidence of skill-based behaviour, where there is no conscious monitoring of the behaviour; rather, it is 'automatic' and 'recognition primed' (Klein, Calderwood and MacGregor 1989) with key world information enabling the appropriate actions to be selected directly, without conscious reference to the schema. This type of behaviour is typically demonstrated by those with considerable experience with the scenario under investigation (Bartlett 1995), which reflects the expert participant pool used in their study. Plant and Stanton (2015) did not offer explanation of the cognitive processes that could account for the other examples of CC behaviour, reflecting the early stages of the theory of the CC.

The aim of this paper is to demonstrate the benefits of Neisser's (1976) PCM as an effective framework to understand the interaction between driver and semiautonomous vehicle for on-road studies, as well as the challenges of applying the method when VPs are generated concurrently with the event, rather than posthumously in a post-incident interview or accident transcripts. Further, it intends to show how generalised recommendations for improvements in designs can be generated, inspired from different parts of the PCM triad (schema, world, and action), recognising that their effectiveness is ultimately dependent on the interdependency of all three parts of the cycle. Finally, this paper takes the opportunity to further explore the PCM CC introduced by Plant and Stanton (2015), demonstrating how it can aid understanding in the analysis of on-road studies.

9.2 METHOD

9.2.1 PARTICIPANTS

Six participants (five males and one female), aged between 26 and 56, participated in part one of this study. The participants formed two equal groups, undertaking the study on separate days and following the same route and protocol. Ethical approval was granted by the University of Southampton Ethics and Research Governance Office (ERGO number: 21090) for participation.

9.2.2 EQUIPMENT

The study was undertaken in a Mercedes S Class with pilot assist features comprising 'Distronic Plus' and 'Steering Assist'. Distronic Plus consists of two short-range sensors and a long-range radar, used to provide ACC to automatically maintain a safe headway from the vehicle in front by braking when necessary and accelerating again when the traffic conditions permit (Daimler AG 2016). Steering Assist uses stereo cameras to identify lane markings and passes the signal to the electric steering, which maintains the position between lane markings. It keeps the vehicle in the centre of the lane on straight roads and around bends (Mercedes-Benz 2012). This combined function approach (National Highway Traffic Safety Administration 2013) of two SAE Level 2 features, Distronic Plus and Steering Assist, results in a driver experience that is conceptually 'hands and feet free' from control, as there is no need to make inputs to control lateral (steering) or longitudinal (accelerator and brake) locomotion (Banks, Stanton and Harvey 2014a). This feature does not allow 'eyes free' control of the vehicle however (available from SAE Level 3+), as the driver is still required to monitor the road and automation status and capability. Drivers are also not 'mind free', as they must be ready to take manual control of the vehicle at short notice when the car reaches the limits of its operational design (Banks, Stanton and Harvey 2014b; Banks et al. 2018).

The drivers' actions and verbalisations were recorded using two handheld digital video cameras operated by experimenters. One camera focused on the view through the windscreen and recorded the voice of the driver. The second camera was focused over the right shoulder of the driver, allowing a view of the steering wheel, the hands of the driver, and the Distronic display. A head-mounted GoPro video camera was used to identify the broad direction of gaze. For redundancy, a head-mounted microphone, connected to a digital Dictaphone, was used to capture driver verbalisations. Salmon et al. (2017) demonstrated that driver behaviour was comparable with and without concurrent VPs.

9.2.3 PROCEDURE

Prior to the experiment, all participants were trained on the 'think-aloud' technique and were requested to read a participant information sheet and complete a consent form. A safety driver conducted a 20-min training session on a test track to familiarise each participant with the Mercedes S Class and the Distronic Plus and Steering Assist features. While drivers who purchase a vehicle with SAE Level 2 capability will not be required to undergo training prior to use, it is reasonable to expect that they will have trialled the features during a test drive, or on familiar routes, before using the combined features in earnest. The safety driver was present throughout the study to provide continued advice on the route and verify that conditions were suitable for activation of the automation features.

Following directional instruction from the safety driver, the participant then drove two predetermined routes of approximately 20-min duration. One route comprised predominantly motorway, whereas second route urban roads through a small town. Order of driving condition was counterbalanced such that half the drivers drove each route manually and then in semi-autonomous mode, and the other half of drivers drove in semi-autonomous mode followed by manual mode. Throughout each route, the drivers were required to use the 'think-aloud' method in order to generate VP data for PCM analysis (Stanton et al. 2013). While further papers will focus on comparisons between manual and automated driving, for this paper, our interest is on how the PCM can provide insights into how to improve interaction design following incidents during semi-autonomous mode. As such, incidents during manual mode will not be described.

9.2.4 DATA ANALYSIS

VP data was transcribed and entered into NVivo software for qualitative data analysis. The content was coded according to Neisser's (1976) PCM, using the criteria in Figure 9.1. During coding, verbalisations of driver frustration, confusion, or panic, which demonstrated a lack of synergy between driver and vehicle, were identified. The video data for each incident was then examined in detail to populate the PCM with observation and context data, following the process adopted by Plant and Stanton (2012). These elements were identified with [square brackets] to distinguish them from the VP strings in Figures 9.3, 9.4 and 9.6, and used to populate diagrams based on Figure 9.1. Any inferences made, particularly regarding schema content, were verified with the participant in question in a follow-up session in person. VP strings, observational, and inferred data were numbered according to the sequence of occurrence (e.g. 1, 2, 3). However, if a VP segment for an incident contained evidence of more than one 'schema', these were distinguished by a decimalised numbering system for clarity (e.g. Schema 1-1.1, 1.2; Schema 2-2.1, 2.2). As distinguishing between schemata hinders the reader from understanding the sequential order of VP strings, the full VP transcript from each instance is also supplied. The combination of observed, inferred, and VP content is presented both in diagram and in table form. The former is used to emphasise the interrelation between the three triads, while the latter to emphasise the sequence of content over time.

9.3   RESULTS AND DISCUSSION: THREE CASE STUDIES OF DRIVER FRUSTRATION

This section presents three case studies relating to frustrations originating in the drivers' schema, drivers' actions, and drivers' observation of the world. These demonstrate not only the flexibility of the PCM framework to explore driver–vehicle interaction from on-road studies, but also its utility in generating insights that can inform the development of design recommendations for semi-autonomous vehicles. This section demonstrates how each VP can easily be allocated according to one of the three PCM triads (world, action, and schema). By exploring how each triad links with the others, and if the sequence follows the traditional PCM cycle or CC proposed by Plant and Stanton (2015), the analysis is encouraged to embrace a 'systems perspective'. The inclusion of world data, in particular, allows non-optimal behaviour to be understood in terms of a context beyond the user. The interaction between all three triads emphasises their interdependency.

9.3.1   CASE STUDY 1: 'THAT WAS SCARY ....' – THE RISK OF AN INAPPROPRIATE SCHEMA

The first case study relates to TD processing from the 'schema' triad of PCM during use of the Distronic Plus in the Mercedes S Class on a UK motorway. It occurred when the participant was in the centre lane behind two cars (one red and one white) and a lorry. The red car directly ahead of the Mercedes S Class car indicated and moved to the right-hand lane to overtake the lorry (Figure 9.2a). The participant decided that they also wanted to manually overtake the lorry, so activated the indicator. During this time, the white car directly behind the lorry also indicated and moved to the right-hand lane, proceeding to overtake the lorry. While waiting for a clear gap in the right-hand lane to pull out, the participant noticed their car accelerating (Figure 9.2b). In confusion and panic, they braked suddenly to counter the acceleration that they attributed to an error in the automation (Figure 9.2c). They then continued with the (manual) overtake procedure and verbalised their fear that the automated car had been on course to drive into the back of the lorry in the

FIGURE 9.2   Screenshots from the scenario show the Distronic Plus is correctly representing the outside world but is yet to reduce the headway to the standard setting following two cars behind a lorry changing lanes in quick succession. A manual lane change is initiated by the driver as the car automatically increases speed to reduce headway, resulting in panic breaking.

central lane. Figure 9.2 shows three screenshots of the incident describing the scene. Figure 9.2a depicts the scene just after the red car has moved into the right lane, and before the white car starts to indicate. The Distronic Plus dashboard clearly shows the white car has been sensed (car ahead icon is present) but the headway has not yet been reduced to the prescribed setting (gap evident between headway line and the car ahead icon). Following is the VP transcript excerpt for the incident:

> And there's vehicles all around me. It feels quite heavy traffic. So, we've dropped down to – another bit of input, it won't – okay, just given it. I'm thinking about doing an over-take now. So, I get past this lorry and I'll try indicating. Check behind me. But we're speeding up – oh, no. Blimey! Brake. And I didn't trust it. And so, I'm pulling out now. And – no, that was scary. So, I think I'm going to have put that back on again. We're doing 60, hands off the wheel. But generally, if I hadn't had grabbed then it would have ploughed into that lorry.

The VPs during the incident were broken down and numbered in order of occurrence, then categorised by relevance to schema, action, or world in Figure 9.3. Two different schemata were identified: Schema 1 – relating to undertaking a manual overtake from semi-automated drive, and Schema 2 – relating to undertaking a manual overtake from manual drive. Here, it can be seen that the lead up to the incident (VP 1.1–1.9 in Figure 9.3) is relatively calm with attention being focused mainly on the 'world' in terms of outside traffic conditions, the speed status, and the alert (dashboard icon) to remind the driver to keep their hands on the wheel. Before the distance between the Mercedes S Class and the white car could be reduced to maintain the set headway, the white car also moved into the right-hand lane, extending the

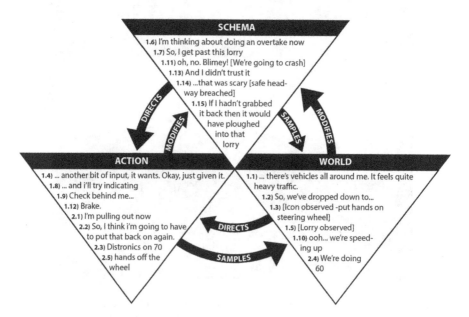

FIGURE 9.3 Verbal protocols from case study 1 arranged according to Neisser's (1976) Perceptual Cycle Model.

headway between the Mercedes S Class and the lorry ahead (Figure 9.2b). Both the video and audio data provide no evidence that the driver paid attention to 'world' data relating to the status of headway maintenance, shown on the dashboard. Reductions in awareness can increase the occurrence of mode errors (Sarter and Woods 1995; Stanton, Dunoyer and Leatherland 2011) and automation surprises (Sarter, Woods and Billings 1997). The Distronic Plus accelerated to bridge the gap between the Mercedes S Class and the lorry ahead.

This acceleration, however, was noticed by the participant, who sensed the change in the 'world' – VP 10, Figure 9.3, *ooh.. we're speeding up.* Without checking the headway status to help interpret the situation, the participant assessed the situation on incomplete data from the world, triggering an alternate 'schema' that dictated the 'action' of sudden braking (Figure 9.2c) as a means of preventative action to avoid an accident (VP 1.11–1.12, Figure 9.3). These, and the statements *And I didn't trust it...* (VP 1.13, Figure 9.3) and *that was scary [safe headway breached]* (VP 1.14, Figure 9.3) highlight the fear and lack of trust experienced during the incident.

There was clearly a 'mismatch' between the 'expected' and 'observed' behaviour of the vehicle experienced by the driver. After resuming the manual overtake in manual drive (VP 2.1–2.5, Figure 9.3), the reflection (VP 1.15, Figure 9.3) '... If I hadn't grabbed it back then, it would have ploughed into that lorry...!' clearly indicates that the participants' existing schema could not reconcile the automated acceleration of the vehicle at that point in their manual manoeuvre, with correct functioning of the Distronic Plus. To understand the reason for this, it is necessary to consider the context of the incident in terms of a manual overtake of a large vehicle, and how this differs from that of a small vehicle. When overtaking a small vehicle, if the lane to the right is clear and there is visibility ahead of the vehicle for re-entry into the lane, it is appropriate to accelerate towards the vehicle to adopt 'overtake position', allowing the overtake to occur immediately (Roadcraft 2013). In the scenario described, traffic was present in the right-hand lane, preventing an immediate start to the manoeuvre, and the large vehicle obscured the view of the road ahead. In this situation, it is prudent to drop back to a following positioning until a gap in the right-hand lane is imminent. This also allows greater visibility of the road ahead of the vehicle and avoids positioning your own vehicle in its blind spot for any extended period (Roadcraft 2013).

If the driver had been in manual control of the longitudinal locomotion, it is unlikely they would have accelerated at the point chosen by the Distronic Plus. As they were about to engage in a manual manoeuvre, it is not surprising that a schema for manual control was in place that interfered with appropriate interpretation of events. A key issue relates to the automatic systems' lack of awareness that the driver had changed their goal from 'automated cruise' to 'vehicle overtake'. Although the driver did activate an indicator, which would necessitate manual lateral control, this input is not integrated with the performance of the Distronic Plus, preventing a more appropriate rule to be modelled by the automated system (e.g. wait until lane change is complete before adjusting speed to maintain set headway). In the same way, the driver was unaware of the intention of the Distronic Plus to reduce the leading headway between the Mercedes S Class and the lorry in front in order to keep the prescribed setting. A key factor for the incident clearly arose out of a conflict of goals between the vehicle and the driver, and obstacles to communicating each other's intentions. We are not suggesting that there was a serious safety breach with Mercedes Distronic Plus; rather, there was a mismatch between the expectations of the driver and the behaviour of the vehicle. Ideally, this apparent mismatch could be addressed in the design of vehicle automation. There may also have been a 'gap' between the driver's 'device model' of the function of ACC as a feature that maintains headway, rather than speed (which is more applicable to the functioning of standard cruise control) (Norman 1986). This incident reflects the findings relating to ACC by Kazi et al. (2007), and Stanton, Dunoyer, and Leatherland (2011), who identified cognitive mismatch leading to mode confusion (Woods 1988; Baxter, Besnard and Riley 2007), particularly where insufficient monitoring of the system was observed. A poor conceptual model from the driver impacted levels of trust with ACC, supporting Lee and See's (2004) view that appropriateness of trust should be considered in relation to the context of the environment and goals of the user. Norman (1990) also emphasises that ACC is not considered to be sufficiently intelligent to cope with a high demand situation. The changing goals of the driver observed in this study could be considered a high demand situation for the current capabilities of semi-autonomous vehicles. With interrelated nature between schema, action, and world information, recognition that the poor conceptual model of the driver was likely to have resulted from a poor 'device image' by the system, points to design strategies to improve the partnership between driver and automated vehicle (Chapanis 1999; Banks et al. 2018).

9.3.1.1 Evidence of Counter Cycle in Case Study 1

To better understand where examples of CC are found, the same data has been arranged in Table 9.1, showing the sequence of verbalisations over time as well as by PCM category. For clarity, the grey shading represents VPs belonging to a second schema introduced in the case study. CC examples are highlighted in black for both schemas and inferred, or video observed, data (rather than VP data) is distinguished by [square brackets]. Traditional PCM cycle flows in direction of schema–action–world–schema. CC PCM will conversely flow schema–world–action–schema.

From Table 9.1, three examples of CC sequence between VPs show two different types of activity. An example of each type will be described. The first, between VP1.3 *[icon observed]* and VP1.4 *Another bit of input it wants. OK just given it*, reflects the world–action relationship found frequently by Plant and Stanton (2015), suggesting skill-based behaviour. However, although this participant had considerable experience driving, their experience with the interface and interaction design was limited. Another interpretation may be that the icon advising steering wheel input was sufficiently intuitive to be self-explanatory or easily learnt, so reference to the schema was unnecessary. The second example is between VP1.12 *Brake* and VP 1.13 *And I didn't trust it*, which provide an example of actions influencing the schema. Video footage showed the driver focusing 'eyes out' during this area to ensure safety, so no observations are related to the icons or interface that could have helped add 'observed/inferred' content to the 'world' column to complete the traditional direction of the PCM cycle and link to ideas relating to trust. The cognitive processes between these two triads are not clear, but one hypothesis to consider is

TABLE 9.1

Verbal Protocols for Case Study 1 Shown in Table Format to Demonstrate Sequence of Verbalisations

Schema	Action	World
		1.1 … there's vehicles all around me. It feels like quite heavy traffic.
		1.2 So we've dropped down to…
		1.3 [Icon observed – put hands on steering wheel]
	1.4 … another bit of input it wants. Okay, just given it.	
		1.5 [Lorry observed]
1.6 I'm thinking about doing an overtake now		
1.7 So, I get past this lorry		
	1.8 … and I'll try indicating	
	1.9 Check behind me…	
		1.10 Ooh we're speeding up
1.11 Oh no. Blimey! [We're going to crash]		
	1.12 Brake.	
1.13 And I didn't trust it		
2.1 And so I'm pulling out now		
1.14 And – no, that was scary [safe headway breached]		
2.2 So, I think I'm going to have to put that back on again.		
2.3 Distronics on 70		
	2.4 We're doing 60	
2.5 Hands on wheel		
1.15 If I hadn't grabbed it back then it would have ploughed into that lorry		
Key: Schema 1 content	*Schema 2 content*	**Example of counter cycle**

that, having undertaken avoiding action, the driver is justifying this behaviour by reflecting on their changed schema of the automation following unexpected vehicle behaviour.

9.3.2 Case Study 2: 'Oh, I've just done the *Distronic* again ….' – Impeding Intended Actions

The second case study provides an example of BU processing, relating to the 'action' triad in the PCM. It depicts the accidental activation of Distronic Plus in a Mercedes

S Class when leaving a car park on an urban road. Following is the VP transcript excerpt for the incident:

> You can see there's a van coming in sight, because we're coming more away to the left, then to the right, and over to me... Oh, I've just done the distronic again, when I'm meant to indicate right. I've done that twice now. Absolutely ridiculous levels of complexity in the controls of this car.

The VPs are depicted in Figure 9.4, where the driver observes other road users in the 'world', before describing their own 'actions' when manoeuvring around the car park (VP 1–3, Figure 9.4). The participants' schema identifies the need to indicate using the stalk control (VP 4, Figure 9.4) and video footage verified the action *[stalk pushed]* (VP5, Figure 9.4), informing other road users the intention to turn and prompting the seeking out of 'world' information to verify the action had been completed. They noticed from the dashboard interface that the Distronic system had been activated (VP 6, Figure 9.4) and stated in frustration *Oh I've just done the Distronic again, when I'm meant to indicate right, I've done that twice now...* (VP 6–7, Figure 9.4), before complaining about the complexity of the car interface (VP 8, Figure 9.4).

By viewing the controls, it is clear why the frustration occurs: activation of Distronic mode and activation of the indicator both require a stalk to the left of the steering wheel to be nudged and are designed to be activated 'eyes out'. The stalks are difficult to distinguish by feel, are activated by the same hand, and are in a similar location with similar visual appearance. Similar to the indicator stalk, the Distronic plus can be activated both by tilting upwards and by tilting downwards

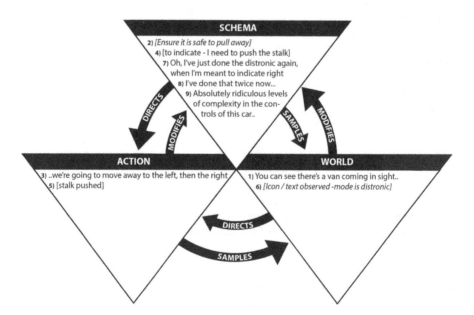

FIGURE 9.4 Verbal protocols from case study 2 arranged according to Neisser's (1976) Perceptual Cycle Model.

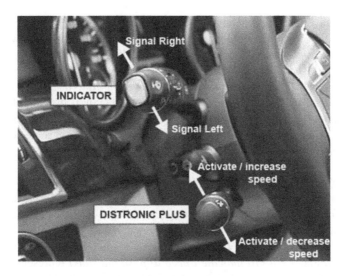

FIGURE 9.5   Showing similar proximity and design of indicator stalk and Distronic Plus stalk.

(these same movements are then used to increase/decrease the ACC speed setting). Figure 9.5 shows the close proximity, and similar style and method of activation of disparate controls.

Confusion between controls with different functions can have serious consequences. In SAE Level 2 automation, the role of changing lanes is achieved manually and can occur while still in Distronic mode. This means that control confusion could result in the absence of indicating when intended (as in this case study). In a motorway context, this would prevent other road users from understanding the driver's intentions, resulting at best in surprise if a manoeuvre is completed without the control error being identified, or at worst a collision if a car speeds up or overtakes a vehicle that suddenly changes lanes. In addition, if the car is already in Distronic mode and the same stalk is adjusted, this would have the impact of unintentionally changing the speed value for ACC, compromising drivers' situation awareness of vehicle state (Stanton, Dunoyer and Leatherland 2011; Stanton, et al. 2017). While not demonstrated in this particular example, the reverse control confusion could also occur (evident by other participants in the study) where the driver's intention is to activate Distronic plus, but they accidentally activate the indicator. Confusion to other road users may still occur when a manoeuvre does not occur, but the consequences are less serious in this case. If the driver had not checked world information however, and incorrectly assumed ACC was then active, this schema would be triggered, providing incorrect expectations of vehicle behaviour. This could influence driver actions, such as failing to manually operate brakes, or operating brakes late when approaching vehicles (resulting in a potential for collisions). Accidental deactivation of combined SAE Level 2 features has also been observed in Banks et al. (2018) due to torque input via the steering wheel in a Tesla Model S. Mode errors can have potentially disastrous outcomes (Stanton, Dunoyer and Leatherland 2011; Reason

1990; Norman 1990; Woods, et al. 1994). The problem found between control stalks is similar to that found in aviation cockpit design. After a series of runway crashes of the Boeing B-17, it was found that certain cockpit controls (flaps and landing gear) were confused with each other, due partly to their proximity and similarity of shape. Design principles, such as shape coding, were used to disambiguate controls to improve safety (Chapanis 1999). Automotive manufacturers often implement new features, and particularly optional extra ones, by overlaying their functionality onto extant controls for good engineering and cost reasons; however in some contexts, the usability cost of this reuse can be very high.

9.3.2.1   Evidence of Counter Cycle in Case Study 2

The data for case study 2 has been arranged in Table 9.2, showing the sequence of verbalisations and inferred content over time, as well as by PCM category. A far simpler incident, VPs from a single schema are evident so no grey shading is required. A single CC example is shown between VP 3 *We're going to move away to the left, then the right* and VP 4 *[To indicate I need to push the stalk]* representing another example of action to schema processing. It is noteworthy that this CC example relates to inferred content, distinguished by [square brackets]. From a cognitive processing perspective, it is reasonable to see how the intention to take action triggers a 'subschema' relating to the process for indicating. The nature of the indicating task being carried out 'eyes free' explains why no VP describing viewing the indicator in

---

TABLE 9.2

Verbal Protocols for Case Study 2 Shown in Table Format to Demonstrate Sequence of Verbalisations

Schema	Action	World
		1 You can see there's a van coming in sight because…
2 [Ensure it is safe to pull away]		
	3 We're going to move away to the left, then the right.	
4 [To indicate I need to push the stalk]		
	5 [Stalk pushed]	
		6 [Icon/text observed – mode is distronic]
7 Oh, I've just done the distronic again, when I'm meant to indicate right.		
8 I've done that twice now.		
9 Absolutely ridiculous levels of complexity in the controls of this car.		
Key:	**Example of counter cycle**	

---

the 'world' before activating it, was evident. Since the driver will be highly skilled in using the indicator to pull away, the schema they are accessing is likely to be implicit, impeding verbalisation. However, it is clear that without the inferred and observed content in VP 4–6, the sequence from VP 3 *Were going to move away to the left, then the right* and VP 7 *Oh, I've just done the distronic again, when I'm meant to indicate right* fails to illuminate either the control in question used in the 'action' triad or the feedback mechanism in the 'world' triad to focus strategies for reducing error or improving design.

9.3.3   CASE STUDY 3: 'I THINK IT'S GREEN NOW, … NO IT'S NOT!' – INEFFECTIVE WORLD INFORMATION

This final case study considers the impact of the 'world' triad in the PCM, with driver VPs depicted in Figure 9.6. It highlights the influence of environmental factors that limit how information in the world can be perceived. This specific driver was more experienced than some of the other participants, in terms of both a conceptual understanding of Distronic Plus and its method of activation. However, they found themselves in the position of non-optimal driving on a UK motorway due to the performance of the interface in a typical context of use (bright sunshine). Following is the VP transcript excerpt for the incident.

> Again re-engage, just drive closer behind the lorry, increase my speed to 70. Green steering wheel which looks like we have. Oh it's gone grey and now it's gone grey again. It's very difficult to see with this sunlight in my sunglasses. Definitely both on. I think it's green now, no it's not. It is really difficult to see with the… I'll take my sunglasses off because they're making it harder for me to see. Lots of traffic so just keeping the set distance for the vehicle in front. The speed is 58 miles an hour. Still no steering. It's turned green now. I'll just take my hands off. Spending too much time looking at the green steering wheel rather than the traffic around me.

Figure 9.6 represents this interaction with two schemas relating to: Schema 1 – engaging automation and monitoring (mode status and other road users), and Schema 2 – troubleshooting issues with interpreting the mode status. The participant activates Distronic Plus, which initiates the ACC, allowing them to be 'feet free' (VP 1.1, Figure 9.6). To also be 'hands free', they identify from the 'world', the interface icon *green steering wheel* (VP 1.2, Figure 9.6), that conditions for auto-steering are available. Unlike the participant in case study 1, whose attention on the road ahead led to an 'eyes out' focus, this participant spends a considerable amount of attention 'eyes down' trying to determine the status of the vehicle automation features – e.g., *oh its gone grey….* (VP 1.4, Figure 9.6), *definitely both on* (1.6, Figure 9.6), *I think its green now, …no its not* (1.7, Figure 9.6), *Still no steering* (1.11, Figure 9.6), *its turned green now* (1.12, Figure 9.6), with only one out of eight VPs in this triad focused on other road users (VP 1.17, Figure 9.6). Their awareness of the imbalance of their attention with the world is clear in VP 2.5 (Figure 9.6), *Spending too much time looking at the green steering wheel rather than the traffic*. This represents an example of BU processing, whereby the visual issues trigger the second schema, relating to difficulties with visibility. Figure 9.6 shows that most of the verbalisations categorised

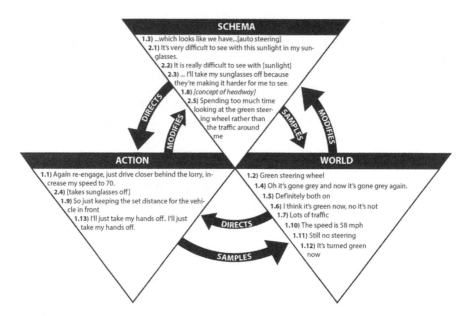

FIGURE 9.6 Verbal protocols from case study 3 arranged according to Neisser's (1976) Perceptual Cycle Model.

in the 'schema' triad are focused on the understanding that bright outside light conditions and associated aids such as sunglasses, while improving visibility 'eyes out', diminish visibility 'eyes in' when viewing the dashboard interface, *It's very difficult to see with this sunlight in my sunglasses* (VP 2.1, Figure 9.6) and *It is really difficult to see with the [sunlight]* (VP 2.2, Figure 9.6). This influences the driver's actions to remove them ...*I'll take my sunglasses off* (VP 2.3, Figure 9.6) to improve visibility 'eyes down' (despite the negative impact on visibility 'eyes out').

The diagram in Figure 9.7 shows a representation of the Mercedes interface with the auto-steering icon in active (green) and inactive (grey) modes, placed side by side for comparison (the interface shows a single icon that lights up green when auto-steering conditions are met). While the green coding of the icon follows western colour coding conventions, it does not employ redundancy to take account of issues with colour as a primary distinguisher of mode. While this standard ergonomics advice for interface design is intended to take into account colour blindness, the benefits would also be reaped in circumstances such as bright sunlight, when colour changes are more difficult to distinguish (Stanton et al. 2009). The consequences of misunderstanding auto-steering mode status are also not reflected in the hierarchy of information on the interface. Figure 9.7 shows the auto-steering icon as considerably smaller than primary information such as speed and rpm. It could be argued that automation status is one of the primary system states, which should be communicated in semi-autonomous vehicles and warrants far greater prominence than currently shown by the industry. The current, relatively low-salience implementation of this autonomous mode display in the instrument cluster is, however, consistent with this feature being treated as an 'optional extra', and therefore an add-on to an

FIGURE 9.7 Mercedes S Class interface highlighting the mode status shown (with 'off' mode duplicated by the side of the green 'on' steering wheel icon) by icon colouring and minimal priority compared to other types of car status information (see colour version of this figure on the online resource).

instrument cluster design optimised for multiple versions of the vehicle which may not offer this functionality.

9.3.3.1 Evidence of Counter Cycle in Case Study 3

The data for case study 3 has been arranged in Table 9.3, showing the sequence of verbalisations and inferred content over time, as well as by PCM category. Similar to case study 1, the VPs refer to two different schemas and have been shaded to reflect this, with Schema 1 represented with a white background and Schema 2 with a grey background. As before, examples of CC activity are highlighted in black, and inferred, or video observed, data (rather than VP data) is distinguished by [square brackets].

Two examples of CC activity are seen in case study 3. The world–action sequence in Schema 1, between VP 1.7 *lots of traffic* and VP 1.8 *so just keeping a set distance for the vehicle in front,* is another example of skill-based behaviour, where driving experience sees a direct link between observing in the world the busyness of the road and subsequent action to prioritise headway. The implicit schema relates to safety and a greater likelihood for traffic incidence in busy traffic conditions than more sparsely populated roads. The other example of CC processing is from action–world for Schema 2. Observed data VP 2.4 *[removes sunglasses]* and schema data VP 2.5 *Spending too much time looking at the green steering wheel rather than the traffic around me* clearly identifies the balance of visual attention is inappropriate for the driving task for SAE Level 2 (where monitoring the road ahead and other road users is essential). As with case study 1, this reflection directly follows an action that was not anticipated had the original expectation of the interface allowed the driver to decipher the automation mode and capability in typical environmental conditions. Unlike case study 1, however, this reflection represents a genuine limitation of the interface, rather than a misunderstanding of automation function.

9.3.4 Implications for Interaction Design

Semi-automated cars (anything below SAE Level 5) require input from drivers where the road, infrastructure, surroundings, or environmental conditions prevent full

TABLE 9.3

Verbal Protocols for Case Study 3 Shown in Table Format to Demonstrate Sequence of Verbalisations

Schema	Action	World
	1.1 Again re-engage, just drive closer behind the lorry, increase my speed to 70	
		1.2 Green steering wheel
1.3 Which looks like we're having … [auto-steering]		
	1.4 oh it's gone grey now and its gone grey again	
2.1 *Its really difficult to see with this sunlight in my sunglasses.*		
	1.5 Definitely both on	
	1.6 I think it's green now, no it's not	
2.2 *It is really difficult to see with the [sunlight]*		
2.3 *I'll take my sunglasses off because they're making it harder for me to see*		
	2.4 *[removes sunglasses]*	
		1.7 lots of traffic
	1.8 so just keeping a set distance for the vehicle in front	
		1.9 The speed is 58 mph
		1.10 Still no steering
		1.11 It's turned green now
	1.12 I'll just take my hands off	
2.5 *Spending too much time looking at the green steering wheel rather than the traffic around me*		
Key: Schema 1 content	*Schema 2 content*	**Example of counter cycle**

automation (also where the driver chooses to drive manually). High levels of synergy between driver and automation are required for performance and safety, but this synergy relies on trust and understanding on both sides (Walker, Stanton and Salmon 2016; Stanton et al. 2017). Hancock, Hancock, and Warm (2009) argued that individual case representations are increasingly relevant for the design of human–machine systems, and generalisations can be derived. From the case studies described in this paper, the goals to be achieved through interaction design can be summarised as follows:

- **Design for conflicting intentions between driver and automation**: For example, when the driver begins or ends a manual manoeuvre, there needs to be a means for this to be communicated to the existing assistive technologies so alternate models of autonomous behaviour can be activated, or where necessary, autonomous assistance is disabled until the manual manoeuvre is complete (case study 1).

- **Design for reassurance:** Such as when an atypical change in longitudinal or lateral locomotion is initiated by the automation, the action and its reason need to be successfully communicated to the driver to avoid unnecessary, or potentially risky, manual intervention by the driver. Visual dashboard displays may be insufficient if the driver is focused 'eyes out' of the windscreen, observing potential hazards, so head-up displays or alerts may be more appropriate. Eriksson and Stanton (2017) concluded that visual and vocal displays should be used to convey semantics to the driver to support cognitive processing and decision selection, a key stage of automation as defined by Parasuraman, Sheridan, and Wickens (2000) (case study 1).
- **Design for appropriate mental models:** Ensuring the function of any assistive automation is effectively understood by the driver and does not conflict directly with their existing mental models, to promote a positive transfer of knowledge. Without an appropriate schema or mental model of automation function, the predictability of vehicle behaviour and appropriate human response is diminished (Revell and Stanton 2014, 2015, 2017, 2018). At worst, safety is compromised, and at best, user trust and commercial adoption is diminished (Banks et al. 2018; Walker, Stanton and Salmon 2016) (case studies 1 and 2).
- **Design for context:** For example, recognise that the expected behaviour by a car when following a large vehicle will differ from that when following a standard vehicle. Interaction design should either appropriately set the drivers' expectations in a range of contexts, or the automation should be programmed to adjust to differing expectations based on context (case study 1).
- **Design for the variety of human capabilities and the variety of environmental conditions of use:** The interface elements should be designed to accommodate the visual, hearing, cognitive, reach, and dexterity capabilities of the potential user population (Waller et al. 2015). Where automation sensors (radar etc.) experience limitations in sensing the road, drivers also have limitations in sensing the low contrast or diminutive display icons in non-ideal weather conditions (e.g. bright sunshine). Digital displays provide the flexibility to offer redundancy in presentation, and traditional ergonomic guidelines should be adhered to and tested against the variety of relevant capabilities in the user population (case study 3).
- **Design an adequately prominent mode state:** While occurring from different causes in different contexts, and relating primarily to different triads of the PCM, each case study ultimately resulted in mode errors. This supports the extensive body of literature investigating human interaction with automated systems (Andre and Degani 1997; Degani, Shafto and Kirlik 1995; Leveson 2004; Norman 1983; Rushby, Crow and Palmer 1999; Sarter and Woods 1995; Sheridan and Parasuraman 2005). The importance of an explicit mode state and feedback which are detectable and interpretable by

the user population's capabilities when mode changes cannot be understated (case studies 1–3).

- **Disambiguate controls of different function:** Disambiguate two very different actions by changing the design of controls – e.g., by using different shapes and activation of controls (Chapanis 1999), or by positioning controls in distinct locations, or requiring different hands or methods of activation (case study 2).

9.3.5 Evaluation of Applying PCM to On-Road Concurrent VP Dialogue

Three case studies presenting concurrent VPs during an on-road study of non-optimal interaction between a driver and semi-autonomous vehicle were successfully explored within the framework of the PCM. While this method of thematically analysing VPs takes a deductive rather than grounded theory approach, the benefits include providing a theoretical framework that directs the analyst to systematically seek out links between triads, and ensures consideration of context-based interdependency from a system's perspective. While other usability methods have much to contribute, traditional usability questionnaires (e.g. the system usability scale, the system acceptance scale) provide only high-level understanding of the user experience and cannot target where change is needed. Other more in-depth methods such as product walk throughs are difficult to implement in a fast-moving dynamic demonstration. The authors acknowledge that for more simplistic case studies such as 2 and 3, observations of behaviour without the PCM framing would also have successfully captured the issues identified. However, the PCM framework may be more appropriate for more complex sociotechnical systems that it can be easily applied and reveal both complex and more simple causes of non-optimal system behaviour, highlights its advantage over a method that can only cope with more basic issues.

Both the traditional direction of the PCM, proposed by Neisser (1976), and the CC, presented by Plant and Stanton (2015), were evident in the VPs.

Using inferred content (through observation of video footage, or post-study discussions) ultimately had an impact on whether a sequence of VPs followed a traditional direction of the PCM or CC. For example, in case study 3, VP 1.7–1.9 demonstrates traditional BU followed by TD processing. However, without the 'inferred' content, *[concept of headway]*, VP 1.8, Figure 9.6, could have be classed as CC BU processing, going straight from *lots of traffic* (world) to *just keeping the set distance from the vehicle in front* (action). This change in direction is compatible with Plant and Stanton's (2015) description of skilled behaviour, where schema content is not verbalised as it is implicitly, rather than explicitly, understood. The schema element remains part of the cognitive processing, but cannot be captured directly from the transcript. The pace of the changing road context also impacted the ability of the driver to express all perceptions, thoughts, and actions, particularly during manual manoeuvres. In case study 2, four out of nine elements in Figure 9.4 were inferred, with three of these taken from video footage. The process of indicating was completely omitted in the VPs (VPs 4 and 5, Figure 9.4), *To indicate, I need to push the stalk* (schema) and *Stalk pushed* (action), are both key aspects to understanding the incident where the incorrect stalk was operated, so warranted inclusion. It is important to note, therefore, that VPs alone may be insufficient for deep understanding of non-optimal interaction. Framing dialogue using the PCM, and considering whether sequential strings follow the traditional cycle, CC provides a systematic way by which the analyst or practitioner can deeply understand the interaction and gain insights for focus. These insights could be used to inform interventions to improve user experience and safety through design, training, or legislation.

9.4 CONCLUSIONS

This paper set out to highlight the benefit of applying Neisser's (1976) PCM to explain non-optimal interaction between drivers and semi-autonomous vehicles. Three naturalistic case studies were presented that highlight how driver frustration and lack of synergy can occur when interacting with semi-autonomous vehicles, and the potential impact on safety. Seven design recommendations were generated and the benefits of gaining deep understanding through the PCM, in contrast to categorisation by codes, were discussed. Further evidence of Plant and Stanton's (2015) CC was found and aided the interpretation of non-optimal interaction. The ability to interrogate the evidence from the perspective of 'world', 'schema', and 'action' proved useful in deriving generalisable recommendations for interaction design, to improve safety and break the cycle of frustration for drivers when interacting with semi-autonomous vehicles.

ACKNOWLEDGEMENTS

This work was supported by Jaguar Land Rover and the UK-EPSRC Grant EP/-011899/1 as part of the jointly funded 'Towards Autonomy: Smart and Connected Control (TASCC) Programme'.

REFERENCES

Andre, A., and A. Degani. 1997. "Do you know what mode you're in? An analysis of mode error in everyday things." In *Human-Automation Interaction: Research and Practice*, by M. Mouloua and J. M. Koonce, 19–28. Mahwah, NJ: Lawrence Erlbaum Associates.

Banks, V. A., A. Eriksson, J. O'Donoghue, and N. A. Stanton. 2018. "Is partially automated driving a bad idea? Observations from an on-road study." *Applied Ergonomics*, 68 138–145. DOI: 10.1016/j.apergo.2017.11.010.

Banks, V. A., and N. A. Stanton. 2015. "Discovering driver-vehicle coordination problems in future automated control systems: Evidence from verbal commentaries." *Procedia Manufacturing*, 3 2497–2504. DOI: 10.1016/j.promfg.2015.07.511.

Banks, V. A., and N. A. Stanton. 2016. "Keep the driver in control: Automating automobiles of the future." *Applied Ergonomics*, 53, Part B 389–395. DOI: 10.1016/j.apergo. 2015.06.020.

Banks, V. A., K. L. Plant, and N. A. Stanton. 2018. "Driver error or designer error: Using the Perceptual Cycle Model to explore the circumstances surrounding the fatal Tesla crash on 7th May 2016." *Safety Science*, 108 278–285. DOI: 10.1016/j.ssci.2017.12.023.

Banks, V. A., N. A. Stanton, and C. Harvey. 2014a. "Sub-systems on the road to vehicle automation: Hands and feet free but not "mind" free driving." *Safety Science* 62 505–514. DOI: 10.1016/j.ssci.2013.10.014.

Banks, V. A., N. A. Stanton, and C. Harvey. 2014b. "What the drivers do and do not tell you: using verbal protocol analysis to investigate driver behaviour in emergency situations." *Ergonomics, 57, 3* 332–342. https://doi.org/10.1080/00140139.2014.884245.

Bartlett, F. C. 1995. *Remembering: A Study in Experimental and Social Psychology.* Cambridge, UK: Cambridge University Press.

Baxter, G., D. Besnard, and D. Riley. 2007. "Cognitive mismatches in the cockpit: Will they ever be a thing of the past?" *Applied Ergonomics, 38,* 4 417–423. DOI: 10.1016/j. apergo.2007.01.005.

Carsten, O., K. Kircher, and S. Jamson. 2013. "Vehicle-based studies of driving in the real world: The hard truth?" *Accident Analysis & Prevention, 58* 162–174. DOI: 10.1016/j. aap.2013.06.006.

Chapanis, A. 1999. *The Chapanis Chronicles: 50 Years of Human Factors Research, Education, and Design.* Santa Barbara, CA: Aegean Publishing.

Daimler AG. 2016. *DISTRONIC PLUS.* Accessed September 2017. http://techcenter.mercedes-benz.com/en/distronic_plus/detail.html.

de Winter, J. C.F., P. M. van Leeuwen, and R. Happee. 2012. "Advantages and disadvantages of driving simulators: A discussion." *Proceedings of Measuring Behavior 2012. 8th International Conference on Methods and Techniques in Behavioral Research.* Utrecht, The Netherlands: Noldus Information Technology. 47–50.

Degani, A., M. Shafto, and A. Kirlik. 1995. "Mode usage in automated cockpits: Some initial observations." *IFAC Proceedings Volumes, 28,* 15 345–351. DOI: 10.1016/S1474-6670(17)45256-6.

Endsley, M. R. 2017. "Autonomous driving systems: A preliminary naturalistic study of the tesla model S." *Journal of Cognitive Engineering and Decision Making* 225–238. DOI: 10.1177/1555343417695197.

Eriksson, A., and N. A. Stanton. 2017. "The chatty co-driver: A linguistics approach applying lessons learnt from aviation incidents." *Safety Science 99* 94–101. DOI: 10.1016/j. ssci.2017.05.005.

Hancock, P. A., G. M. Hancock, and J. S. Warm. 2009. "Individuation: the N = 1 revolution." *Theoretical Issues in Ergonomics Science, 10,* 5 481–488. DOI: 10.1080/ 14639220903106387.

Heikoop, D. D., J. C.F. de Winter, B. van Arem, and N. A. Stanton. 2017. "Effects of platooning on signal-detection performance, workload, and stress: A driving simulator study." *Applied Ergonomics, 60* 116–127. DOI: 10.1016/j.apergo.2016.10.016.

Kazi, T. A., N. A. Stanton, G. H. Walker, and M. S. Young. 2007. "Designer driving: Drivers' conceptual models and level of trust in adaptive cruise control." *International Journal of Vehicle Design, 45,* 3 339–360. https://doi.org/10.1504/IJVD.2007.014909.

Klein, G. A., R. Calderwood, and D. MacGregor. 1989. "Critical decision method for eliciting knowledge." *IEEE Transactions on Systems, Man, and Cybernetics, 19,* 3 462–472. DOI: 10.1109/21.31053.

Lee, J. D., and K. A. See. 2004. "Trust in automation: Designing for appropriate reliance." *Human Factors: The Journal of the Human Factors and Ergonomics Society, 46,* 1 50–80. DOI: 10.1518/hfes.46.1.50_30392.

Leveson, N. 2004. "A new accident model for engineering safer systems." *Safety Science, 42,* 4 237–270. DOI: 10.1016/S0925-7535(03)00047-X.

Mercedes-Benz. 2012. "DISTRONIC PLUS with Steering Assist - Mercedes-Benz original." *YouTube.* 10 December. Accessed September 29, 2017. https://www.youtube. com/watch?v=lpIddjyHHVs.

National Highway Traffic Safety Administration. 2013. *Preliminary Statement of Policy Concerning Automated Vehicles.* Washington, DC: U.S. Department of Transportation.

Naujoks, F., C. Purucker, and A. Neukum. 2016. "Secondary task engagement and vehicle automation – Comparing the effects of different automation levels in an on-road experiment." *Transportation Research Part F: Traffic Psychology and Behaviour, 38* 67–82. DOI: 10.1016/j.trf.2016.01.011.

Neisser, U. 1976. *Cognition and Reality: Principles and Implications of Cognitive Psychology.* San Francisco, CA: W. H. Freeman.

Norman, D. A. 1986. "Cognitive engineering." In *User Centered System Design: New Perspectives on Human-computer Interaction*, by D. A. Norman, 29–61. Boca Raton, FL: CRC Press. https://doi.org/10.1201/9780367807320.

Norman, D. A. 1983. "Design rules based on analyses of human error." *Communications of the ACM, 26*, 4 254–258. DOI: 10.1145/2163.358092.

Norman, D. A. 1990. "The "problem" with automation: inappropriate feedback and interaction, not "overautomation". *Philosophical Transaction of the Royal Society of London, 327* 585–593.

Parasuraman, R., T. B. Sheridan, and C. D. Wickens. 2000. "A model for types and levels of human interaction with automation." *IEEE Transactions on Systems, Man, and Cybernetics - Part A: Systems and Humans, 30*, 3 286–297. DOI: 10.1109/3468.844354.

Plant, K. L., and N. A. Stanton. 2016. *Distributed Cognition and Reality: How Pilots and Crews Make Decisions.* London, UK: CRC Press. https://doi.org/10.1201/9781315577647.

Plant, K. L., and N. A. Stanton. 2015. "The process of processing: exploring the validity of Neisser's perceptual cycle model with accounts from critical decision-making in the cockpit." *Ergonomics, 58*, 6 909–923. DOI: 10.1080/00140139.2014.991765.

Plant, K. L., and N. A. Stanton. 2012. "Why did the pilots shut down the wrong engine? Explaining errors in context using Schema Theory and the Perceptual Cycle Model." *Safety Science, 50*, 2 300–315. DOI: 10.1016/j.ssci.2011.09.005.

Rasmussen, J. 1983. "Skills, rules, and knowledge; signals, signs, and symbols, and other distinctions in human performance models." *IEEE Transactions on Systems, Man, and Cybernetics, 13*, 3 257–266. DOI: 10.1109/TSMC.1983.6313160.

Reason, J. 1990. *Human Error.* Cambridge, UK: Cambridge University Press.

Revell, K. M.A., and N. A. Stanton. 2014. "Case studies of mental models in home heat control: Searching for feedback, valve, timer and switch theories." *Applied Ergonomics, 45*, 3 363–378. DOI: 10.1016/j.apergo.2013.05.001.

Revell, K. M.A., and N. A. Stanton. 2018. "Mental model interface design: putting users in control of home heating." *Building Research and Information, 46*, 3 251–271. DOI: 10.1080/09613218.2017.1377518.

Revell, K. M.A., and N. A. Stanton. 2017. "Mind the gap – Deriving a compatible user mental model of the home heating system to encourage sustainable behaviour." *Applied Ergonomics, 57* 48–61. DOI: 10.1016/j.apergo.2016.03.005.

Revell, K. M.A., and N. A. Stanton. 2015. "When energy saving advice leads to more, rather than less, consumption." *International Journal of Sustainable Energy, 36*, 1 1–19. DOI: 10.1080/14786451.2014.999071.

Roadcraft. 2013. *Roadcraft: The Police Driver's Handbook.* London, UK: The Stationery Office.

Rushby, J., J. Crow, and E. Palmer. 1999. "An automated method to detect potential mode confusions." *Gateway to the New Millennium. 18th Digital Avionics Systems Conference. Proceedings.* St Louis, US: IEEE. 4.B.2. DOI: 10.1109/DASC.1999.863725.

SAE J3016. 2016. Taxonomy and Definitions for Terms Related to Driving Automation Systems for On-Road Motor Vehicles. Accessed November 3, 2020. https://www.sae.org/standards/content/j3016_201806/.

Salmon, P. M., G. J.M. Read, N. A. Stanton, and M. G. Lenné. 2013. "The crash at Kerang: Investigating systemic and psychological factors leading to unintentional non-compliance at rail level crossings." *Accident Analysis & Prevention, 50* 1278–1288. DOI: 10.1016/j.aap.2012.09.029.

Salmon, P. M., M. G. Lenne, G. H. Walker, N. A. Stanton, and A. Filtness. 2014. "Exploring schema-driven differences in situation awareness between road users: an on-road study of driver, cyclist and motorcyclist situation awareness." *Ergonomics*, *57*, 2 191–209. DOI: 10.1080/00140139.2013.867077.

Salmon, P. M., N. Goode, A. Spiertz, M. Thomas, E. Grant, and A. Clacy. 2017. "Is it really good to talk? Testing the impact of providing concurrent verbal protocols on driving performance." *Ergonomics*, *60*, 6 770–779. DOI: 10.1080/00140139.2016.1214752.

Sarter, N. B., and D. D. Woods. 1995. "How in the world did we ever get into that mode? Mode error and awareness in supervisory control." *Human Factors: The Journal of the Human Factors and Ergonomics Society*, *37*, 1 5–19. DOI: 10.1518/001872095779049516.

Sarter, N. B., D. D. Woods, and C. E. Billings. 1997. "Automation Surprises." In *Handbook of Human Factors and Ergonomics*, by G. Salvendy, 1926–1943. New York: John Wiley and Sons.

Sheridan, T. B., and R. Parasuraman. 2005. "Human-Automation Interaction." *Reviews of Human Factors and Ergonomics*, *1*, 1 89–129. DOI: 10.1518/155723405783703082.

Stanton, N. A., A. Dunoyer, and A. Leatherland. 2011. "Detection of new in-path targets by drivers using Stop & Go Adaptive Cruise Control." *Applied Ergonomics*, *42*, 4 592–601. DOI: 10.1016/j.apergo.2010.08.016.

Stanton, N. A., and G. H. Walker. 2011. "Exploring the psychological factors involved in the Ladbroke Grove rail accident." *Accident Analysis & Prevention*, *43*, 3 1117–1127. DOI: 10.1016/j.aap.2010.12.020.

Stanton, N. A., and P. Marsden. 1996. "From fly-by-wire to drive-by-wire: Safety implications of automation in vehicles." *Safety Science*, *24* 35–49. DOI: 10.1016/S0925-7535(96)00067-7.

Stanton, N. A., P. M. Salmon, D. Jenkins, and G. Walker. 2009. Human Factors in the Design and Evaluation of Central Control Room Operations. Boca Raton, FL: CRC Press. https://doi.org/10.1201/9781439809921.

Stanton, N. A., P. M. Salmon, G. H. Walker, E. Salas, and P. A. Hancock. 2017. "State-of-science: Situation awareness in individuals, teams and systems." *Ergonomics*, *60*, 4 449–466. DOI: 10.1080/00140139.2017.1278796.

Stanton, N. A., P. M. Salmon, L. A. Rafferty, G. H. Walker, C. Baber, and D. P. Jenkins. 2013. Human Factors Methods: A Practical Guide for Engineering and Design. London, UK: CRC Press. https://doi.org/10.1201/9781315587394.

Stapel, J. C.J., F. A. Mullakkal-Babu, and R. Happee. 2017. "Driver behavior and workload in an on-road automated vehicle." *Proceedings Road Safety & Simulation International Conference 2017*. The Hague, Netherlands. 17–19.

Walker, G. H., N. A. Stanton, and P. M. Salmon. 2016. "Trust in vehicle technology." *International Journal of Vehicle Design*, *70*, 2 157–182. DOI: 10.1504/IJVD.2016.074419.

Waller, S., M. Bradley, I. Hosking, and P.J. Clarkson. 2015. "Making the case for inclusive design." *Applied Ergonomics*, *46*, Part B 297–303. DOI: 10.1016/j.apergo.2013.03.012.

Woods, D. D. 1988. "Coping with complexity: the psychology of human behaviour in complex systems." In *Tasks, Errors, and Mental Models*, by L. P. Goodstein, H. B. Andersen and S. E. Olsen, 128–148. London, UK: Taylor & Francis, Inc.

Woods, D. D., L. J. Johannesen, R. I. Cook, and N. B. Sarter. 1994. *Behind Human Error: Cognitive Systems, Computers, and Hindsight*. Ohio, US: Crew Systems Ergonomics Information Analysis Center.

10 Semi-Automated Driving Has Higher Workload and Is Less Acceptable to Drivers than Manual Vehicles

An On-Road Comparison of Three Contemporary SAE Level 2 Vehicles

Jisun Kim and Kirsten M. A. Revell
University of Southampton

Patrick Langdon
Edinburgh Napier University

Michael Bradley
University of Cambridge

Ioannis Politis
The MathWorks Inc.

*Simon Thompson, Lee Skrypchuk,
and Jim O'Donoghue*
Jaguar Land Rover

Joy Richardson and Neville A. Stanton
University of Southampton

CONTENTS

10.1   INTRODUCTION

Despite the proliferation of SAE Level 2 semi-autonomous vehicles (AVs) (SAE International 2018) currently on the road, drivers' workload and acceptance in a real-world setting have not been fully understood (Eriksson, Banks and Stanton 2017). One of the reasons is that simulators and closed test tracks have widely been chosen to investigate the issues (Heikoop et al. 2019; Endsley 2017; de Winter et al. 2014). This is partly due to the risks associated with non-professional drivers operating prototype vehicles on public roads. However, the simulated environments and test tracks are limited in emulating the complexity of real road conditions (Eriksson, Banks and Stanton 2017).

A number of on-road studies have suggested evidence of reduction in workload associated with autonomous driving compared to manual driving (Stapel, Mullakkal-Babu and Happee 2019), but the opposite effect has also been presented (Banks and Stanton 2016). Therefore, a consensus about workload in the real-world autonomous driving remains to be seen. Moreover, workload seems to vary across complexity in driving environments (Stapel, Mullakkal-Babu and Happee 2019), and drivers' prior experience in automated driving (Solís-Marcos, Ahlström and Kircher 2018). This represents a need for further investigation of drivers' workload in differing on-road settings, and drivers' experience to help develop comprehensive understanding about their workload. Furthermore, a rationale for examining perceived workload is that it could negatively affect acceptance of AVs (Nees 2016). Acceptance has been explored to account for the influencing factors: to identify acceptance of the general public and to understand acceptance after experiencing AVs as drivers, or passengers. However, these attempts hardly created evidence-based insights into acceptance of AVs in comparison to manual driving in a real-world setting. Gaining this understanding would help reveal if barriers to acceptance need to be overcome in order to encourage a widespread adoption of SAE Level 2 semi-AVs (SAE International 2018), over the existing norm to opt for manual vehicles.

Advantages of AVs are well established. For example, they could perform driving with a higher degree of precision in comparison to human drivers whose performance could be negatively influenced by distraction, fatigue, and emotions (Clamann, Aubert and Cummings 2017; Van Brummelen et al. 2018; Singh 2015). For the benefits of SAE Level 2 semi-AVs to be realised (SAE International 2018), negative attitudes of the public need to be overcome to increase uptake. Improving driver acceptance of AVs is a key step towards this goal (Liu et al. 2019), given that acceptance is a major antecedent to drivers' intention to use, or actual use of the system (Venkatesh et al. 2003; Venkatesh and Bala 2008; Venkatesh and Davis 2000; Ghazizadeh, Lee and Boyle 2012).

Workload is the level of effort required from the operator to interact with the system (Moller et al. 2009). It is one of the key criteria to measure driver–AV interaction (de Winter et al. 2014). It is defined as 'the cost incurred by human operators to achieve a specific level of performance', rather than determined by demands of the objective task. Individuals' perception of workload may differ as they perceive the task parameters differently (Hart and Staveland 1988). Drivers' workload in AVs has been explored widely in simulators, as reviewed in de Winter et al.'s (2014) study, as well as in on-road settings (Banks and Stanton 2016; Heikoop et al. 2019; Eriksson, Banks and Stanton 2017; Solís-Marcos, Ahlström and Kircher 2018; Stapel, Mullakkal-Babu and Happee 2019). However, the focus of this study is to investigate workload in the real-world settings; thus, a review of on-road studies will be provided.

Previous studies have shown varying results concerning drivers' perceived workload in automated and manual driving conditions (Stapel, Mullakkal-Babu and Happee 2019; Solís-Marcos, Ahlström and Kircher 2018; Banks and Stanton 2016). Banks and Stanton (2016) report higher perceived workload associated with driving in driver-initiated automated control system than in manual driving. In particular, the automation led to a significant increase in mental and temporal demands, and frustration. The increase in perceived workload could be explained by the additional requirements imposed on drivers to monitor the surroundings, the system's behaviour, and to confirm the vehicle reacted appropriately to ensure safety (Banks and Stanton 2016). In contrast to the Banks and Stanton's (2016) findings, automated driving induced lower perceived workload than manual driving among automation-experienced drivers (Solís-Marcos, Ahlström and Kircher 2018; Stanton, Young and McCaulder 1997; de Winter et al. 2014). Stapel, Mullakkal-Babu, and Happee (2019) found that inexperienced drivers reported workload to be similar to manual driving. Although the actual workload of attentive monitoring was underestimated by the drivers, it could increase objective workload in an automated driving condition. The rationale behind the increase was that partial automation does not expand spare capacity as much as the drivers expect for secondary tasks, or interaction with in-vehicle interfaces. Indeed, the effects of automation can actually reduce attentional capacity (Young and Stanton 2002). Young and Stanton (2007) showed that novice and intermediate drivers rating of workload was lower than that of advanced drivers, who reported similar levels to manual driving. This is presumably because they were monitoring and supervising the automation properly. Furthermore, in this study, the effect of traffic complexity was found to increase perceived and objective workload (Stapel, Mullakkal-Babu and Happee 2019). Low ratings of perceived workload were reported in SAE Level 2 automated driving (SAE International 2018) in a real traffic environment. This may demonstrate the possibility of lower perceived workload associated with automated driving than with manual driving. The low ratings were accompanied by relatively low heart rate and respiratory rate, which were similar to those in resting conditions. However, the level of stress of the participants may have been lower than that of the general public as they were experienced drivers with advanced driver training. Therefore, a need for investigation with less experienced drivers was present (Heikoop et al. 2019). In a similar vein, evidence was found that perceived workload could be low in non-critical control transactions in highly automated driving in both on-road and simulator conditions. This argument was made based on the median values for perceived workload measured through the NASA-Task Load Index (NASA-TLX) in both conditions, which were lower than the scale's halfway point. In addition, they were lower than the average NASA-TLX workload scores reported in de Winter et al.'s (2014) study showing that unweighted mean self-reported workload was highest in manual (43.5%), followed by adaptive cruise control (ACC) (38.6%) and lowest in highly automated driving (22.7%) (Eriksson, Banks and Stanton 2017). Individual aspects of perceived workload in the NASA-TLX were investigated in SEA Level 2 automation for experienced and novice drivers, in which an additional task was involved. The effort scores were significantly lower in automated than in manual driving, and the frustration scores were higher among the novice than among experienced drivers. It was assumed that the additional task as well as task switching induced the drivers' mental demands. Monitoring in automated driving was thought to be more disruptive to performing the additional task than that in manual driving (Solís-Marcos, Ahlström and Kircher 2018).

Acceptance is behavioural responses and attitudes about a product formulated after use (Schade and Schlag 2003; Schuitema, Steg and Forward 2010). Acceptance of AVs has been studied in three main approaches. Firstly, acceptance about autonomous driving technology has been understood based on theories of human behaviour (Rahman et al. 2017). Related theories are the Technology Acceptance Model that posits perceived usefulness and ease of use as determinants of system usage (Davis 1989), and the Unified Theory of Acceptance and Use of Technology, an integrated model formulated across eight relevant models elucidating users' acceptance of new information technologies (Venkatesh et al. 2003). Secondly, public opinion about acceptance of, and willingness to purchase, purchasing AVs was investigated using survey method (Liu et al. 2019; Kyriakidis, Happee, and de Winter 2015; Becker and

Axhausen 2017). Thirdly, acceptance of AVs was examined empirically on the basis of driving tasks in simulation (Rahman et al. 2017), or on-road environments. The focus of this study is to investigate acceptance in a naturalistic environment; therefore, a review of related studies conducted on the road will be provided.

Previous studies presented unchanged or improved acceptance of AVs after experiencing the technology as a driver or a passenger. Drivers' acceptance did not show a significant difference assessed before and after driving a Level 2 AV on the open road. In the study, ACC and lane keeping assist features were mentioned as sources of stress (Stanton and Young, 2005; 2010). In addition, engaging and monitoring lane keeping assist features was seen to be more challenging than ACC (Biondi et al. 2017). Acceptance of a SAE Level 3 AV was enhanced after direct experience of the vehicle as a passenger in the test bed, containing driving scenarios close to the real-world conditions. It was verified through prior and post hoc questionnaires, and the results also confirmed that the participants' trust, perceived usefulness, ease of use, and behaviour intention to use were improved (Xu et al. 2018). Positive attitude towards acceptance was reported among the participants who had physical experience of a Level 4 AV as a passenger in realistic traffic environments on campuses. Acceptance, in conjunction with perceived safety, was verified as a significant antecedent of intention to use. Zoellick et al. (2019) also revealed that positive attitudes such as acceptance, perceived safety, trust, and intention to use were correlated.

10.1.1 RESEARCH GAP AND AIM

Levels of workload associated with autonomous driving seem to vary across driving conditions such as modes, environments, and drivers' experience. Generally, workload tended to be lower in automated than in manual driving, as reviewed in de Winter et al.'s study (2014). However, their review only included workload assessed in Level 1 (ACC) and Level 3 (highly automated driving) autonomous driving mostly in simulation environments. Among the previous studies reported on perceived workload in Level 2 AVs, conflicting results were reported. Subjective workload was higher in semi-automated than in manual mode (Banks and Stanton 2016), and monitoring was assumed to impose higher workload (Solís-Marcos, Ahlström and Kircher 2018). On the contrary, subjective workload was lower in semi-automated than in manual driving (Stapel, Mullakkal-Babu and Happee 2019). Similarly, the likelihood of reduction in workload in semi-automated driving was described (Heikoop et al. 2019; Eriksson, Banks and Stanton 2017). Therefore, little consensus has been reached about the degree of workload in Level 2 AVs, and few explanations have been furnished to account for what possible interactions have resulted in the discrepancies. Even human factors experts are divided on the issues and their relative importance (Kyriakidis et al. 2019). Given that workload can be affected by driving modes (manual and automated), environments (simulator and on-road), and conditions (complex and monotonic), comparing the levels of workload in different driving scenarios seems meaningful. With regard to acceptance, little evidence has been provided about drivers' acceptance based on their direct experience with SAE Level 2 semi-AVs in naturalistic environments (SAE International 2018).

Therefore, this study aims to extend the current knowledge on drivers' workload and acceptance of Level 2 AVs in the real-world settings. This will be achieved through a comparative investigation into drivers' experience of Level 2 AVs on public roads in both manual and semi-autonomous modes for monotonous and complex conditions. Qualitative understanding about drivers' behaviours and thoughts in semi-automated driving will further explain the quantified results. Findings of this study give insights into automotive product manufacturers to enhance AVs that help reduce drivers' workload and improve acceptance.

This study explores two research questions as follows:

1. How will workload and acceptance of SAE Level 2 semi-automated driving compare to manual driving, in naturalistic conditions?
2. How could experience of SAE Level 2 semi-automated driving from a driver's viewpoint explain differences in workload and acceptance compared to manual driving?

10.2   METHOD

This explorative study was designed to investigate drivers' perceived workload and acceptance of SAE Level 2 AVs in various naturalistic conditions. The focus of investigation was placed on the difference between manual and automated driving experience in various conditions (in manual and automated modes, in highway and urban environments, in three AVs). Drivers' workload and acceptance were examined in quantitative and qualitative manners. In order to respond to the first research question, survey technique was applied to quantify drivers' subjective workload and acceptance in manual and automated modes. For data collection, standard questionnaires were used to assess perceived workload and acceptance. The selected scales were the NASA-TLX (Hart and Staveland 1988) and the acceptance of advanced transport telematics scale (Van Der Laan, Heino and De Waard 1997). These scales were chosen because they were used to measure the perceptions extensively in the area of AV research (Forster et al. 2018). The participants were asked to fill in the forms shortly after they finished a driving task in each condition. In order to address the second research question, a think-aloud method was applied to further the understanding about the differences in workload and acceptance between manual and automated driving to identify what could influence drivers' workload and acceptance.

10.2.1   Experiment Design

Experiment conditions were designed as follows. Participants conducted driving tasks: (1) in two driving modes: manual and automated modes; (2) in two different environments: highway and urban environments; and (3) in three different SAE Level 2 AVs (SAE International 2018). The selected routes were located on the M40 around Coventry, and public urban roads in Coventry area, UK. The chosen vehicles were (1) Jaguar I-Pace 2018, (2) Mercedes S350 2014, and (3) Tesla Model S 2016. All three vehicles were equipped with ACC, steering assist, and 'stop and go' feature, which brings the vehicles to a halt when the lead vehicle is slowing down its speed and stopping, and resumes autonomously a few seconds after the lead vehicle begins to move. The Tesla was equipped with an automated lane change assist feature that enabled autonomous lane change with the driver's initiation by indicating (Jaguar Land Rover 2020; Mercedes-Benz 2014; Tesla 2016). Table 10.1 details the 12 study conditions, Table 10.2 summarises the autonomous features of the vehicles used in the study, and Figure 10.1 indicates the route used.

10.2.2 PARTICIPANTS

Eight healthy participants holding a full UK driving licence participated in the study. They were comprised of one female and seven male drivers whose mean age was 44.13 years (SD = 8.41), who had varying levels of knowledge about, and experience in using AV features. They were recruited only within the project team affiliated in the University of Southampton, the University of Cambridge, and Jaguar Land Rover to conform with the company's insurance requirements. The study was approved by the University of Southampton Ethics and Research Governance Office (ERGO number: 49792).

10.2.3 PROCEDURE

Experiment procedures consisted of three sections. Firstly, the recruited participants were given time to familiarise themselves with the experiment by reading a participant information sheet, which included information about required driving tasks, designed routes, and vehicles. They had opportunities to ask questions about the study, and they signed the consent form as an expression of agreement to take part in the study. Secondly, they received related training, including practice sessions about autonomous features, such as ACC and steering assist on the designed route at the Jaguar Land Rover test track offered by the company's safety driver. Thirdly, in the

TABLE 10.1
Study Condition (A Total of 12 Different Conditions)

Environment	Mode	Vehicle	Data
1. Highway	1. Manual	1. I-PACE	Data collected from 8 participants:
		2. Mercedes	1. Standard questionnaires to measure perceived
		3. Tesla	workload (NASA-TLX) and acceptance
	2. Automated	1. I-PACE	(system acceptance scale) after each trial
		2. Mercedes	2. Think-aloud protocol (video and audio)
		3. Tesla	during each trial
2. Urban	1. Manual	1. I-PACE	
		2. Mercedes	
		3. Tesla	
	2. Automated	1. I-PACE	
		2. Mercedes	
		3. Tesla	

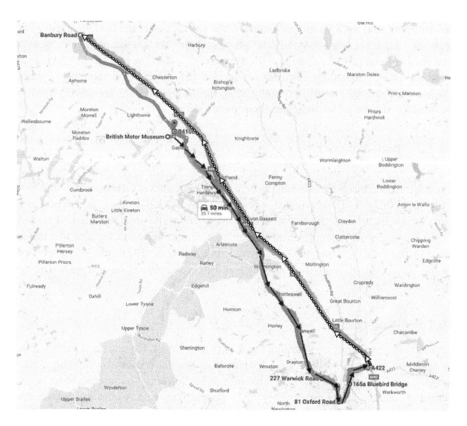

FIGURE 10.1   Map indicating the highway route (dotted line) and urban route (solid line).

experiment, they drove the three vehicles: on the highway and urban roads, in manual and automated modes. The distances and approximate durations were 16.1 miles (approximately 15 min) for the highway and 14.5 miles (approximately 16.6 min) for the urban route. The urban environment included junctions, roundabouts, traffic lights, and pedestrian crossings. The participants were asked to engage autonomous features as much as they could, only when they felt safe to do so. A safety driver, located in the front passenger seat, monitored the drivers' and vehicles' statuses and behaviours, and the traffic condition. They advised or intervened when necessary to ensure the vehicle was safely driven both for the driver and for the other road users.

10.2.4   DATA ANALYSIS

Descriptive and inferential statistics techniques were applied for questionnaire data analysis. Box plots (shown in Figures 10.2 and 10.3) were used to visualise overall tendencies of the data sets showing the median, upper and lower hinges, and extremes (Mcgill, Tukey and Larsen 1978). Additionally, statistical differences between manual and automated modes were estimated using the Wilcoxon signed-rank test. The method was chosen because it allows the analyses of differences between ordinal data

TABLE 10.2

Autonomous Features of the Vehicles

Vehicle	Autonomous Feature	Required Conditions for Activation	Alert Information Modality
Jaguar I-PACE (2019)	ACC	Minimum speed of 20 mph Can be activated at stationary when a vehicle in front is detected Operated by driver's ACC button push	Visual only when activated
	Steering assist	Lane markings of both sides detected Operated autonomously when the conditions are met	Visual only when activated
	Stop and Go	The vehicle in front gradually stops and resumes while it is detected	N/A
Mercedes S350 (2014)	ACC	Minimum speed of 20 mph Can be activated at stationary when a vehicle in front is detected Operated by driver's ACC stalk control	
	Steering Assist	Lane markings of both sides detected Operated autonomously when the conditions are met	Visual only
	Stop and Go	The vehicle in front gradually stops and resumes while it is detected	N/A
Tesla Model S (2016)	ACC	Minimum speed of 5 mph Can be activated at stationary when a vehicle in front is detected Operated by driver's ACC stalk control	Visual and auditory when de/activated
	Steering Assist	Lane markings of both sides detected for initial activation Once activated, it can be maintained if one of the types information is received (either one of the lane markings, the vehicle in front) Operated by driver's ACC stalk control when available	Visual and auditory when de/activated
	Stop and Go	A vehicle in front gradually stops and resumes while it is detected	N/A
	Auto lane change	Initiated by drivers' indicating, vehicle makes a lane change when no vehicle or obstacles are in the target lane	N/A

sets collected from a small sample (Field 2017). For workload data, the NASA-TLX unweighted raw scores were taken into analysis because raw and weighted workload scores have shown strong correlations in a number of studies (Grier 2016). Moreover, comparisons between raw-TLX and the original version revealed that raw-TLX was either more sensitive, or less sensitive, or equally sensitive. Thus, either can be applicable (Hart 2006). Representative scores for acceptance were generated according to the guidance of the scale (Van Der Laan, Heino and De Waard 1997).

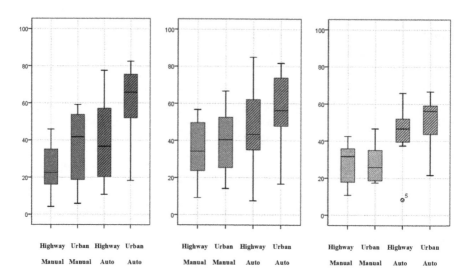

FIGURE 10.2 **Left to right:** (a) workload in the I-PACE, (b) workload in the Mercedes, (c) workload in the Tesla. [*Note*: Y-axis=score, ▨ : 25%~75%, ——: Median line, ◓: Outlier].

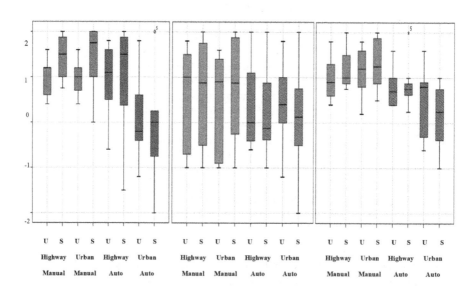

FIGURE 10.3 Left to right: (a) acceptance (usefulness and satisfying) in the I-PACE, (b) acceptance (usefulness and satisfying) in the Mercedes, (c) acceptance (usefulness and satisfying) in the Tesla. [*Note*: Y-axis=score, ▨ : 25%~75%, ——: Median line, ◓, ★: Outlier, U = Usefulness subscore, S = Satisfying subscore].

Secondly, qualitative data was analysed for relevant instances that might have had an effect on drivers' subjective workload, and acceptance of automated driving in each condition (in highway and urban environments, in three vehicles) will be described. Situations that required prompt actions from the drivers, including a takeover during supervised automated driving, were selectively observed because these situations could have led to an increase in workload. This is because the drivers needed to conduct necessary driving tasks that consisted of multiple activities within a short period of time. This process required them to allocate mental and physical resources which were limited to meet the demand (Stapel, Mullakkal-Babu and Happee 2019; Hart and Staveland 1988). In addition, takeover situations during automation deserve attention because they are linked to driving safety (Choi et al. 2020). Furthermore, monitoring and supervision of automation, and intermittent responses to takeover requests were not required in manual driving. For the analysis, examples of participants who rated their automated driving experience less favourably (represented by highest or second highest workload, and lowest acceptance in comparison to the scores for the other vehicles rated by the same driver to avoid participants' different tendencies to rate scores, e.g. extreme response style) were chosen. Autonomous features and interfaces varied across the vehicles and will be described individually.

10.3   RESULTS AND DISCUSSIONS

This section describes data from the questionnaire survey and think-aloud protocol to answer the two research questions established in Section 10.1.1. Firstly, comparisons of overall trends in workload, and acceptance data in the defined conditions (see Table 10.1), will be provided. Additionally, relevant instances from SAE Level 2 semi-automated driving (SAE International 2018) will be assessed in order to promote our understanding about an increase or a decrease in perceived workload and acceptance of the participants. As the focus of this study is to offer insights about drivers' interaction with SAE Level 2 semi-AVs (SAE International 2018), the results will be presented centring on the driving task.

10.3.1   COMPARISONS BETWEEN MANUAL AND AUTOMATED DRIVING

Firstly, workload (unweighted means) tended to be higher in automated mode in all the three vehicles considering the medians, the range of quartiles, and outliers (see Figure 10.2). This excluded the case of participant 5 who rated a lower score for the automated mode in the Tesla, highway environment. Statistical differences in workload between automated and manual modes were detected as follows. In the highway environment, they were found to be significantly higher in automated mode in the I-Pace ($p = 0.021$; $Z = -2.313$) and the Tesla ($p = 0.017$; $Z = 2.380$) only, but not the Mercedes. In the urban environment, workload was identified to be statistically higher than in manual mode in all vehicles: I-Pace ($p = 0.012$; $Z = -2.521$), Mercedes ($p = 0.025$; $Z = -2.240$), and Tesla ($p = 0.017$; $Z = -2.383$).

Higher workload in automated driving may be explained as follows. In this study, the drivers were required to pay attention to the traffic conditions at all times and put their hands close to the steering wheel in automated mode in case they needed to intervene. In this condition, they needed to monitor the outside environment and information from the cluster display, and to understand the vehicle's mental model, automation status, and behaviour to see if the vehicle reacted appropriately, otherwise to take control back. This process seemed to increase workload, especially mental demand, which is consistent with Banks and Stanton's (2016), as well as Stapel, Mullakkal-Babu and Happee's (2019) study results. Interpretations of the autonomous status, and the behaviour in relation to the traffic situation, and the decision-making based on the information are not needed in manual driving in which automation does not replace the relevant roles (Koo et al. 2015). In the process, having an understanding about the vehicles' mental model was essential; however, it usually takes time for the driver to develop it (Banks and Stanton 2016; Beggiato et al. 2015). Therefore, from the drivers' point of view, the interpretation as well as development of mental model needed to be coordinated while driving in automated mode. The identified higher workload in the automated driving seems dissimilar to previous findings that workload was mostly lower in automated driving as described in Heikoop et al.'s (2019) and Eriksson, Banks, and Stanton's (2016) studies, and also as reviewed in de Winter et al.'s study (2014). The disparity may be clarified that most studies reviewed in de Winter et al.'s paper adopted simulation as a method that cannot reflect the complex real-world conditions (Endsley 2017). Thus, the drivers in the simulation did not need to react to changes in road conditions as frequently as the complex actual road environment. Consequently, in this study, the workload may have been heightened, especially due to increased mental demand whose statistical differences were detected in more cases than other dimensions of workload (I-PACE: $p=0.011$, $Z=-2.533$; Tesla: $p=0.017$, $Z=-2.385$ in highway environment; I-PACE: $p=0.011$, $Z=-2.533$; Mercedes: $p=0.018$, $Z=-2.371$; Tesla: $p=0.018$, $Z=-2.371$ in urban environment, the Wilcoxon signed-rank test). In addition, the disparity could be interpreted as greater risk perception (Carsten and Jamson 2011) and actual physical risk of accidents (Heikoop et al. 2019). Hence, the drivers of this study might have been more sensitive to structures on the road (e.g. kerb and vehicles running on the opposite side of the road) while monitoring. Moreover, higher mental workload in automated mode may present that cognitive and/or perceptual activity (e.g. thinking, deciding, calculating, remembering, looking, searching) was required more than in manual driving (Hart and Staveland 1988). This may explain the substantial task demand from monitoring automation as described in Stapel, Mullakkal-Babu, and Happee's (2019) study. Furthermore, the disparity in workload between manual and automated driving may have resulted from significant differences in frustration (I-PACE: $p=0.028$, $Z=-2.200$; Mercedes: $p=0.041$, $Z=-2.043$; Tesla: $p=0.027$, $Z=-2.205$ in highway environment; I-PACE: $p=0.017$, $Z=-2.395$; Mercedes: $p=0.012$, $Z=-2.524$; Tesla: $p=0.035$, $Z=-2.111$ in urban environment, the Wilcoxon signed-rank test), which was increased in automated mode. Frustration represents 'how insecure, discouraged, irritated, and annoyed versus secure, gratified, content, and complacent', the participant felt (Hart and Stavelend 1988). This may demonstrate that the participants might have felt less secure or content about autonomous driving than about manual driving. This may highlight the gap between the vehicles operation in relation to diminished control compared to manual driving in which lateral and longitudinal control was under drivers' control (Koo et al. 2015).

Secondly, acceptance explained by usefulness and satisfying tended to be lower in automated driving than in manual driving in all three vehicles, considering the medians, the range of quartiles, and outliers (see Figure 10.3). This excluded (1) the I-Pace highway satisfying score and (2) the Mercedes urban usefulness score. However, the exclusions did not seem to considerably affect the general tendencies described earlier. For the case of the I-PACE, although the median value was higher in automated mode, the minimum and the first quartile were positioned much lower than those in the highway setting. For the case of the Mercedes, the maximum score was greater in automated mode, but the third quartile, median, the first quartile, and the minimum scores were rated lower in automated mode. Statistical differences in acceptance between automated and manual modes were seen in the urban environment in the I-PACE (usefulness: $p = 0.035$, $Z = -2.106$; satisfying: $p = 0.017$, $Z = -2.388$) and the Tesla (usefulness: $p = 0.034$, $Z = -2.117$; satisfying: $p = 0.017$; $Z = -2.388$), but not the Mercedes.

The lower level of acceptance in automated driving may be explained by the monitoring and supervising tasks, and intermittent interventions that limited the usefulness of the automation features. They may have had little opportunity to experience automated driving as more useful, effective, or assisting than manual driving. In manual driving, e.g., actions could be taken based on the outcome of their decision-making without observation of the vehicle's autonomous behaviour.

10.3.2 The Effects of Complexity in the Driving Condition

The discrepancies in workload between manual and automated driving seem to be greater in the urban environment (see Figures 10.2 and 10.3). This could be interpreted that the drivers were engaged in more complex situations, including roundabouts, pedestrians, traffic lights, and the road infrastructure in the urban environment (Stapel, Mullakkal-Babu and Happee 2019). The current generation of Level 2 AVs were likely to reach the limitations due to road complexity, including sharp bends. Regarding longitudinal control (ACC and Stop & Go features), the drivers needed to slow down and stop manually by pressing the brake pedal when necessary. This led to cancellation of the autonomous features; therefore, the drivers needed to activate ACC again when they wanted to. Furthermore, regarding lateral control (Steering Assist), if the vehicle did not adjust the steering angle appropriately when relevant lane markings were missing or they were not detected, the driver had to intervene at the right time. Therefore, in the urban conditions, the drivers' active monitoring of the road ahead, the automation status by checking the interfaces, and the vehicle's behaviour were required to decide when and how to intervene. The processes associated with the disengagement and re-engagement seemed to be connected to the increase in workload. These actions could be interpreted that the driver attempted to maintain and regain the sufficient level of situation awareness to close longitudinal and lateral loops (de Winter et al. 2014; Endsley 2017). The higher workload reported in this study is consistent with the Stapel, Mullakkal-Babu, and Happee's (2019) findings that workload increased in complex road conditions. Also, participants did not seem to find the system more useful or satisfying to conduct autonomous driving in the urban environment than in the highway.

10.3.3   THE EFFECTS OF DRIVERS' PRIOR EXPERIENCE

There are outliers in the case of workload in the Tesla, highway automated driving (see Figure 10.2), and acceptance in the cases of the I-PACE, urban automated, and Tesla highway automated driving. Participant 5 was not a daily driver of the vehicles under test; however, he had more expert experience with the vehicles. Therefore, an understanding and expectation of the system's actions, and decision-making would become more natural than other drivers (Thompson, Personal communication, 2019). This enhanced ability to maintain situation awareness may have led to a lower degree of workload. Moreover, the participant's acceptance scores were higher than those of other participants in all conditions. This may be explained that experienced drivers could build a more appropriate mental model of the system's limitations and functionality needed for safe usage of AVs than novice drivers (Stapel, Mullakkal-Babu and Happee 2019). Thus, these outliers could support the effects of learning that mental model of vehicle automation, and in turn, acceptance, could be enhanced through the process.

10.3.4   QUALITATIVE INVESTIGATION OF INSTANCES WHICH MAY HAVE INFLUENCED DRIVERS' WORKLOAD AND ACCEPTANCE IN AUTOMATED DRIVING

Accounts will be given to help enhance designing driver–AV interaction by understanding the differences in workload and acceptance between automated and manual driving. Verbal descriptions of participants with highest/lowest values of these results were investigated for each vehicle type, and the situations in highway and urban environments were considered with reference to the existing literature.

10.3.5   CONSIDERATIONS FOR DESIGNING DRIVER–AUTONOMOUS VEHICLE INTERACTION IN HIGHWAY ENVIRONMENT

Regarding the I-PACE, participant 2 marked the highest score on workload and the lowest score on acceptance (usefulness and satisfying) as compared with scores they rated for other conditions (see Figures 10.2 and 10.3). This may be explained by a number of instances when the vehicle required the driver to place their hands on the wheel in response to a hands-on-wheel warning (Jaguar Land Rover 2019) while monitoring. For example, the vehicle in front was slowing down its speed by braking, and the participant noticed it. At the same time, hands-on-wheel alert was given while they were anticipating the host vehicle to decrease its speed in response to the vehicle in front. They needed to place their hands on the steering wheel and wiggle to show full attentiveness so the alert could be deactivated. There were two instances of this during the run. Additionally, the driver was trying to make a lane change to leave the highway. Shortly before one mile to the exit, a hands-on-wheel alert was issued.

They had to react to the alert while they were actively making preparations for exiting. Responding to the alert while attentively monitoring the traffic, and making a decision whether to intervene or not, may have increased workload. This is because the related situations required mental (looking/searching, thinking, and deciding) and physical efforts. In addition, the driver could have experienced time pressure (Hart and Staveland 1988).

Concerning the Mercedes, participant 1 rated the highest score on workload and the lowest score on acceptance (usefulness and satisfying) in comparison to the scores they marked on other vehicles (see Figures 10.2 and 10.3). This may have resulted from the instances that they could not easily check the status of steering assist due to the reflection of direct sunlight which was mentioned three times during the drive. Consequently, they were not sure when the vehicle had stopped steering and they were required to put their hands on the wheel accordingly. This case could represent an issue regarding interface design and driver–AV communication. The vehicle only provided visual information for the steering assist status (an icon toggling between green and grey according to the status of steering assist) without auxiliary warning signal, same as the majority of SAE Level 2 AVs. As a result, when the information on the cluster display was not visible, the driver could not acquire the relevant information for the decision-making for the following action.

Additionally, participant 1 disengaged automation mistakenly by pulling down the ACC stalk when intending to operate the indicator. The disengagement of the automation could have led to a serious consequence if the driver had not taken control back immediately. This layout does not seem to facilitate an efficient use of both functions, as it seemed to require a careful selection of the levers according to their intention. This arrangement was not effective, especially when the drivers' cognitive resources were allocated to monitoring tasks, e.g. looking at the road ahead, and the system concurrently.

With regard to the Tesla, participant 4 rated the highest score on workload and the lowest score on acceptance (satisfying) compared to the scores they rated on other conditions (see Figures 10.2 and 10.3). This may have been attributed to the instances as follows. They addressed confusion about using the cruise control lever when they were trying to decrease the ACC speed but they instead turned the function off, or turned the indicator on by mistake. Participant 4 also mentioned that they did not trust the automation when the car started to approach the right lane and there was a car behind in the target lane, while attempting to change lane using the auto lane change function. This may be analysed that the sensors for blind spot monitoring were not able to detect the car behind. As a result, they had to cancel the lane change function to stay in the current lane. The driver expressed that they did not feel confident about the system when the car did not slow down its speed as they expected when there was a car cutting in front. Consequently, they reduced the speed manually by pressing the brake pedal to keep a safe distance. It could be defined as a technical limitation that the related sensors did not detect the vehicle in front in time, or the driver's sensitiveness. Furthermore, they identified that the car accelerated abruptly in ACC when they were exiting the highway, because the targeted vehicle used for the ACC was lost due to the lane change. However, they did not know why the vehicle behaved in the way. An insufficient understanding about the ACC feature could lead to a surprise to the driver in the situation (Banks and Stanton 2016; Stanton, Dunoyer and Leatherland 2011; Sarter, Woods and Billings 1997). When there is no car detected in front, AVs in ACC could speed up when the current speed is slower than the ACC set speed. The maximum ACC speed was marked on the top of the cluster display, but it could not be easily remembered by the driver unless trained well especially when focused on other tasks. Relevant training seems helpful as the benefits were described that trained drivers had a better perceived understanding about the system and confidence in the usage of it than untrained drivers. In particular, the trained ones were able to achieve a more instinctive knowledge about when the system would work or not (Krampell, Solís-Marcos and Hjälmdahl 2020).

10.3.6 CONSIDERATIONS FOR DESIGNING DRIVER–AUTONOMOUS VEHICLE INTERACTION IN URBAN ENVIRONMENT

SAE Level 2 semi-automated driving experience in urban environments instances described in this section includes situations in which SAE Level 2 AVs currently on the market could reach their limits due to complexity in road conditions in the environments. Moreover, the current autonomous systems are developed only for motorway use. However, it seems beneficial to pay attention to drivers' interaction with the vehicles in complex scenarios to consider for future development of the technology.

Regarding the I-PACE, participant 2 rated the highest score on workload and the lowest score on acceptance (usefulness and satisfying) compared to the scores they rated for other conditions (see Figures 10.2 and 10.3). The driver stated that lane markings were not detected in time, four times, and they had to manually steer during these instances. The cases present that lane detection did not offer a stable support presenting the limitation of steering assist of the current generation of SAE Level 2 AVs. This technical instability seemed to lead to the circumstances that required the driver's cognitive and perceptual activities. However, dropout of lane detection is common, and it is not perceived as risky (Dikmen and Burns 2017). During the instances, the driver had to keep monitoring the road ahead, and the interface to check the steering assist status to gather the relevant information to control lateral position. On the basis of the information, the driver made a decision to take control back when the steering assist was not active. This is advised by SAE International; drivers of Level 2 AVs must supervise the automation constantly and be ready to intervene whenever needed to ensure safety (SAE International 2018). As with most SAE Level 2 AVs, the vehicle did not offer auxiliary warning signals (e.g. auditory or tactile) when the steering assist changed its status. Without auditory warning signals, the driver had to take their eyes off the road and look down the interface to check the status regularly. Auditory or tactile signals may be helpful to alert the driver in similar cases when they are fixing their gaze at the road ahead (Eriksson et al. 2019).

Concerning the Mercedes, participant 6 scored the highest on workload and the lowest on acceptance (usefulness, satisfying) compared to the scores rated for other conditions (see Figures 10.2 and 10.3). This may be explained by certain instances as follows. The vehicle zoomed off to accelerate up to the ACC set speed while there was no vehicle in front. It was after stopping at red light while passing a junction. Hence, the driver had to slow down quickly. The driver had the knowledge that the car could speed up if the set speed were higher than the current speed in ACC; however, they were not aware of the speed at the moment the vehicle was accelerating. They recurred the last stored speed once the vehicle was accelerating abruptly. Although the surge of the vehicle was a normal behaviour, the driver did not feel comfortable. Moreover, the vehicle came to a halt autonomously following the behaviour of the vehicle in front; however, it did not resume itself when traffic got moving again (relating to Stop & Go). In contrast, it came to a standstill itself, and resumed autonomously following the traffic. These inconsistent behaviours of the vehicle caused confusion to the driver. Messages showing required actions would be helpful for the driver to decide what to do in similar situations. Additionally, the driver noticed that the steering assist went off by looking at the icon, which had turned grey without auditory warning. The lack of warning and difficulty in distinguishing the icon mode was problematic for the participant during the urban drive. Supporting text may deliver clearer messages, especially for drivers with colour weakness (Jenny and Kelso 2007).

Regarding the Tesla, participant 8 measured the second highest workload and the lowest acceptance (usefulness, satisfying) compared to the scores marked on other vehicles (see Figures 10.2 and 10.3). This may have been attributed by instances as follows. The driver addressed that the targeted vehicle in front, used for steering control, was lost, and the driver had to intervene by steering abruptly. This kind of instance was recorded twice. Drivers of AVs need to be aware that this instance could occur, and to check the status in windy road, and to be ready to takeover. However, this process could increase mental demand induced by the effort to remember, look, and decide (Hart and Staveland 1988). Additionally, the driver tried to engage steering assist; however, it was not successful because the system was not available. The driver was not aware that they could not activate steering assist while the system was not ready, although the relevant information (steering assist not available) was provided on the cluster display. Nevertheless, steering assist is generally available on straight roads with clear lane markings, such as highway environments. Further, when steering assist was activated, it was cancelled by the driver's manual steering input, but they did not understand why it was deactivated. A lack of understanding about conditions for engaging automation was reported. Relevant driver training about when and where relevant information is provided on the interface displays, and conditions for activation/deactivation of autonomous features.

As the current SAE Level 2 AVs have been developed only for highway use, a number of instances in the urban environment were identified due to its complexity in traffic situation, or road structure and furniture. The complexity seemed to lead to limitations of automation more frequently in the environment that required drivers' manual interventions. Nevertheless, it is meaningful to look at the driving experience in urban environment to further develop the technology that can drive the vehicle in a wider range of scenarios.

10.3.7 Recommendations for Designing Driver–Autonomous Vehicle Interaction

The following recommendations are made to improve driver experiences with SAE Level 2 AVs.

Firstly, mode awareness (Revell et al. 2020; Stanton, Dunoyer and Leatherland 2011) was addressed, including engagement and disengagement, which might relate to issues of situation awareness. In order for drivers to use the system appropriately, there were various things to remember, including conditions for activation and deactivation of ACC and steering assist, and limitations of the automation. Thus, they had to constantly pay attention to the road and the system to intervene when necessary (SAE International 2018). Therefore, it seems that clear presentation about the automation status, including a significant information for longitudinal and lateral control, is needed, such as the lead vehicle, lane markings, and the ACC speed (Revell et al. 2018). This could also help reduce mode confusion which was reported as one of the issues of partially automated driving (Banks et al. 2018). Provision of multimodal information for changes in automation status (such as a mixture of visual and auditory) seems more effective especially for takeover requests (Yoon, Kim and Ji 2019). Furthermore, the addition of vibrotactile stimulus could be considered because it was confirmed to shorten steer-touch times (Petermeijer et al. 2017). For example, it would be helpful to convey a takeover request while drivers are fixing their gaze at the road ahead and are not able to check the cluster information easily.

Secondly, an issue relevant to manoeuvrability was observed that may be linked to usability. The incorrect usage of the ACC and indicating levers was addressed in the cases of the Mercedes and Tesla with participants 1 and 4. The positions of the levers for two different features mattered: one for longitudinal and the other for lateral control. This seemed to require a careful selection of the lever according to their intentions (Revell et al. 2020). This arrangement was not effective especially when the drivers' cognitive resources were allocated to monitoring tasks, e.g. looking at the road ahead, and the system concurrently. Therefore, it seems reasonable to put the ACC control, e.g., in a different place, or in a different shape. This could in turn enable intuitive usage as buttons with different functions located close to each other could hinder ease of use, as observed in the cases of the Mercedes and Tesla (Clarkson 2008).

Thirdly, issues related to drivers' insufficient knowledge were identified, which may be associated with driver–AV communication (Kazi et al. 2007). The cases of participants 4 and 6 showed that a certain level of knowledge of an autonomous feature (ACC) was essential for comfortable and safe autonomous driving. It needs to be remembered that AVs can accelerate and decelerate quickly to match the ACC set speed. Thus, the drivers need to be ready either to check the speed, knowing where to look, or to adjust the speed accordingly (Stanton and Young 2005). Inappropriate usage for activation led to confusion of participant 8. However, this could have been avoided if the driver had been able to build knowledge on appropriate usage considering the vehicles' automation status which showed its availability. Driver training may be effective for drivers to make the most of autonomous features (Stanton et al. 2007), and it was suggested as a useful tool to develop drivers' mental model (Krampell, Solís-Marcos and Hjälmdahl 2020). Even though essential information was provided on the interface displays, it may not have been acquired as effectively as it was supposed to be from a driver's point of view. Drivers' information behaviour needs to be considered, including information seeking, retrieval, and usage (Wilson 2000) for the training.

In conclusion, these problems seem to be mainly related to maintaining the appropriate level of situation awareness, and understanding of the vehicle's mental model, and the system that could assist the processes, apart from the issues caused by technical limitations. Considering the challenges of collaboration between the human and automation in relation to the handover of authority between the human and the system in the Level 2 autonomy (SAE International 2018), the driver's role seems critical based on the observations made in the naturalistic setting. In the current system, it seems that the driver needs to identify when to give authority to the vehicle, and when to take it back, and even when the vehicle has the authority, the driver still had to monitor the system as well as the environment. If the vehicle requires the human intervention (such as the Level 2 and 3 systems), the interface should be able to help reduce the workload by providing relevant information to keep them in the loop. Further, this may lead to enhancement of perceived usefulness and satisfaction about the use of the vehicle.

10.3.8 OVERALL SUMMARY

This study was conducted to assess drivers' perceptual responses to naturalistic driving of SAE Level 2 AVs (SAE International 2018), and to identify events that may have affected the reactions. Drivers conducted automated and manual driving tasks in highway and urban environments, and in three different SAE Level 2 AVs: I-PACE, Mercedes, and Tesla. Interpretations were synthesised based on comparisons of subjective workload and acceptance among the driving modes, and environments. The main findings were the drivers' workload was higher and acceptance was lower in automated than in manual driving both in highway and in urban environments, apart from a few exceptions.

Higher workload and lower acceptance in automated mode may be construed as follows. Firstly, the drivers were required to pay attention to the traffic conditions and be ready to intervene at all times in automated mode. Secondly, they needed to observe the vehicle's behaviour to see whether it reacted appropriately in relation to the surroundings, if not to intervene. This process led to immediate actions to ensure safety. Once automation was deactivated, drivers needed to think about conditions required to be met to put the vehicle back into autonomous mode successfully. The processes for monitoring and supervising may have led to an increase in workload, especially frustration and mental demand compared to manual driving. Further, they were not perceived as useful and satisfying as much as manual driving.

Some instances that could have influenced the perceptions were described. They seemed to be related to the issues about awareness of the mode (manual or automated), automation status (activation and the status of ACC and steering assist: targeted vehicles and lane markings), and mental model (vehicles' behaviour and the intention behind it). The more the systems relied on the driver's memory, the more the system could be perceived to be demanding. Moreover, if the drivers were not aware of necessary information, the events could have been seen as automation surprise (Stanton, Dunoyer and Leatherland 2011; Sarter, Woods and Billings 1997). It is suggested to enable a clear communication between AV and the driver in relation to mode limitation. The interface needs to be designed to better support to shape the driver's mental model to narrow the gap between the driver's expectation and the vehicle's behaviour and the information conveyed through the interfaces.

10.4 CONCLUSIONS

This study has provided empirical evidence collected in naturalistic settings that workload could be higher in semi-automated driving compared to manual driving. It is suggested that the greater mental demand in semi-automated driving tasks is likely to be due to the situation awareness requirements, such as monitoring and supervising semi-AVs. Additionally, frustration was reported to be higher in semi-automated driving. These findings lend support to the more recent literature on workload reported by drivers of semi-AVs. The research also offers explanations for the higher workload. Also, in terms of practical contributions, the findings from this study offer possible design strategies for automotive manufacturers to help reduce workload, and ultimately enhance acceptance of vehicle automation.

ACKNOWLEDGEMENTS

This work was supported by Jaguar Land Rover and the UK-EPSRC Grant EP/-N011899/1 as part of the jointly funded Towards Autonomy: Smart and Connected Control (TASCC) Programme.

REFERENCES

Banks, V. A., and N. A. Stanton. 2016. "Keep the driver in control: Automating automobiles of the future." *Applied Ergonomics*, *53*, Part B 389–395. DOI: 10.1016/j. apergo.2015.06.020.

Banks, V. A., A. Eriksson, J. O'Donoghue, and N. A. Stanton. 2018. "Is partially automated driving a bad idea? Observations from an on-road study." *Applied Ergonomics*, *68* 138–145. DOI: 10.1016/j.apergo.2017.11.010.

Becker, F., and K. W. Axhausen. 2017. "Literature review on surveys investigating the acceptance of automated vehicles." *Transportation*, *44* 1293–1305. DOI: 10.1007/s11116-017-9808-9.

Beggiato, M., M. Pereira, T. Petzoldt, and J. Krems. 2015. "Learning and development of trust, acceptance and the mental model of ACC. A longitudinal on-road study." *Transportation Research Part F: Traffic Psychology and Behaviour*, *35* 75–84. DOI: 10.1016/j.trf.2015.10.005.

Biondi, F., R. Goethe, J. Cooper, and D. Strayer. 2017. "Partial-autonomous Frenzy: Driving a Level-2 Vehicle on the Open Road." In Engineering Psychology and Cognitive Ergonomics: Cognition and Design, by D. Harris EPCE 2017. Lecture Notes in Computer Science, vol *10276*. Cham, Switzerland: Springer. 329–338. https://doi. org/10.1007/978-3-319-58475-1_25.

Carsten, O., and A. H. Jamson. 2011. "Driving simulators as research tools in traffic psychology." In Handbook of Traffic Psychology, by B. E. Porter, 87–96. Elsevier Inc. https://-doi.org/10.1016/B978-0-12-381984-0.10007-4.

Choi, D., T. Sato, T. Ando, T. Abe, M. Akamatsu, and S. Kitazaki. 2020. "Effects of cognitive and visual loads on driving performance after take-over request (TOR) in automated driving." *Applied Ergonomics*, *85* 103074. DOI: 10.1016/j.apergo.2020.103074.

Clamann, M., M. Aubert, and M. Cummings. 2017. "Evaluation of Vehicle-to-Pedestrian Communication Displays for Autonomous Vehicles." *96th Annual Transportation Research Board Meeting.* Washington DC, USA.

Clarkson, P.J. 2008. "Human capability and product design." In Product Experience, by H. N.J. Schifferstein and P. Hekkert, 165–198. https://doi.org/-10.1016/B978-008045089-6.50009-5. Elsevier Ltd.

Davis, F. D. 1989. "Perceived usefulness, perceived ease of use, and user acceptance of information technology." *MIS Quarterly, 13,* 3 319–340. DOI: 10.2307/249008.

de Winter, J. C.F., R. Happee, M. H. Martens, and N. A. Stanton. 2014. "Effects of adaptive cruise control and highly automated driving on workload and situation awareness: A review of the empirical evidence." *Transportation Research Part F: Traffic Psychology and Behaviour, 27,* Part B 196–217. DOI: 10.1016/j.trf.2014.06.016.

Dikmen, M., and C. Burns. 2017. "Trust in autonomous vehicles: The case of Tesla Autopilot and Summon." *2017 IEEE International Conference on Systems, Man, and Cybernetics (SMC).* Banff, Canada: IEEE. 1093–1098. DOI: 10.1109/SMC.2017.8122757.

Endsley, M. R. 2017. "Autonomous driving systems: A preliminary naturalistic study of the tesla model S." *Journal of Cognitive Engineering and Decision Making* 225–238. DOI: 10.1177/1555343417695197.

Eriksson, A., V. A. Banks, and N. A. Stanton. 2017. "Transition to manual: Comparing simulator with on-road control transitions." *Accident Analysis & Prevention, 102* 227–234. DOI: 10.1016/j.aap.2017.03.011.

Eriksson, A., S. M. Petermeijer, M. Zimmermann, J.C.F. de Winter, K. J. Bengler, and N. A. Stanton. 2019. "Rolling out the red (and green) carpet: Supporting driver decision making in automation-to-manual transitions." *IEEE Transactions on Human-Machine Systems, 49,* 1 20–31. DOI: 10.1109/THMS.2018.2883862.

Field, A. 2017. Discovering Statistics Using IBM SPSS Statistics, 5th edition. Thousand Oaks, CA: SAGE Publications.

Forster, Y., S. Hergeth, F. Naujoks, and J. F. Krems. 2018. "How usability can save the day - Methodological considerations for making automated driving a success story." *AutomotiveUI - Proceedings of the 10th International Conference on Automotive User Interfaces and Interactive Vehicular Applications.* Toronto, Canada: Association for Computing Machinery. 278–290. https://doi.org/10.1145/3239060.3239076.

Ghazizadeh, M., J. D. Lee, and L. N. Boyle. 2012. "Extending the technology acceptance model to assess automation." *Cognition, Technology & Work* 39–49. DOI: 10.1007/s10111-011-0194-3.

Grier, R. A. 2016. "How High is High? A Meta-Analysis of NASA-TLX Global Workload Scores." *Proceedings of the Human Factors and Ergonomics Society Annual Meeting.* Los Angeles, CA, USA: SAGE Publications. 1727–1731. https://doi.org/10.1177/1541931215591373.

Hart, S. G. 2006. "Nasa-Task Load Index (NASA-TLX); 20 Years Later." *Proceedings of the Human Factors and Ergonomics Society Annual Meeting* 904–908. https://doi.org/10.1177/154193120605000909.

Hart, S. G., and L. E. Staveland. 1988. "Development of NASA-TLX (task load index): Results of empirical and theoretical research." *Advances in Psychology, 52* 139–183. DOI: 10.1016/S0166-4115(08)62386-9.

Heikoop, D. D., J. C.F. de Winter, B. van Arem, and N. A. Stanton. 2019. "Acclimatizing to automation: Driver workload and stress during partially automated car following in real traffic." *Transportation Research Part F: Traffic Psychology and Behaviour, 65* 503–517. DOI: 10.1016/j.trf.2019.07.024.

JaguarLandRover.2019."All-ElectricJaguarI-Pace-TheArtofPerformance."AccessedNovember 4, 2020. https://media.jaguar.com/2018/jaguar-i-pace-art-electric-performance.

Jaguar Land Rover. 2020. *Jaguar Owner Information.* Accessed April 10, 2020. https://www.ownerinfo.jaguar.com/model/4K/2019/document/29937_en_GBR.

Jenny, B., and N. V. Kelso. 2007. "Color design for the color vision impaired." *Cartographic Perspectives* 61–67. DOI: 10.14714/CP58.270.

Kazi, T. A., N. A. Stanton, G. H. Walker, and M. S. Young. 2007. "Designer driving: drivers' conceptual models and level of trust in adaptive cruise control." *International Journal of Vehicle Design*, 45, 3 339–360. DOI: 10.1504/IJVD.2007.014909.

Koo, J., J. Kwac, W. Ju, M. Steinert, L. Leifer, and C. Nass. 2015. "Why did my car just do that? Explaining semi-autonomous driving actions to improve driver understanding, trust, and performance." *International Journal on Interactive Design and Manufacturing*, 9 269–275. DOI: 10.1007/s12008-014-0227-2.

Krampell, M., I. Solís-Marcos, and M. Hjälmdahl. 2020. "Driving automation state-of-mind: Using training to instigate rapid mental model development." *Applied Ergonomics*, 83 102986. DOI: 10.1016/j.apergo.2019.102986.

Kyriakidis, M., R. Happee, and J.C.F. de Winter. 2015. "Public opinion on automated driving: Results of an international questionnaire among 5000 respondents." *Transportation Research Part F: Traffic Psychology and Behaviour* 127–140. DOI: 10.1016/j.trf.2015.04.014.

Kyriakidis, M., J. C.F. de Winter, N. Stanton, T. Bellet, B. van Arem, K. Brookhuis, M. H. Martens, and et al. 2019. "A human factors perspective on automated driving." *Theoretical Issues in Ergonomics Science 20*, 3 223–249.

Liu, H., R. Yang, L. Wang, and P. Liu. 2019. "Evaluating initial public acceptance of highly and fully autonomous vehicles." *International Journal of Human-Computer Interaction*, 35, 11 919–931. DOI: 10.1080/10447318.2018.1561791.

Mcgill, R., J. W. Tukey, and W. A. Larsen. 1978. "Variations of box plots." *The American Statistician*, 32, 1 12–16. DOI: 10.1080/00031305.1978.10479236.

Mercedes-Benz. 2014. *S-Class Operator's Manual.* Accessed 2020. https://www.mercedes-benz.co.uk/passengercars/being-an-owner/mb-guides.html.

Moller, S., K. Engelbrecht, C. Kuhnel, I. Wechsung, and B. Weiss. 2009. "A taxonomy of quality of service and Quality of Experience of multimodal human-machine interaction." *2009 International Workshop on Quality of Multimedia Experience.* San Diego, CA, USA. 7–12. DOI: 10.1109/QOMEX.2009.5246986.

Nees, M. A. 2016. "Acceptance of self-driving cars: An examination of idealized versus realistic portrayals with a self- driving car acceptance scale." *Proceedings of the Human Factors and Ergonomics Society Annual Meeting, 60,* 1 1449–1453. DOI: 10.1177/1541931213601332.

Petermeijer, S. M., P. Bazilinskyy, K. Bengler, and J. de Winter. 2017. "Take-over again: Investigating multimodal and directional TORs to get the driver back into the loop." *Applied Ergonomics, 62* 204–215. DOI: 10.1016/j.apergo.2017.02.023.

Rahman, M. M., M. F. Lesch, W. J. Horrey, and L. Strawderman. 2017. "Assessing the utility of TAM, TPB, and UTAUT for advanced driver assistance systems." *Accident Analysis & Prevention, 108* 361–373. DOI: 10.1016/j.aap.2017.09.011.

Revell, K., J. Richardson, P. Langdon, M. Bradley, I. Politis, S. Thompson, L. Skrypchuk, and et al. 2018. ""That was scary…" exploring driver-autonomous vehicle interaction using the Perceptual Cycle Model." In Contemporary Ergonomics and Human Factors. Abingdon-on-Thames, UK: Taylor and Francis.

Revell, K., J. Richardson, P. Langdon, M. Bradley, I. Politis, S. Thompson, L. Skrypchuk, and et al. 2020. "Breaking the cycle of frustration: Applying Neisser's perceptual cycle model to drivers of semi-autonomous vehicles." *Applied Ergonomics, 85* 103037. DOI: 10.1016/j.apergo.2019.103037.

SAE International. 2018. *SAE International Releases Updated Visual Chart for Its "Levels of Driving Automation" Standard for Self-Driving Vehicles.* 11 December. Accessed June

11, 2019.vhttps://www.sae.org/news/press-room/2018/12/sae-international-releases-updated-visual-chart-for-its-%E2%80%9Clevels-of-driving-automation%E2%80%9D-standard-for-self-driving-vehicles.

Sarter, N. B., D. D. Woods, and C. E. Billings. 1997. "Automation surprises." In Handbook of Human Factors and Ergonomics, by G. Salvendy, 1926–1943. New York: John Wiley and Sons.

Schade, J., and B. Schlag. 2003. "Acceptability of urban transport pricing strategies." *Transportation Research Part F: Traffic Psychology and Behaviour, 6*, 1 45–61. DOI: 10.1016/S1369–8478(02)00046-3.

Schuitema, G., L. Steg, and S. Forward. 2010. "Explaining differences in acceptability before and acceptance after the implementation of a congestion charge in Stockholm." *Transportation Research Part A: Policy and Practice, 44*, 2 99–109. DOI: 10.1016/j.tra.2009.11.005.

Singh, S. 2015. Critical reasons for crashes investigated in the national motor vehicle crash causation survey. Washington, DC: U.S. Department of Transportation.

Solís-Marcos, I., C. Ahlström, and K. Kircher. 2018. "Performance of an additional task during level 2 automated driving: An on-road study comparing drivers with and without experience with partial automation." *Human Factors, 60*, 6 778–792. DOI: 10.1177/0018720818773636.

Stanton, N. A., and M. S. Young. 2005. "Driver behaviour with adaptive cruise control." *Ergonomics, 48*, 10 1294–1313. DOI: 10.1080/00140130500252990.

Stanton, N. A., and M. S. Young. 2010. "A proposed psychological model of driving automation." *Theoretical Issues in Ergonomics Science, 1*, 4 315–331. DOI: 10.1080/14639220052399131.

Stanton, N. A., M. Young, and B. McCaulder. 1997. "Drive-by-wire: The case of driver workload and reclaiming control with adaptive cruise control." *Safety Science, 27*, 2 149–159. DOI: 10.1016/S0925-7535(97)00054-4.

Stanton, N. A., A. Dunoyer, and A. Leatherland. 2011. "Detection of new in-path targets by drivers using Stop & Go Adaptive Cruise Control." *Applied Ergonomics, 42*, 4 592–601. DOI: 10.1016/j.apergo.2010.08.016.

Stanton, N. A., G. H. Walker, M. S. Young, T. Kazi, and P. M. Salmon. 2007. "Changing drivers' minds: The evaluation of an advanced driver coaching system." *Ergonomics, 50*, 8 1209–1234. DOI: 10.1080/00140130701322592.

Stapel, J., F. A. Mullakkal-Babu, and R. Happee. 2019. "Automated driving reduces perceived workload, but monitoring causes higher cognitive load than manual driving." *Transportation Research Part F: Traffic Psychology and Behaviour, 60* 590–605. DOI: 10.1016/j.trf.2018.11.006.

Tesla. 2016. *Tesla Model S Owner's manual.* Accessed 2020. https://www.tesla.com/sites/default/files/model_3_owners_manual_north_america_en.pdf.

Van Brummelen, J., M. O'Brien, D. Gruyer, and H. Najjaran. 2018. "Autonomous vehicle perception: The technology of today and tomorrow." *Transportation Research Part C: Emerging Technologies, 89* 384–406. DOI: 10.1016/j.trc.2018.02.012.

Van Der Laan, J. D., A. Heino, and D. De Waard. 1997. "A simple procedure for the assessment of acceptance of advanced transport telematics." *Transportation Research Part C: Emerging Technologies, 5*, 1 1–10. DOI: 10.1016/S0968-090X(96)00025-3.

Venkatesh, V., and F. D. Davis. 2000. "A theoretical extension of the technology acceptance model: Four longitudinal field studies." *Management Science, 46*, 2 169–332. DOI: 10.1287/mnsc.46.2.186.11926.

Venkatesh, V., and H. Bala. 2008. "Technology acceptance model 3 and a research agenda on interventions." *Decision Sciences, 39*, 2 273–315. DOI: 10.1111/j.1540-5915.2008.00192.x.

Venkatesh, V., M. G. Morris, G. B. Davis, and F. D. Davis. 2003. "User acceptance of information technology: Toward a unified view." *MIS Quarterly, 27*, 3 425–478. DOI: 10.2307/30036540.

Wilson, T. D. 2000. "Human information behavior." *Informing Science: The International Journal of an Emerging Transdiscipline, 3*, 2 49–56. DOI: 10.28945/576.

Xu, Z., K. Zhang, H. Min, Z. Wang, X. Zhao, and P. Liu. 2018. "What drives people to accept automated vehicles? Findings from a field experiment." *Transportation Research Part C: Emerging Technologies, 95* 320–334. DOI: 10.1016/j.trc.2018.07.024.

Yoon, S. H., Y. W. Kim, and Y. G. Ji. 2019. "The effects of takeover request modalities on highly automated car control transitions." *Accident Analysis & Prevention, 123* 150–158. DOI: 10.1016/j.aap.2018.11.018.

Young, M. S., and N. A. Stanton. 2002. "Malleable attentional resources theory: A new explanation for the effects of mental underload on performance." *Human Factors: The Journal of the Human Factors and Ergonomics Society* 365–375. DOI: 10.1518/0018720024497709.

Young, M. S., and N. A. Stanton. 2007. "What's skill got to do with it? Vehicle automation and driver mental workload." *Ergonomics, 50*, 8 1324–1339. DOI: 10.1080/00140130701318855.

Zoellick, J. C., A. Kuhlmey, L. Schenk, D. Schindel, and S. Blüher. 2019. "Amused, accepted, and used? Attitudes and emotions towards automated vehicles, their relationships, and predictive value for usage intention." *Transportation Research Part F: Traffic Psychology and Behaviour, 65* 68–78. DOI: 10.1016/j.trf.2019.07.009.

11 The Iconography of Vehicle Automation – A Focus Group Study

Joy Richardson, Kirsten M. A. Revell,
Jisun Kim, and Neville A. Stanton
University of Southampton

CONTENTS

11.1 INTRODUCTION

As partially automated (SAE level 2) vehicles have been available for several years and conditionally automated (SAE level 3) vehicles are currently in production (Gasser and Westhoff 2012; Kim et al. 2020), dashboard displays are becoming more sophisticated and required to communicate more complex information. This automation is formed from combining a system of longitudinal control (such as adaptive

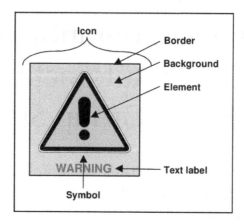

FIGURE 11.1 Composition of an icon (Carney, Campbell and Mitchell 1998).

cruise control (ACC)) providing brake and throttle input with system(s) of lateral control (such as following lanes or a lead vehicle) providing steering. It is essential that the driver is aware of the mode of the vehicle in order to remain safe, and this is most commonly communicated on the dashboard or other in-vehicle display pictorially (Stanton, Dunoyer and Leatherland 2011).

The use of pictures to convey meaning is one of the oldest forms of communication; prior to the invention of written languages, these pictures would be used to record history and tell stories (Horton 1994; Moser 1998). As shown in Figure 11.1, icons are a way of using pictures to deliver a specific message and are formed of several elements which can include a border, a background and text but are focused on the pictorial element, known as the symbol (Carney, Campbell and Mitchell 1998). Studies have shown that the well-designed icons can be recognised more quickly and accurately than textual displays (Horton 1994; Carney, Campbell and Mitchell 1998; Green 1993; Baber and Wankling 1992). They also have the benefit of consuming less space than text, of particular importance on the small and busy real estate of a screen (Green 1993; Baber and Wankling 1992), and, if well designed, can be universally understood and language independent (McDougall, de Brujin and Curry 2000; Chanwimalueng and Rapeepisarn 2013; Zwaga and Mijksenaar 2000).

Carney, Campbell, and Mitchell (1998) proposed that icons can be classified into three types: pictorial representations of the object or action they represent with meaning easily derived and little effort required to learn; concept-related icons based on an image or a property of an actual object or action – these can be context specific so are more difficult to learn – and arbitrary icons which are only meaningful through convention and rely on particular knowledge, which can be cultural. The latter of these are the most difficult type of icon to learn. Others have categorised types of icon as abstract or concrete (Lin 1994) or consider abstract and concrete as opposite ends of a scale (McDougall, de Brujin and Curry 2000). However, all agree that the easiest to understand are those which closest represent real items, as the visual metaphor is clear without the need for prior learning or experience (Carney, Campbell and Mitchell 1998; Lin 1994; McDougall, de Brujin and Curry 2000).

Previous studies, evaluating the automation systems and human–machine interfaces across a range of SAE Level 2 vehicles, identified some mode confusion whereby the driver has been unsure of whether they are in control of the vehicle or if it is automated, despite the mode being displayed on the vehicle's dashboard (Stanton, Dunoyer and Leatherland 2011; Revell et al. 2018; Kim et al. 2020). During these studies across different vehicles, it was observed that the icon indicating the activation or deactivation of automated modes could be green, white, or blue, but the icon remained a pictorial image of a steering wheel. It is clear from examining the interfaces from different manufacturers that the levels of functionality are being interpreted very differently. It is possible this could be due to the lack of standardisation in associated displays and imagery.

The International Organisation for Standardisation (ISO) is an independent non-governmental organisation whose responsibility is to make international standards. They have a voluntary membership of 162 standard bodies worldwide and create standards across a wide range of sectors, from agriculture and healthcare, to transport (ISO 2014, 2016). ISO standardised signs (small icons) and symbols (larger images), both of which will be referred to as icons in this paper, have been commonly used to communicate messages to automobile drivers on the vehicle dashboard for many years, and the ISO aim to maximise the potential of icons through worldwide standardisation (Zwaga and Mijksenaar 2000). This can be seen to originate with the standardisation of traffic light colours in the 1920s (Priest, Wilson and Salas 2005), which are reflected in the use of those same colours on the dashboard, through to the development of icon recommendations in the 1970s (Green 1993; Saunby, Farber and DeMello 1988). Due to new technologies, these icons are now also being displayed on additional screens within the vehicle, commonly located in the centre of the vehicle or as a head-up display. These combined screens are known as in-vehicle information systems (IVIS), and while they may also provide entertainment, comfort, and navigation (Harvey and Stanton 2013), an additional purpose is to provide information to the driver in order to increase safety during driving, thus protecting road users and reducing costs associated with accidents, collision, and congestion (Carney, Campbell and Mitchell 1998). ISO 2575 outlines all the standardised icons in use and is revised every five years (ISO 2017); it also forms the basis for British Standard BS ISO 2575 (BSI 2010). Currently, there are over 300 of these standards falling into 18 categories, from lighting and signalling, to vehicle handling. The basic shape and colours of these icons are standardised, which is important in helping ensure the messages they convey are recognised and understood (ISO 2013). These standardised icons are designed to assist the driver in interpreting dashboard messages, even when driving an unfamiliar vehicle.

Within ISO standard 2575, icons supporting the communication of aspects of automated driving are few. Section J relates to vehicle handling and cruise control, and covers icons for such capabilities, including parking assist and hill descent control. The section contains icon J.09 to indicate when ACC is active, this system accelerates or decelerates a vehicle to maintain a set speed and distance to the vehicle in front, and forms the basis for many of the more advanced automated driving systems available in SAE Level 2 vehicles. Icon J.10 indicates if the ACC system has failed (see Figure 11.2). However, there are a number of key characteristics related to

FIGURE 11.2   Icons relating to ACC (ISO 2010).

automated driving which are absent. In March 2018, a draft for the new version of the standard was released to the public for comments and is due for publication shortly. This draft does not contain any additional icons to support automated driving (BSI 2018). There is an additional, more recent, draft resolution which has recently been circulated to ISO international committee stakeholders, which does look at introducing icons which could be used to indicate modes in SAE Level 2 vehicles.

In addition to prescribing the shape of icons, the ISO standard 2575 also states how colours should be used as they have the following meanings attached. Importantly, an icon may be shown in more than one colour to convey a change in operation of that system (ISO 2010).

- Red – Immediate or imminent danger to persons or equipment.
- Yellow/amber – Caution, malfunction, damage likely, or potential hazard.
- Green – Safe, normal operation.
- Blue – Only used in relation to headlights (main beam or high beam).
- White – May be used when no other condition applies.

Benefits of standardisation have been shown to include opening products up to wider international markets by helping overcome language and cultural barriers, providing a method of communication which can surpass them (ISO 2013; Stuart-Buttle 2006; ISO 2014; Green 1993; Chong, Clauer and Green 1990). They can improve the image of the original equipment manufacturers and reduce exposure to liability from new technologies (Priest, Wilson and Salas 2005). Most importantly, standards can increase safety (Sherehiy, Karwowski and Rodrick 2006; Priest, Wilson and Salas 2005; ISO 2014), which, in the context of driving, can save lives. Road traffic collisions are the eighth most common cause of death (World Health Organization 2018), accounting for 1.25 million deaths per year worldwide (Gorea 2016) and 1793 deaths in 2017 in Great Britain (Department for Transport 2018). Since their initiation in 1968, the US safety standards are thought to have saved the lives of thousands of road users (Priest, Wilson and Salas 2005). This is due to standardisation making systems more understandable (Priest, Wilson and Salas 2005), thereby reducing confusion and failure (Horton 1994). Equally, the impact of the lack of standardisation in IVIS icons has led to poor design, including the use of multiple different icons across manufacturers, which are attempting to communicate the same message (Carney, Campbell and Mitchell 1998). This combination of divergence and underdeveloped design can cause confusion or distraction (Horton 1994; Baber and Wankling 1992;

Stanton, Dunoyer and Leatherland 2011). Poorly designed icons make a system more difficult to use (Horton 1994), increasing workload, errors, and stress (Priest, Wilson and Salas 2005), all of which can increase the likelihood of confusion and therefore of being involved in a road traffic accident by delaying or creating an incorrect input from the driver (Frank, Koenig and Lendholt 1973).

Human factors methods have been used to help design icons. Green (1981) and Chong, Clauer, and Green (1990) both used design workshops with members of the public designing icons for use in IVIS. In both cases, the participants were asked to hand-draw their own ideas for a variety of alerts such as oil level, coolant pressure, and a variety of cruise control functions. While these icons may have lacked in aesthetics and legibility, both studies had similar conclusions, agreeing that the ideas created by the end user were a good point from which to develop candidate icons for future testing, and had the benefit of being quite different from that which would have been developed by designers. In addition to these design methods, the ISO have produced test 9186, Part 1: Method for Testing Comprehensibility (BSI 2014). This has been developed due to the increasing volume of information related via icons to the public, in order to ensure that only one symbol is used internationally to convey each meaning. The test can be used to quantify the comprehensibility of an icon across a representative sample of the end user population, to include those of different age, sex, education level, cultural or ethnic background, and physical ability. When associated with an icon to be used internationally, it requires participants from a minimum of three countries, preferably from different cultural backgrounds. The participants are shown an image of the icon with a short description of where it could be found and then asked two questions: (1) 'What do you think the symbol means?' and (2) 'What action should you take in response to this symbol?' The results are then classified as:

- 1 – Correct,
- 2 – Wrong,
- 2a – Wrong and response is the opposite of the intended meaning,
- 3 – Don't know,
- 4 – No response.

An icon must receive 66% correct answers to be acceptable, but this is raised to 85% in a safety-critical situation, such as for automation mode (BSI 2014; Zwaga and Mijksenaar 2000; Foster, Koyama and Adams 2010; Arcia et al. 2019). Where results are poor, analysis of the results can be used to redesign the icon, resulting in an iterative approach (Zwaga and Mijksenaar 2000; Foster, Koyama and Adams 2010).

No literature has assessed the implications of the lack of standardisation, and much inconsistency in icons, related to automated driving. Misinterpretation of the icons relating to automation involves risks, which at their extreme could cause no one to be in control of the vehicle or for a dangerous unnecessary intervention to take place. Therefore, a focus group was held to determine whether these safety-critical icons could be recognised and understood by the driving public. A demographically diverse range of participants in terms of age, gender, type of vehicle used, annual mileage, and driving experience were recruited. They were asked to interpret the meaning of a variety of displays from the three different manufacturers of SAE Level 2 vehicles used in previous studies, an SAE Level 3 vehicle, for comparison, and from two high-fidelity simulators where studies had previously been conducted (Politis et al. 2018).

11.2 METHOD

11.2.1 Participants

Ethical approval was gained via the University of Southampton Ethics and Research Governance Office (ERGO number: 48777). Seven participants were recruited across three age groups: 18–34, 35–56, and 56–80+ to ensure an inclusive sample meeting the group size (5–10) commonly recommended in the literature (Caplan 1990; Krueger and Casey 2015; Morgan 1997a). Participants were recruited via posters around the university campuses, posts on social media, direct emailing to a list of people who had previously expressed interest in taking part in research undertaken by this team, and to ensure age group 3 was captured, contacting the local branch of 'The University of the Third Age', an organisation who describe themselves as a group 'which brings together people in their "third age" to develop their interests and continue their learning' (University of the Third Age 2019). These age groups were chosen in order to align with other work within the same project (Clark, Stanton and Revell 2019). Time constraints allowed only a single focus group to be held, which, according to Morgan (1997c), is acceptable as long as the results are interpreted cautiously. Demographic information for the participants is shown in Table 11.1.

11.2.2 Design

The focus group lasted two hours, in line with the recommendations of Krueger and Casey (2015) and Morgan (1997b). Demographic information collected included age, gender, annual mileage, years holding a full driving license, any advanced driver qualifications (none reported for any participant), and experience in using a range of 17 advanced driver assistance systems (ADAS). The ADAS categories were carefully worded to represent systems used by all manufacturers, and standard explanations

TABLE 11.1
Participant Demographics

Participant	Age Group	Gender	Years Since Passing Test	Annual Mileage	ADAS Score
1	2	M	25	1,000	4
2	1	F	6	2,000	3
3	1	F	13	10,000	11
4	2	M	22	30,000	2
5	3	F	54	10,000	2
6	3	M	56	13,000	6
7	1	M	3	2,000	3

were provided for each. Participants were asked to select one of three radio buttons to describe their experience of each ADAS as either, 'no experience', 'I have tried this but I'm not experienced', or 'I regularly use this feature and consider myself an experienced user'.

The event started with an introduction to the subject of the workshop and focus group in line with common practice (Krueger and Casey 2015; Morgan 1997b). The concepts of signs and symbols which are currently commonly displayed on a vehicle dashboard during highly automated driving and automated to manual takeover were explained and then demonstrated using a manufacturer promotional video. The session then commenced and was formed of three exercises which followed the convention of progressing from structured to less structured (Morgan 1997c; Cooper and Baber 2004).

11.2.3 EQUIPMENT

Multiple strategies were used for data capture:

- Two Sony HandyCam video cameras were mounted on tripods close to the group in order to obtain footage and audio recording of the group discussions, which was subsequently transcribed.
- A GoPro video camera was mounted high up above the group in order to obtain footage of the group moving and placing the materials on the table.
- Printed copies of icons (compiled in Table 11.2).

TABLE 11.2
Mode Icons Used in the Focus Group

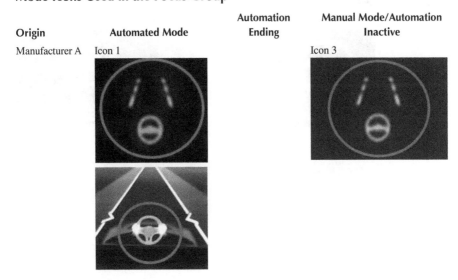

Origin	Automated Mode	Automation Ending	Manual Mode/Automation Inactive
Manufacturer A	Icon 1		Icon 3

(Continued)

TABLE 11.2 (*Continued*)
Mode Icons Used in the Focus Group

Origin	Automated Mode	Automation Ending	Manual Mode/Automation Inactive
Manufacturer B	Icon 1		Icon 2
Manufacturer C	Icon 1		Icon 2
Manufacturer D	Icon 1	Icon 2	Icon 3
Simulators	Icon 1 Icon 2		Icon 3

- Workbooks containing images of the icons, accompanied by notes written by the participants, and comments from the group written by the facilitators on flip charts during the exercises.

11.2.4 PROCEDURE

The first exercise was a written task. In this procedure, the participants were presented with a workbook containing images of icons from the aforementioned vehicles and simulators, each with an adjacent area for writing their response. The icons were shown in context within the vehicle, accompanied by an isolated enlarged version shown in Table 11.2, and they were asked the same four questions for each image.

  i. What is the meaning of this icon?
  ii. Why is the icon being presented?
  iii. What do you like about this icon?
  iv. What do you dislike about this icon?

These questions were devised in line with the recommendations of Cooper and Baber (2004).

The group was then given some further guidance from the facilitator about the types of messages these icons could be conveying, and the participants were asked to categorise the icons by placing the images under categories placed on the table. These categories were as follows:

  i. What the vehicle is sensing.
  ii. What automation capability is possible.
  iii. What driver actions are required.
  iv. What mode the vehicle was in.

The participants then discussed their choices as a group and rearranged the icons on the table until they came to a consensus as to their meaning. The facilitator ensured that all participants were able to engage with the discussions. The event finished with the third exercise, an unstructured session allowing for any comments and recommendations to be made, which were captured by a facilitator on a flip chart.

11.2.5 METHOD OF ANALYSIS

The scripts from the written task were categorised by icon and the comments compared; they were sorted into 'correct', 'wrong', 'wrong and response is the opposite of the intended meaning', and 'don't know', in line with the ISO 9186 Graphical Symbols Test (BSI 2014). This was repeated with the printed icons from the second task. The audio from videos of the study was transcribed, categorised, and sorted in the same manner. For the final task, the flip chart pages were sorted into themes such as colour, standardisation, and recommendation.

As colour was of such a concern to the participants, a comparison was made between the interpretations of the meaning of colours from the participants'

TABLE 11.3

Comparisons of Meanings of Colours From Manufacturers, Simulators, Focus Group, ISO, and Recommendations

Source	Red	Amber/Yellow	Green	Blue	White/Grey
Manufacturer A	-	-	Auto Active	-	Auto Not Active
Manufacturer B	-	-	Auto Active	-	Auto Not Active
Manufacturer C	Auto Ending Imminent	-	-	Auto Active	Auto Not Active
Simulators		Manual Mode Auto Not Active	Auto Active	Auto Active	-
Focus Group	Not Active Unsafe	Caution Standby Normal	Auto Active Safe	Unsure Ambiguous Hybrid	Ambiguous
ISO 2575	Danger Immediate or Imminent	Caution	Safe Normal	High Beam	Used When No Other Condition Applies
Green (1995)	Critical Warning Action Required	Caution	-	-	-
Horton (1994)	Danger	Caution	Safe	Information	-

comments, how these were used in the icons used in the focus group, standards recommended by authors, and the recommendations from ISO 2575. The results are given in Table 11.3.

11.3   RESULTS

11.3.1   EXERCISE ONE

11.3.1.1   Icons Indicating Automation Mode Active

Each manufacturer, and the simulators, has icons to indicate that the automation system is active. Most of the participants were able to interpret these icons accurately, but some participants had more unusual opinions: 'the car is driving correctly within the lane' (e.g. Table 11.2, Manufacturer A, Icon 1), 'pavement is irregular, or narrow roads, slow down'. Some participants expressed confusion: 'unclear', 'is car driving itself?', 'no idea'. One participant interpreted Manufacturer A, Icon 2 to mean the opposite to that intended: 'you are holding the steering wheel, feedback to driver that car recognises they are doing as asked' (Table 11.2, Manufacturer A, Icon 2).

11.3.1.2   Icons Indicating Manual Mode or Automation Ending/Inactive

All of the manufacturers have icons indicating that the automation is inactive or ending, and the simulator has an icon to indicate manual mode. Some of the more unusual interpretations were 'indentation in pavement, take care of kerb', 'road narrows, slow down' (e.g. Table 11.2, Manufacturer A, Icon 1), and 'speed control active', and some were confused: 'unclear. Could be anything or nothing', 'unsure, maybe AI inactive'. Some participants again made opposing interpretations: 'don't use steering wheel', 'steering is in auto mode for the rest of the journey', 'AI is in control' (e.g. Table 11.2, Manufacturer C, Icon 2).

11.3.1.3   Colour

There were many comments about the colour of the icons and what the colours meant:

Green (e.g. Table 11.2, Manufacturer A, Icon 1) – 'I like colour coding' (green meaning active), 'active as it's green', 'the steering wheel is green which is positive confirmation', and 'green resonates as something being active'.

Orange/amber (e.g. Table 11.2, Simulators, Icon 2) – 'amber resonates as caution/be alert'.

Red (e.g. Table 11.2, Manufacturer C, Icon 2) – 'could be red to give you a clear on-off indication' (referring to an icon which greys out when inactive).

11.3.1.4   Size and Text Labels

The participants were able to see the icons in context within the IVIS, and so were able to determine how large they were. There were several comments regarding size: 'icon so small on dash', 'it's small', 'very small – easily missed', 'maybe a bit small', 'small, meaning it is unclear', 'icon is quite small in cluster'. Those icons that were accompanied with a text label, e.g. Table 11.2, Simulators, Icon 1, were more easily understood than those without, and it was suggested by one participant that text should be added to those icons which were more ambiguous; however, some participants found the text annoying or too long.

11.3.2   EXERCISE TWO

The various interpretations were discussed by the group in detail, and this revealed some opposing views, e.g. Table 11.2, Manufacturer A, Icon 2:

Participant A, 'I think if my hands were off the wheel and I saw that, it would be very obvious that I need to put my hands on the wheel'.

Participant B, 'If my hands were on the wheel, I'd take them off the wheel'.

Also, some ambiguity, e.g. Table 11.2, Manufacturer C, Icon 2:

Participant C, 'I don't know whether it's available or not. I wouldn't be sure if it's available'.

Participant D, 'I'm not sure that it's not available either'.

There were also many discussions regarding the use and meaning of colour. A number of icons used green colouring in the automated mode, while one was blue. Orange and white/grey were used to indicate no automation/manual mode, and one icon used red to indicate the automated mode was ending (see Table 11.2). The participants again had opinions as to what the colours should represent:

Green – 'Because it's green, it's suggesting automated mode', 'Green is automated active', 'So green means safe'.

Orange/amber – 'Kind of a standby thing', 'Orange so it's a caution', 'warning', 'orange means normal'.

Red – 'not active', 'red means unsafe', 'and red that was not active'.

Blue – 'Blue is a big question mark', 'Blue suggests hybrid'.

White – 'Not being green or any colours (e.g. greyed out/white) also suggests to me that it's not on an automated mode'.

There was also a discussion around standardisation and how this is helpful in aiding understanding. One participant commented, 'Most things are standardised… like traffic lights. They become standardised… so that everyone ends up having it the same way'.

11.3.3 EXERCISE THREE

The final task was a less structured, open discussion, but raised some of the same concerns as the previous tasks.

The use of colour in icons was still of significance, with participants commenting: 'consistency of colour across manufacturers is important' and 'colour coding should be consistent – stick to red/orange/green. Blue and grey are ambiguous'. With further comments around standardisation: 'don't like icons – need to be standardised' and 'icons should be standardised'.

11.3.3.1 ADAS Experience

From the form completed before the study, the lowest score was 2 and the highest was 11. As the maximum possible score was 34, this range of scores indicates that the participants had a low to moderate amount of experience in using ADAS (all scores can be seen in Table 11.1). Those participants with the lowest amount of experience using ADAS were also more likely to interpret the icons incorrectly during both the unguided individual task, and following the extra guidance during the second task, making interpretations such as 'pavement is irregular', or 'narrow roads, slow down' for the first task and 'road narrows, slow down' in the second task. Equally, those with more ADAS experience were more able to identify the icons correctly, even from the initial unguided task.

11.4 DISCUSSION

It can be seen that the majority of icons from the manufacturers and from the simulator, designed to indicate mode, are based on the representation of a steering wheel. One manufacturer has opted to use the image of a car. These icons can be classed as concept-related icons; they are based on an image of an actual object and sometimes an associated action (Carney, Campbell and Mitchell 1998). These are moderately difficult to learn, and it is therefore not surprising that some participants found them challenging to interpret without guidance or previous experience. Where participants had made more unusual interpretations of the icons, it can be seen that they were examining them more closely as a pictorial-type icon, seeing the image in front of them as a literal representation of the road ahead; these participants also had less ADAS experience. Those with more experience of ADAS were more likely able to identify the modes indicated by the icons correctly. While no participants have these automation capabilities in their own vehicles, it may be possible to hypothesise that those with more ADAS experience were more used to the concept-related icon, and were therefore able to interpret these new icons more easily.

Some participants interpreted icons identifying the automation mode inversely, effectively believing a car was not in automated mode when it was, but more concerning was the understanding that the car was in automated mode when it was not. This ambiguity could cause a driver to omit making driving inputs; causing an accident; and resulting in damage to the vehicle and harm to the driver, passengers, and other road users. There would also be a negative economic impact to the driver in terms of increased insurance premiums, vehicle repair or replacement, any casualties from medical bills, and missed employment, and to society as a whole from emergency services and highway repair (Gorea 2016). In addition, accidents related to an automation system can cause negative media coverage, such as that received by Tesla following a fatal crash in 2016 (NTSB 2020), leading to public distrust in the manufacturer or similar systems and therefore also having a negative economic impact on the original equipment manufacturers.

Colour of icons was important to the participants and was raised by them in all three exercises. Colour can increase the likelihood icons are noticed (Young 1991), and when used well can aid communication. However, when used carelessly, they can increase confusion (Horton 1994). The colours used in the icons during this focus group were white/grey, green, blue, amber/yellow, and red, as given in Table 11.2. Participants made many comments about how they interpret the meaning of colours. They referred to the use of colour in IVIS as they experience it in their own vehicles, but also made comparisons to traffic lights.

These meanings of colours stated by the participants were, what they considered, a conventional interpretation. This aligns closely with the meaning attributed to colour in ISO 2575. The use of new colours (white and blue), or colours not meeting their expectation, was deemed confusing by the participants. It is interesting to compare the opinions of the focus group and the ISO standards with those of authors who have written recommendations for the use of colour in icon design from an IT perspective (Horton 1994) and from a human factors perspective (Green 1995). Table 11.3 shows a summary of the colours and actual meanings from the icons used in the focus group, the focus group participants, ISO 2575, and those of the authors of recommendations. From Table 11.3, it is clear that colour use is inconsistent between the manufacturers and simulators. But they also appear misaligned from the opinions of the focus group. The participants' opinions generally match the ISO standards, which they will have experienced in their own vehicles, and with those listed in the recommendations.

It is acknowledged that this paper was limited in its scope by relying on a single focus group, and the participants having low to moderate ADAS experience. However, recommendations for future work could include repeating the exercises with more focus groups and recruiting some participants with higher experience of ADAS, perhaps comparing novice and expert users.

Carney, Campbell, and Mitchell (1998), Green (1995), and Zwaga and Mijksenaar (2000) have highlighted the importance of a human factors-led approach to icons in IVIS, and the studies conducted by Green (1979) and Chong, Clauer, and Green (1990) had user-led approach to the initial stages of design. After some polishing, these designs could be scored using the ISO 9186 test as part of a human factors and user-led iterative design process.

11.5 CONCLUSION

Due to a lack of standardisation in the icons relating to automated driving, manufacturers are independently designing icons, causing inconsistency (Carney, Campbell and Mitchell 1998). A focus group was conducted in order to determine if a range of icons, identifying mode in an automated vehicle, could be understood by driving members of the public. During the initial exercise, the participants were asked to write their interpretations and meanings of a number of icons used in semi-autonomous vehicles and simulators. They reported a wide range of ideas about the meaning of the icons. Those participants with more experience of ADAS in their own vehicles were more likely to have correct or similar ideas as to the meaning of the icons. Those with less experience of ADAS attributed a wider range of ideas related to the way they interpreted the icons, seeming to make more literal interpretations. Before the second exercise, the participants received additional guidance about the possible meanings of the same icons. The participants were able to make a correct interpretation of more icons than in the previous task, suggesting that these could not be considered pictorial or concrete types as some learning was required in order to identify them (Carney, Campbell and Mitchell 1998; Lin 1994; McDougall, de Brujin and Curry 2000). However, they were also in agreement that several icons remained ambiguous, suggesting these icons were arbitrary or abstract. The final exercise allowed the participants to openly discuss the icons they had seen. This section showed that colour and standardisation were considered important when trying to understand icons. The participants considered it essential that colours were only used in what they considered the conventional manner; their opinions on the use of colour aligned well with the colour standards outlined in ISO 2575.

Following these exercises and discussions, it is recommended that ISO standards should be created for the icons used in IVIS, indicating automated and manual modes, in order to reduce driver confusion. These icons should be designed using the existing human factors methodologies such as using the target population to help develop candidate icons, and tested using the ISO 9186 to ensure they are simple, clear, concise, and universally understood.

ACKNOWLEDGEMENTS

This work was supported by Jaguar Land Rover and the UK-EPSRC Grant EP/N011899/1 as part of the jointly funded 'Towards Autonomy: Smart and Connected Control (TASCC) Programme'. The authors also acknowledge the support of Simon Thompson at Jaguar Land Rover for his support in preparing this paper.

REFERENCES

Arcia, A., L. V. Grossman, M. George, M. R. Turchioe, S. Mangal, and R. M.M. Creber. 2019. "Modifications to the ISO 9186 Method for Testing Comprehension of Visualizations:

Successes and Lessons Learned." *2019 IEEE Workshop on Visual Analytics in Healthcare.* Vancouver, Canada: IEEE. 41–47. doi: 10.1109/VAHC47919.2019.8945036.

Baber, C., and J. Wankling. 1992. "An experimental comparison of text and symbols for in-car reconfigurable displays." *Applied Ergonomics, 23,* 4 255–262. DOI: 10.1016/0003-6870(92)90153-M.

BSI. 2010. "BS ISO 2575: Road vehicles - Symbols for controls, indicators and tell-tales." Accessed November 4, 2020. https://www.iso.org/standard/68409.html.

BSI. 2014. "BS ISO 9186-1: 2014, Graphical symbols — Test methods — Part 1: Method for testing comprehensibility." Accessed November 4, 2020. https://www.iso.org/standard/59226.html.

BSI. 2018. "BS ISO 2575. Road vehicles. Symbols for controls, indicators and tell-tales." Accessed November 4, 2020. https://www.iso.org/standard/68409.html.

Caplan, S. 1990. "Using focus group methodology for ergonomic design." *Ergonomics, 33,* 5 527–533. DOI: 10.1080/00140139008927160.

Carney, C., J. L. Campbell, and E. A. Mitchell. 1998. *In-Vehicle Display Icons and Other Information Elements: Literature Review.* Literature Review, US Department of Transportation, (No. FHWA-RD-98-164). Turner-Fairbank Highway Research Center, McLean, VA.

Chanwimalueng, W., and K. Rapeepisarn. 2013. "A study of the recognitions and preferences on abstract and concrete icon styles on smart phone from Easterners and Westerners' point of view." *2013 International Conference on Machine Learning and Cybernetics.* Tianjin, China: IEEE. 1613–1619. doi: 10.1109/ICMLC.2013.6890858.

Chong, M., T. Clauer, and P. Green. 1990. *Development of candidate symbols for automobile functions.* Technical Report, The University of Michigan Transportation Research Institute.

Clark, J. R., N. A. Stanton, and K. Revell. 2019. "Directability, eye-gaze, and the usage of visual displays during an automated vehicle handover task." *Transportation Research Part F: Traffic Psychology and Behaviour, 67* 29–42. DOI: 10.1016/j.trf.2019.10.005.

Cooper, L., and C. Baber. 2004. "Focus Groups." In *Handbook of Human Factors and Ergonomics Methods,* by N. A. Stanton, A. Hedge, K. Brookhuis, E. Salas and H. Hendrick, 321–328. Boca Raton, FL: CRC Press.

Department for Transport. 2018. "Reported road casualties in Great Britain: Quarterly provisional estimates year ending June 2018." 8 November. Accessed October 1, 2019. https://assets.publishing.service.gov.uk/government/uploads/system/uploads/attachment_data/file/754685/quarterly-estimates-april-to-june-2018.pdf.

Foster, J. J., K. Koyama, and A. Adams. 2010. "Paper and on-line testing of graphical access symbols in three countries using the ISO 9186 comprehension test." *Information Design Journal, 18,* 2 107–117. DOI: 10.1075/idj.18.2.02fos.

Frank, D., N. Koenig, and R. Lendholt. 1973. "Identification of symbols for motor vehicle controls." *SAE Transactions, 82* 2153–2161. www.jstor.org/stable/44717612.

Gasser, T. M., and D. Westhoff. 2012. *BASt-study: Definitions of Automation and Legal Issues in Germany.* 25 July. *Proceedings of the 2012 road vehicle automation workshop.* Automation Workshop, Irvine, CA.

Gorea, R. 2016. "Financial impact of road traffic accidents on the society." *International Journal of Ethics Trauma & Victimology, 2,* 1 6–9. DOI: 10.18099/ijetv.v2i1.11129.

Green, P. 1979. "Development of pictographic symbols for vehicle controls and displays." *1979 Automotive Engineering Congress and Exposition.* SAE International. DOI: 10.4271/790383.

Green, P. 1981. "Displays for automotive instrument panels: production and rating symbols." *HSRI Research Review, 12,* 1 1–12.

Green, P. 1993. "Design and evaluation of symbols for automobile controls and displays." In *Automotive Ergonomics,* by B. Peacock and W. Karwowski, 237–268. Washington, DC: Taylor & Francis.

Green, P. 1995. *Measures and methods used to assess the safety and usability of driver information systems.* Technical Report FHWA-RD-94-088, McLean, VA: Federal Highway Administration.

Harvey, C., and N. A. Stanton. 2013. *Usability Evaluation for In-Vehicle Systems.* London, UK: CRC Press.

Horton, W. K. 1994. *The Icon Book: Visual Symbols for Computer Systems and Documentation.* New York: John Wiley & Sons, Inc.

ISO. 2010. "ISO 2575:2010, Road vehicles — Symbols for controls, indicators and tell-tales." Accessed November 4, 2020. https://www.iso.org/standard/54513.html.

ISO. 2013. The International Language of ISO Graphical Symbols. Accessed November 4, 2020. https://www.iso.org/news/2013/10/Ref1787.html.

ISO. 2014. "Economic Benefits of Standards." Accessed November 4, 2020. https://www.iso.org/publication/PUB100403.html.

ISO. 2016. "ISO 17488:2016, Road vehicles — Transport information and control systems — Detection-response task (DRT) for assessing attentional effects of cognitive load in driving." Accessed November 4, 2020. https://www.iso.org/standard/59887.html.

ISO. 2017. "ISO 2575:2010/AMD 7:2017, Road vehicles — Symbols for controls, indicators and tell-tales — Amendment 7." Accessed November 4, 2020. https://www.iso.org/standard/71505.html.

Kim, J., K. Revell, P. Langdon, M. Bradley, I. Politis, S. Thompson, L. Skrypchuk, and et al. 2020. "Drivers' Interaction with, and Perception Toward Semi-autonomous Vehicles in Naturalistic Settings." *International Conference on Intelligent Human Systems Integration. Advances in Intelligent Systems and Computing.* Cham, Switzerland: Springer. 20–26. https://doi.org/10.1007/978-3-030-39512-4_4.

Krueger, R. A., and M. A. Casey. 2015. *Focus Groups: A Practical Guide for Applied Research.* Thousand Oaks, CA: Sage Publications.

Lin, R. 1994. "A study of visual features for icon design." *Design Studies, 15,* 2 185–197. DOI: 10.1016/0142-694X(94)90024-8.

McDougall, S. J.P., O. de Brujin, and M. B. Curry. 2000. "Exploring the effects of icon characteristics on user performance: The role of icon concreteness, complexity, and distinctiveness." *Journal of Experimental Psychology: Applied, 6,* 4 291–306. DOI: 10.1037/1076-898X.6.4.291.

Morgan, D. L. 1997a. *Focus Groups as Qualitative Research (2nd ed.).* California: Sage Publications. https://doi.org/10.4135/9781412984287.

Morgan, D. L. 1997b. *The Focus Group Guidebook.* California: Sage Publications.

Morgan, D. L. 1997c. *Planning Focus Groups.* California: Sage Publications.

Moser, S. 1998. *Ancestral Images: The Iconography of Human Origins.* Stroud, UK: Sutton Publishing Limited.

NTSB. 2020. *NTSB Opens Public Docket for 2 Ongoing Tesla Crash Investigations.* 11 February. https://www.ntsb.gov/news/press-releases/Pages/NR20200211.aspx.

Politis, I., P. Langdon, D. Adebayo, M. Bradley, P.J. Clarkson, L. Skrypchuk, A. Mouzakitis, and et al. 2018. "An Evaluation of Inclusive Dialogue-Based Interfaces for the Takeover of Control in Autonomous Cars." *Proceedings of the 2018 Conference on Human Information Interaction & Retrieval - IUI'18: 23rd International Conference on Intelligent User Interfaces.* Tokyo, Japan: ACM Press. 601–606. https://doi.org/10.1145/3172944.3172990.

Priest, H. A., K. A. Wilson, and E. Salas. 2005. "National standardization efforts in ergonomics and human factors." In *Handbook of Standards and Guidelines in Ergonomics and Human Factors,* by W. Karwowski, https://doi.org/10.1201/9781482289671. Boca Raton, FL: CRC Press.

Revell, K., J. Richardson, P. Langdon, M. Bradley, I. Politis, S. Thompson, L. Skrypchuk, and et al. 2018. ""That was scary..." exploring driver-autonomous vehicle interaction using the Perceptual Cycle Model." *Contemporary Ergonomics and Human Factors.* Abingdon-on-Thames: Taylor and Francis.

Saunby, C. S., E. I. Farber, and J. DeMello. 1988. *Driver Understanding and Recognition of Automotive ISO Symbols.* Technical Paper 880056. https://doi.org/10.4271/880056, SAE International.

Sherehiy, B., W. Karwowski, and D. Rodrick. 2006. "Human Factors and Ergonomics Standards." In *Handbook of Human Factors and Ergonomics*, by G. Salvendy. New Jersey: John Wiley & Sons, Inc.

Stanton, N. A., A. Dunoyer, and A. Leatherland. 2011. "Detection of new in-path targets by drivers using Stop & Go Adaptive Cruise Control." *Applied Ergonomics*, *42*, 4 592–601. DOI: 10.1016/j.apergo.2010.08.016.

Stuart-Buttle, C. 2006. "Overview of National and International Standards and Guidelines." In *Handbook of Standards and Guidelines in Ergonomics and Human Factors*, by W. Karwowski. Boca Raton, FL: CRC Press.

University of the Third Age. 2019. *About the U3A.* Accessed September 16, 2019. https://www.u3a.org.uk/about.

World Health Organization. 2018. *Global status report on road safety 2018.* Accessed February 26, 2020. https://www.who.int/violence_injury_prevention/road_safety_status/2018/en/.

Young, S. L. 1991. "Increasing the noticeability of warnings: Effects of pictorial, color, signal icon and border." *Proceedings of the Human Factors and Ergonomics Society Annual Meeting*, 35, 9 580–584. DOI: 10.1518/107118191786754662.

Zwaga, H. J., and P. Mijksenaar. 2000. "The development and standardization of warning symbols; the role of design and human factors." *Proceedings of the Human Factors and Ergonomics Society Annual Meeting*, 44, 28 782–785. DOI: 10.1177/1541931200044028104.

Part IV

HMI Simulator

12 Customisation of Takeover Guidance in Semi-Autonomous Vehicles

*James W. H. Brown, Kirsten M. A. Revell,
Joy Richardson, and Jediah R. Clark*
University of Southampton

Nermin Caber and Theocharis Amanatidis
University of Cambridge

Patrick Langdon
Edinburgh Napier University

Michael Bradley
University of Cambridge

Simon Thompson and Lee Skrypchuk
Jaguar Land Rover

Neville A. Stanton
University of Southampton

CONTENTS

12.1   INTRODUCTION

This necessity for customisation is not just limited to physical ergonomics; there are multiple additional aspects of usability. Shackel's (1991) framework describes the interaction between the user, environment, tool, and the task. In the context of transition of control from a semi-autonomous vehicle to a driver with a traditional fixed interface design, the user, the environment, and the task are all variables. Considering just the driver as a variable, different drivers will have varying dimensions, such as motivations, experience levels, alertness, and preferences (Harvey et al. 2011). Environmental variances might include darkness levels, sound levels, and distraction levels. The task also varies in difficulty, e.g. in terms of road conditions, traffic density, and visibility levels. A fixed interface represents a compromise for the array of differing variables. The Usability and EC Directive 90/270 specifically describes user characteristics which represent their motivation, skills, and knowledge (Stanton and Baber 1992). The consideration of these characteristics might improve acceptance levels of any in-vehicle technology. Therefore, a customised interface, matched by the user to their own specific needs and preferences, may result in an increased use and acceptance. Driving is considered to be a complex task (Regan, Lee and Young 2008); it is also safety-critical, and the performance may affect the safety of the vehicle occupants and those around them. This provides perhaps the strongest motivation for matching the human–machine interface (HMI) to the needs and requirements of the driver. Green (1995) stated that lateral deviations can result from drivers experiencing high attentional demands and/or fatigue. Green also described how drivers tend to reduce their speed to compensate for the attentional demand, in order to retain a safety margin. Both the variation in speed and lateral deviation can have safety implications (Hoogendoorn, van Arerm and Hoogendoom 2014; Miura 2014).

There is increasing support for the adaption of the interface to match the driver's needs and preferences (Harvey et al. 2011). An interface that allows customisation by the driver could provide the method by which this is achieved. Customisation of modalities, intensities, and locations of information may not only match the driver's subjective preferences, but there is a significant literature to suggest the information presented prior to takeover requests (TORs) can have a positive effect upon performance (Eriksson et al. 2019; Melcher et al. 2015; Naujoks et al. 2019; Petermeijer, Cieler and de Winter 2017). The SAE definition of a TOR is, *'notification by the ADS to a human driver that he/she should promptly begin or resume performance of the dynamic driving task'* (SAE J3016 2016).

Zhang et al. (2019) carried out a meta-analysis of 129 studies of takeover times (TOTs) in SAE Level 2 automated vehicles. They found that a shorter mean TOT can be associated with multiple potential influences. Increasing scenario urgency, lack of a handheld secondary task device, lack of a visual-based non-driving task, receiving haptic or audio modality TORs compared to just visual TORs, and previous experience are all associated with shorter mean TOTs (Zhang et al. 2019). In the case of a handheld device, the ability to display a TOR on the screen is likely to reduce TOT (Yoon, Kim and Ji 2019). Providing the ability to add or remove audio and haptic modalities enables drivers to ensure that their performance remains safe, even when they may be listening to background music and focusing on a secondary task. Borojeni et al. (2016) describe TORs as a twin-phase process; drivers need to be disengaged from their secondary task with sensory cues and then provided with information on action that needs to be taken. Customisation allows these sensory cues to be modified to suit the drivers' unique environmental requirements and information modified to suit preferences. McDonald et al. (2019) reviewed the literature on vehicle-to-driver transitions of control in automation Levels 2–4, to identify the factors that influence driver performance in terms of TOT and a range of additional metrics. They stated that engagement in secondary task, trust level, driving environment, and TOR modality have a significant effect on both lateral and longitudinal control post-takeover. All of these factors could potentially be optimised by providing the driver with a customisable interface.

Employing audio and visual modalities simultaneously was found by Naujoks, Mai, and Neukum (2014) to both reduce TOT and improve lateral control performance. Toffetti et al. (2009) compared an interface that featured visual and acoustic TOR modalities with an additional vocal modality. Their findings suggested that in the event of time-critical TORs, the addition of the vocal modality improved reaction time; however, in non-critical TORs, the reaction time was lower for just the acoustic modality. This demonstrates how as the task changes, the performance of the driver for a given interface may also change, and hence the rationale for a driver will be able to customise their interface specifically to the task. Yoon, Kim, and Ji (2019) examined the effect of information modality on takeover performance under four varying conditions of secondary task. They found that purely visual TORs resulted in drivers taking longer to put their hands on the steering wheel, and this was considered to be due to lower visual priority being given. A customisable interface would allow the driver to mitigate this by adding an additional modality. Indeed, they found that hands-on time decreased when multiple modalities were employed. An additional disadvantage of visual-only TORs was found to be an increase in hands-on time due to drivers having to fixate on the visual alert prior to re-fixing their gaze on the road. This highlights the potential advantage of allowing customisation of visual

TOR alerts. Placing them in a head-up display (HUD) would potentially reduce the fixation duration issue, additionally, providing the driver the ability to place them on their secondary device would ensure that they are visible in a short duration. Wintersberger et al. (2018) specifically examined the placement of visual TORs on secondary task devices; they found that takeover performance was better and TOTs reduced in comparison with presenting the TORs on the vehicle's interface. They also found that trust levels were higher for device-based TORs. Van den Beukel, van der Voort, and Eger (2016) examined driver performance using three combinations of TOR modalities; they found that in scenarios where driver intervention was required, a sound and icon-based visual resulted in best performance, followed by an illumination and haptics TOR, and finally an acoustic-only TOR. However, in the event that a driver has impaired hearing and/or close vision, they may find their performance significantly impaired, as the modalities do not suit their disabilities. A customisable interface that provided a choice of modality would make the vehicle more inclusive and improve safety through TOR performance. Holländer and Pfleging (2018) examined the effects of different forms of pre-TOR alerts; they state that the type of non-driving-related activity will determine the best modality (or set of modalities) for the TOR. This represents a strong argument for customisation. A fixed interface might be optimised for a given scenario; however, it may be suboptimal in another. A driver reading a book might want an audio alert, but if they had a child asleep in the car, they might opt for a haptic alert instead. Customisation allows for the adaption to circumstances.

Eriksson et al. (2019) examined the effect of augmented reality on takeover performance; they examined four HMI variations, each providing a unique set of information to the drivers. Their findings indicated that takeover performance was improved when drivers were provided with additional artificial reality-based information. This lends some weight to the argument that it may be advantageous to provide an interface that can display a wealth of information in various modalities, intensities, and locations, and allow the driver to reduce the interface to the level and forms of information that they desire. The authors do, however, state that linguistic theory suggests that communication is optimised when information relevance is high, and density is correctly balanced. An overly dense interface may actually impede the driver's performance. Therefore, a good approach may be to limit the maximum density of some modalities, primarily visual, to prevent this, while still allowing customisability. Wandtner, Schömig, and Schmidt (2018) explored the effects of a predictive HMI that provided the driver with a preview of an upcoming TOR; they found that drivers self-regulated their secondary tasks, reducing them when they became aware that a TOR was imminent. It is conceivable that a customisable interface could allow the driver to specify the level of pre-alert. This would allow, e.g., for a driver to set the time interval to the time it would take them to write an average email. In this way, they could be writing emails during automation and know that when the pre-alert occurs, they will likely have the time to complete one, but no time to start another. This is in preference to a pre-alert that results in a driver attempting to complete their secondary task when the TOR occurs. Indeed, Wandtner, Schömig, and Schmidt (2018) found that drivers who fail to disengage from their secondary task when they carry out a takeover, exhibit significantly impaired performance in terms of TOT, and lateral control.

While there are clear benefits to customisation, limits should potentially be imposed in terms of minimum and maximum levels of information across modalities. Clearly, it should not be possible to remove all informational elements of a TOR, but some modalities when used exclusively may result in reduced takeover performance. Indeed, Melcher et al. (2015) concluded that multimodal TORs are necessary, regardless of the TOR urgency. Politis, Brewster, and Pollick (2017) stated that in critical takeovers, unimodal visual TORs presented on an in-car interface detrimentally affected response accuracy and lateral deviation, and increased TOTs. They also noted that language-based warnings were less effective than abstract warnings when presented on the vehicles' interface. Both of these effects were resolved by placing the information on the secondary task. Politis, Brewster, and Pollick (2014) suggest the number of modalities should be limited, unless the TOR is time-critical; they recommend a combination of audio and visual modalities for secondary task TOR alerts. McDonald et al. (2019) state that if ecological alerts are presented at the wrong time, or they are of long durations, driver decision-making can be affected; however, if they are presented optimally, TOT can be reduced. Integrating HMI customisations with other vehicle systems may yield additional benefits. Wintersberger et al. (2018) found that presenting TORs at secondary task boundaries resulted in lower stress levels than those presented within tasks. The ability to time a TOR or TOR pre-alert based on the secondary task duration could potentially improve both usability and performance/safety. Indeed, the authors found that both reaction times and lane change performance were degraded when TORs were presented within task on the in-vehicle information system (IVIS).

There is a wide breadth of literature on the effects of HMI-based information on takeover performance. Much of it examines specific conditions and combinations of modalities/HMI designs, and the findings in each case allow the most appropriate HMIs to be selected for those conditions. However, conditions are not constant, and driver needs and preferences are not constant. Vehicles already have many customisable systems, such as climate control, seat position, and media. These provide drivers with the ability to improve their levels of comfort based on preferences. The issue of resuming control from an automated vehicle is safety-critical, and taking into account the fact that drivers have different preferences and conditions are non-constant, it is logical to consider that a customisable HMI might provide a solution to improve safety, usability, and efficiency.

12.2 METHOD

12.2.1 PARTICIPANTS

A total of 68 participants were recruited by Jaguar Land Rover (JLR) via an external agency, and three cancellations occurred during the study, resulting in data being collected for a total of 65 participants. Selection criteria specified an equal split of three age ranges (18–34 (22), 35–56 (23), and 57–82 (20)) (mean age=44.7, SD=16.175), and an equal gender split (33 females, 32 males). An additional inclusion criterion was for 10% of participants to own premium vehicles, defined as newer than seven years old, and manufactured by Audi, BMW, JLR, Mercedes Benz, MINI, Porsche,

Tesla, Volvo, or VW. All participants held a full UK driver's license, and were in good health, with corrected vision acceptable. The University of Southampton Ethics and Research Governance Office provided ethical approval for the study (ERGO Number: 41761.A3).

12.2.2 EXPERIMENTAL DESIGN

The study consisted of four trials, with a repeated-measures condition defined as time out of the loop (OOTL) (i.e. one participant=two short time OOTL and two long time OOTL conditions). Trials were counterbalanced to ensure that order effects were controlled. There were four counterbalance conditions consisting of SSLL, LLSS, SLSL, and LSLS – where S=short (1 min) and L=long (10 min). Counterbalance conditions were equally distributed according to age category, gender, and premium car ownership. A secondary task in the form of a tablet-based Tetris-style game was employed during periods of automation. At the end of each trial, participants had the opportunity to customise their HMI settings, as shown in Figure 12.1.

The driving environment consisted of a three-lane motorway with gentle curves; traffic conditions were designed to be congested, resulting in an average speed of approximately 40 mph. Road conditions were dry with good visibility.

Recorded performance data from STISIM included speed, lateral lane position, headway, throttle, brake, and steering. Data was time-stamped to allow performance 25 s post-takeover to be analysed. The times at which HMI elements were presented – and control transition button pressed – were logged, allowing protocol timings and TOTs to be calculated. Data was logged for the secondary Tetris task, including scores and button presses. Customisation settings were saved for each trial to reveal patterns in preference to subsequent trials. Subjective data on workload (NASA-TLX), usability (system usability scale), technology acceptance (technology

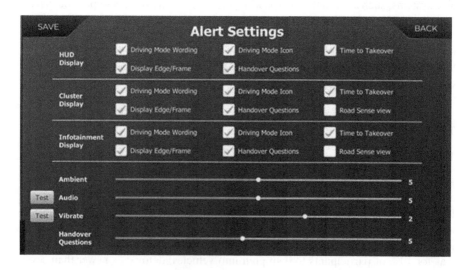

FIGURE 12.1 The customisation panel that appeared on the central console between trials.

acceptance scale), and automation trust were collected after each trial to provide insight into how these varied between trials. A final post-trial interview was administered at the end of the study to capture data on preferences and opinions relating to the HMI and customisation.

12.2.3 EQUIPMENT

The study employed a fixed-base simulator at a JLR research facility (Figure 12.2). The simulator cabin was designed to mimic a generic JLR vehicle; LCD monitors were built in to provide multiple in-car interfaces, including a touchscreen central console, an instrument cluster, and HUD. The HUD monitor was mounted horizontally and displayed an inverted image; this was reflected onto a sheet of angled glass fitted in the drivers' view. The in-car interfaces were highly adaptable and featured a range of graphics and text that could be displayed. The central console provided additional functionality through the display of a customisation screen that allowed the HMI graphics and text on the central display, cluster, and HUD to be modified within specific parameters when between trials. A Range Rover steering wheel was modified to include two additional buttons, mounted adjacent to the drivers' thumb positions. The steering system was connected to a Sensodrive direct drive system to provide force feedback and turn the steering wheel automatically when the automation was activated. The drivers' seat was modified to provide haptics cues using Arduino-controlled motors fitted with eccentric weights. Ambient lighting was provided through the use of Arduino-controlled RGB LEDs fitted under the top of the cluster, within the drivers' footwell, and under the central console. The main road view was provided via a large LCD screen mounted directly in front of the simulator cabin. The rear-view mirror was depicted graphically on the main road view screen and placed in the correct position from the drivers' point of view. Door

FIGURE 12.2 The simulator environment with the HUD reflector screen, cluster, and central control visible.

mirrors were provided via small LCDs mounted on the cabin in the traditional positions. STISIM Drive simulation software was employed to generate the driving environment, mirrors, and sound. A control desk and PC behind the simulator provided the experimenters with the 'Wizard of Oz'-style ability to supply responses to the takeover protocols. Audio sounds and vocalisations were generated automatically by the automation system and played through speakers mounted in the simulator cabin. A Microsoft Surface tablet running a Tetris-style game provided the secondary task during automation; this tablet was networked with the automation system to provide a visual alert to the driver as part of the takeover protocol.

12.2.4 HMI DESIGN AND CUSTOMISATION

Between trials, participants had the opportunity to customise their HMI using an alert settings screen (Figure 12.1) that appeared on the central console between trials. This provided the ability to hide or display aspects on the HUD, cluster, and central console (Figure 12.3), including wording regarding driving mode, a driving mode icon, a time to takeover readout, takeover question wording, and a coloured edge frame that indicated driving mode. A 'road sense' view could additionally be displayed on the cluster or central console, which graphically indicated the presence of vehicles around the drivers' car.

Analog settings included the ability to adjust the brightness of ambient light which indicated automation mode based on the colour. Audio volume could be adjusted that would affect both audio tones and vocalisations generated by the automation system. The vibrate setting allowed the intensity of the haptics to be adjusted. The takeover questions setting allowed the driver to specify the number of questions they would like the system to ask as part of the takeover protocol.

When a participant completed making their adjustments, the settings were saved, and the customised HMI was employed in the HMI of the next trial.

12.2.5 PROCEDURE

After arriving at the JLR facility, participants were welcomed and presented with a participant information sheet informing them of the details of the study. Their

FIGURE 12.3 The three visual aspects of the HMI: on the left is the central console, bottom-right is the cluster, and the HUD is shown on the top-right.

right to halt the study at any time was explained; they were then provided with an informed consent form which they had to read, initial, and sign in order for the study to continue. On completion of the consent form, participants were given a bibliographical form to complete to capture demographics data. They were then introduced to the simulator. The main controls were explained, and they were informed of the functionality of the customisation screen and the effects it has on the HMI. The customisation screen provided dynamic updates of the HMI as changes were made, allowing the participant to see the effects, and to test audio levels, light, and haptic intensities. Participants then took part in a test run, where they experienced three takeovers after 60-s OOTL intervals. After completion of the test run, the customisation settings were reset to defaults and the study started. Trials started with the participant being asked to accelerate onto the motorway, join the middle lane, keep up with traffic, and follow the HMI's instructions. After a period of approximately one minute, the HMI indicated to the participant that automation was available and informed them via text, icon, vocalisation, and two flashing green steering wheel buttons. Participants then activated automation by pressing the two steering wheel buttons; this was followed by the HMI, indicating that automation was now in control of the vehicle. Participants then picked up the secondary task tablet and started to play the Tetris secondary task. After a period of either 1 or 10 min, dependent upon counterbalancing, the automation indicated via the HMI that the driver was required to get ready to take control. Participants were expected to put aside their secondary task, and follow the instructions presented by the HMI. The takeover protocol consisted of a set of questions designed to raise situation awareness. These questions could be presented in vocal, word, and icon form, dependent upon the HMI customisations. Participants responded vocally to each question, the answers to which were judged by an experimenter taking the part of the automation system using the Wizard of Oz approach. Incorrect or missed questions were repeated twice before moving to the next. Once the questions were answered by the participant, the HMI indicated for them to take control, which they did by pressing the two green buttons on the steering wheel. This constituted one takeover, and the process was repeated twice more. After completion, the participant was asked to pull safely to the hard shoulder and stop the vehicle. The customisation screen was then displayed on the central console, and the participant was asked to make adjustments to the HMI for the next trial. Once the participant was happy with their customisations, these were saved, and they were presented with three questionnaires in electronic form. Completion of the questionnaires marked the end of a trial. Four trials took place in total, two featuring 1-min OOTL durations and two featuring 10-min OOTL durations. Once all four trials were complete, the participant filled out a post-trial interview questionnaire.

12.2.6 ANALYSIS

The sample size of 65 participants meets the standard for automobile research and adheres to the rules of thumb suggested for multivariate analysis. Analysis will consist of subjective measures such as workload (NASA-TLX), usability (system usability scale), acceptance (system acceptance scale), trust in automation, customisation settings, and preference data being recorded using a questionnaire upon completing the study. Quantitative measures related to performance will be collected (including lateral positioning, lateral velocity, headway, and reaction times). This data will be analysed using analysis of variance techniques (ANOVA, and regression analyses where appropriate). Data will also be collected from visual and audio recordings which will then be repurposed to signal detection theory analysis to determine the accuracy of behavioural predictions using operator event sequence diagrams. Visual and audio recordings will also be analysed using thematic analysis to code the major themes that emerge regarding drivers' experiences in the study.

Performance data, consisting of speed, throttle, steering, brake, headway, and lateral lane position, was recorded at 20 Hz; however, this was not to millisecond accuracy. To mitigate this issue, the method adopted by Eriksson and Stanton (2017) was employed and data was binned into temporal portions of 0.333 s each. The 25-s post-takeover data set was therefore represented by 75 bins, each containing approximately six elements of data for each channel. Some pre-processing was necessary to remove data where participants had activated or deactivated automation at incorrect times, resulting in periods of less than 25 s of manual control, post-automation. Three participant's data were missing due to unforeseen circumstances and a technical issue. Some channels also required some filtering prior to analysis; all channels were checked for anomalous values, such as consistent zero values caused by technical issues. Steering and lane position data was filtered to remove data that included lane changes; this was done by removing all data within takeovers that exhibited lane position values that indicated a lane change. Speed and lane position data was also normalised as the automation did not maintain a consistent lateral position within lane and the speed prior to takeover was variable in response to position within traffic. Data for steering and lane position were converted to absolute values for analysis. Data was visualised using time series graphs, with mean values and standard deviations plotted for each channel. Negative log-transformed p values were also used to compare data from trials 1 and 4; these were plotted to reveal significant differences that may have been a result of the customisation. The red dashed line represents an alpha of 0.05, while the blue dotted line represents a Bonferroni-corrected alpha of 0.00067. Finally, Cohen's D was also employed to compare trials 1 and 4 and provide further insight into the effects of customisation.

12.3 RESULTS

12.3.1 Speed

Speed values were provided from STISIM in mph. At the point of takeover, not all participants' vehicles were travelling at the same speed due to traffic variations and other variables. Speed values were therefore normalised.

On examination of mean speeds in Figure 12.4, the gradient of the slope for trial 4 indicates the lowest level of acceleration between 5 and 10 s, which is good for both economy and safety and could be a result of the drivers' heightened awareness causing them to apply throttle earlier and therefore requiring less application later. Participants in trial 1 allowed their speed to drop by a mean value of approximately 1.5 mph in the first 5 s post-takeover; this then required them to accelerate more, between 3 and 11 s,

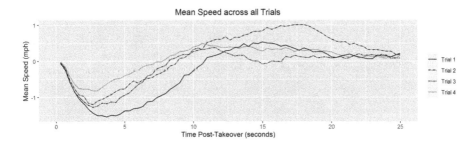

FIGURE 12.4 The mean speed per trial across all participants. Speed was normalised; therefore, the graph shows the change relative to the start speed.

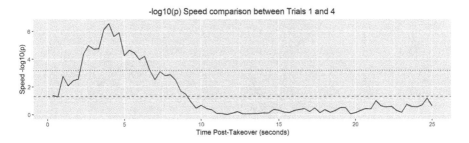

FIGURE 12.5 The significance of a speed comparison between trials 1 and 4 that represent default settings and iterated customised settings.

to regain the lost speed. They surpass the automation's set speed at 11 s and have to decelerate from 15 s. By around 7.5 s in trial 4, participants have passed the speed at which the automation was driving, and they have to decelerate. In trial 4, participants match the speed at which the automation was driving at around 25 s, while participants in all other trials match this speed at approximately the same time.

Examining the $-\log_{10}(p)$ values for speeds in the first and last trials (Figure 12.5), the highest indicated significance at around 4 s post-takeover, likely due to the drivers accelerating later in trial 1 than they do in trial 4. This is statistically significant between approximately 2 and 7 s. Speeds between trials then match at around 10 s.

12.3.2 THROTTLE

Figure 12.6 shows the mean throttle percentage applied by participants within each trial, by time. It is clear that throttle usage was similarly applied in all trials; however, less throttle was applied around the time of takeover in trial 1 (approximately 7%) compared to trial 4 (10%).

The $-\log_{10}(p)$ values comparing trials 1 and 4 (Figure 12.7) indicate the effect of the participants in trial 4 applying more throttle at the point of takeover, as seen in the mean throttle percentage graph. The spike at around 24 s was potentially due to participants in trial 1 gently accelerating from around 22 s, whereas in trial 4 they tended to maintain a more consistent level of throttle.

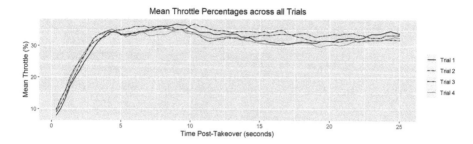

FIGURE 12.6 The mean throttle percentage applied by participants within each trial, by time.

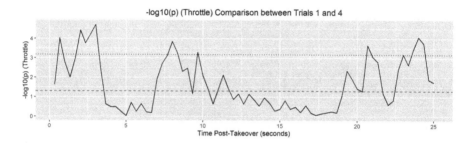

FIGURE 12.7 The significance of the throttle values, comparing trial 1 with trial 4.

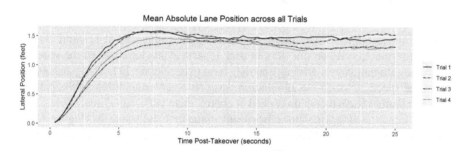

FIGURE 12.8 Absolute changes in lane position for each trial. The values represent mean deviations from the starting position.

12.3.3 LANE POSITION

Mean absolute lane position (Figure 12.8) indicates that in all trials, drivers deviate laterally by a mean of approximately 1.4 ft. The deviations do not appear to correlate with trial order. Trials 3 and 4 appear to result in the least lateral deviation; however, this is not a significant finding.

Figure 12.9 indicates that absolute lateral position values in trials 1 and 4 were significant primarily between 1 and 4 s post-takeover. This is reflected in the mean lane position graph that indicates a greater lateral deviation occurring faster in trial 1 compared to trial 4.

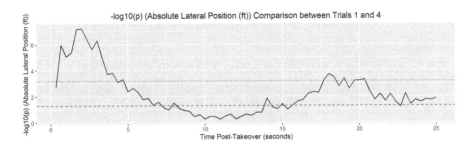

FIGURE 12.9   The significance of a comparison of absolute lateral position data in trials 1 and 4.

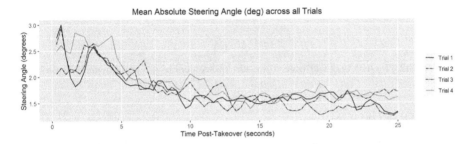

FIGURE 12.10   The mean absolute steering angles across all trials.

12.3.4   STEERING ANGLE

Steering values (Figure 12.10) start at nonzero, likely due to the automation handing over while cornering. There appears to be no correlation between trial number and steering angle. Values are largely similar across trials with all exhibiting higher initial values of steering angle, peaking at between 2.5° and 3.2° before gradually reducing by approximately 10 s post-takeover to between 1.25° and 2°.

12.3.5   TAKEOVER TIME

TOTs representing the durations between the start of the takeover protocol and the participants transitioning into manual control were calculated for trials 1 and 4. This enabled the effect of customisation on protocol TOT to be identified. In all cases, data for the final takeover of control was utilised as participants would be at their most experienced with the system at this point.

Trial 1 used the default settings and trial 4 employed the final iteration of customisation settings.

Examination of Figure 12.11 reveals a longer TOT for trial 1 of approximately 30 s; when using default HMI settings, the spread of TOTs was also narrow with an interquartile range of approximately 4 s. When using customised HMIs, participants' median TOT was less, at approximately 24 s, but with a greater spread (interquartile range of approximately 12 s).

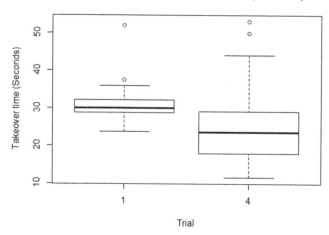

FIGURE 12.11   The takeover time from start of protocol to the point manual control was resumed.

12.4   DISCUSSION

The experiment was designed to examine the post-takeover performance effects of a customisable TOR interface. Post-takeover performance can reveal behaviours that might compromise driver safety. By examining longitudinal and lateral performance data from baseline HMI settings and from those that have been iteratively customised, the resultant performance effects are identified. Clark and Feng (2017) state that an effective takeover is one that results in the least deviation in speed and lateral position; therefore, any significant differences in performance data will highlight the potential areas of improvement or degradation due to customisation. Due to the level of customisation available and the multimodal nature of the HMI, many of the advantages of multimodality (McDonald et al. 2019; Naujoks, Mai and Neukum 2014; Toffetti et al. 2009) are potentially available to drivers; however, this is tempered by the subjectivity associated with customisation. For example, an interface that is subjectively 'good looking' is not necessarily usable; the usability–subjectivity relationship is not mutually exclusive but there can be a bias. It is entirely dependent upon the drivers' choice, and this is the rationale for examining not only how the HMI might have been improved, but also its effect on post-takeover performance, particularly with regard to safety. There is a difficulty associated with exploring the effects of customisation with respect to the literature on performance, as usually the HMI parameters represent independent variables, whereas in this experiment, complete customisation of HMI parameters was allowed in the three trials following the control trial. This experiment included the use of a handheld device that displayed a TOR, similar to what was investigated by Yoon, Kim, and Ji (2019); they stated that TOTs were likely to be reduced, which suggests a raised level of readiness. However, these alerts were available in all trials and were not customisable; therefore, their effect was not able to be measured. They do represent an important factor though, as Borojeni et al. (2016) describe the importance of disengaging participants from their secondary task. They may have contributed to improving performance overall. Zhang et al. (2019) highlighted the advantage of multimodality in reducing TOT; indeed, Politis, Brewster, and Pollick (2017) warned that in critical takeovers, unimodal TORs could result in performance degradation in terms of lateral control and accuracy of response; however, no participants customised their settings to the point where the interface became unimodal.

12.4.1 Speed and Throttle

Speed and throttle input are clearly closely coupled; the results showed that while participants in all trials allowed their speed to drop immediately post-takeover, in trial 4, participants' speed dropped less than in trial 1. Examining the $-\log_{10}(p)$ values indicates that this is a statistically significant difference between 2 and 7 s post-takeover. The fundamental reason for the speed drop in all trials is considered to be due to participants resuming control when they have little or no throttle applied, and this correlates with the throttle data. Throttle data shows a similar behaviour of throttle input in all trials; however, the throttle is applied earlier in trial 4 than in trial 1, resulting in the significantly greater mean throttle percentage values for trial 4 and subsequent reductions in speed lost post-takeover. The reduced speed reduction in trial 4 could be considered to be an improvement in longitudinal performance, as Hoogendoorn, van Arerm, and Hoogendoom (2014) describe how speed variability may influence traffic stability. Variations in speed may cause string instability; small decelerations can be amplified in such a way to cause large amplitude traffic shockwaves (Wu, Bayen and Mehta 2018). These shockwaves can result in stop/go traffic conditions with resultant safety consequences (Kim and Zhang 2008). Minimising the reduction in speed post-takeover is therefore considered advantageous as it has the potential to mitigate the generation of traffic shockwaves and their effects. Results therefore indicate that customisation has multiple benefits with respect to longitudinal performance. Regarding usability, McDonald et al. (2019) linked longitudinal performance with secondary task engagement, driving environment, TOR modality, and trust. It might be that participants are optimising their interfaces in terms of TOR modality and the ability to select the desired functionality may have improved trust levels. However, it is difficult to discern whether trust may also have been gained through iterative exposure to the customisations and systems. Eriksson et al. (2019) state that when applying linguistics theory to HMI design, the best communication is achieved when information is relevant, and the density is not overly high. Customisation may be providing the participants with the ability to match these with their desires and needs, thus improving their performance by increasing readiness, resulting in more throttle being applied at the point of takeover.

Borojeni et al. (2016) posited the theory that drivers should be disengaged from their secondary task prior to being presented with information on required actions. Van der Heiden, Iqbal, and Janssen (2017) also advocate the priming of drivers. In this experiment, participants were provided with pre-warnings at fixed durations prior to takeover. Additionally, the takeover protocol provided participants with alerts and time to prepare to take control, and yet they failed to apply enough throttle to match the speed at takeover successfully. This is potentially due to unfamiliarity with the car's required throttle positions or possibly an unwillingness to apply much throttle without experiencing the effect it has once manual control is resumed. It suggests that a method by which drivers can visualise the automation's throttle input and match it prior to taking manual control might prevent the drop in speed.

Telpaz et al. (2015) found that when presenting participants with haptic cues, they started to accelerate earlier on manual takeover; they considered that this was potentially due to participants being more prepared. The customisations that we provided included haptics as part of the takeover cue, but this could be turned off as well as increased in intensity. However, with iterative customisations, participants will be changing settings holistically to match what they consider to be improving preparedness and performance. Indeed, Harvey et al. (2011) state that an IVIS should account for a wide range of different user characteristics, in addition to other aspects such as driving environments. The ability to customise allows drivers to adapt their interface specifically to their needs.

12.4.2 LANE POSITION AND STEERING ANGLE

Lane position and steering angle are also tightly coupled; the data revealed a significant difference in steering angle almost immediately post-takeover due to customisation. This appeared to be caused by participants in trial 1 applying a considerable steering input prior to correcting it back to a more nominal level, whereas participants in trial 4 maintained a relatively stable steering input. In terms of absolute lateral position, the findings indicated that there was a statistically significant effect of customisation on lateral position in the period immediately post-takeover. Trial 1 using the default settings caused a more accelerated change in lateral position; indeed, it represented the greatest rate of change of position of all trials and resulted in a peak lateral position change significantly more than trial 4. While the difference is significant, the magnitude of difference was small; however, even small lateral movement can place a vehicle outside a lane if it is close to a boundary. Green (1995) describes how lateral deviations can result from fatigue and attentional demand; therefore, it is possible that the customisable interface may play a role in reducing fatigue and/or reducing attentional demand. These findings are in line with what might be expected, as a driver who has iteratively customised their interface is likely to have adjusted it closely to their preferences and characteristics and therefore improved the usability of the interface (Stanton and Baber 1992). The increase in usability may result in improved acceptance (Stanton and Baber 1992), but also performance improvement, through increased readiness and situation awareness. One example might be a customisation of volumes to enable the participant to hear the TOR information more easily. Hickson et al. (2010) specifically noted how hearing impairment can degrade driving performance. Wintersberger et al. (2018) reported significant differences in lateral positioning when employing the Daimler lane change performance metric depending upon whether IVIS-based TORs were issued between tasks or within tasks. This experiment employed a secondary task with a Tetris-style game; however, transitions between games were not deliberately aligned with TORs. We additionally provided pre-warnings, and this may have resulted in participants self-limiting their gameplay in expectation of upcoming TORs.

12.4.3 TAKEOVER TIME

The TOTs comparing default settings (trial 1) with iteratively customised settings (trial 4) indicated that customised settings result in a faster TOT but with a larger spread of values. A major contributory factor to this finding is the customisable number of questions within the protocol; by reducing the number of questions, the TOT is subsequently reduced. The spread may also be partially explained by changes in other settings. Vogelpohl et al. (2017) described the HMI as an influential factor in TOT, specifically the modalities employed as part of the TOR. However, it remains a useful finding as it indicates the benefit of providing drivers with the ability to adjust the number of protocol questions to their requirements. The unique nature of the protocol TOT renders it somewhat difficult to compare with the existing literature as it is not just a response; it includes multiple processes. The takeovers within this experiment were also ostensibly driver-paced, as there was no impending hazard or requirement to resume manual control, other than the TOR itself. The driver-paced nature may contribute to the takeover quality as the participants would not feel rushed. Vogelpohl et al. (2017) stated that when more time is provided for takeovers, the takeover quality is higher and TOT is lower. Zhang et al. (2019) concluded that a low TOT should not, in all cases, be considered as a design goal. Drivers take more time if it is available; however in the case of a customisable HMI and a driver-paced takeover, a balance is naturally found.

The performance of participants utilising an iteratively customised TOR interface (trial 4) clearly demonstrates significant performance improvements, in terms of both lateral control and longitudinal control. The customisable interface appears to provide drivers with not only a subjectively more pleasing experience, but driver readiness appears to be improved, manifesting as earlier throttle response and less steering variability immediately post-takeover. An optimal TOR will result in a seamless transition of control (Clark and Feng 2017), and customisable interface appears to contribute towards achieving this goal.

12.4.4 LIMITATIONS

All trials did indicate a trend towards a lateral movement to the left on examination of normalised, non-absolute values over time. This is postulated to have been due to the automation software placing the vehicle in an unnaturally right-biased position within lane. In some trials, participants initiated lane changes that would have skewed the lane position and steering angle data. A threshold was used to exclude data sets that included lane changes.

12.5 CONCLUSIONS

Providing drivers with the ability to select the takeover information that they desire in terms of modalities, densities, and location enables them to satisfy their perceived requirements and desires. The benefits of attending to the users' and tasks' characteristics are well established in the literature; however, the customisations will be made based on the drivers' subjective decisions. This raises the question of whether the customisations that are made are beneficial in terms of driver performance, in addition to being perceived as improvements. Examination of speed and lateral position as representative metrics for takeover performance, as well as correlated control inputs of throttle and steering, determined that customisation had no negative effects, and may actually improve performance. Lane position was significantly improved by customisation, albeit with a small magnitude, and examination of speed post-takeover revealed some benefit to customisation as participants appeared to have a higher level of readiness, in terms of throttle position and speed. This can have beneficial effects in terms of third-party safety through the reduction of traffic shockwaves. Additionally, if the drivers' readiness levels are higher for throttle, it is conceivable that they are also higher for other aspects of control such as steering and braking. A driver with higher readiness levels will likely react faster to situations that require driver inputs, again representing a positive outcome of customisability in terms of safety.

ACKNOWLEDGEMENTS

This work was supported by Jaguar Land Rover and the UK-EPSRC Grant EP/N011899/1 as part of the jointly funded 'Towards Autonomy: Smart and Connected Control (TASCC) Programme' HI:DAVe Project.

REFERENCES

Borojeni, S. S., L. Chuang, W. Heuten, and S. Boll. 2016. "Assisting Drivers with Ambient Take-Over Requests in Highly Automated Driving." *AutomotiveUI - Proceedings of the 8th International Conference on Automotive User Interfaces and Interactive Vehicular Applications*. Ann Arbor, MI, USA: Association for Computing Machinery. 237–244. https://doi.org/10.1145/3003715.3005409.

Clark, H., and J. Feng. 2017. "Age differences in the takeover of vehicle control and engagement in non-driving-related activities in simulated driving with conditional automation." *Accident Analysis & Prevention, 106* 468–479. DOI: 10.1016/j.aap.2016.08.027.

Eriksson, A., and N. A. Stanton. 2017. "Driving performance after self-regulated control transitions in highly automated vehicles." *Human Factors, 59*, 8 1233–1248. DOI: 10.1177/0018720817728774.

Eriksson, A., S. M. Petermeijer, M. Zimmermann, J.C.F. de Winter, K. J. Bengler, and N. A. Stanton. 2019. "Rolling out the red (and green) carpet: Supporting driver decision making in automation-to-manual transitions." *IEEE Transactions on Human-Machine Systems, 49*, 1 20–31. DOI: 10.1109/THMS.2018.2883862.

Green, P. 1995. *Measures and methods used to assess the safety and usability of driver information systems*. Technical Report FHWA-RD-94-088, McLean, VA, USA: Federal Highway Administration.

Harvey, C., N. A. Stanton, C. A. Pickering, M. McDonald, and P. Zheng. 2011. "A usability evaluation toolkit for in-vehicle information systems (IVISs)." *Applied Ergonomics, 42*, 4 563–574. DOI: 10.1016/j.apergo.2010.09.013.

Hickson, L., J. Wood, A. Chaparro, P. Lacherez, and R. Marszalek. 2010. "Hearing impairment affects older people's ability to drive in the presence of distracters." *Journal of the American Geriatrics Society, 58*, 6 1097–1103. DOI: 10.1111/j.1532-5415.2010.02880.x.

Holländer, K., and B. Pfleging. 2018. "Preparing drivers for planned control transitions in automated cars." *Proceedings of the 17th International Conference on Mobile and Ubiquitous Multimedia.* Cairo, Egypt: Association for Computing Machinery. 83–92. https://doi.org/10.1145/3282894.3282928.

Hoogendoorn, R., B. van Arerm, and S. Hoogendoom. 2014. "Automated driving, traffic flow efficiency, and human factors: Literature review." *Transportation Research Record: Journal of the Transportation Research Board, 2422,* 1 113–120. DOI: 10.3141/2422-13.

Kim, T., and H. M. Zhang. 2008. "A stochastic wave propagation model." *Transportation Research Part B: Methodological, 42,* 7 619–634. DOI: 10.1016/j.trb.2007.12.002.

McDonald, A. D., H. Alambeigi, J. Engström, G. Markkula, T. Vogelpohl, J. Dunne, and N. Yuma. 2019. "Toward computational simulations of behavior during automated driving takeovers: A review of the empirical and modeling literatures." *Human Factors: The Journal of the Human Factors and Ergonomics Society, 61,* 4 642–688. DOI: 10.1177/0018720819829572.

Melcher, V., S. Rauh, F. Diederichs, H. Widlroither, and W. Bauer. 2015. "Take-over requests for automated driving." *Procedia Manufacturing, 3* 2867–2873. DOI: 10.1016/j.promfg. 2015.07.788.

Miura, S. 2014. "Direct adaptive steering – Independent control of steering force and wheel angles to improve straight line stability." *5th International Munich Chassis Symposium 2014.* Wiesbaden, Germany: Springer. 75–89. https://doi.org/10.1007/978-3-658-05978-1_8.

Naujoks, F., C. Mai, and A. Neukum. 2014. "The effect of urgency of take-over requests during highly automated driving under distraction conditions." *5th International Conference on Applied Human Factors and Ergonomics.* Krakow, Poland: AHFE. 431–438.

Naujoks, F., K. Wiedermann, N. Schömig, S. Hergeth, and A. Keinath. 2019. "Towards guidelines and verification methods for automated vehicle HMIs." *Transportation Research Part F: Traffic Psychology and Behaviour, 60* 121–136. DOI: 10.1016/j.trf.2018.10.012.

Petermeijer, S. M., S. Cieler, and J. C.F. de Winter. 2017. "Comparing spatially static and dynamic vibrotactile take-over requests in the driver seat." *Accident Analysis & Prevention, 99* 218–227. DOI: 10.1016/j.aap.2016.12.001.

Politis, I., S. A. Brewster, and F. Pollick. 2014. "Evaluating multimodal driver displays under varying situational urgency." *Proceedings of the SIGCHI Conference on Human Factors in Computing Systems.* Toronto, Canada: Association for Computing Machinery. 4067–4076. https://doi.org/10.1145/2556288.2556988.

Politis, I., S. Brewster, and F. Pollick. 2017. "Using multimodal displays to signify critical handovers of control to distracted autonomous car drivers." *International Journal of Mobile Human-Computer Interaction, 9,* 3 1–16. DOI: 10.4018/ijmhci.2017070101.

Regan, M. A., J. D. Lee, and K. Young. 2008. *Driver Distraction.* Boca Raton, FL: CRC Press.

SAE J3016. 2016. *Taxonomy and Definitions for Terms Related to Driving Automation Systems for On-Road Motor Vehicles.* https://www.sae.org/standards/content/j3016_201806/ (accessed 3 November 2020).

Shackel, B. 1991. "Usability – Context, framework, definition, design and evaluation." In *Human Factors for Informatics Usability,* by B. Shackel and S. J. Richardson, 21–37. New York: Cambridge University Press.

Stanton, N. A., and C. Baber. 1992. "Usability and EC Directive 90270." *Displays, 13,* 3 151–160. DOI: 10.1016/0141-9382(92)90017-L.

Telpaz, A., B. Rhindress, I. Zelman, and O. Tsimhoni. 2015. "Haptic seat for automated driving: Preparing the driver to take control effectively." *AutomotiveUI: Proceedings of the 7th International Conference on Automotive User Interfaces and Interactive Vehicular Applications.* Nottingham, UK: Association for Computing Machinery. 23–30. https://doi.org/10.1145/2799250.2799267.

Toffetti, A., E. S. Wilschut, M. H. Martens, A. Schieben, A. Rambaldini, N. Merat, and F. Flemisch. 2009. "CityMobil: Human factor issues regarding highly automated vehicles on eLane." *Transportation Research Record: Journal of the Transportation Research Board, 2110,* 1 1–8. DOI: 10.3141/2110-01.

van den Beukel, A. P., M. C. van der Voort, and A. O. Eger. 2016. "Supporting the changing driver's task: Exploration of interface designs for supervision and intervention in automated driving." *Transportation Research Part F: Traffic Psychology and Behaviour, 43* 279–301. DOI: 10.1016/j.trf.2016.09.009.

van der Heiden, R. M.A., S. T. Iqbal, and C. P. Janssen. 2017. "Priming Drivers before Handover in Semi-Autonomous Cars." *Proceedings of the 2017 CHI Conference on Human Factors in Computing Systems.* Denver, US: Association for Computing Machinery. 392–404. https://doi.org/10.1145/3025453.3025507.

Vogelpohl, T., M. Vollrath, M. Kühn, T. Hummel, and T. Gehlert. 2017. *Übergabe von hochautomatisiertem Fahren zu manueller Steuerung.* Research Report, Berlin, Germany: GDV.

Wandtner, B., N. Schömig, and G. Schmidt. 2018. "Secondary task engagement and disengagement in the context of highly automated driving." *Transportation Research Part F: Traffic Psychology and Behaviour, 58* 253–263. DOI: 10.1016/j.trf.2018.06.001.

Wintersberger, P., A. Riener, C. Schartmüller, A.K. Frison, and K. Weigl. 2018. "Let me finish before I take over: Towards attention aware device integration in highly automated vehicles." *AutomotiveUI: Proceedings of the 10th International Conference on Automotive User Interfaces and Interactive Vehicular Applications.* Toronto, Canada: Association for Computing Machinery. 53–65. https://doi.org/10.1145/3239060.3239085.

Wu, C., A. M. Bayen, and A. Mehta. 2018. "Stabilizing traffic with autonomous vehicles." *2018 IEEE International Conference on Robotics and Automation.* Brisbane, Australia: IEEE. 6012–6018. doi: 10.1109/ICRA.2018.8460567.

Yoon, S. H., Y. W. Kim, and Y. G. Ji. 2019. "The effects of takeover request modalities on highly automated car control transitions." *Accident Analysis & Prevention, 123* 150–158. DOI: 10.1016/j.aap.2018.11.018.

Zhang, B., J. de Winter, S. Varotto, R. Happee, and M. Martens. 2019. "Determinants of take-over time from automated driving: A meta-analysis of 129 studies." *Transportation Research Part F: Traffic Psychology and Behaviour, 64* 285–307. DOI: 10.1016/j.trf.2019.04.020.

13 Effects of Interface Customisation on Drivers' Takeover Experience in Highly Automated Driving

Jisun Kim, Kirsten M. A. Revell,
James W. H. Brown, Joy Richardson,
and Jediah R. Clark
University of Southampton

Nermin Caber
University of Cambridge

Michael Bradley
University of Cambridge

Patrick Langdon
Edinburgh Napier University

Simon Thompson and Lee Skrypchuk
Jaguar Land Rover

Neville A. Stanton
University of Southampton

CONTENTS

13.1   INTRODUCTION

Highly automated driving (HAD) provides benefits to drivers which have not been enabled in lower levels of automated driving. Freedom to disengage from driving, and availability to engage in non-driving-related secondary tasks are examples. However, HAD still involves control transitions between the driver and the vehicle as it is not entirely automated. They include situations when drivers must drive, when all the required conditions for the autonomous features are not met (SAE International 2018). In this transition period, drivers need to rapidly shift their attention to manual driving and understand the situation by acquiring related information to be able to take control back safely. This process occurs after automated driving in which drivers' level of situation awareness could be decreased as a result of a lack of interaction with the automated system (Li et al. 2019; Yoon, Kim and Ji 2019). In the takeover situation, regaining situation awareness comfortably and rapidly is vital for the driver's safety and subjective experience (Miller, Sun and Ju 2014; Lv et al. 2018).

Takeovers can be affected by situational circumstances, the driver's state, as well as their differences in takeover abilities. Accordingly, it is proposed that future systems need to be adjusted to better suit the driver, and the context to promote drivers' experience, and safer transitions of control (Walch et al. 2017). Moreover, individual variance in takeover times was observed that could hardly be represented by the mean or the median. This justifies a need for an inclusive design approach to developing automation settings (Eriksson and Stanton 2017).

Customisation has drawn attention because it helps improve users' subjective experience regarding the use of products by offering them an opportunity to actively influence the technology. For instance, users can transform the product or the system to better suit their preferences and needs, rather than consuming the content passively (Marathe and Sundar 2011). Further, users are able to adapt service interfaces, and select required information to match their needs (Nguyen and Colman 2010) based on their own preferences and interests (Macías and Paternò 2008; Jorritsma, Cnossen and van Ooijen 2015). It is considered a secondary activity for users to adapt devices and interfaces to conduct a task more efficiently. In turn, this is expected to improve user experience (Marathe and Sundar 2011). Moreover, it has been gaining more importance as a strategy to meet customers' specific needs, and to enhance customer satisfaction (Ma, Wang and Li 2019). Harvey et al. (2011) suggest adaptive interfaces as one of the possible solutions to minimising usability problems associated with managing multiple functions in in-vehicle information systems. Customisable human–machine interface (HMI) could be beneficial to improve users' experience, particularly in takeover situations because they need to rebuild situation awareness and make a decision within a short period of time. The process can be influenced by individual factors (such as long-term memory, information processing mechanisms, abilities, prior experience, and training) (Endsley 1995). Also, customisable HMI could embrace individual differences in perceptions (such as speed perception, reaction times, and time constant) that could affect performance (Butakov and Ioannou 2015).

13.1.1 Driver Experience during Takeover

Understanding drivers' subjective experience is of importance for designing autonomous driving information system that could be more widely accepted by drivers. User experience with vehicle information systems (including modality, format, and information provision) could influence perceived usability and acceptance about the system (Petersson, Fletcher and Zelinsky 2005). Hence, it is worth making an effort to investigate subjective experience to enhance autonomous driving systems that could be adopted more by the public. The background for this claim is that users' acceptance of the system predicts actual usage and purchase of the technology (Davis 1989; Venkatesh et al. 2003). Acceptance could in turn lead to realisation of potential benefits of autonomous driving (Fraedrich and Lenz 2016). A number of factors have been confirmed as antecedents of the acceptance, such as perceived usefulness and trust (Choi and Ji 2015). Usability, trust, and usefulness have been identified as factors affecting the experience of autonomous transition of control (Linehan et al. 2019). Furthermore, perceived workload could have an impact on drivers' acceptance (Nees 2016), which is constructed through users' interaction with the product/system (Hart and Staveland 1988). In this context, this study will focus on drivers' perceived workload, usability, trust, and acceptance. The definitions and importance of the factors will be described as follows.

Workload represents 'the cost of accomplishing mission requirements for the human operator' (Hart 2006). It matters because users of the system often cannot accomplish tasks expected to be done with accuracy in a timely and reliable manner. In addition, the human cost is excessively high to maintain performance (Hart 2006). Workload of a task imposed on an operator results from objectives of the task, structure, duration, and resources provided by the human and system (Hart and Staveland 1988). In HAD, workload could be reduced because physical activity to operate the steering wheel and pedals, and cognitive activity related to manual driving are reduced. Additionally, HAD could increase workload when the driver needs to monitor the automation and stay vigilant (de Winter et al. 2014).

Usability in HMI in autonomous vehicles is of importance because drivers' control and autonomy are mediated by the HMI. Usability is understood to be affected by intuitiveness, corresponding to perceived ease of use, as explained by Davis's (1989) technology acceptance model. Intuitiveness could lower the barrier to using the technology and increase its accessibility, and intuitiveness can be assessed through a usability test. Usability is identified as 'the extent to which a product can be specified users to achieve specified goals with effectiveness, efficiency and satisfaction in a specific context of use' (ISO 1998). It is also defined as a characteristic of the 'interaction between a product, a user, and the task, or set of tasks, that he or she is trying to complete' (Jordan 1998). Understanding of the users and their characteristics is essential to investigate usability (Jordan 1998).

Acceptance is the key antecedent for benefits of new automotive technologies to be realised. It is drivers' willingness to incorporate an in-vehicle system in their driving, or intention to utilise the system if not available. It is a multidimensional concept, involving dimensions such as willingness to pay and intention to use (Xu et al. 2018; Adell, Varhelyi and Nilsson 2014). Acceptance of technology is often explained by the technology acceptance model, which is extensively adopted to acceptance behaviour of information technology. The model postulates that intention to use a technology is determined by perceived usefulness and ease of use (Venkatesh et al. 2003; Davis, Bagozzi and Warshaw 1989; Choi and Ji 2015). Intention to use a technology (e.g. highly autonomous vehicles) is worth investigating rather than actual use of it at an early stage (Choi and Ji 2015).

Trust is considered as a significant factor predicting the use of autonomous vehicles (Kaur and Rampersad 2018). It is defined as

> willingness of a party to be vulnerable to the actions of another party based on the expectation that the other will perform a particular action important to the trustor, irrespective of the ability to monitor and control that other party

> *Mayer, Davis, and Schoorman (1995).*

Performance, process, purpose, and foundation are identified as aspects of trust in the automation domain. Performance is based upon automation consistency, stability, and desirability of performance. Process relates to users' knowledge about the system's methods/rules or algorithms which governs the systems to function. Purpose represents the underlying intentions of the system. A foundation of trust helps make other dimensions of trust possible (Lee and Moray 1992).

13.1.2 RELATED WORK

The effects of HMI information settings on drivers' subjective experience in takeover situations in related studies were determined. Yoon, Kim, and Ki (2019) identified that multiple modalities of takeover request (TOR) (a combination of two or three modalities) were perceived to be more useful, satisfactory, and trustworthy. They then suggested to consider the inclusion of multimodal alert system by providing evidence that visual unimodal TOR may not be sufficient to alert driver to retake control as quickly as possible. The drivers conducted takeover after engaging in non-driving-related tasks with varying degrees in simulated HAD environment. Lu et al.'s (2019) study results showed that provision of monitoring request (MR) followed by TOR could lead to a better subjective experience than TOR issued in the absence of MR. The study was motivated by the assumption that the presentation of uncertainty in a traffic situation and MRs could increase drivers' situation awareness and cognitive/mental preparedness. The results presented that the participants' workload was lower, acceptance and trust were higher, and their performance was better in the MR plus TOR condition. Körber, Prasch, and Bengler's (2018) research demonstrated that post hoc explanation of TOR did not influence drivers' trust and acceptance but it helped enhance their understanding about the system. In the experiment, they engaged in a visually and manually demanding non-driving-related task during the conditional automated driving in the simulator environment. Eriksson et al. (2019) evaluated the effects of augmented visual presentations on drivers' reaction to TOR and subjective experience in a simulated conditional automated driving environment. Significantly higher usefulness was reported for the conditions with augmented reality overlay that could support decision-making for lane change and braking scenarios than the condition without any augmented effects (baseline condition). Although there was not a significant difference in perceived workload among the conditions, workload in augmented conditions seemed to be lower than the baseline condition. Interpretation was added that information that facilitates the decision-making process could lead to the improvement of satisfaction and usefulness about the takeover experience. Li et al. (2019) identified that prompting TOR with information about vehicle status along with the reason to take over led to greater positive attitudes, lower subjective workload, and better performance. Further, TOR with vehicle status information also produced positive impacts on performance, attitudes, and workload. Both conditions produced better results in comparison to baseline TOR, which did not inform the drivers about the vehicle status or reason for takeover. It was interpreted that improved situation awareness of the drivers being in better informed condition about the status and reasons resulted in better takeover performance and attitudes.

In summary, some benefits of multimodal HMI have been suggested as an efficient tool for alerting drivers during takeover situations, which helped improve their subjective experience (Yoon, Kim and Ji 2019). In addition, situation awareness could be enhanced by being better informed, e.g. with information about vehicle status and context of takeover, which could lead to a better experience about the takeover and performance (Li et al. 2019; Lu et al. 2019). However, despite the significance and need for customisation described previously, few attempts have been made to introduce customisable HMI that could facilitate subjective takeover experience.

Therefore, this study aims to investigate the effect of customisable multimodal HMI on drivers' subjective experience in HAD situation. The HMI was chosen as the medium because successful takeover can hardly be achieved without effective communication between the driver and the vehicle. During the process, both parties should be able to work collaboratively as a team based on an understanding about each other's 'abilities, strengths, and weaknesses'. Moreover, vehicle interfaces play a significant role as the communication channel (Miller, Sun and Ju 2014) that directs attention, offers feedback and feedforward about changes in system status, and promotes drivers' timely intervention (van den Beukel, van den Voort and Eger, 2016).

13.2 METHOD

13.2.1 Participants

A total of 68 participants were recruited by Jaguar Land Rover (JLR) via an external agency, three cancellations occurred during the study, resulting in data being collected for a total of 65 participants. Selection criteria specified an equal split of three age ranges (18–34 (22), 35–56 (23), and 57–82 (20)) and an equal gender split (33 females, 32 males). An additional inclusion criterion was for 10% of participants to own premium vehicles, defined as newer than seven years old, and manufactured by Audi, BMW, JLR, Mercedes Benz, MINI, Porsche, Tesla, Volvo, or VW. All participants held a full UK driver's license, and were in good health, with corrected vision acceptable. The University of Southampton Ethics and Research Governance Office provided ethical approval for the study (ERGO Number: 41761.A3).

13.2.2 Experimental Design

The study consisted of four trials, with a repeated-measures condition defined as time out of the loop (OOTL) (i.e. one participant = two short time OOTL and two long time OOTL conditions). Trials were counterbalanced to ensure that order effects were controlled. There were four counterbalance conditions consisting of SSLL, LLSS, SLSL, LSLS – where S = short (1 min) and L = long (10 min). Counterbalance conditions were equally distributed according to age category, gender, and premium car ownership. A secondary task in the form of a tablet-based Tetris-style game was employed during periods of automation. At the end of each trial, participants had the opportunity to customise their HMI settings as shown in Figures 13.1 and 13.2.

The driving environment consisted of a three-lane motorway with gentle curves; traffic conditions were designed to be congested, resulting in an average speed of approximately 40 mph. Road conditions were dry with good visibility.

Recorded performance data from STISIM included speed, lateral lane position, headway, throttle, brake, and steering. Data was time-stamped to allow performance 25 s post-takeover to be analysed. The times at which HMI elements were presented –

FIGURE 13.1 The customisation panel that appeared on the central console between trials.

FIGURE 13.2 Three visual aspects of the HMI: on the left is the central console, bottom-right is the cluster, and the HUD is shown on the top-right.

and control transition button pressed – were logged, allowing protocol timings and takeover times to be calculated. Data was logged for the secondary Tetris task, including scores and button presses. Customisation settings were saved for each trial to reveal patterns in preference to subsequent trials. Subjective data on workload (NASA-Task Load Index (NASA-TLX)), usability (system usability scale), technology acceptance (technology acceptance scale), and automation trust were collected after each trial to provide insight into how these varied between trials. A final, post-trial interview was administered at the end of the study to capture data on preferences and opinions relating to the HMI and customisation.

13.2.3 EQUIPMENT

The study employed a fixed-base simulator at a JLR research facility. The simulator cabin was designed to mimic a generic JLR vehicle. LCD monitors were built in to provide multiple in-car interfaces, including a touchscreen central console, an instrument cluster, and head-up display (HUD). The HUD monitor was mounted horizontally and displayed an inverted image, which was reflected on a sheet of angled glass fitted in the drivers' view. The in-car interfaces were highly adaptable and featured a range of graphics and text that could be displayed. The central console provided additional functionality through the display of a customisation screen that allowed the HMI graphics and text on the central display, cluster, and HUD to be modified within specific parameters when between trials. A Range Rover steering wheel was modified to include two additional buttons, mounted adjacent to the drivers' thumb positions. The steering system was connected to a Sensodrive direct drive system to provide force feedback and turn the steering wheel automatically when the automation was activated. The drivers' seat was modified to provide haptic cues using Arduino-controlled motors fitted with eccentric weights. Ambient lighting was provided through the use of Arduino-controlled RGB LEDs fitted under the top of the cluster, within the drivers' footwell, and under the central console. The main road view was provided via a large LCD screen mounted directly in front of the simulator cabin. The rear-view mirror was depicted graphically on the main road view screen and placed in the correct position from the drivers' point of view. Door mirrors were provided via small LCDs mounted on the cabin in the traditional positions.

STISIM Drive simulation software was employed to generate the driving environment, mirrors, and sound. A control desk and PC behind the simulator provided the experimenters with the 'Wizard of Oz'-style ability to supply responses to the takeover protocols. Audio sounds and vocalisations were generated automatically by the automation system and played through speakers mounted in the simulator cabin. A Microsoft Surface tablet running a Tetris-style game provided the secondary task during automation; this tablet was networked with the automation system to provide a visual alert to the driver as part of the takeover protocol.

13.2.4 HMI Design and Customisation

Between trials, participants had the opportunity to customise their HMI using an alert settings screen (Figure 13.1) that appeared on the central console between trials. This provided the ability to hide or display aspects on the HUD, cluster, and central console, including wording regarding driving mode, a driving mode icon, a time to takeover readout, takeover question wording, and a coloured edge frame that indicated driving mode. A 'road sense' view could additionally be displayed on the cluster or central console; this graphically indicated the presence of vehicles around the drivers' car.

Analog settings included the ability to adjust the brightness of ambient light, which indicated automation mode based on the colour. Audio volume could be adjusted that would affect both audio tones and vocalisations generated by the automation system. The vibrate setting allowed the intensity of the haptics to be adjusted. The takeover questions setting allowed the driver to specify the number of questions they wanted the system to ask as part of the takeover protocol.

When a participant completed making their adjustments, the settings were saved, and the customised HMI was employed in the HMI of the next trial.

13.2.5 Procedure

After arriving at the JLR facility, participants were welcomed and presented with a participant information sheet informing them of the details of the study. Their right to halt the study at any time was explained; they were then provided with an informed consent form which they had to read, initial, and sign in order for the study to continue. On completion of the consent form, participants were given a bibliographical form to complete to capture demographics data. They were then introduced to the simulator. The main controls were explained, and they were informed of the functionality of the customisation screen and the effects it had on the HMI. The customisation screen provided dynamic updates of the HMI as changes were made, allowing the participant to see the effects, and to test audio levels, light, and haptic intensities. Participants then took part in a test run, where they experienced three takeovers after 1-min OOTL intervals. After completion of the test run, the customisation settings were reset to defaults and the study started. Trials started with the participant being asked to accelerate onto the motorway, join the middle lane, keep up with traffic, and follow the HMI's instructions. After a period of approximately 1 min, the HMI indicated to the participant that automation was available and informed them via text, icon, vocalisation, and two flashing green steering wheel buttons. Participants then activated automation by pressing the two steering wheel buttons; this was followed by the HMI indicating that automation was now in control of the vehicle. Participants then picked up the secondary task tablet and started to play the Tetris secondary task. After a period of either 1 or 10 min, dependent upon counterbalancing, the automation indicated via the HMI that the driver was required to get ready to take control. Participants were expected to put aside their secondary task, and follow the instructions presented by the HMI. The takeover protocol consisted of a set of questions designed to raise situation awareness. These questions could be presented in vocal, word, and icon form, dependent upon the HMI customisations. Participants responded vocally to each question, the answers to which were judged by an experimenter taking the part of the automation system using the Wizard of Oz approach. Incorrect or missed questions were repeated twice before moving to the next. Once the questions were answered by the participant, the HMI indicated for them to take control, which they did by pressing the two green buttons on the steering wheel. This constituted one takeover, the process was repeated twice more, and after completion, the participant was asked to pull safely to the hard shoulder and stop the vehicle. The customisation screen was then displayed on the central console, and the participant was asked to make adjustments to the HMI for the next trial. Once the participant was happy with their customisations, these were saved, and they were presented with three questionnaires in electronic form. Completion of the questionnaires marked the end of a trial. Four trials took place in total, two featuring 1-min OOTL durations and two featuring 10-min OOTL durations. Once all four trials were complete, the participant filled out a post-trial interview questionnaire.

13.2.6 Analysis

Standard questionnaires were used to assess drivers' subjective attitudes and perceptions about takeover experience in HAD environment. The forms were administered immediately after they finished each trial in the simulator. Data was collected from 68 participants who took part in this study. Forms from three participants were excluded for analysis because they were not completed as the trials could not be finished. Missing responses were replaced by the median values.

13.2.6.1 Workload

Participants' perceived workload was assessed by the NASA-TLX. As workload is a multidimensional construct, the scale was comprised of six questions regarding mental, physical, and temporal demand, as well as performance, effort, and frustration. Each item ranged from very low to very high, divided into 20 intervals. Averaged raw scores (raw TLX) were used to gain each participant's overall workload score (Hart 2006). This scale has been used to evaluate drivers' subjective workload in autonomous driving situations in a wide range of studies (de Winter et al. 2014).

13.2.6.2 Usability

The system usability scale was selected because it is a useful tool to examine subjective usability considering appropriateness of systems, or tools to a purpose viewed in relation to the context that they are used. The scale is composed of items demonstrating a wide range of aspects of usability such as complexity, the need for training, and support. It consists of ten questions including five positively and five negatively worded questions (Brooke 1996).

13.2.6.3 Acceptance

The acceptance of advanced transport telematics scale was adopted to measure the participants' acceptance about takeover experience. It consists of two subscales assessing system usefulness, reflecting practical aspects, and satisfaction, representing pleasantness. Subscales are comprised of five and four items, respectively. Scores from the two subscales were calculated after verifying reliability was above the suggested value (Cronbach's $\alpha > 0.65$) (Van Der Laan, Heino and De Waard 1997).

13.2.6.4 Trust

The scale of trust between people and automation was used to assess trust in the interfaces during the takeover. It consists of five questions regarding the level of distrust, and seven questions regarding the level of trust. Items were scored on a Likert scale ranging from 1 (not at all) to 7 (extremely) (Jian, Bisantz and Drury 1998). Mean values were used as representative scores because this method has been employed in relevant studies (Gold et al. 2015; Blömacher, Nöcker and Huff 2018; Feldhütter et al. 2016).

13.2.6.5 Data Analysis

Perceived workload, usability, acceptance, and trust of takeover experience through customisable interfaces were investigated based on comparisons between trials. The purpose was to identify the participants' perception and attitude towards the takeover experience before and after customising HMI settings. Data sets measured after the first trial with the default settings, and last (fourth) trial, carried out with the last setting after three customisations between trials, were used because the settings for the last trial were more optimised than those for the second and third trials. Wilcoxon signed-rank test was used to test the differences in subjective experience: perceived workload, usability, acceptance, and trust of customisable interfaces in autonomous vehicles between the first and the last trials. It was chosen because it allows a comparison of two sets of data collected from the same participants (Field 2017) when the data is measured on an ordinal scale (McCrum-Gardner 2008). IBM SPSS Statistics 24 was used for the analysis.

13.3 RESULTS AND DISCUSSIONS

Comparisons between trial 1 and trial 4 were made in order to interpret the effects of customisation of the HMI on the drivers' takeover experience. Generally, enhancement in all the criteria of subjective experience was identified. For instance, perceived workload was lower in trial 4 than in trial 1; usability, acceptance, and trust were higher in trial 4 than in trial 1. The differences were all statistically significant. This could be regarded as benefits of customisation of the HMI, which enabled the drivers to create a driving environment that could suit their needs better than the default settings (without customisation).

13.3.1 WORKLOAD

The level of overall perceived workload decreased in trial 4 in comparison to trial 1, and the difference was statistically significant (see Table 13.1). This could be explained as the customised interface may have helped reduce the amount of effort required during the takeover situation in trial 4. Reduction in demand was observed in all six aspects of workload, and the differences were all statistically significant.

The biggest difference was shown in mental demand followed by effort (see Table 13.2). The decreased mental demand means that the participants perceived that takeovers in trial 4 required less cognitive activity involving information searching (e.g. automation status from the interface, and the road ahead), and processing, and decision-making (when and to what extent they needed to operate the pedals and steering wheel). Additionally, the reduced effort represents that they perceived the takeover in trial 4 less mentally and physically demanding than in trial 1 (Hart and Staveland 1988).

The reduction in perceived workload may be interpreted that the drivers found the information settings were more optimal, in terms of modalities (visual, auditory, and tactile), the amount of visual information on each display, and the intensities of auditory and tactile information. As a result, they could conduct takeover tasks more effectively, requiring less effort and focusing better on takeover tasks in the environment in which the information was delivered in the way they preferred over the default settings.

13.3.2 USABILITY

Usability scores were generated according to the suggested guidelines (Brooke 1996). Average usability scores for both trials were above the 'acceptable' level (Brooke 2013). The average score for trial 4 was greater than that for trial 1 (see Table 13.3), and the difference was statistically significant. This increase is noteworthy because the average scores for the trials showed a change in grade, from a 'good' to 'excellent' level (Brooke 2013). This can be interpreted that the participants perceived the assistance from the customised HMI settings during the takeover as more usable than from the default HMI settings. For example, they perceived that they needed less support and training, and that the system was clearer or easier to use after they customised the interface (Brooke 1996). This may be construed as they felt they were assisted more effectively by the HMI alert system regarding the procedure for the takeover – including where to look and what to do at which stage.

TABLE 13.1

Perceived Workload Scores between Trials

Trial	N	Mean	Std. Deviation	Minimum	Maximum	Wilcoxon Signed-Rank Test Sig. (2-Tailed)
1	65	5.7920	3.5999	1.00	14.67	$Z = -4.930$
4	65	3.7811	3.0505	1.00	12.67	$p = 0.000$

TABLE 13.2

Perceived Workload of Each Element, and the Averages for Trial 1 and Trial 4

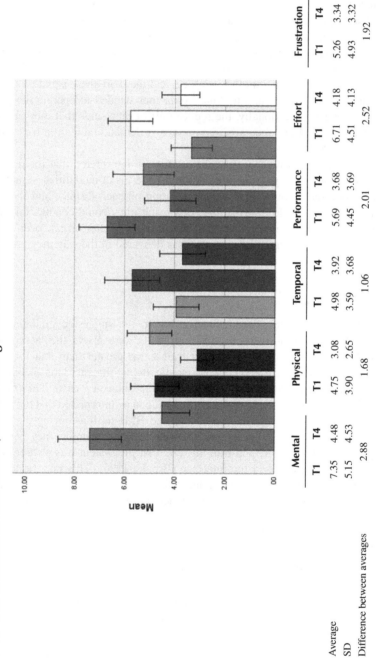

	Mental		Physical		Temporal		Performance		Effort		Frustration		Average	
	T1	T4	T1	T4	T1	T4	T1	T4	T1	T4	T1	T4	T1	T4
Average	7.35	4.48	4.75	3.08	4.98	3.92	5.69	3.68	6.71	4.18	5.26	3.34	5.79	3.78
SD	5.15	4.53	3.90	2.65	3.59	3.68	4.45	3.69	4.51	4.13	4.93	3.32	3.60	3.05
Difference between averages	2.88		1.68		1.06		2.01		2.52		1.92		2.01	

TABLE 13.3
System Usability Scores between Trials

Trial	N	Mean	Std. Deviation	Minimum	Maximum	Wilcoxon Signed-Rank Test Sig. (2-Tailed)
1	65	80.8308	14.43122	50.00	100.00	$Z = -4.180$
4	65	86.2538	13.74293	45.00	100.00	$p = 0.000$

TABLE 13.4
Acceptance Scores between Trials

Subscale	Trial	N	Cronbach's Alpha	Mean	Std. Deviation	Min.	Max.	Wilcoxon Signed-Rank Test Sig. (2-Tailed)
Usefulness	1	65	0.786	1.2031	0.59633	0.00	2.00	$Z = -2.796$
	4	65	0.817	1.3415	0.62322	-0.60	2.00	$p = 0.005$
Satisfying	1	65	0.869	1.1654	0.73058	-0.75	2.00	$Z = -3.873$
	4	65	0.930	1.4346	0.75684	-0.75	2.00	$p = 0.000$

It was clearer and easier to understand the situation regarding the surroundings and the vehicle through protocol questions, and to make an appropriate decision in a timely and appropriate manner. It may have been aided through the customised interface which presented the relevant information where the participants could find it easily, or redundant information which could be removed to avoid information overload. It was followed by actions to take, e.g. when and how to take control back by pressing the green buttons, and controlling speed and lane position based on the established situation awareness.

13.3.3 ACCEPTANCE

Overall scores for each subscale (usefulness and satisfying subscale) were calculated after verifying the Cronbach's alpha values were above the acceptable level (0.65) (Van Der Laan, Heino and De Waard 1997). Mean values for the usefulness and satisfying subscores of trial 4 were higher than those of trial 1 (see Table 13.4). Statistically significant differences in both subscores were found between the trials. This means that the participants' attitudes towards the customisable interface settings in automated vehicles were enhanced following experience of the customised system. For instance, their attitudes were more approach-oriented than avoidance-oriented, and they found the system more favourable than unfavourable after customising the HMI settings (Van Der Laan, Heino and De Waard 1997).

It can be interpreted that there were alterations in their opinion about the system, and they were more positive about the customised HMI settings. This means that the effects of the customised HMI settings during the takeover were more appealing and accepted by the participants. The significant increase in acceptance over the trials is meaningful in the circumstance that public's opinion about acceptance of automated vehicles may tend not to be positive (König and Neumayr 2017; Li et al. 2019). The differences may represent that having hands-on experience of driving in the HMI settings customised by themselves helped them show more positive attitudes about it, rather than negative perceptions. This may have been achieved because the drivers were able to choose type (e.g. textual or graphical; static or dynamic), the amount, or the places they preferred in the manner they could receive and understand effectively and comfortably.

13.3.4 TRUST

Significant increase in the participants' overall trust score was found in trial 4 compared to trial 1 (see Table 13.5). This could be due to the level of trust in the system during the takeover improved with experience of it: for instance, drivers tended to gain more confidence, and find the system more dependable and reliable. Also the level of distrust reduced: e.g., they were inclined to perceive the system as less deceptive, or unclear about the intent, action, or outputs of automation after experiences of the system with customised interfaces (Jian, Bisantz and Drury 1998). The increase in trust may be due to the participants being able to change the HMI settings to a way that they could perceive the information with accuracy and reliability. This could be construed that in the customised settings, they were able to gain the necessary information to rebuild situation awareness because it was easier and more comfortable to understand than the default settings. This helped them identify what was going on during the takeover situation (e.g. limits or strengths of the automation), and thus, they were able to decide at which point, and what actions they needed to take, and what they could let the automation do. The customised settings may have been more helpful in the relevant establishment of situation awareness and decision-making processes, which led to an increase in perceived accuracy and reliability of the system (National Research Council 1997, cited in Jian, Bisantz and Drury 1998). It could also be explained that the participants could understand and operate the system more appropriately without experiencing malfunctions or failures of drivers or systems that could negatively affect trust in automation (Hoff and Bashir 2014; Hartwich et al. 2019).

TABLE 13.5
Trust Scores between Trials

Trial	N	Mean	Std. Deviation	Minimum	Maximum	Wilcoxon Signed-Rank Test Sig. (2-tailed)
1	65	5.1513	0.983	3.250	7.000	$Z = -5.056$
4	65	5.7387	1.129	3.330	7.000	$p = 0.000$

13.4 CONCLUSION

This study was conducted to observe the effects of customisation of HMI on drivers' subjective takeover experience in a HAD situation. It was posited that offering customisable HMI could help alleviate drivers' perceived workload, and enhance usability, acceptance, and trust about the takeover experience. The participants completed four trials of highly autonomous driving, which included three takeovers, in a simulator environment. Between the trials, they had a chance to customise the HMI alert settings. The effects of customisation were identified by comparing the participants' subjective experience (workload, usability, acceptance, and trust) between trial 1 (default settings) and trial 4 (final customised settings). The results demonstrated that noticeable differences in the experience between the trials were found. For all the aspects of the subjective experience, positive changes were strengthened in trial 4 compared to trial 1, showing statistical significance. These findings could imply that autonomous vehicles with customisable interfaces could help enhance drivers' subjective experience by reducing workload, and improving usability, acceptance, and trust. This may have been explained that the participants could create the driving environment by customising the multimodal HMI settings as the way they could receive, and process the information more effectively helped improve situation awareness and readiness. The increase in acceptance is significant because it is a strong predictor of intention to purchase, and actual usage of the system (Davis 1989; Venkatesh et al. 2003). This positive attitude is further verified by the improvement in usability and trust that have been confirmed as antecedents of acceptance (Choi and Ji 2015). Findings from the present study help clarify the effect of customisation of the HMI by producing the empirical evidence that strengthens the need to introduce customisable platforms to boost user experience.

Limitations of this study are as follows. The positive changes in the attitudes and perceptions may be partly explained by learning effects (Benndorf, Rau and Sölch 2019; Mäkinen et al. 2006) that resulted from repeated exposures to the equivalent experimental settings. For instance, it may have been easier for the participants to find necessary information from the HMI by knowing where to look, to predict what to do next more accurately, and to operate the interfaces during takeovers more easily. This in turn may have made them feel that they found the system more effective, and less complex, and needed less support over trials. However, the participants conducted a practice run before actual trials which helped them understand how to operate the system sufficiently. The learning effect could be minimised through this process.

This study extends current knowledge by producing empirical evidence demonstrating that drivers' subjective experience could be enhanced through the effects of a system customised by the user, based on their own preferences and needs. The findings are reassuring in the circumstance that the necessity for customisable vehicle interface to accommodate varying levels of abilities of drivers (Politis et al. 2017), and related concerns coexist. Furthermore, they are meaningful because they may serve as momentum to consider offering users higher level of control in customising interfaces in autonomous vehicles with confidence. They are expected to give insights to automobile manufacturers to create more user-centred HMI, which helps better promote drivers' experience of HAD.

ACKNOWLEDGEMENTS

This work was supported by Jaguar Land Rover and the UK-EPSRC Grant EP/N011899/1 as part of the jointly funded 'Towards Autonomy: Smart and Connected Control (TASCC) Programme' HI:DAVe Project.

REFERENCES

Adell, E., A. Varhelyi, and L. Nilsson. 2014. "The definition of acceptance and acceptability." In *Driver Acceptance of New Technology*, by T. Horberry, M. A. Regan and A. Stevens, 11–21. DOI: 10.1201/9781315578132-2. London, UK: CRC Press.

Benndorf, V., H. A. Rau, and C. Sölch. 2019. "Minimizing learning in repeated real-effort tasks." *Journal of Behavioral and Experimental Finance, 22* 239–248. DOI: 10.1016/j.jbef.2019.04.002.

Blömacher, K., G. Nöcker, and M. Huff. 2018. "The role of system description for conditionally automated vehicles." *Transportation Research Part F: Traffic Psychology and Behaviour, 54* 159–170. DOI: 10.1016/j.trf.2018.01.010.

Brooke, J. 1996. "SUS: A 'Quick and Dirty' usability scale." In *Usability Evaluation in Industry*, by P. W. Jordan, B. Thomas, I. L. McClelland and B. Weerdmeester, 189–194. DOI: 10.1201/9781498710411. London, UK: Taylor & Francis Group.

Brooke, J. 2013. "SUS: A retrospective." *Journal of Usability Studies, 8,* 2 29–40.

Butakov, V., and P. Ioannou. 2015. "Driving Autopilot with Personalization Feature for Improved Safety and Comfort." *Proceedings of the 2015 IEEE 18th International Conference on Intelligent Transportation Systems.* IEEE. 387–393. https://doi.org/10.1109/ITSC.2015.72.

Choi, J. K., and Y. G. Ji. 2015. "Investigating the importance of trust on adopting an autonomous vehicle." *International Journal of Human–Computer Interaction, 31,* 10 692–702. DOI: 10.1080/10447318.2015.1070549.

Davis, F. D. 1989. "Perceived usefulness, perceived ease of use, and user acceptance of information technology." *MIS Quarterly, 13,* 3 319–340. DOI: 10.2307/249008.

Davis, F. D., R. P. Bagozzi, and P. R. Warshaw. 1989. "User acceptance of computer technology: A comparison of two theoretical models." *Management Science, 35,* 8 903–1028. DOI: 10.1287/mnsc.35.8.982. de Winter, J. C.F., R. Happee, M. H. Martens, and N. A. Stanton. 2014. "Effects of adaptive cruise control and highly automated driving on workload and situation awareness: A review of the empirical evidence." *Transportation Research Part F: Traffic Psychology and Behaviour, 27,* Part B 196–217. DOI: 10.1016/j.trf.2014.06.016.

Endsley, M. R. 1995. "Toward a Theory of Situation Awareness in Dynamic Systems." *Human Factors: The Journal of the Human Factors and Ergonomics Society, 37,* 1 32–64. DOI: 10.1518/001872095779049543.

Eriksson, A., and N. A. Stanton. 2017. "Takeover time in highly automated vehicles: Noncritical transitions to and from manual control." *Human Factors: The Journal of the Human Factors and Ergonomics Society, 59,* 4 689–705. DOI: 10.1177/0018720816685832.

Eriksson, A., S. M. Petermeijer, M. Zimmermann, J.C.F. de Winter, K. J. Bengler, and N. A. Stanton. 2019. "Rolling out the red (and green) carpet: Supporting driver decision making in automation-to-manual transitions." *IEEE Transactions on Human-Machine Systems, 49,* 1 20–31. DOI: 10.1109/THMS.2018.2883862.

Feldhütter, A., C. Gold, A. Hüger, and K. Bengler. 2016. "Trust in automation as a matter of media influence and experience of automated vehicles." *Proceedings of the Human Factors and Ergonomics Society Annual Meeting, 60,* 1 2024–2028. DOI: 10.1177/1541931213601460.

Field, A. 2017. *Discovering Statistics Using IBM SPSS Statistics*, 5th edition. Thousand Oaks, CA: SAGE Publications.

Fraedrich, E., and B. Lenz. 2016. "Societal and individual acceptance of autonomous driving." In *Autonomous Driving*, by M. Maurer, J. Gerdes, B. Lenz and H. Winner, 621–640. DOI: 10.1007/978-3-662-48847-8_29. Berlin, Germany: Springer.

Gold, C., M. Körber, C. Hohenberger, D. Lechner, and K. Bengler. 2015. "Trust in automation – before and after the experience of take-over scenarios in a highly automated vehicle." *Procedia Manufacturing*, 3 3025–3032. DOI: 10.1016/j.promfg.2015.07.847.

Hart, S. G. 2006. "Nasa-Task Load Index (NASA-TLX); 20 Years Later." *Proceedings of the Human Factors and Ergonomics Society Annual Meeting*, 904–908. DOI: 10.1177/154193120605000909.

Hart, S. G., and L. E. Staveland. 1988. "Development of NASA-TLX (Task Load Index): Results of empirical and theoretical research." *Advances in Psychology*, 52 139–183. DOI: 10.1016/S0166-4115(08)62386-9.

Hartwich, F., C. Witzlack, M. Beggiato, and J. F. Krems. 2019. "The first impression counts – A combined driving simulator and test track study on the development of trust and acceptance of highly automated driving." *Transportation Research Part F: Traffic Psychology and Behaviour*, 65 522–535. DOI: 10.1016/j.trf.2018.05.012.

Harvey, C., N. A. Stanton, C. A. Pickering, M. McDonald, and P. Zheng. 2011. "In-vehicle information systems to meet the needs of drivers." *International Journal of Human– Computer Interaction*, 27, 6 505–522. DOI: 10.1080/10447318.2011.555296.

Hoff, K. A., and M. Bashir. 2014. "Trust in automation: Integrating empirical evidence on factors that influence trust." *Human Factors: The Journal of the Human Factors and Ergonomics Society*, 407–434. DOI: 10.1177/0018720814547570.

ISO. 1998. "ISO 9241-11: Ergonomic requirements for office work with visual display terminals (VDTs) — Part 11: Guidance on usability." https://www.iso.org/standard/16883.html (accessed 20 November 2020).

Jian, J.Y., A. M. Bisantz, and C. G. Drury. 1998. "Towards an empirically determined scale of trust in computerized systems: Distinguishing concepts and types of trust." *Proceedings of the Human Factors and Ergonomics Society Annual Meeting*, 42, 5 501–505. DOI: 10.1177/154193129804200512.

Jordan, P. W. 1998. *An Introduction to Usability*. Bocan Raton, FL: CRC Press.

Jorritsma, W., F. Cnossen, and P. M.A. van Ooijen. 2015. "Adaptive support for user interface customization: A study in radiology." *International Journal of Human-Computer Studies*, 77 1–9. DOI: 10.1016/j.ijhcs.2014.12.008.

Kaur, K., and G. Rampersad. 2018. "Trust in driverless cars: Investigating key factors influencing the adoption of driverless cars." *Journal of Engineering and Technology Management*, 48 87–96. DOI: 10.1016/j.jengtecman.2018.04.006.

König, M., and L. Neumayr. 2017. "Users' resistance towards radical innovations: The case of the self-driving car." *Transportation Research Part F: Traffic Psychology and Behaviour*, 44 42–52. DOI: 10.1016/j.trf.2016.10.013.

Körber, M., L. Prasch, and K. Bengler. 2018. "Why do I have to drive now? Post hoc explanations of takeover requests." *Human Factors*, 60, 3 305–323. DOI: 10.1177/0018720817747730.

Lee, J., and N. Moray. 1992. "Trust, control strategies and allocation of function in human-machine systems." *Ergonomics*, 35, 10 1243–1270. DOI: 10.1080/00140139208967392.

Li, S., P. Blythe, W. Guo, A. Namdeo, S. Edwards, P. Goodman, and G. Hill. 2019. "Evaluation of the effects of age-friendly human-machine interfaces on the driver's takeover performance in highly automated vehicles." *Transportation Research Part F: Traffic Psychology and Behaviour*, 67 78–100. DOI: 10.1016/j.trf.2019.10.009.

Linehan, C., G. Murphy, K. Hicks, K. Gerling, and K. Morrissey. 2019. "Handing over the keys: A qualitative study of the experience of automation in driving." *International Journal of Human-Computer Interaction*, 35, 18 1681–1692. DOI: 10.1080/10447318.2019.1565482.

Lu, Z., B. Zhang, A. Feldhütter, R. Happee, M. Martens, and J. C.F. de Winter. 2019. "Beyond mere take-over requests: The effects of monitoring requests on driver attention, take-over performance, and acceptance." *Transportation Research Part F: Traffic Psychology and Behaviour, 63* 22–37. DOI: 10.1016/j.trf.2019.03.018.

Lv, C., D. Cao, Y. Zhao, D. J. Auger, M. Sullman, H. Wang, L. M. Dutka, L. Skrypchuk, and A. Mouzakitis. 2018. "Analysis of autopilot disengagements occurring during autonomous vehicle testing." *IEEE/CAA Journal of Automatica Sinica, 5,* 1 58–68. DOI: 10.1109/JAS.2017.7510745.

Ma, S., Y. Wang, and D. Li. 2019. "The influence of product modularity on customer perceived customization: The moderating effects based on resource dependence theory." *Emerging Markets Finance and Trade, 55,* 4 889–901. DOI: 10.1080/1540496X.2018.1506328.

Macías, J. A., and F. Paternò. 2008. "Customization of Web applications through an intelligent environment exploiting logical interface descriptions." *Interacting with Computers, 20,* 1 29–47. DOI: 10.1016/j.intcom.2007.07.007.

Mäkinen, T. M., L. A. Palinkas, D. L. Reeves, T. Pääkkönen, H. Rintamäki, J. Leppäluoto, and J. Hassi. 2006. "Effect of repeated exposures to cold on cognitive performance in humans." *Physiology & Behavior, 87,* 1 166–176. DOI: 10.1016/j.physbeh.2005.09.015.

Marathe, S., and S. S. Sundar. 2011. "What drives customization?: Control or identity?" *Proceedings of the SIGCHI Conference on Human Factors in Computing Systems.* Vancouver, Canada: Association for Computing Machinery. 781–790. https://doi.org/10.1145/1978942.1979056.

Mayer, R. C., J. H. Davis, and F. D. Schoorman. 1995. "An integrative model of organizational trust." *The Academy of Management Review, 20,* 3 709–734. DOI: 10.2307/258792.

McCrum-Gardner, E. 2008. "Which is the correct statistical test to use?" *The British Journal of Oral & Maxillofacial Surgery, 46,* 1 38–41. DOI: 10.1016/j.bjoms.2007.09.002.

Miller, D., A. Sun, and W. Ju. 2014. "Situation awareness with different levels of automation." *2014 IEEE International Conference on Systems, Man, and Cybernetics.* San Diego, US: IEEE. 688–693. DOI: 10.1109/SMC.2014.6973989.

Nees, M. A. 2016. "Acceptance of self-driving cars: An examination of idealized versus realistic portrayals with a self-driving car acceptance scale." *Proceedings of the Human Factors and Ergonomics Society Annual Meeting, 60,* 1 1449–1453. DOI: 10.1177/1541931213601332.

Nguyen, T., and A. Colman. 2010. "A feature-oriented approach for web service customization." *2010 IEEE International Conference on Web Services.* Miami, US: IEEE. 393–400. DOI: 10.1109/ICWS.2010.64.

Petersson, L., L. Fletcher, and A. Zelinsky. 2005. "A framework for driver-in-the-loop driver assistance systems." *Proceedings. 2005 IEEE Intelligent Transportation Systems.* Vienna, Austria: IEEE. 771–776. DOI: 10.1109/ITSC.2005.1520146.

Politis, I., P. Langdon, M. Bradley, L. Skrypchuk, A. Mouzakitis, and P.J. Clarkson. 2017. "Designing autonomy in cars: A survey and two focus groups on driving habits of an inclusive user group, and group attitudes towards autonomous cars." *International Conference on Applied Human Factors and Ergonomics - Advances in Design for Inclusion.* Springer. 161–173. https://doi.org/10.1007/978-3-319-60597-5_15.

SAE International. 2018. *SAE International Releases Updated Visual Chart for Its "Levels of Driving Automation" Standard for Self-Driving Vehicles.* 11 December. Accessed June 11, 2019. https://www.sae.org/news/press-room/2018/12/sae-international-releases-updated-visual-chart-for-its-%E2%80%9Clevels-of-driving-automation%E2%80%9D-standard-for-self-driving-vehicles.

van den Beukel, A. P., M. C. van der Voort, and A. O. Eger. 2016. "Supporting the changing driver's task: Exploration of interface designs for supervision and intervention in automated driving." *Transportation Research Part F: Traffic Psychology and Behaviour, 43* 279–301. DOI: 10.1016/j.trf.2016.09.009.

Van Der Laan, J. D., A. Heino, and D. De Waard. 1997. "A simple procedure for the assessment of acceptance of advanced transport telematics." *Transportation Research Part C: Emerging Technologies*, 5, 1 1–10. DOI: 10.1016/S0968-090X(96)00025-3.

Venkatesh, V., M. G. Morris, G. B. Davis, and F. D. Davis. 2003. "User acceptance of information technology: Toward a unified view." *MIS Quarterly*, 27, 3 425–478. DOI: 10.2307/30036540.

Walch, M., K. Mühl, J. Kraus, T. Stoll, M. Baumann, and M. Weber. 2017. "From car-driver-handovers to cooperative interfaces: Visions for driver–vehicle interaction in automated driving." In Automotive User Interfaces. Human–Computer Interaction Series, by G. Meixner and C. Müller, 273–294. DOI: 10.1007/978-3-319-49448-7_10. Cham, Switzerland: Springer.

Xu, Z., K. Zhang, H. Min, Z. Wang, X. Zhao, and P. Liu. 2018. "What drives people to accept automated vehicles? Findings from a field experiment." *Transportation Research Part C: Emerging Technologies*, 95 320–334. DOI: 10.1016/j.trc.2018.07.024.

Yoon, S. H., Y. W. Kim, and Y. G. Ji. 2019. "The effects of takeover request modalities on highly automated car control transitions." *Accident Analysis & Prevention*, 123 150–158. DOI: 10.1016/j.aap.2018.11.018.

14 Accommodating Drivers' Preferences Using a Customised Takeover Interface

Nermin Caber
University of Cambridge

Patrick Langdon
Edinburgh Napier University

Michael Bradley
University of Cambridge

Theocharis Amanatidis
University of Cambridge

James W.H. Brown
University of Southampton

Simon Thompson
Jaguar Land Rover

Joy Richardson
University of Southampton

Lee Skrypchuk
Jaguar Land Rover

Kirsten M. A. Revell and Jediah R. Clark
University of Southampton

Ioannis Politis
The MathWorks Inc.

P. John Clarkson
University of Cambridge

Neville A. Stanton
University of Southampton

CONTENTS

14.1   INTRODUCTION

With 3,000 people dying in road accidents daily, driving is a high-risk activity and is being undertaken by an ever-increasing number of people (World-Health-Organization 2015). A reduction in accidents and fatalities could be achieved through autonomous driving systems as they show the potential to improve road safety and accommodate drivers of a wide range of abilities (Fagnant and Kockelman 2015). The Society of Automotive Engineers (SAE) categorises these systems into six levels of driving automation ranging from SAE Level 0 (no automation) up to SAE Level 5 (full automation) (SAE J3016 2016). The first SAE Level 3 autonomous system, Traffic Jam Pilot, was released in 2018 by Audi (McNamara 2017) and, when regulatory issues are resolved, will be the first conditionally automated car on our roads (Bishop 2019; Hetzner 2017). The implications are substantial. It will mark the first time that the driver is legally allowed to take both hands and feet off the controls while driving and engage with a secondary task, such as watching a movie.

Automation systems at SAE Levels 3 and 4 do not require permanent monitoring but do request human interaction when their limitations are reached. The absence of permanent monitoring is perceived as one of the major benefits of autonomous driving (Fagnant and Kockelman 2015). However, early on Bainbridge (1983) warned that automation can exacerbate problems with the human operator instead of eliminating them, and today, wide-ranging concerns have arisen about the interaction between these systems and human drivers especially during mode changes between autonomous and manual driving (Kyriakidis et al. 2017). SAE Levels 0 and 5 are the only levels that do not require such mode changes. Human factor experts are concerned specifically about effects of automation on reducing situation awareness and mental workload, as a result of so-called time out of the loop (OOTL) (Endsley 2010; Merat et al. 2012; Stanton, Young and McCaulder 1997; Stanton and Young 2005; Young and Stanton 1997). This can lead to longer reaction times (Young and Stanton 2007). These concerns have led to extensive research in the field of driving control transition with a focus on the transition from autonomous to manual mode. The prompt to take control from the car is designated as a takeover request (TOR). The main areas of this study have been the TOR interface design and the time needed for a safe transition (takeover time).

Unimodal visual TOR alerts produce higher response times and, consequently, a poorer driving performance than multimodal interfaces (Politis, Brewster and Pollick 2015). Similarly, visual overlays on the surroundings do not reduce initial reaction times for the TOR; however, they do support driver decision-making processes about the necessary action after taking control of the vehicle (Eriksson et al. 2019). The location of the visual warning has proved to be of major importance; displaying the visual alert on the respective side task screen can reduce response time significantly (Politis, Brewster and Pollick 2017). All these findings have led to extensive research in multimodal approaches to the TOR human–machine interface in order to lower response times further. Here, ambient displays addressing drivers' peripheral vision have proved beneficial as they lead to shorter reaction times (Borojeni et al. 2016). Auditory cues, in addition to visual ones, give rise to reduced driver hands-on time (Naujoks, Mai and Neukum 2014). An experiment on the effect of vibrotactile stimuli as TOR warnings revealed its high effectiveness (Petermeijer, Cieler and de Winter 2017). However, the integration of further modalities, such as mobile phone alerts and additional brake jerks, did not reduce the response time (Melcher et al. 2015). To summarise, multimodal interfaces for TORs have proven beneficial for driving performance as well as response time (Eriksson and Stanton 2017; Petermeijer et al. 2017; Politis, Brewster and Pollick 2017). The benefits of a multimodal approach lead to the first two of the three research questions addressed in this chapter:

1. *What interface do participants rely on the most when taking control from the car?*
2. *What groupings can be identified within the participants' use of interface variables? What different preferences do drivers have for TOR interfaces?*

With regard to the takeover time, two time durations were researched: takeover request lead time (TORlt) and takeover reaction time (TOrt). TORlt is defined as the time frame between the issuance of the TOR and the critical event, such as a construction site ahead. TOrt is the time required by the driver to take control from the automated system (Eriksson and Stanton 2017). Melcher et al. (2015) argue that a TORlt of 10 s is reasonable to allow a comfortable takeover for all drivers. This is supported by studies researching the effect of age on TOrt which show that older drivers did not take longer to takeover than younger drivers (Körber et al. 2016). The literature review by Eriksson and Stanton (2017) shows that the vast majority of TORlt picked for experimental studies was in the range 0–10 s. However, there are some studies (Eriksson and Stanton 2017; Merat et al. 2014) which report higher TOrt than the TORlt of 10 s mentioned above. This means that some drivers would not be able to resume control on time, e.g., before arriving at the critical event. Another important factor influencing the TOrt is the OOTL time. A study showed that an OOTL time of 60 min increased TOrt significantly when compared to an OOTL time of 10 min (Bourrelly et al. 2019). As a consequence, Eriksson and Stanton (2017) argue for tailored interfaces to accommodate the different capabilities among drivers, which may result as a consequence of wide age ranges (Green 2008), and to mitigate the effect of long OOTL times of 1 h or more. This leads to the final research question addressed in this chapter:

3. *How does the OOTL time, between long and short, influence participants' TOR customisation settings?*

14.1.1 User-Tailorable Interfaces

Three different approaches are applied to enable user-tailorable interfaces: adaptation, personalisation, and customisation. Adaptation is defined as: 'As a response to some stimuli, a system alters something in such a way that the result of the alteration corresponds to the most suitable solution in order to fulfil some specific needs' (García-barrios, Mödritscher and Gütl 2005). In broader terms, the system adjusts specific characteristics based on the interaction between the system itself and the environment (Supulniece 2012).

Personalisation can be seen as a subtype of adaptation in which the system adapts towards a specific user (García-barrios, Mödritscher and Gütl 2005). This means that a personalised system gathers data about the user who is interacting with the system and adjusts product characteristics accordingly (Arora et al. 2008).

Customisation, on the other hand, allows the driver to decide on system characteristics and to make appropriate changes on his/her own (Arora et al. 2008). The customisation of car interiors started in the 1920s with the implementation of the seat-sliding mechanism, allowing differently sized drivers to adapt the distance to driving controls, such as pedals. This process continued over the decades (Akamatsu, Green and Bengler 2013). While the early era of customisation mainly focused on anthropometry in order to accommodate a wide range of different human bodies, today the main focus lies on mentally coping with the two simultaneous tasks, driving and interacting with in-vehicle information systems.

This challenge is met by adapting the car human–machine interface, often referred to as adaptive interfaces, pursued in projects such as COMUNICAR and AIDE (Amditis et al. 2006). For example, specific notifications are delayed during periods of high workload to avoid an overload situation and consequently increase driving safety (Bellotti et al. 2005; Piechulla et al. 2003; Wright et al. 2017). To address individual driving-related preferences and behaviours, personalised advanced driver assistance systems modelling driving behaviour were proposed (Miyajima et al. 2007; Politis et al. 2017). Adaptive interfaces show positive effects on user attitude and behaviour in web portals (Kalyanaraman and Sundar 2006). Higher levels of tailoring led to increased attentiveness along with perceived relevance, involvement, and interactivity.

Car manufacturers are reacting and, accordingly, adapting and tailoring the car's interior, including the interface, to drivers' preferences (Caber, Langdon and Clarkson 2019; Kew 2018; Pollard 2018; Porsche-AG 2018; Volkswagen-AG 2017). In modern high-end cars, drivers can assign themselves a driver profile which saves a range of preferences and settings. This profile can be retrieved when entering the car and then makes the appropriate adjustments (Caber, Langdon and Clarkson 2018; Nica 2015).

Despite the recent surge in interest and consequently research and development in the field of adaptation in the automotive sector (Amditis et al. 2010), their actual application has remained limited. Particularly, the often indirect approach to predict user state based on user performance and input, along with limited understanding on how to translate collected data into useful adaptations, has proven difficult (Feigh, Dorneich and Hayes 2012). A suboptimal adaptation implementation which selects average, not-tailored, settings automatically for specific drivers could influence their experience of a user-tailorable TOR interface as less useful or effective and therefore negatively impact performance (Schnelle et al. 2017; Yi et al. 2019). As a consequence, it was decided to use customisation for a first evaluation of the potential of user-tailorable TORs and to allow participants to select settings.

14.1.2 PURPOSE

Thus, in this chapter, we designed an experiment to examine a user-tailorable TOR interface. A medium-fidelity driving simulator, comprising a range of multimodal user-tailorable interfaces, was used to conduct a study with 65 participants. The idea was that by tailoring the takeover process and therewith keeping the right level of workload and situation awareness at the time of transition (Piechulla et al. 2003; Wright et al. 2017; Young and Stanton 1997), we can create a smoother transition and avoid inappropriate driver actions (Oliver and Pentland 2000). The goal was to examine the effects of user-tailored TOR interfaces on takeover time by varying levels of OOTL times, short (1 min) and long (10 min), and to thereby explore drivers' need for user-tailorable interfaces. Since cluster-aided algorithms, which use identified clusters as input, can improve prediction performance (Yi et al. 2019), a cluster analysis on the collected TOR interface settings was performed. The goal is to thereby inform the design of future adaptive systems for TOR interfaces. There is scarce evidence in the literature that a tailored approach to TORs has been tested.

14.2   METHOD

14.2.1   EQUIPMENT AND DRIVING SIMULATOR

The medium-fidelity, right-hand drive simulator was composed of an artificially imitated car body, equipped with two car seats, three displays, a steering wheel, and an 85-inch TV screen in front of it (see Figure 14.1). The 'STISIM Drive®' was

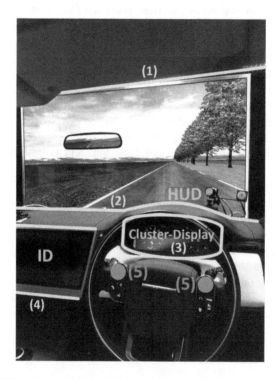

FIGURE 14.1   LED placement for driving mode indication (yellow): (1) Top of the windshield and behind sun visor, (2) on the dashboard at the bottom of the windshield, (3) surrounding instrument cluster-display, and (4) top and bottom of centre screen; (5) automation buttons (green); displays (blue).

employed to simulate a motorway with variable curve radii. The cluster-display and head-up display (HUD) depicted several types of driving related information in addition to the options given for customisation. The HUD displayed time and speed, while the cluster display additionally depicted engine revolution, fuel level, battery status, outside temperature, and fuel consumption. The resemblance of a conventional cluster-display with all this information was expected to increase ecological validity of the experienced interface. While in autonomous mode, participants played a Tetris video game on a 10-inch tablet. LED strips were placed at various locations in the driver's field of vision, as observable in Figure 14.1, to enable ambient lighting.

14.2.2 STUDY INTERFACE DESIGN

To understand the design of the interface used in the study, it is helpful to consider the elements of which it comprises, to discuss the underlying rationale, and to describe the takeover process in more detail.

The TOR interface design consisted of the interface elements: HUD, instrument cluster-display, infotainment display (ID), secondary task display, ambient lighting, audio cues, vocalisations, and haptics. Except the secondary task display, all of them were customisable by the user.

The motivation for the applied TOR interface design was derived from the *response-based* concept presented in Politis et al. (2018) which required drivers to answer driving-related questions prior to taking control from the car. The goal of this design was to increase situation awareness, thereby improving driving performance (Walker, Stanton and Salmon 2015). In contrast to the definition of system monitored takeover in Walch et al. (2015), this TOR interface design does not monitor the driver after, but during, the takeover process. The main reason for using a dialogue-based interaction was that research showed higher trust and usability (Forster, Naujoks and Neukum 2017) along with lower visual workload (Naujoks et al. 2016) for speech-based interfaces.

Subsequently, the takeover process is described in detail. Firstly, drivers were prompted verbally by the interface to 'Get ready to drive' using a computer-generated female voice. This was multimodally augmented using a vibrating seat and by showing the driver the auditory message along with an animated icon which depicted hands moving to a steering wheel. It was the only message also being depicted on the secondary task monitor as the driver was expected to put down the secondary task and resume driving position. The animated icon was not depicted on the secondary task display.

Secondly, drivers were required to answer questions relating to the driving environment and specific car features. The questions included information such as current speed, traffic situation ahead, and fuel level, and were spoken by the same voice and depicted on all screens except the secondary task one. The recognition of correct and incorrect responses was performed using the Wizard of Oz method. In case a question was answered incorrectly, it was repeated once to allow the driver a second attempt.

After answering all questions, the driver was prompted to take control of the car by using the vocalisation 'Press Green Buttons To Take Control Now', visual cues, pulsing steering wheel buttons along with an animated icon, and by seat vibration.

FIGURE 14.2   Driving mode icons.

By pressing both green automation buttons (see Figure 14.1), the driver took over control and received the auditory and visual confirmation 'You Are in Control', along with a vibration felt through the seat.

Ambient lighting in conjunction with colour coding and icons (Figure 14.2) was used to indicate the current driving mode. Amber indicated a manual mode, while blue indicated an autonomous mode. These colours were chosen to minimise any negative visual effects on drivers with reduced colour perception (Persad, Langdon and Clarkson 2007).

14.2.3   Selectable Customisation Settings

Figure 14.3 shows the customisation settings selectable by the driver. The first three sections were distributed vertically on the display and extended horizontally as collections of checkboxes. The sections are 'HUD Display', 'Cluster-Display', and 'Infotainment Display', and they share mostly the same six binary (on/off) options: 'Driving Mode Wording', 'Driving Mode Icon', 'Time to Takeover', 'Display Edge/Frame', 'Handover Questions', and 'Road Sense View'.

Due to the size of the HUD and its technology, 'Road Sense View' was not an option and 'Display Edge/Frame' had no function for the HUD despite being depicted on the customisation interface. Participants were informed accordingly.

'Driving Mode Wording' affects whether the mode, 'Manual Mode' or 'Autonomous Mode', is stated underneath the driving mode icon (see Figure 14.2). The option 'Driving Mode Icon' enables the driver to turn on/off the icon itself on the particular interface.

Settings 'Time to Takeover' and 'Handover Questions' display or hide the respective information – i.e. the time left till the TOR is issued and the current question asked.

The specific mode colour is used to illuminate the edge/frame of the particular interface monitor when option 'Display Edge/Frame' is selected.

'Road Sense View' is a road and object mimic display depicting the own car and detected surrounding vehicles to the driver when in autonomous mode, similar to Tesla's 'Enhanced Autopilot' interface (Lambert 2018).

The next four groupings were distributed vertically on the display and extended horizontally as sliders. These are 'Ambient', 'Audio', 'Vibrate', and 'Handover Questions', and they offer ordinal settings allowing participants to pick a specific intensity.

For 'Ambient' and 'Audio', participants were able to select 11 different levels ranging from 0 (off) to 10 (maximum intensity). The LED strips ('Adafruit DotStar

Digital LED Strip') offered 32 brightness degrees which were broken down to 11 levels from 0 (brightness degree: 0) up to 10 (brightness degree: 30) for the ambient lighting. For the audio settings, standard computer speakers were used and adjusted by testing different volume levels for audio cues and vocalisations on 10 participants aged 25–59.

For the vibrate settings, trials with the same participants were conducted to define intensity levels for the seat vibration which was powered by an electric actuator with an acceleration of approximately 50–75 m/s^2, an amplitude of 0.3 mm, and a voltage range of 9–16 V. This resulted in four different levels: 0 (off), 1 (9 V), 2 (12 V), and 3 (16 V).

The option 'Handover Questions' allowed participants to select the number of questions (1–10) they are asked before the prompt to take control appears.

14.2.4 EXPERIMENTAL DESIGN

The experimental design was a single independent variable (OOTL time) driving simulator study. Dependent variables were subjective measures and one objective measure, total duration of takeover. The experiment applied a within-subjects design with two levels for the independent variable: short (1 min) and long (10 min) OOTL time. They were presented in a counterbalanced sequence (S = short; L = long): SSLL, LLSS, SLSL, and LSLS. This means that every participant experienced four trials, two with a short and two with a long OOTL time. Each trial consisted of three TORs, and the OOTL time was fixed within a trial. Varying OOTL times were expected to encourage setting changes due to different levels of disengagement from the driving task and, in consequence, requiring varying levels of interface salience. The following dependent variables were measured and examined:

- **Customisation settings:** Selected settings of all interfaces, binary (on/off) and ordinal (scale positions), were gathered after each trial to enable cluster analyses of interfaces and participants.
- **Takeover time:** Total duration in seconds from the first request to prepare for takeover to the actual pressing of the takeover buttons.
- **Post-task questionnaire:** A specially designed questionnaire allowed participants to evaluate the experienced interface. For the purpose of this chapter, the following questions are relevant:
  - Would you be happy for your final settings to be implemented in a car? If No, could you briefly explain why?
  - Please rank in order the elements that you relied on for information on the automation system? (1 = least reliance, 7 = most reliance) Please write down any comments.
  - Do you think that, as you became accustomed to the automation system, you might further adjust the customisations to suit you? Please write down any comments.
  - What is your view on the level of customisation? (Too few options; Correct number of options; Too many options) Please write down any comments.

14.2.5 Procedure

14.2.5.1 Pre-Trial

Participants were welcomed, informed about safety procedures, and familiarised with the purpose of the experiment. All pre-trial documentations, i.e. consent form, attendance list, and demographics form, were completed by the participants before entering the simulator. Here, they were asked to adjust the steering wheel and seat to their preference and had the interfaces, including the secondary task, explained to them. The secondary task was a Tetris video game with increasing difficulty over playtime. This task was chosen as it has been applied in several studies to impose different levels of cognitive workload (Dyke et al. 2015; Grabisch et al. 2006; Miller et al. 2011) and because it requires the same cognitive resources as driving (visual–spatial processing and manual response). After a short training trial, namely 1-min OOTL time, to accustom participants to the interface and simulator, they were familiarised with the customisation settings and told that changes would not affect their settings in the first experimental trial where every participant experienced the same default set of customisation (see Figure 14.3). Participants did not know OOTL time prior to the trial. The OOTL times, short and long, were presented in counterbalanced sequences.

14.2.5.2 Trial

The simulation started on the hard shoulder of a motorway, and participants were instructed to drive onto the motorway, to comply with the highway code, and to drive as they would in a real car. After a short drive of approximately 10 s, participants reached the end of a traffic congestion enforcing a speed between 30 and 40 mph. Thirty seconds after the trial's start, the simulator would offer autonomous mode which participants were asked to turn on by pressing both green buttons on the steering wheel (see Figure 14.1). Depending on the OOTL time, participants were

FIGURE 14.3 Customisation settings.

engaged in the secondary task before prompted to get ready to drive, to engage with the verbal protocol, and finally to take control by pressing both green buttons.

After resuming manual control, participants drove for1 min, a time selected based on assuming an inverse to Young and Stanton (2020) who show decaying attentional resources after 1 min of automated driving, prior to the automation system offering autonomous mode again. For each trial, this procedure was repeated three times, resulting in three TORs per trial. At the end of each trial, the customisation settings were shown on the centre screen and participants were asked whether they would like to make changes to the interface which they then would experience in the next trial. It was clearly stated that changes do not need to be made if the participant is happy with the current layout. The final interface settings selected by participants are intended to be used in a follow-up on-road study. Participants were informed accordingly.

14.2.5.3   Post-Trial

After completion of all four trials, lasting approximately 160 min, participants completed the post-task questionnaire and took a break with some refreshments before starting their journey home.

14.2.6   HYPOTHESES

14.2.6.1   Hypothesis 1

Participants were expected to mostly rely on the cluster-display, audio cues, and vocalisations due to familiarity (Gough, Green and Billinghurst 2006) with current car interfaces, which widely use the former two systems (Akamatsu, Green and Bengler 2013) and due to research showing higher usability of speech-based vocalisations (Forster, Naujoks and Neukum 2017).

14.2.6.2   Hypothesis 2

The non-normally distributed results for reaction time, driving performance, and subjective workload (Eriksson and Stanton 2017) demonstrate considerable differences among drivers for the takeover. Therefore, a large amount of variability in chosen customisation settings in order to increase or decrease salience was expected between participants.

14.2.6.3   Hypothesis 3

A significant effect of OOTL time on customisation settings was expected due to research showing increased drowsiness for longer OOTL times (Bourrelly et al. 2019). This was believed to influence drivers' interface requirements resulting in higher intensity characteristics for longer OOTL times.

14.2.6.4   Hypothesis 4

The OOTL times, short (1 min) and long (10 min), were expected to significantly affect the takeover time needed. This also tests whether the findings in Bourrelly et al. (2019) are observable for short and long OOTL times.

14.2.7   DATA ANALYSIS

The data analysis was performed using 'IBM SPSS® Statistics 25', and the significance threshold was set at 0.05. The statistical methods to be applied were chosen based on Field (2017).

14.2.7.1   Binary Settings

To measure to what extent participants explored specific settings, the number of changes of each setting over all participants and over all trials was counted. In addition, the number of changes for each setting over all participants between the first and the last trial was counted. The second number was subtracted from the first to get the measurement of exploration. One change constitutes the adjustment from 'On' to 'Off', or vice versa.

The effect of OOTL time sequence on the binary settings was analysed using between-subjects, Pearson chi-square because of their categorical character. Since the Pearson chi-square test needs to be performed for each binary setting independently, only statistical results for significant observations are reported.

The nonparametric Wilcoxon signed-rank test is applied to detect significant effects of OOTL time on the binary settings since OOTL time is a within-subject variable. Again, only statistically significant results are reported due to the number of tests performed.

14.2.7.2   Ordinal Settings and Takeover Time

The Kolmogorov–Smirnov (K-S) test was used to test the ordinal settings and takeover time for normal distribution. No test was performed for the ordinary setting 'Vibrate'. Because of its small number of levels, a normal distribution is not expected.

Because of the interval character of the ordinal settings and the expected non-normal distribution of takeover time (Eriksson and Stanton 2017), the effect of OOTL time sequence and age on these was tested using the Kruskal–Wallis test. When significance was detected, potential linear trends were assessed using a Jonckheere–Terpstra test.

For the effect of OOTL time on ordinal settings and takeover time, the Wilcoxon signed-rank test was applied. For the same reasons as before, only statistically significant results are reported.

14.2.7.3   Cluster Analysis

Hierarchical agglomerative cluster analysis (Everitt et al. 2011; Langdon et al. 2003) was used to group participants and interfaces based on similarities in their final customisation profiles. The decision to apply cluster analysis is based on its wide application to divide human participants into homogenous groups as used in market research and psychiatry (Everitt et al. 2011). The measured similarity can be interpreted as the psychological distance participants perceived between option settings (Langdon et al. 2003; Nosofsky 1985).

The squared Euclidean distance was used as the similarity measure due to its known suitability and wide acceptance to assess differences in human participants

(Clatworthy et al. 2005; Nosofsky 1985). All values were standardised by variable in the range 0–1 to account for different measurement scales among the settings. After exploring different linkage methods, namely single, complete, and Ward's, it was decided to apply Ward's linkage as the underlying intergroup proximity measure since it revealed clearly observable clusters. In order to decide on the cut-off distance, a scree-like plot indicating the distance on the y-axis against the number of groups on the x-axis was applied.

14.2.7.4 Post-Task Questionnaire

We performed a Friedman test to evaluate participants' interface reliance rankings. Dunn–Bonferroni post hoc tests were used to determine which interfaces show different reliance rankings.

14.2.8 PARTICIPANTS

Participants consisted of 32 female and 33 male UK drivers who were recruited by an external agency and aged between 21 and 75 ($M = 44.7$, SD = 16.2). Adhering to the principles of inclusive design (Langdon and Thimbleby 2010), participants were recruited and divided into three different age bands: (1) 18–34 years ($N = 21$, 11 females), (2) 35–56 years ($N = 21$, 10 females), and (3) 57–82 years ($N = 19$, 9 females). The definition of these age bands was decided on a principal basis using statistical analysis (Langdon and Stanton, Personal communication, 2017). An equal distribution of gender was pursued within each age group. The participants had driving experience of 2 up to 57 years ($M = 24.27$, SD = 16.22) and drove between 4,300 and 36,000 miles per year ($M = 11,793$, SD = 5,512).

The study complied with the American Psychological Association Code of Ethics and was approved by the Department of Engineering Research Ethics Committee at the University of Cambridge. Informed consent was obtained from each participant.

14.3 RESULTS

For overview reasons, the following sections have been structured by theme. The resulting implications for the hypotheses will be tied together in the discussion.

14.3.1 CUSTOMISATION SETTINGS

Overall, 63 participants (97%) decided to deviate from the default settings in their final profile at the end of all trials. All 63 of these profiles are unique in their combination. To calculate the variability between participants, all setting options were standardised and added up. This summarised each customisation profile into one overall variable with a potential minimum value of 0.11 and a maximum value of 20. Subsequently, the mean and standard deviation ($M = 11.94$, SD = 4.19) along with the range ($R = 16.37$) were calculated for this variable. The coefficient of variation for this variable is 35.1% (standard deviation divided by mean), which indicates the dispersion of drivers' preferences over all interface settings.

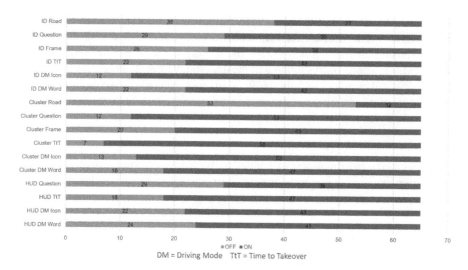

FIGURE 14.4   Selected binary (on-off) interfaces in final customisation profiles.

14.3.1.1   Binary

The frequencies for the selected binary (on/off) interface settings are shown in Figure 14.4. The majority of participants, 53 (82%) for the instrument cluster-display setting and 38 (58%) for the ID setting, decided to keep the road view option turned off in their final customisation profile. These findings indicate that the majority of participants did not appreciate the road view option.

Over all participants, and for all trials, the default, pre-selected binary options (see Figure 14.3) were changed by 4.57 levels on average, while the non-default settings, i.e. 'Road Sense View' for cluster-display and ID, were adjusted by 37 levels on average. Consequently, participants explored non-default settings far more than the default settings.

Between-subjects, Pearson chi-square tests did not reveal any significance of OOTL time sequence (SSLL, LLSS, SLSL, and LSLS) on binary (on/off) interface settings ($p > 0.05$). Wilcoxon signed-rank tests did not detect any within-subjects effects of OOTL time on binary settings. This means that neither different OOTL time sequences nor different OOTL times resulted in distinct binary settings.

14.3.1.2   Ordinal

An overview of descriptive statistics, mean, and standard deviation, and the K-S test for normality are given for all ordinal settings in Table 14.1; all ordinal settings are non-normally distributed at a significance level of $p < 0.001$. Concerning the setting 'Questions', it was discovered that 42 participants (65%) decreased the number of questions by 112 levels in total. As can be seen in Figure 14.5, the distribution is right-skewed ($z = 2.75$, $p < 0.01$) and significantly deviates from a normal distribution. Univariate Kruskal–Wallis and Wilcoxon signed-rank tests did not reveal any significant effect of OOTL time sequence and OOTL time, respectively, on any

TABLE 14.1

Descriptive Statistics for Ordinal Customisation Settings

Ordinal Setting	Mean	SD	K-S Test
Ambient	3.95	2.41	$D(65)=0.191$
Audio	4.94	1.1	$D(65)=0.355$
Haptic	1.75	1.13	—
Questions	3.42	1.81	$D(65)=0.163$

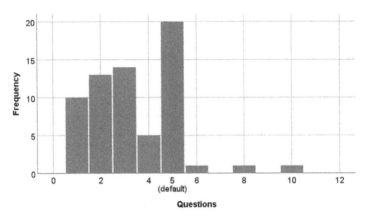

FIGURE 14.5 Frequencies of final settings for number of questions.

ordinal setting. Ergo, participants preferred fewer questions than the default of five, independent of the OOTL time itself or its sequence.

14.3.1.3 Cluster Analyses

Firstly, a hierarchical agglomerative cluster analysis was performed to group participants based on customisation settings. The scree-like plot did not reveal a clear result. However, it did show observable changes in distance for grouping stages two to four. Therefore, an in-depth analysis of the selected settings for two, three, and four clusters was completed. In this analysis, it was explored whether there is a clear distinction between the distinctive clusters.

In order to differentiate different interface clusters by one specific measure, the mean interface density was calculated for each cluster. Here, a density of one defines the maximum possible – i.e. all options are adjusted to their highest levels. Accordingly, a density of zero constitutes the sparsest interface – i.e. all options are turned to their minimum. As there was a clear distinction apparent for four clusters and a higher number of clusters allows for more accurate targeting, it was decided to define four clusters instead of two or three. The four major clusters are defined as heavy, medium-high, medium-low, and light. Table 14.2 summarises the corresponding values for each cluster of customisation profiles. The heavy interface

TABLE 14.2

Density Matrix Showing the Mean Interface Densities for the Four Identified Customisation Clusters

Interface	N	Mean Interface Density		
		Binary Settings (%)	Ordinal Settings (%)	Overall Settings (%)
Heavy	25	88.25	47.76	80.28
Medium-high	13	66.35	41.49	62.43
Medium-low	11	47.73	33.05	44.28
Light	16	34.38	42.04	35.96

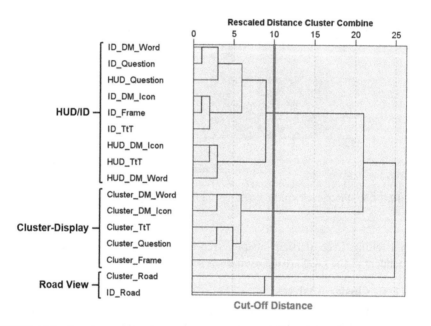

FIGURE 14.6 Dendrogram for cluster analysis of binary interface settings.

accommodates 25 participants and is thereby the biggest cluster. Hence, participants show a wide variability of interface preferences. While some prefer a comprehensive and intense interface with a lot of information about the systems' current state, others favour a more minimalistic approach with interfaces just depicting basic, necessary information.

A second cluster analysis was performed to group the interface variables in order to identify participants' perceived psychological similarities between interfaces. As in the previous cluster analysis, squared Euclidean distance as the similarity measure and Ward's linkage as the intergroup proximity measure were applied. For the binary (on/off) settings, no standardisation was utilised. The resulting dendrogram for binary (on/off) settings can be seen in Figure 14.6. A factor analysis was performed, from which a scree-like plot was generated. It was concluded to keep three clusters by using the point of inflexion as a criterion. The three clusters observable are HUD/ID, cluster-display, and road view. This means that ID and HUD were perceived as similar interfaces, while instrument cluster-display was regarded as a separate group. The road view setting was perceived as separate from the other customisation options independent of the interface element it was shown on. For the ordinal settings, no interesting observations were made about interface clusters.

14.3.2 Takeover Time

A Kruskal–Wallis test showed no significant effect of OOTL time sequence on takeover time ($H(3) = 2.503$, $p = 0.475$). However, a significant effect of age on takeover time ($H(2) = 10.145$, $p = 0.006$) was detected. Furthermore, a significant linear trend between age and takeover time ($J = 972$, $z = 3.241$, $p = 0.001$, $r = 0.402$) was discovered (see Figure 14.7). A Wilcoxon signed-rank test showed no significant effect of OOTL time on takeover time ($Z = -1.389$, $p = 0.165$). The mean required takeover time per question was $M = 7.276$ s (SD = 2.307). With a mean number of questions of 3.42, this makes an average takeover time of 24.88 s. Consequently, neither OOTL time nor different sequences of it influenced the takeover time in the experiment. Age, on the other hand, affected the takeover time, showing that older people tend to take longer for a takeover.

14.3.3 Post-Task Questionnaire

A big majority of 63 participants (96.92%) would be happy for their final settings to be implemented in a real car. Either of the two participants who replied negatively stated 'Didn't think I performed too good', referring more to their own performance than the actual interface.

A Friedman test showed a significant difference of participants' reliance on the presented interfaces (χ^2 (6) = 83.187, $p < 0.001$). Dunn–Bonferroni post hoc tests

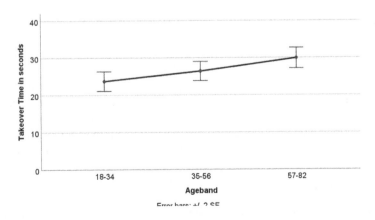

FIGURE 14.7 Means of takeover time by age band with two standard error, error bars.

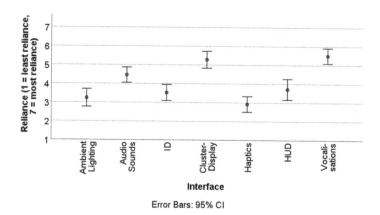

FIGURE 14.8 Mean error bars of reliance by interface with 95% confidence interval.

revealed significant differences between vocalisations and haptics ($p < 0.001$), vocalisations and ambient lighting ($p < 0.001$), vocalisations and ID ($p < 0.001$), vocalisations and HUD ($p < 0.001$), cluster-display and haptics ($p < 0.001$), cluster-display and ambient lighting ($p < 0.001$), cluster-display and ID ($p < 0.001$), cluster-display and HUD ($p = 0.001$), audio sounds and haptics ($p = 0.001$), and audio sounds and ambient lighting ($p < 0.05$). All p-values are after Bonferroni adjustments. This means that participants clearly relied more on vocalisations and cluster-display than on most other interfaces. Additionally, they were closely followed by audio sounds which showed a high value of reliance. An overview is given in Figure 14.8, which shows the means and error bars for each interface.

Participants were divided on whether they would further adjust the customisation settings as they became accustomed to the system with 50% answering yes or no, respectively.

Overall, 71% of the participants were satisfied with the level of customisation offered by the interface. Only 5% thought there were too few options. Representative examples of the comments given in the questionnaire are shown in Table 14.3.

14.4 DISCUSSION

This chapter investigated drivers' experience of a customised takeover process by analysing participants' customisation profiles, identifying emerged groupings for interface elements and participants, and examining the effect of different OOTL times and OOTL time sequences on customisation and takeover time. The results indicate a high diversity in preference to numerous unique customisation profiles indicating widely shared appreciation of a tailored TOR interface. Our findings are discussed with respect to our hypotheses, driver experience, and possible limitations.

14.4.1 HYPOTHESES

Hypothesis 1 is accepted since results demonstrate a high reliance on cluster-display, audio cues, and vocalisations. Participants relied the most on cluster-display and

TABLE 14.3
Summary of Comments in the Post-Task Questionnaire

Question	Comments
Do you think that as you became accustomed to the automation system you might further adjust the customisations to suit you? Please write down any comments.	'lighting and vibration – would increase if I was feeling fatigued or on a long journey' 'After the first batch of changes were made, I was happy with them and did not need to change them again' 'Because I think each person is different and the settings might best being able to change slightly'
What is your view on the level of customisation? Please write down any comments.	'It was good to have the choice and increased comfort and security'. 'Although as mentioned previously, I don't feel that there should be an option to turn them all off completely'. 'I would like the ability to organise the instrument panel to my liking' 'Lots of choice, possibly too many'

speech-based vocalisations, followed closely by audio sounds. While according to the reliance data, cluster-display and vocalisations were more relied on by participants than all other interfaces, audio cues did not show a higher participant reliance than HUD.

High reliance on the cluster-display and audio cues can be attributed to familiarity and, consequently, habit, as was expected in the hypothesis. The high reliance on audio cues can furthermore be explained by the fact that participants achieve the required information without needing to look up at any other interface or interrupt the secondary task. However, since the information conveyed by a specific auditory cue needs to be remembered, identified, and recalled at the moment of interaction, it could result in a heightened demand of mental processing and therefore higher cognitive workload.

Vocalisations, on the other hand, offer similar benefits but present the information in a more direct way and therefore may require less attention in working memory. This aligns with the benefits for speech-based vocalisation presented in other studies (Forster, Naujoks and Neukum 2017; Naujoks et al. 2016).

In summary, the conducted study gives a comprehensive overview of drivers' reliance on different interfaces during a TOR. The gained insights can be used in future automotive interface design to support the decision-making process concerning information location.

Hypothesis 2 is accepted since the analysis discovered a high percentage of unique customisation profiles, a substantial standard deviation, and a large range within the final settings which illustrates the diverse interface preferences between participants. Considering that the maximum possible customisation level was 20 and the calculated range was 16.37 underlines the inherent differences in participants' interface preferences.

Within these profiles, four distinguishable participant clusters were discovered. The most popular grouping was the one labelled *Heavy* and was characterised by a high-density TOR interface. On the one hand, this could be attributed to drivers' preferences for high-density interfaces which provide a lot of information and a considerable amount of alertness for the TOR. On the other hand, the already high default interface density of 81% may have motivated participants to preserve a dense interface. An explanation may be that participants assume that the default settings are optimised by the interface designers, resulting in hesitation to deselect them. This is supported by the fact that non-default settings were explored more than default settings.

Furthermore, the results revealed three different interface clusters: HUD/ID, cluster-display, and road view. Participants seem to perceive a high psychological similarity between the interfaces HUD and ID despite their spatial difference. This could be explained by less familiarity, usage, and experience in comparison to the cluster-display. In contrast to HUD and ID which were released in top-of-the-range cars in the late 1980s, cars have been equipped with cluster-displays since the 1920s (Akamatsu, Green and Bengler 2013).

The road view settings for cluster-display and ID were grouped together and were experienced as separate entities from the other cluster-display and ID options. This may relate to the lack of prior experience with such a road view setting in previous cars or to the fact that they were the only options not included in the default customisation profile. As non-default settings, participants may have seen them as additional, technical gadgets not necessarily cueing for change.

These findings show the uniqueness of drivers and their preferences, and therefore support the argument for a user-tailorable TOR interface over a single 'one-size-fits-all' offering. The different clusters identified could be used as an initial grouping for an adaptive interface more in line with personalisation, which then can be fine-tuned to the specific driver based on takeover time, OOTL time, age, and driver's interaction with the settings.

Hypothesis 3 is rejected since there was no significant effect of OOTL time on the preferred customisation settings, neither for binary (on/off) nor for ordinal. A possible reason could be that by experiencing both OOTL times, participants decided to choose a medium customisation profile which suited both levels of the independent variable. Furthermore, the difference between the OOTL times might not have been sufficient to trigger different customisation profiles. An OOTL time of 10 min may not be sufficient to cause increased inattention in drivers along with the demand for a more salient interface.

Furthermore, the TOR interface design with its multimodal approach combining visual, auditory, and tactile interfaces, as advocated in multiple studies (Borojeni et al. 2016; Eriksson et al. 2019; Melcher et al. 2015; Naujoks, Mai and Neukum 2014; Petermeijer et al. 2017; Politis, Brewster and Pollick 2015, 2017), may have been perceived as salient enough even with a lighter interface density.

The insight that participants potentially assume an optimised default interface may be another explanation. This could make participants expect that the default interface covers a range of scenarios, short and long OOTL times, in the best possible way, making major adjustments due to changes in the scenario seem irrelevant.

A further interesting observation was that participants generally seem to prefer fewer than five questions independent of OOTL time. This could be an indication that participants did not see the necessity for an extended takeover process as they did not feel fatigue after the longer OOTL time. Support for this explanation comes from some participants who reported that it might be beneficial to have more questions on a long journey.

Nevertheless, Bourrelly et al. (2019) show how longer OOTL times, in this case 60 min, can cause increased driver drowsiness. Therefore, it is recommended to test a user-tailorable interface for an OOTL time of 60 min in order to compare the findings with the current study. If changes in customisation settings based on OOTL time can be found, it can be a good factor for implementation in an adaptive interface, which intensifies interface salience with increasing OOTL time.

Hypothesis 4 is rejected as our results did not show any effect of OOTL time on takeover time. As mentioned above, the long OOTL time of 10 min in this study may not have been sufficient to result in the deterioration of takeover performance (Bourrelly et al. 2019).

A statistical analysis of age revealed a significant effect of age on takeover time and a linear tendency between them. This indicates that as the age increased, the takeover time rose proportionally. All this cannot be attributed to the number of questions since there was no effect of age on the number of questions. A possible explanation, in accordance with inclusive design, may be the decline in cognitive capabilities with increasing age, slowing down response times, and potentially resulting in a more risk-averse strategy (Langdon and Thimbleby 2010). This contradicts findings in other studies which reported no effect of age on takeover time (Körber et al. 2016). However, Körber et al. (2016) did show differences in drivers' reactions depending on their age. The differing results may be attributed to the differences in study design since Körber et al. (2016) looked into critical takeovers, whereas this study researched non-critical takeovers. Moreover, the response-based interface design presented here required substantially more actions from drivers and let them define their takeover pace themselves. These findings accordingly indicate that drivers indeed want to take longer for a safe takeover, aligning with the suggestions and findings in Eriksson and Stanton (2017).

As a sufficient takeover time should minimise the risk of drivers taking over control before situation awareness is fully built up (Eriksson and Stanton 2017), a wide range of capabilities will have to be accommodated by future TOR interface design. Consequently, the user-tailorable approach which allows drivers to define their takeover pace and interface preferences, therefore accommodates diverging driver capabilities, is a reasonable concept for non-critical takeover interfaces.

As a consequence, for the implementation of an adaptive interface for non-critical TORs, age might be an important factor in order to issue the TOR with sufficient lead time. Based on the findings in this study, the lead time should increase proportionally to age allowing older drivers more time to go through the TOR protocol, build up situation awareness, and finally, safely take over control from the car.

14.4.2 Driver Experience

The high number of unique customisation profiles, the high approval rate concerning the implementation of the settings in a real car, and the great satisfaction with the number of customisation options offered, demonstrate the power of tailored interfaces to accommodate a wide range of individual capabilities, needs, desires, and requirements. At least for one of the two participants who did not approve of the implementation, the dissatisfaction with the interface settings may relate more to the participant's own perceived performance as being low, rather than their actual appreciation of customisation. These findings align with the beneficial effects of tailored interfaces on user experience in web portals (Kalyanaraman and Sundar 2006).

Moreover, there was a substantial indecisiveness as to whether participants would further adjust the settings once they became accustomed to the TOR design. Such indecisiveness could have been caused by ambiguous understanding of the question or by a lack of understanding regarding the possibilities of customisation. This outcome may also indicate two different types of drivers. Some participants seem to be satisfied with an initial adjustment which they then keep. Other participants appear to desire ongoing situation-dependent changes, such as journey length and fatigue. The latter group of drivers could clearly benefit from an adaptive interface which adjusts automatically based on personal preferences, non-driving activity undertaken by driver, and external conditions, such as traffic density, daytime, OOTL time, journey length, drowsiness, and weather.

Two types of drivers can also be identified in their judgement of customisation level. While some participants are concerned there may be too many options, other participants aim to go even a step further and change the location of certain information elements on the specific interface.

The discovered variation in preferences concerning customisation level and ongoing adaptation is a further point of consideration for prospective adaptive interfaces. A possibility might be to initially only offer the customisation of basic and/or most important settings and increase the range incrementally, as long as the user keeps interacting with the settings. Moreover, customisation of information elements and predefined interaction methods within an interface can be part of an initial system set-up allowing drivers an even greater degree of customisation.

14.4.3 Limitations of the Study

In this study, participants were actively asked whether they would like to make changes to the interface. Although the invitation was subtle and participants were told that there is no necessity of change, it could have motivated them to adjust settings with which they would not have bothered in everyday life. Therefore, there might have been a significant impact on customisation usage in comparison to the real-world usage rates.

Hierarchical agglomerative cluster analysis as an exploratory method raises further possible limitations. Even if applied to the same data, hierarchical clustering methods can result in different groupings, making the results inconclusive. Furthermore, the failure to recover non-spherical clusters and correctly choosing the number of clusters may be additional limiting factors.

The application of a driving simulator understandably cannot reach the same level of ecological validity as an on-road study. Accordingly, the choice of settings and drivers' preferences in a real-world scenario could differ substantially from the ones observed in this simulator study.

The short OOTL times of 1 and 10 min may not have sufficed to cause different levels of disengagement in drivers. Therefore, whether, and to what extent, a long OOTL time causes different interface settings could not be addressed appropriately in this study.

14.5 CONCLUSION AND FUTURE WORK

User-tailorable TOR interfaces enable future drivers to interact and be alerted by the system not only based on their personal preferences but also external conditions. This chapter investigated the customisation of TOR interfaces as a first step towards a fully adaptive TOR interface design. Participants' positive experience, and high appreciation, of customisation suggests that a tailored approach should play a significant role in future TOR interface design. The diverse participant clusters identified show that there is not only a demand but a need for tailorable TOR interfaces.

This need for tailorable TOR interfaces is additionally supported by the difference in takeover time depending on driver's age. A sufficient lead time is recommended to enable a safe transition for older drivers.

The cluster-display proves to be an important interface which most of the drivers still use as a means for information gathering. However, vocalisations and audio cues have been identified as additional important information sources. The conducted study adds to the available literature on speech-based vocalisations and highlights drivers' reliance on it.

Potentially important factors for future TOR interface design, such as OOTL time and age, identified in the literature and this study, can be used to transform the customisable approach presented here into an adaptive one. This removes the necessity for drivers to implement their own settings, making it more comfortable to use and potentially safer.

In a subsequent study, the user-tailorable approach will be tested with an on-road study to enhance ecological validity. An additional experiment investigating the effect of a longer OOTL time, such as, 60 min, will also be considered to clarify the consequences of longer periods OOTL on customisation settings. The results gained from this study show there is a desire and a need for drivers to adjust TOR settings to better suit their preferences and capabilities. Optimal settings for drivers are likely to be linked to driver acceptance and safe performance during takeover, warranting further exploration.

ACKNOWLEDGEMENTS

This work was supported by Jaguar Land Rover and the UK-EPSRC Grant EP/N011899/1 as part of the jointly funded 'Towards Autonomy: Smart and Connected Control (TASCC) Programme'.

REFERENCES

Akamatsu, M., P. Green, and K. Bengler. 2013. "Automotive technology and human factors research: past, present, and future." *International Journal of Vehicular Technology* *2013* 1–27. DOI: 10.1155/2013/526180.

Amditis, A., A. Polychronopoulos, L. Andreone, and E. Bekiaris. 2006. "Communication and interaction strategies in automotive adaptive interfaces." *Cognition, Technology & Work, 8* 193–199. DOI: 10.1007/s10111-006-0033-0.

Amditis, A., L. Andreone, K. Pagle, G. Markkula, E. Deregibus, M. Romera Rue, F. Bellotti, and et al. 2010. "Towards the automotive HMI of the future: Overview of the AIDE-integrated project results." *IEEE Transactions on Intelligent Transportation Systems, 11,* 3 567–578. DOI: 10.1109/TITS.2010.2048751.

Arora, N., X. Dreze, A. Ghose, J. D. Hess, R. Iyengar, B. Jing, Y. Joshi, and et al. 2008. "Putting one-to-one marketing to work: personalization, customization, and choice." *Marketing Letters, 19* 305–321. DOI: 10.1007/s11002-008-9056-z.

Bainbridge, L. 1983. "Ironies of Automation." *Analysis, Design and Evaluation of Man–Machine Systems. Proceedings of the IFAC/IFIP/IFORS/IEA Conference.* Baden-Baden, Germany: Elsevier Ltd. 129–135. https://doi.org/10.1016/B978-0-08-029348-6.50026-9.

Bellotti, F., A. De Gloria, R. Montanari, N. Dosio, and D. Morreale. 2005. "COMUNICAR: Designing a multimedia, context-aware human-machine interface for cars." *Cognition, Technology & Work, 7* 36–45. DOI: 10.1007/s10111-004-0168-9.

Bishop, R. 2019. *Is 2020 The Year For Eyes-Off Automated Driving?* October 1. Accessed April 21, 2020. https://www.forbes.com/sites/richardbishop1/2019/10/01/is-2020-the-year-for-eyes-off-automated-driving/.

Borojeni, S. S., L. Chuang, W. Heuten, and S. Boll. 2016. "Assisting Drivers with Ambient Take-Over Requests in Highly Automated Driving." *AutomotiveUI - Proceedings of the 8th International Conference on Automotive User Interfaces and Interactive Vehicular Applications.* Ann Arbor, MI, USA: Association for Computing Machinery. 237–244. https://doi.org/10.1145/3003715.3005409.

Bourrelly, A., C. Jacobé de Naurois, A. Zran, F. Rampillon, J. Vercher, and C. Bourdin. 2019. "Long automated driving phase affects take-over performance." *IEEE Intelligent Transport Systems, 13,* 8 1249–1255. DOI: 10.1049/iet-its.2019.0018.

Caber, N., P. Langdon, and P.J. Clarkson. 2018. "Intelligent Driver Profiling System for Cars – A Basic Concept." *International Conference on Universal Access in Human-Computer Interaction. Virtual, Augmented, and Intelligent Environments.* Cham, Switzerland: Springer. 201–213. https://doi.org/10.1007/978-3-319-92052-8_16.

Caber, N., P. Langdon, and P.J. Clarkson. 2019. "Designing Adaptation in Cars: An Exploratory Survey on Drivers' Usage of ADAS and Car Adaptations." *International Conference on Applied Human Factors and Ergonomics - Advances in Human Factors of Transportation.* Cham, Switzerland: Springer. 95–106. https://doi.org/10.1007/978-3-030-20503-4_9.

Clatworthy, J., D. Buick, M. Hankins, J. Weinman, and R. Horne. 2005. "The use and reporting of cluster analysis in health psychology: A review." *British Journal of Health Psychology, 10* 329–358. DOI: 10.1348/135910705X25697.

Dyke, F. B., A. M. Leiker, K. F. Grand, M. M. Godwin, A. G. Thompson, J. C. Rietschel, C. G. McDonald, and M. W. Miller. 2015. "The efficacy of auditory probes in indexing cognitive workload is dependent on stimulus complexity." *International Journal of Psychophysiology, 95,* 1 56–62. DOI: 10.1016/j.ijpsycho.2014.12.008.

Endsley, M. R. 2010. "Level of automation effects on performance, situation awareness and workload in a dynamic control task." *Ergonomics, 42,* 3 462–492. DOI: 10.1080/001401399185595.

Eriksson, A., and N. A. Stanton. 2017. "Takeover time in highly automated vehicles: Noncritical transitions to and from manual control." *Human Factors: The Journal of the Human Factors and Ergonomics Society, 59,* 4 689–705. DOI: 10.1177/0018720816685832.

Eriksson, A., S. M. Petermeijer, M. Zimmermann, J.C.F. de Winter, K. J. Bengler, and N. A. Stanton. 2019. "Rolling out the red (and green) carpet: Supporting driver decision making in automation-to-manual transitions." *IEEE Transactions on Human-Machine Systems, 49,* 1 20–31. DOI: 10.1109/THMS.2018.2883862.

Everitt, B. S., S. Landau, M. Leese, and D. Stahl. 2011. Cluster Analysis, 5th Edition. Chichester, UK: John Wiley & Sons, Ltd. https://doi.org/10.1002/9780470977811.

Fagnant, D. J., and K. Kockelman. 2015. "Preparing a nation for autonomous vehicles: Opportunities, barriers and policy recommendations." *Transportation Research Part A: Policy and Practice, 77* 167–181. DOI: 10.1016/j.tra.2015.04.003.

Feigh, K. M., M. C. Dorneich, and C. C. Hayes. 2012. "Toward a characterization of adaptive systems: A framework for researchers and system designers." *Human Factors: The Journal of the Human Factors and Ergonomics Society, 54* 1008–1024. DOI: 10.1177/0018720812443983.

Field, A. 2017. *Discovering Statistics Using IBM SPSS Statistics*, 5th edition. Thousand Oaks, CA: SAGE Publications.

Forster, Y., F. Naujoks, and A. Neukum. 2017. "Increasing anthropomorphism and trust in automated driving functions by adding speech output." *IEEE Intelligent Vehicles Symposium (IV)* 365–372. DOI: 10.1109/IVS.2017.7995746.

García-barrios, V. M., F. Mödritscher, and C. Gütl. 2005. "Personalisation versus Adaptation? A User-centred model approach and its application." *Proceedings of the International Conference on Knowledge Management (I-KNOW).* 120–127.

Gough, C. A.D., R. Green, and M. Billinghurst. 2006. "Accounting for user familiarity in user interfaces." *CHINZ'06: Proceedings of the 7th ACM SIGCHI New Zealand Chapter's International Conference on Computer-Human Interaction: Design Centered HCI.* Christchurch, New Zealand: ACM Press. 137–138. https://doi.org/10.1145/1152760.1152778.

Grabisch, M., H. Prade, EA. Raufaste, and P. Terrier. 2006. "Application of the choquet integral to subjective mental workload evaluation." *IFAC Proceedings Volumes, 39,* 4 135–140. DOI: 10.3182/20060522-3-FR-2904.00022.

Green, P. A. 2008. "Motor vehicle-driver interfaces." In *Human Computer Interaction Handbook: Fundamentals, Evolving Technologies, and Emerging Applications, Human Factors and Ergonomics,* by A. Sears and J. A. Jacko, 701–719. New York: Lawrence Erlbaum Association.

Hetzner, C. 2017. *German industry welcomes self-driving vehicles law.* Accessed November 19, 2019. https://europe.autonews.com/article/20170515/ANE/170519866/german-industry-welcomes-self-driving-vehicles-law.

Kalyanaraman, S., and S. S. Sundar. 2006. "The psychological appeal of personalized content in web portals: Does customization affect attitudes and behavior?" *Journal of Communication, 56,* 1 110–132. DOI: 10.1111/j.1460-2466.2006.00006.x.

Kew, O. 2018. *This is the bigger, lighter, techier new BMW 3 Series.* Accessed October 3, 2018. https://www.topgear.com/car-news/paris-motor-show/bigger-lighter-techier-new-bmw-3-series.

Körber, M., C. Gold, D. Lechner, and K. Bengler. 2016. "The influence of age on the takeover of vehicle control in highly automated driving." *Transportation Research Part F: Traffic Psychology and Behaviour, 39* 19–32. DOI: 10.1016/j.trf.2016.03.002.

Kyriakidis, M., J. C.F. de Winter, N. Stanton, T. Bellet, B. van Arem, K. Brookhuis, M. H. Martens, and et al. 2017. "A human factors perspective on automated driving." *Theoretical Issues in Ergonomics Science* 1–27. DOI: 10.1080/1463922X.2017.1293187.

Lambert, F. 2018. *Tesla plans to release Autopilot 'On Ramp/Off Ramp' feature in version 9.0 update next month.* Accessed March 24, 2019. https://electrek.co/2018/08/02/tesla-autopilot-on-ramp-off-ramp-feature-version-9-update/.

Langdon, P., and H. Thimbleby. 2010. "Inclusion and interaction: Designing interaction for inclusive populations." *Interacting with Computers, 22* 439–448. DOI: 10.1016/j.intcom.2010.08.007.

Langdon, P., R. Japikse, P.J. Clarkson, and K. Wallace. 2003. "The effectiveness of Psychological distance in an empirical study of guideline clustering for aerospace design." *Proceedings of ICED 03, the 14th International Conference on Engineering Design.* Stockholm, Sweden: The Design Society. 589–590.

McNamara, P. 2017. *How did Audi make the first car with Level 3 autonomy?* July 12. Accessed May 2, 2018. https://www.carmagazine.co.uk/car-news/tech/audi-a3-level-3-autonomy-how-did-they-get-it-to-market/.

Melcher, V., S. Rauh, F. Diederichs, H. Widlroither, and W. Bauer. 2015. "Take-over requests for automated driving." *Procedia Manufacturing, 3* 2867–2873. DOI: 10.1016/j.promfg.2015.07.788.

Merat, N., A. H. Jamson, F. C.H. Lai, and O. Carsten. 2012. "Highly automated driving, secondary task performance, and driver state." *Human Factors: The Journal of the Human Factors and Ergonomics Society, 54* 762–771. DOI: 10.1177/0018720812442087.

Merat, N., A. H. Jamson, F. C.H. Lai, M. Daly, and O. M.J. Carsten. 2014. "Transition to manual: Driver behaviour when resuming control from a highly automated vehicle." *Transportation Research Part F: Traffic Psychology and Behaviour, 27* 274–282. DOI: 10.1016/j.trf.2014.09.005.

Miller, M. W., J. C. Rietschel, C. G. McDonald, and B. D. Hatfield. 2011. "A novel approach to the physiological measurement of mental workload." *International Journal of Psychophysiology, 80,* 1 75–78. DOI: 10.1016/j.ijpsycho.2011.02.003.

Miyajima, C., Y. Nishiwaki, K. Ozawa, T. Wakita, K. Itou, K. Takeda, and F. Itakura. 2007. "Driver modeling based on driving behavior and its evaluation in driver identification." *Proceedings of the IEEE, 95,* 2 427–437. DOI: 10.1109/JPROC.2006.888405.

Naujoks, F., C. Mai, and A. Neukum. 2014. "The Effect of Urgency of Take-Over Requests During Highly Automated Driving Under Distraction Conditions." *5th International Conference on Applied Human Factors and Ergonomics.* Krakow, Poland: AHFE. 431–438.

Naujoks, F., Y. Forster, K. Wiedemann, and A. Neukum. 2016. "Speech improves human-automation cooperation in automated driving." *Mensch und Computer 2016 – Workshopband* http://doi.org/10.18420/muc2016-ws08-0007.

Nica, G. 2015. *How to Use BMW Personal Profiles for the Perfect Driving Position Every Time.* Accessed September 25, 2019. https://www.autoevolution.com/news/how-to-use-bmw-personal-profiles-for-the-perfect-driving-position-every-time-video-101615.html.

Nosofsky, R. 1985. "Overall similarity and the identification of separable-dimension stimuli: A choice model analysis." *Perception & Psychophysics, 38* 415–432. DOI: 10.3758/BF03207172.

Oliver, N., and A. P. Pentland. 2000. "Driver behavior recognition and prediction in a SmartCar." *Proceedings Volume 4023. Enhanced and Synthetic Vision 2000. Presented at the AeroSense 2000.* Orlando, US: Society of Photo-Optical Instrumentation Engineers (SPIE). 280–290. https://doi.org/10.1117/12.389351.

Persad, U., P. Langdon, and P.J. Clarkson. 2007. "Characterising user capabilities to support inclusive design evaluation." *Universal Access in the Information Society, 6* 119–135. DOI: 10.1007/s10209-007-0083-y.

Petermeijer, S. M., S. Cieler, and J. C.F. de Winter. 2017. "Comparing spatially static and dynamic vibrotactile take-over requests in the driver seat." *Accident Analysis & Prevention, 99* 218–227. DOI: 10.1016/j.aap.2016.12.001.

Petermeijer, S. M., P. Bazilinskyy, K. Bengler, and J. de Winter. 2017. "Take-over again: Investigating multimodal and directional TORs to get the driver back into the loop." *Applied Ergonomics, 62* 204–215. DOI: 10.1016/j.apergo.2017.02.023.

Piechulla, W., C. Mayser, H. Gehrke, and W. König. 2003. "Reducing drivers' mental workload by means of an adaptive man–machine interface." *Transportation Research Part F: Traffic Psychology and Behaviour, 6,* 4 233–248. DOI: 10.1016/j.trf.2003.08.001.

Politis, I., S. Brewster, and F. Pollick. 2015. "Language-based multimodal displays for the handover of control in autonomous cars." *AutomotiveUI - Proceedings of the 7th International Conference on Automotive User Interfaces and Interactive Vehicular Applications.* Nottingham, UK: Association for Computing Machinery. 3–10. https://doi.org/10.1145/2799250.2799262.

Politis, I., S. Brewster, and F. Pollick. 2017. "Using multimodal displays to signify critical handovers of control to distracted autonomous car drivers." *International Journal of Mobile Human-Computer Interaction, 9*, 3 1–16. DOI: 10.4018/ijmhci.2017070101.

Politis, I., P. Langdon, M. Bradley, L. Skrypchuk, A. Mouzakitis, and P.J. Clarkson. 2017. "Designing autonomy in cars: A survey and two focus groups on driving habits of an inclusive user group, and group attitudes towards autonomous cars." *International Conference on Applied Human Factors and Ergonomics - Advances in Design for Inclusion.* Springer. 161–173. https://doi.org/10.1007/978-3-319-60597-5_15.

Politis, I., P. Langdon, D. Adebayo, M. Bradley, P.J. Clarkson, L. Skrypchuk, A. Mouzakitis, and et al. 2018. "An Evaluation of Inclusive Dialogue-Based Interfaces for the Takeover of Control in Autonomous Cars." *Proceedings of the 2018 Conference on Human Information Interaction & Retrieval - IUI'18: 23rd International Conference on Intelligent User Interfaces.* Tokyo, Japan: ACM Press. 601–606. https://doi.org/10.1145/3172944.3172990.

Pollard, T. 2018. BMW's new iDrive OS 7.0: hands-on test. April 17. Accessed May 2, 2018. https://www.carmagazine.co.uk/car-news/tech/bmw-idrive-os-70-what-you-need-to-know-about-the-new-2018-idrive/.

Porsche-AG. 2018. "Porsche Mission E Cross Turismo: the electric athlete for an active lifestyle - Porsche Great Britain." *Porsche HOME: Porsche Mission E Cross Turismo: the electric athlete for an active lifestyle - Porsche Great* Britain. Accessed April 1, 2019. https://www.porsche.com/uk/aboutporsche/pressreleases/pcgb/?lang=none&pool=international-de&id=482584.

SAE J3016. 2016. *Taxonomy and Definitions for Terms Related to Driving Automation Systems for On-Road Motor Vehicles.* Accessed November 3, 2020. https://www.sae.org/standards/content/j3016_201806/.

Schnelle, S., J. Wang, H. Su, and R. Jagacinski. 2017. "A driver steering model with personalized desired path generation." *IEEE Transactions on Systems, Man, and Cybernetics: Systems, 47*, 1 111–120. DOI: 10.1109/TSMC.2016.2529582.

Stanton, N. A., and M. S. Young. 2005. "Driver behaviour with adaptive cruise control." *Ergonomics, 48*, 10 1294–1313. DOI: 10.1080/00140130500252990.

Stanton, N. A., M. Young, and B. McCaulder. 1997. "Drive-by-wire: The case of driver workload and reclaiming control with adaptive cruise control." *Safety Science, 27*, 2 149–159. DOI: 10.1016/S0925-7535(97)00054-4.

Supulniece, I. 2012. "Conceptual Aspects of User-Oriented Adaptive Systems." *ICIS 2012 Proceedings. Presented at the International Conference on Information Systems (ICIS).* 116–124.

Volkswagen-AG. 2017. "Personalisation – individual settings on demand Arteon, T-Roc, Golf & Co. "recognise" their drivers." *Volkswagen Newsroom.* Accessed April 1, 2019. https://www.volkswagen-newsroom.com:443/en/press-releases/personalisation-individual-settings-on-demand-arteon-t-roc-golf-and-co-recognise-their-drivers-853.

Walch, M., K. Lange, M. Baumann, and M. Weber. 2015. "Autonomous driving: investigating the feasibility of car-driver handover assistance." *AutomotiveUI - Proceedings of the 7th International Conference on Automotive User Interfaces and Interactive Vehicular Applications.* Association for Computing Machinery: Nottingham, UK. 11–18. https://doi.org/10.1145/2799250.2799268.

Walker, G. H., N. A. Stanton, and P. M. Salmon. 2015. *Human Factors in Automotive Engineering and Technology - Human Factors in Road and Rail Transport.* Ashgate, UK: CRC Press.

World-Health-Organization. 2015. *Global Status Report on Road Safety.* Geneva, Switzerland: World Health Organization.

Wright, J., Q. Stafford-Fraser, M. Mahmoud, P. Robinson, E. Dias, and L. Skrypchuk. 2017. "Intelligent scheduling for in-car notifications." *2017 IEEE 3rd International Forum on Research and Technologies for Society and Industry (RTSI).* Modena, Italy: IEEE. 1–6. https://doi.org/10.1109/RTSI.2017.8065957.

Yi, D., J. Su, C. Liu, and W.H. Chen. 2019. "New driver workload prediction using clustering-aided approaches." *IEEE Transactions on Systems, Man, and Cybernetics: Systems, 49* 64–70.DOI: 10.1109/TSMC.2018.2871416.

Young, M. S., and N. A. Stanton. 1997. "Automotive Automation: Investigating the Impact on driver mental workload." *International Journal of Cognitive Ergonomics, 1, 4* 325–336.

Young, M. S., and N. A. Stanton. 2020, July. The decay of malleable attentional resources theory. In *Contemporary Ergonomics 2006: Proceedings of the International Conference on Contemporary Ergonomics (CE2006)*, 4–6 April 2006, Cambridge, UK. 253–257. Abingdon-on-Thames: Taylor & Francis.

Young, M. S., and N. A. Stanton. 2007. "Back to the future: Brake reaction times for manual and automated vehicles." *Ergonomics, 50,* 1 46–58. DOI: 10.1080/00140130600980789.

15 Modelling Automation– Human Driver Interactions in Vehicle Takeovers Using OESDs

*Neville A. Stanton, James W. H. Brown,
Kirsten M. A. Revell, Jediah R. Clark,
and Joy Richardson*
University of Southampton

Patrick Langdon
Edinburgh Napier University

Michael Bradley and Nermin Caber
University of Cambridge

Lee Skrypchuk and Simon Thompson
Jaguar Land Rover

CONTENTS

15.1 INTRODUCTION

Modelling in ergonomics and human factors has a long tradition (Carbonell, Ward and Senders 1968; Moray 2007; Moray, Groeger and Stanton 2017; Senders 1997), going back to the dawn of the discipline, with the invention of the term 'Therblig' to describe a basic human action (Gilbreth 1909). Therbligs comprised 18 elements, including search, find, select, grasp, hold, transport, position, assemble, use, disassemble, inspect, release, delay, plan, and rest. These elements were used by Gilbreth and colleagues to model human manual performance in pursuit of more effective and efficient work patterns. Operator event sequence diagrams (OESDs) (Kurke 1961; Kirwan and Ainsworth 1992) may be considered as a development of the approach, to model human–machine interactions in one pictorial representation. OESDs can be used as part of the interaction design process, in developing new interfaces between humans and technology (Stanton et al. 2013). This may be thought of in terms of the co-evolution of sociotechnical systems, where the human activity and technical process are designed together (Walker et al. 2009), rather than the human activity resulting from tasks left over from automation (Bainbridge 1983). There are many examples of the application of OESDs, such as modelling single pilot operations in commercial aviation (Harris, Stanton and Starr 2015), aircraft landing (Sorensen, Stanton and Banks 2011), air traffic control (Walker et al. 2010), electrical energy distribution (Salmon et al. 2008), maritime collision avoidance (Kurke 1961), process control (Kirwan and Ainsworth 1992), and automatic emergency braking systems in automobiles (Banks, Stanton and Harvey 2014). Despite this wide variety of applications, there is scant validation evidence to support their continued use (Stanton and Young 1999a, b; Stanton 2016). This paper has the twin aims of showing the application of OESDs to the design of the automated vehicle to driver takeover task and providing validation evidence from a high-fidelity driving simulator study.

In one of the very first applications of OESDs, Kurke (1961) showed how this method could be used to design navigation systems on a maritime vessel. The representation was used to compare collision avoidance, using the traditional approach with a new computer-supported approach. Kurke argued that, in the new approach, a computer would automatically calculate the closest point of approach, thus minimising error and reducing collision risk. Comparing the two OESDs side by side allows the analyst to see how the work of the Watch Officer had changed with the introduction of the computer support. This can help to ensure that the system design has been constructed to establish meaningful operational procedures (Kurke 1961), something Gilbreth (1909) himself was also keen to promote. More recently, Harris, Stanton, and Starr (2015) also used OESDs to compare work of two pilots on the flight deck of a civil aircraft with that of a single pilot configuration. Again, the authors invited the reader to spot the difference in a side-by-side comparison of the two versions of the work. The revised work allocation in the reduced crewing option showed that the workload did not reduce by half (although there were substantially fewer cross-checks and communications), and potentially could overload the single pilot (Harris, Stanton and Starr 2015). Both of these examples, and others, show the value of OESDs in modelling future versions of systems and being able to anticipate the work that might be undertaken. Banks, Stanton, and Harvey (2014) have shown how OESDs can be applied to the design of vehicle automation – in particular, to the examination of the interaction between the human driver and the various vehicle subsystems and the road environment. The current paper focuses on the interaction between driver and in-vehicle subsystems during takeover of control.

Other modelling approaches in ergonomics and human factors have made predictions about human error (Embrey 1986; Lane, Stanton and Harrison 2006; Stanton and Harvey 2017; Parnell et al. 2019), time to perform tasks (Card, Newell and Moran 1983; Baber and Mellor 2001; Stanton and Baber 2008), visual sampling behaviour (Senders 1997; Moray, Groeger and Stanton 2017; Clark, Stanton and Revell 2019a), and the structure of work (Annett et al. 1971; Stanton 2006, 2014). The advantage of OESDs is that they make the interactions between various subsystems (including the human operator(s)) explicit, within and between 'swim-lanes' (the columns containing work associated with each of the actors). Each swim-lane contains the activities of that subsystem, and interactions are depicted as connectors across the swim-lane. Like Therbligs (Gilbreth 1909), the approach has a taxonomy of activities, but these can apply to both the social and technical subsystems. OESDs categories include process, decision, delay, display, inspection, operation, receipt, speech, storage, and transport (Kurke 1961; Kirwan and Ainsworth 1992; Stanton et al. 2013). Several of these seem very similar to the Therbligs (such as delay, inspection, and transport), but rather than applying solely to human activities, they can also be applied to the technical aspects of the system. Connectors are used in OESDs to make explicit links between the outputs of one or more subsystem(s) and the inputs into one or more subsystem(s).

Human factors methods are designed to improve product design by understanding or predicting user interaction with devices (Stanton and Young 1999a). These approaches have a long tradition in system design and tend to have greater impact (as well as reduced cost) when applied early on in the design process (Stanton and Young 1999a). Kirwan and Ainsworth (1992) note that the original purpose of OESDs was to represent complex, multiperson tasks. The output of an OESD graphically depicts a task process, including the tasks performed and the interaction between operators over time, using standardised symbols. There are numerous forms of OESDs, ranging from a simple flow diagram representing task order, to more complex analyses of team interaction and communication, and often including a timeline of the scenario under analysis and potential sources of error. OESDs have been used during the design of complex systems, such as nuclear power and petrochemical processing plants (Kirwan and Ainsworth 1992).

Kurke (1961) originally proposed OESDs for human–machine systems interaction design as well as layout of the interfaces. His examples focused on the design of interfaces between human and machines to specify the sequence of operations between these subsystems (Kurke's analysis had swim-lanes for the object to be avoided, the ship, the computer, and the Watch Officer). As such, OESDs may be best suited to scenario analysis where there are discrete sets of tasks to be undertaken. In the present study, we have analysed the takeover of vehicle control by human drivers from automation, which is presented in the next section.

15.1.1 Development of the OESD for Automation– Human Driver Takeover

This section focuses on the first aim of the paper by showing the application of OESDs to the design of the automated vehicle-to-driver takeover task. The rationale for the interaction design is described as well as further detail regarding the elements captured in the diagrams.

Descriptions on the development of OESDs may be found in Kirwan and Ainsworth (1992) and Stanton et al. (2013). In essence, the scenario start and end points are decided, together with the main actors. The analysis presented in the current paper is based on a use-case of vehicle automation takeover scenario on a UK motorway with a SAE Level 4 vehicle (SAE J3016 2016). It is assumed that drivers will drive manually onto the motorway and hand the driving task over to vehicle automation when it becomes available. While vehicle automation is engaged, the driver is free to engage in non-driving tasks (such as reading, emailing, working on a tablet computer). The vehicle would alert the driver of the need to take back control of the vehicle in a planned, non-emergency takeover in a timely manner before the exit junction. These takeovers are described using the task elements from OESDs, as shown in Table 15.1.

The OESDs shown in Figures 15.1–15.5 were developed in workshops with Human Factors and Automotive Engineering experts.

The ten swim-lanes show the different 'actors' under consideration in the design of the takeovers by the human driver from vehicle automation, via the instrument cluster (instruments viewed through the steering wheel), head-up display (HUD – viewed in the windscreen or windshield), centre console (the upper part of the centre of the dashboard), ambient (lighting around the dashboard and vehicle interior), and haptic (vibration through the driver's seat) displays. The arrows are connectors that show the links between the events in the swim-lanes. The takeover protocol presented in Figures 15.1–15.5 was designed to raise the situation awareness of drivers, by presenting them with contextually relevant questions about the vehicle status, other road users, as well as the surrounding environment and infrastructure. This was based on the research evidence that degraded performance of drivers of automated vehicles is, in part, due to poor situation awareness (Stanton et al. 2017). For example, the collisions in the Tesla and Uber vehicles report that the driver was not aware of the environment outside the vehicle (Banks, Plant and Stanton 2018; Stanton et al. 2019). As can be seen in Figure 15.1, it is assumed that the vehicle is under manual control until the system detects that the road is suitable for automation to operate. Then, the system prompts the driver with the message that automation is available, via the four interfaces (cluster, HUD, centre console, and ambient display) should they wish to use it.

If the driver chooses to engage vehicle automation, then they would press two buttons on the steering wheel with their thumbs simultaneously (assuming that their hands are in the ten-to-two clock position on the steering wheel). At this point, the interfaces would display 'Automation Activated' followed by 'The car is in control' (see Figure 15.2). At the same time, the ambient lighting in the car would change from orange (indicating manual driving mode) to blue (indicating automated driving

TABLE 15.1

Key for the OESDs

OESD Task Elements	Description
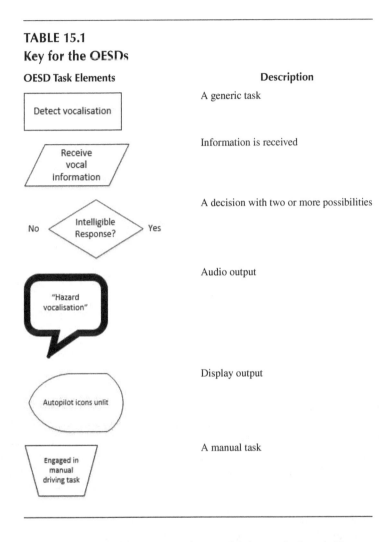	A generic task
	Information is received
	A decision with two or more possibilities
	Audio output
	Display output
	A manual task

mode). Then, the driver is able to engage in non-driving tasks (on a tablet computer in this scenario, as Level 4 SAE is assumed, so there is no need for the driver to monitor the automated driving system). The scenario assumes that there is a planned takeover of driving from automation back to the human driver (such as when their exit from the motorway is coming up, which would have been pre-programmed into the satellite navigation system). The driver is given a 5-, then 2-, then 1-min notice that the takeover process will begin (see Figure 15.2).

Upon the prompt from the automated system that the driver needs to get ready to drive, it is assumed that the driver ceases the non-driving task, puts down the tablet computer, and resumes the driving position (as shown in Figure 15.3). The system then presents a series of questions designed to raise the situation awareness of the driver (such as What speed is the vehicle currently travelling at? What lane are you currently in? What colour is the vehicle in front of you? What is your remaining fuel

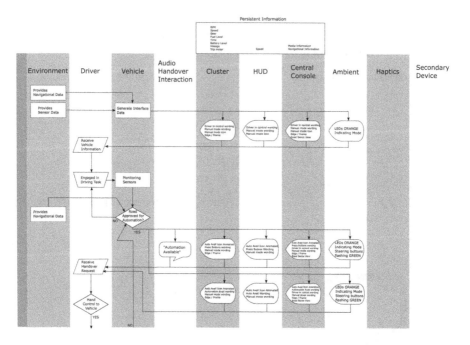

FIGURE 15.1   Human driver in control when automation becomes available.

range? Can you see a bend in the road ahead?). The driver is expected to respond to these questions (which are presented auditorily as well as on all of the visual interfaces). If the answer is correct, then the next question is presented until all questions have been answered. If the answer is incorrect, then the question is repeated for a maximum of two additional times before moving onto the next question. When all questions have been presented, the takeover interaction moves on to that presented in Figure 15.3.

The driver is then requested to take manual control of the vehicle (see Figure 15.4), which will mean placing both hands on the steering wheel and positioning their foot on the accelerator pedal. To transfer control from the automated system to the driver, they need to press two buttons mounted on the steering wheel at the ten-to-two clock position with their thumbs (in the same manner as they do for handing control over to the vehicle automation system).

As Figure 15.5 shows, when control of the vehicle is passed back to the human driver, the ambient lighting changes back from blue to orange (indicating the vehicle is now in manual driving mode) and the words 'Automation deactivated' are presented auditorily as well as on the visual displays. This is followed by the words 'You are in control', which are also presented auditorily and on the visual displays. The human driver is now driving the vehicle.

As well as showing how OESDs can be used to design the interactions between automations and humans, the other aim of this study was to empirically validate the approach. This is introduced in the following section.

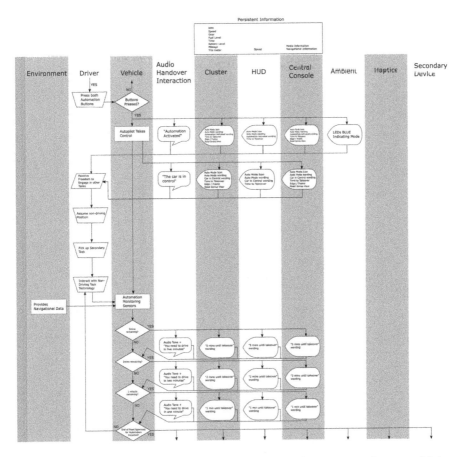

FIGURE 15.2 From automation activated (and driver working on a secondary, non-driving task) up to advice that takeover needs to begin.

15.1.2 VALIDATION OF METHODS

The remainder of this paper focuses on aim two, the validation of OESDs modelling using evidence of actual behaviour from a high-fidelity driving simulator study. Validation of ergonomics and human factors methods has proved very challenging to the community, with scant evidence produced on even the most commonly used approaches (Stanton and Young 1999a, b; Stanton 2016). To paraphrase Stanton and Young (1999a), 'validity is often assumed but seldom empirically tested'. Most ergonomics and human factors methods have theoretical construct and content validity, which helps to convey the credibility of the method to its users. This means that it is often based on a contemporary theory and uses appropriate terminology. With systems theory in ascendancy (Stanton et al. 2017; Stanton, Salmon and Walker 2017), there is perhaps a resurgence of interest in OESDs, as they are able to represent the interactions between subsystems.

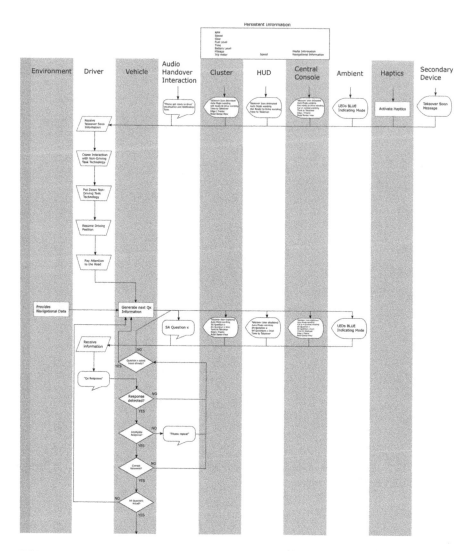

FIGURE 15.3 Driver resumes driving position and engages in takeover protocol.

Nevertheless, establishing concurrent and predictive validity for contemporary applications, such as road vehicle automation, is a priority. Just because a method has 60 years of use, it does not necessarily mean that validation is proven, nor that validity generalisation can be assumed. Therefore, concurrent and predictive validity needs to be formally tested (Annett 2002; Stanton 2002, 2014; Stanton and Young 1999b). Such validations have been undertaken for methods that predict error (Baber and Stanton 1996; Stanton and Stevenage 1998; Stanton et al. 2009) and task time (Baber and Mellor 2001; Stanton and Baber 2008; Harvey and Stanton 2013). A comparison of the reliability and validity of a range of human factors methods has been undertaken by Stanton and Young (1999a, b, 2003). These studies show that the methods vary quite

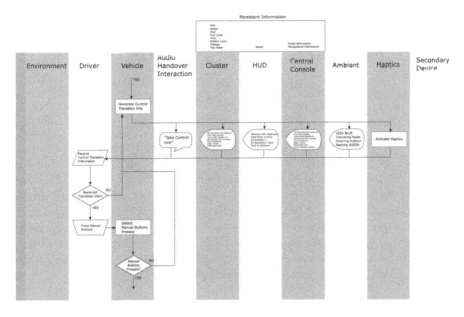

FIGURE 15.4 Human driver requested to take control of driving inputs.

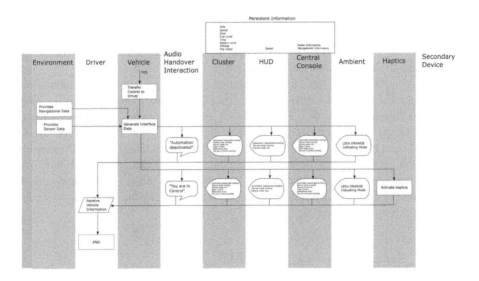

FIGURE 15.5 Human driver resumes manual control.

considerably in their performance. Previous work has made use of the signal detection paradigm for validity testing of ergonomics and human factors methods (Stanton and Young 1999a, b). This approach was used in the current paper, and based on previous work, it was expected that OESDs would achieve reasonable levels of validity, above the threshold for acceptance for use in modelling automation–human interaction.

15.2 METHODS

The methods section covers the recruitment of participants, the design of the study, equipment used, and procedure followed together with approach for data reduction and analysis.

15.2.1 PARTICIPANTS

In this study, 65 participants were recruited in three age bands. Twenty-five were aged between 18 and 34 (mean = 26.6, SD = 4.4), 20 were aged between 35 and 56 (mean = 43.8, SD = 5.8), and 20 were aged between 57 and 82 (mean = 64.8, SD = 5.0). All of the participants were drawn from the general population, had full UK driving licenses, and were in good health, with corrected vision where applicable. Ethical permission for the study was granted by the University of Southampton Research Governance Office (ERGO Number: 41761.A3). Each participant was briefed on the nature of the study and made aware of their right to withdraw at any time. All participants signed a consent form prior to taking part in the study.

15.2.2 STUDY DESIGN

The OESD was used to develop the design of the takeover interface (see Figure 15.2). This meant that the expected behaviour of the human driver and automated system, together with their interaction, can be predicted. To determine the accuracy of this prediction, it was compared with video footage of the actual behaviour of the 65 human drivers who took part in this study.

This study comprised three driver-to-automation handovers, and three automation-to-driver takeovers, repeated over four trials (12 handovers and takeovers in total, although this analysis is focused solely on the automation-to-human takeovers). Two of the trials had shorter out-of-the loop activity (i.e. 1 min), and two of the trials had longer out-of-the-loop activity (i.e. 10 min). These longer (but still relatively short) and shorter out-of-the-loop activity trials were counterbalanced. The last automation-to-human driver takeover for each trial was analysed in this study due to the volume of data, and assuming optimal familiarity during the latter stages of each trial (65 drivers, and 4 takeovers analysed per driver).

15.2.3 EQUIPMENT

The driving simulator was based on a Land Rover Discovery Sport vehicle interior, with a fixed-base running STISIM software (see Figure 15.6). The simulated driving environment comprised a congested, three-lane motorway (to simulate the rush hour in the UK) in dry conditions with good visibility. To help reduce mode error (Sarter and Woods 1995; Stanton, Dunoyer and Leatherland 2011), the cabin ambient lighting displayed two distinct colours. As shown in Figure 15.6, blue was used to indicate that the vehicle was under automated control and orange was used to indicate the vehicle was under manual driver control.

FIGURE 15.6 Four views of the driving simulator, top-left and top-right show the vehicle is under automated control (where the ambient light colour is blue) and the bottom-left and bottom-right panels show that the vehicle is under manual driver control (where the ambient light colour is orange).

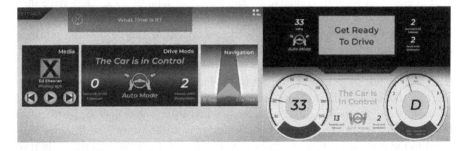

FIGURE 15.7 The three visual aspects of the HMI, on the left is the central console, bottom-right is the cluster, and the HUD is shown top-right.

Interfaces in the vehicle comprised a HUD (showing the car in automated mode but beginning the preparation for the human driver to take control, see Figure 15.7), instrument cluster (showing the icon associated with the car in automated mode, see Figure 15.7), a centre console (showing the car in automated mode, see Figure 15.7), haptic seat (to prompt the driver when the takeover begins), speech input/output (to communicate the takeover questions), and ambient light display (to indicate the driving mode, blue means automation is in control and orange means the human driver is in control). Engagement and disengagement of vehicle automation was undertaken by pressing two green buttons, mounted on the steering wheel, with the driver's thumbs simultaneously. Audio sounds and synthesised speech were generated automatically by the vehicle automation and presented through the in-vehicle speaker system.

In addition, a Microsoft Surface tablet computer was in the cabin loaded with a Tetromino game. This was used as the secondary task to engage the driver when automation was driving the vehicle. In some conditions (if selected by the driver), a visual alert was presented on the tablet to start the takeover process. This was accompanied by a speech alert through the vehicle speakers.

A control desk at the rear of the vehicle was used by the experimenter as a 'Wizard of Oz' environment, to interpret the driver's vocal responses to the speech synthesis questions during the takeover process. If the driver gave an incorrect response, then the question was repeated until a correct response was given. If more than two incorrect responses were given, then the next question was presented until all questions were answered. Then, the driver was requested to resume manual driving. All takeovers were planned (not as a result of an emergency or system failure) and proceeded at the pace of the driver.

15.2.4 PROCEDURE

On arrival at the driving simulation facility, participants were welcomed and presented with a participant information sheet informing them of the details of the study. Their right to halt the study at any time was explained, and they were then provided with an informed consent form which they had to read, initial, and sign in order for the study to continue. On completion of the consent form, participants were given a bibliographical form to complete to capture demographics data. They were then introduced to the simulator. The main driving controls were explained, and they were informed of the functionality of the vehicle automation and the human–machine interfaces (instrument cluster, HUD, centre console, haptic seat, speech input/output, and ambient light display). Participants then took part in a test run, where they experienced three takeovers after 1-min out-of-the-loop intervals. After completion of the test run, the main trials started with the participant being asked to accelerate onto the motorway, join the middle lane, keep up with traffic, and follow the instructions presented on the displays. After a period of approximately 1 min, the displays indicated to the participant that automation was available and informed them via text, icon, vocalisation, and two flashing green steering wheel buttons. Participants then activated automation by pressing the two steering wheel buttons; this was followed by the instruction indicating that the automation system was now in control of the vehicle. Participants then picked up the secondary task tablet and started to engage with the Tetromino secondary task. After a period of either 1 or 10 min, dependent upon counterbalancing, the automation indicated via the displays that the driver was required to get ready to take control. Participants were expected to put aside their secondary task, and follow the instructions presented on the displays. The takeover protocol consisted of a set of questions designed to raise situation awareness. These questions were presented in vocal, word, and icon form. Participants responded vocally to each question, the answers to which were judged by an experimenter taking the part of the automation system using the Wizard of Oz approach. Incorrect or missed questions were repeated twice before moving to the next. When all the questions were answered by the participant, the human–machine interfaces indicated for them to take control, which they did by pressing the two green buttons on the steering wheel. This constituted one takeover; the process was repeated twice more. After completion, the participant was asked to pull safely to the hard shoulder and stop the vehicle. This process was repeated three more times, allowing participants to adjust the takeover displays after each trial. Once the trials were complete, participants were debriefed and thanked for their time.

15.2.5 Data Reduction and Analysis

The video data for each driver during the takeover process was reduced into hits, misses, false alarms, and correct rejections by comparison with the OESD as follows (and in Table 15.2):

- **Hits**: Present in the video and the OESD.
- **Misses**: Present in the video but not the OESD.
- **False alarms**: Not present in video but present in OESD.
- **Correct rejections**: Not present in video and not present in OESD.

The latter category can be difficult to calculate as it could be infinity, but for the purposes of this investigation, it was based on the total number of false alarms generated by all 65 participants, minus the number of false alarms for each individual participant.

Inter-rater reliability testing was conducted on the categorisation scheme for approximately 10% of the video footage between two analysts. Equal-weighted Cohen's kappa was calculated (0.718) showing acceptable agreement between the two independent analysts in their classification of hits, misses, false alarms, and correct rejections (Landis and Koch 1977).

The data for each trial was pooled, and the Matthews correlation coefficient (Phi: Matthews 1975) was calculated using Equation 15.1.

$$\varphi = \frac{\text{Hit} \times \text{CR} - \text{FA} \times \text{Miss}}{\sqrt{(\text{Hit} + \text{FA})(\text{Hit} + \text{Miss})(\text{CR} + \text{FA})(\text{CR} + \text{Miss})}}. \tag{15.1}$$

15.3 RESULTS

The frequency of hits, misses, false alarms, and correct rejections for each of the four takeovers (one for each trial) is presented in Figure 15.8. The box-and-whisker plots

---

TABLE 15.2
Signal Detection Paradigm for OESD

		Behaviour Present in Video	
		Yes	No
Behaviour present in OESD	Yes	Hit	False alarm
	No	Miss	Correct rejection

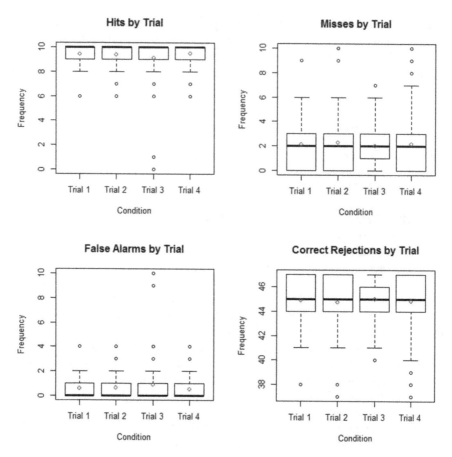

FIGURE 15.8 Hits, misses, false alarms, and correct rejections by trial.

show the median (thick line), range (whisker to whisker), and interquartile range (the box), as well as mean (red circle) and outliers (black circles). As Figure 15.8 shows, the hits are quite high (10 on average) and misses are low (2 on average). Similarly, the false alarms are low (0 on average) and correct rejections are high (45 on average). This is encouraging for the usefulness of OESDs to predict driver–automation takeover activity before empirical studies.

Hit rate (=hits/(hits + misses)) is similarly encouraging, with a median of 1 and a mean very close to 1, as shown in Figure 15.9. A hit rate of 1 is perfect. This means that the OESD process is able to identify those tasks that are observed.

False alarm rate (=false alarms/(false alarms+correct rejections)) is very low, with a median of 0 and a mean around 0.03, as shown in Figure 15.10. This means that the OESD process tends not to predict tasks that do not occur.

Finally, Phi is above 0.8 (the criterion value) for both median and mean values, as shown in Figure 15.11. This means that the OESDs have good predictive validity, at least for driver–automation takeover activity. It should be noted that the outlier in trial three at 0 did not go through any of the takeover protocol and simply

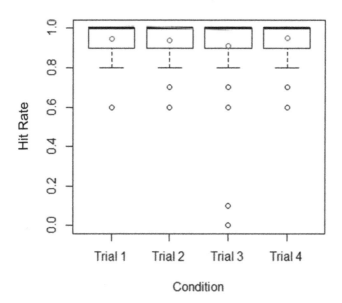

FIGURE 15.9 Hit rate by trial.

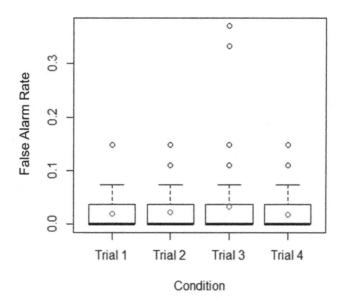

FIGURE 15.10 False alarm rate by trial.

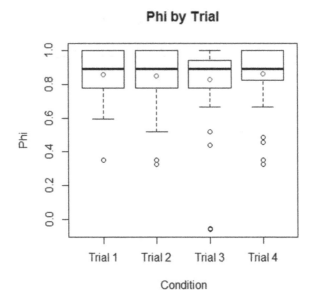

FIGURE 15.11   Matthews correlation coefficient (Phi) by trial.

took manual control of the vehicle. The design of the vehicle allowed drivers to take manual control at any point, but it was anticipated that they would go through the takeover protocol before resuming manual control.

15.4   DISCUSSION

The main finding from this study is that the OESD presented in this paper was able to produce reasonable predictions of the takeover activity conducted by human drivers in all four trials, to an acceptable level (as shown in Figure 15.11). This is very encouraging for the method, as not all ergonomics and human factors methods perform this well (Stanton and Young 1999a, b, 2003). For example, similar values have been reported for the Systematic Human Error Reduction Prediction Approach (Harris et al. 2005) and the Keystroke Level Model (Stanton and Young 1999a). Although OESDs have been in continuous use over the past 60 years, we know of no other formal test of predictive validity (Stanton 2016). This finding has important implications for the use of the method as a modelling activity before empirical studies with human participants. This lends credibility to the comparative analysis used by others for choosing between alternative design solutions (Kurke 1961; Kirwan and Ainsworth 1992; Harris, Stanton and Starr 2015) and modelling future human–automation interactions (Banks et al. 2018). Indeed, on the basis of this study, OESDs could be seen as a possible method for discriminating between alternative approaches for human–automation interaction, where there are too many alternatives for empirical studies.

Moray (2007) has proposed that modelling of human performance in technological systems has long been the goal of the discipline of ergonomics and human factors.

Certainly, this would help raise the credibility of the methods as in the engineering of systems. Some progress has already been made on the empirical validation of some of the methods, such as those that predict task-related human error (Stanton et al. 2009) and task time (Baber and Mellor 2001), which have shown acceptable levels of performance. The prediction of the tasks themselves is what has been presented in the current research. Certainly, the performance of the OESD in the study of takeover has achieved similar levels to those reported for other methods (Stanton et al. 2013; Stanton, Young and Harvey 2014). This means that there is good evidence for their continued use, but future studies should attempt to provide validation data in a range of applications and domains.

Research into takeovers in a wide range of domains has shown that this task is problematic (Clark, Stanton and Revell 2019b), particularly with the failure to transmit appropriate information from one agent to another (Stanton and Harvey 2017). Certainly, takeovers from automated systems to human drivers are no exception (Banks, Plant and Stanton 2018; Salmon, Walker and Stanton 2016). This is a field of endeavour that is currently challenging researchers to facilitate raising the situation awareness of the driver before the takeover is completed (Stanton et al. 2017). Situation awareness is a hotly debated topic in human factors, with views ranging from it being a concept that resides in the mind of humans (Endsley 2015) to that of it being a product of the interaction with a sociotechnical system (Stanton, Salmon and Walker 2015). In vehicle automation, there is an argument for the latter approach (Stanton et al. 2017). Certainly, an automated vehicle needs to be aware of the surrounding vehicles, road environment, and infrastructure, similar to a human driver, albeit in different ways. In the takeover process, some vehicle manufacturers are simply providing a countdown for the time until the vehicle becomes the responsibility of the human driver. This approach does not provide any guarantee about what is transferred in the takeover. The approach presented in this paper took inspiration from human–human takeovers (Clark, Stanton and Revell 2019b), which aim to transfer awareness of the situation from one agent to another. The OESDs can help to make the specification of what is to be transferred more explicit in the design of the interfaces, as was the case in the study reported in this paper.

What the OESD was not able to predict was the occasion when the driver simply took control of the vehicle without going through the formal takeover protocol. This was just one participant, as shown in Figure 15.11 for the third trial. That said, one would not wish to design a takeover protocol that would force drivers to go through every takeover step if they wanted to resume control immediately. The design of the takeover protocol was focused on guiding drivers back into the driving task through a series of situation awareness transactions between the driver and the vehicle (Stanton et al. 2017).

At present, OESDs model normative performance of systems. Armed with this early validation evidence, their use could be expanded to include non-normative performance of systems (Stanton et al. 2019). Typically, such analysis tended to focus on either human (Embrey 1986) or technical (Abedi, Gaudard and Romerio 2019) failures in systems, although not exclusively so (Leveson 2004; Stanton and Harvey 2017). A possible future direction for further development of OESDs could address the potential to model failures in automation and suboptimal responses by human drivers together. The 'miss' data observed in the current study could serve as a starting point for the modelling of alternative human activity, such as taking over control of the vehicle at any point during the takeover protocol.

Another avenue for the development of OESDs could be the addition of time data to both the human and technical activities, to model the minimum, maximum, and median takeover times (Harvey and Stanton 2013). Prediction of task time is quite popular in human factors modelling (Card, Newell and Moran 1983), and there are examples of application in domains such as human–computer interaction (Baber and Mellor 2001), rail (Stanton and Baber 2008) and in-car infotainment systems (Harvey and Stanton 2013). Given the growing literature on automation–human takeover times (see Eriksson and Stanton 2017 for a review), modelling such activities could prove useful in helping to understand how to improve the design and efficiency of the protocols.

There is the broader question of validation of human factors and ergonomics methods more generally, to be addressed by the community (Stanton 2016). These are obvious differences between novice (Stanton and Young 1999a, b, 2003) and expert analysts (Stanton and Baber 2002, 2005), with experts showing better validity data than novices. But there are also domain differences in validity data, suggesting that the classification schemes may have a domain bias (Stanton et al. 2009). The classification scheme for the activities in OESDs has developed over the past 60 years, with developments in technologies and the changing nature of human work. It is reasonable to expect the scheme to continue to evolve, particularly as technologies become more animated and with developments in automation and artificial intelligence (Stanton, Salmon and Walker 2017; Stanton et al. 2017).

In broad terms, this paper has also demonstrated how OESDs can be used to undertake human-centred automation design, focusing on the takeover problem. It is suggested that the approach can be used in the co-evolution of both human and technical subsystems with the sociotechnical systems paradigm. The OESDs are used to guide the task sequences and define what needs to appear on the interfaces in different media. Each of the media had their own 'swim-lane', so that the design could be coordinated. In this way, the OESDs provided a structured approach to guide the design. In each stage of the design, the interaction was simulated in the OESD first, to check that automated and human agents were coordinated. When satisfied that this was the case, the prototype interfaces were built and tested with human participants in a driving simulator. Some iteration between the OESDs and the simulator prototypes was required before the final interfaces were accepted for testing with drivers.

15.5 CONCLUSIONS

This paper sought to apply OESDs to the design of the semi-automated vehicle to driver takeover task and provide validation evidence by comparing theoretical design modelling to behaviours observed from a high-fidelity driving simulator study. The findings from this study demonstrate that the OESD is a useful method for assisting in the specification of interaction design strategies for takeover protocols between road vehicle automation and human drivers. This led to the development of the design of the interfaces and questions to raise the awareness of the driver about their own vehicle status, presence of other vehicles, and road environment. The empirical study in a driving simulator showed that the driver behaviours, as predicted in the OESD, largely matched those observed in videos of the takeover protocol from automation to human drivers. Further research could consider extending the scope of OESDs to predicting error and task time. Conducting studies to provide evidence of validity generalisation in other domains and applications remains an important goal for the future.

ACKNOWLEDGEMENT

This work was supported by Jaguar Land Rover and the UK-EPSRC Grant EP/N011899/1 as part of the jointly funded 'Towards Autonomy: Smart and Connected Control (TASCC) Programme'.

REFERENCES

Abedi, A., L. Gaudard, and F. Romerio. 2019. "Review of major approaches to analyze vulnerability in power system." *Reliability Engineering & System Safety, 183* 153–172. DOI: 10.1016/j.ress.2018.11.019.

Annett, J. 2002. "A note on the validity and reliability of ergonomics methods." *Theoretical Issues in Ergonomics Science, 3,* 2 228–232. DOI: 10.1080/14639220210124067.

Annett, J., K. D. Duncan, R. B. Stammers, and M. J. Gray. 1971. *Task Analysis.* Department of Employment Training Information Paper No. 6, London, UK: Her Majesty's Stationary Office (HMSO).

Baber, C., and N. A. Stanton. 1996. "Human error identification techniques applied to public technology: Predictions compared with observed use." *Applied Ergonomics, 27,* 2 119–131. DOI: 10.1016/0003-6870(95)00067-4.

Baber, C., and B. Mellor. 2001. "Using critical path analysis to model multimodal human–computer interaction." *International Journal of Human-Computer Studies, 54,* 4 613–636. DOI: 10.1006/ijhc.2000.0452.

Bainbridge, L. 1983. "Ironies of Automation." *Analysis, Design and Evaluation of Man–Machine Systems. Proceedings of the IFAC/IFIP/IFORS/IEA Conference.* Baden-Baden, Germany: Elsevier Ltd. 129–135. https://doi.org/10.1016/B978-0-08-029348-6.50026-9.

Banks, V. A., N. A. Stanton, and C. Harvey. 2014. "Sub-systems on the road to vehicle automation: Hands and feet free but not "mind" free driving." *Safety Science 62* 505–514. DOI: 10.1016/j.ssci.2013.10.014.

Banks, V. A., K. L. Plant, and N. A. Stanton. 2018. "Driver error or designer error: Using the Perceptual Cycle Model to explore the circumstances surrounding the fatal Tesla crash on 7th May 2016." *Safety Science, 108* 278–285. DOI: 10.1016/j.ssci.2017.12.023.

Banks, V. A., N. A. Stanton, G. Burnett, and S. Hermawati. 2018. "Distributed Cognition on the road: Using EAST to explore future road transportation systems." *Applied Ergonomics, 68* 258–266. DOI: 10.1016/j.apergo.2017.11.013.

Carbonell, J. R., J. L. Ward, and J. W. Senders. 1968. "A queueing model of visual sampling experimental validation." *IEEE Transactions on Man-Machine Systems, 9,* 3 82–87. DOI: 10.1109/TMMS.1968.300041.

Card, S. K., A. Newell, and T. P. Moran. 1983. *The Psychology of Human-Computer Interaction.* Hillsdale, MI: L. Erlbaum Associates Inc.

Clark, J. R., N. A. Stanton, and K. Revell. 2019a. "Directability, eye-gaze, and the usage of visual displays during an automated vehicle handover task." *Transportation Research Part F: Traffic Psychology and Behaviour, 67* 29–42. DOI: 10.1016/j.trf.2019.10.005.

Clark, J. R., N. A. Stanton, and K. M. Revell. 2019b. "Identified handover tools and techniques in high-risk domains: Using distributed situation awareness theory to inform current practices." *Safety Science, 118* 915–924. DOI: 10.1016/j.ssci.2019.06.033.

Embrey, D. E. 1986. "SHERPA: A systematic human error reduction and prediction approach." *Proceedings of the International Topical Meeting on Advances in Human Factors in Nuclear Power Systems.* Knoxville, US: American Nuclear Society. 184–193.

Endsley, M. R. 2015. "Situation awareness misconceptions and misunderstandings." *Journal of Cognitive Engineering and Decision Making, 9,* 1 4–32. DOI: 10.1177/1555343415572631.

Eriksson, A., and N. A. Stanton. 2017. "Takeover time in highly automated vehicles: Noncritical transitions to and from manual control." *Human Factors: The Journal of the Human Factors and Ergonomics Society, 59,* 4 689–705. DOI: 10.1177/0018720816685832.

Gilbreth, F. B. 1909. *Bricklaying System.* New York: The M.C. Clark Publishing Company.

Harris, D., N. A. Stanton, and A. Starr. 2015. "Spot the difference: Operational event sequence diagrams as a formal method for work allocation in the development of single-pilot operations for commercial aircraft." *Ergonomics, 58,* 11 1773–1791. DOI: 10.1080/00140139.2015.1044574.

Harris, D., N. A. Stanton, A. Marshall, M. S. Young, J. Demagalski, and P. Salmon. 2005. "Using SHERPA to predict design-induced error on the flight deck." *Aerospace Science and Technology, 9,* 6 525–532. DOI: 10.1016/j.ast.2005.04.002.

Harvey, C., and N. A. Stanton. 2013. "Modelling the hare and the tortoise: Predicting the range of in-vehicle task times using critical path analysis." *Ergonomics, 56,* 1 16–33. DOI: 10.1080/00140139.2012.733031.

Kirwan, B., and L. K. Ainsworth. 1992. *A Guide to Task Analysis: The Task Analysis Working Group.* London, UK: CRC Press. https://doi.org/10.1201/b16826.

Kurke, M. I. 1961. "Operational sequence diagrams in system design." *Human Factors: The Journal of the Human Factors and Ergonomics Society, 3,* 1 66–73. DOI: 10.1177/001872086100300107.

Landis, J. R., and G. G. Koch. 1977. "The measurement of observer agreement for categorical data." *Biometrics, 33,* 1 159–174. DOI: 10.2307/2529310.

Lane, R., N. A. Stanton, and D. Harrison. 2006. "Applying hierarchical task analysis to medication administration errors." *Applied Ergonomics, 37,* 5 669–679. DOI: 10.1016/j.apergo.2005.08.001.

Leveson, N. 2004. "A new accident model for engineering safer systems." *Safety Science, 42,* 4 237–270. DOI: 10.1016/S0925-7535(03)00047-X.

Matthews, B. W. 1975. "Comparison of the predicted and observed secondary structure of T4 phage lysozyme." *Biochimica et Biophysica Acta (BBA) - Protein Structure, 405,* 2 442–451. DOI: 10.1016/0005-2795(75)90109-9.

Moray, N. 2007. "Real prediction of real performance." In *People and Rail Systems: Human Factors at the Heart of the Railway,* by J. R. Wilson, B. Norris, T. Clarke and A. Mills, 9–22. London, UK: CRC Press.

Moray, N., J. Groeger, and N. A. Stanton. 2017. "Quantitative modelling in cognitive ergonomics: Predicting signals passed at danger." *Ergonomics, 60,* 2 206–220. DOI: 10.1080/00140139.2016.1159735.

Parnell, K. J., V. A. Banks, K. L. Plant, T. G.C. Griffin, P. Beecroft, and N. A. Stanton. 2019. "Predicting design-induced error on the flight deck: an aircraft engine oil leak scenario." *Human Factors: The Journal of the Human Factors and Ergonomics Society,* DOI: 10.1177/0018720819872900.

SAE J3016. 2016. *Taxonomy and Definitions for Terms Related to Driving Automation Systems for On-Road Motor Vehicles.* https://www.sae.org/standards/content/j3016_201806/ (accessed 3 November 2020).

Salmon, P. M., G. H. Walker, and N. A. Stanton. 2016. "Pilot error versus sociotechnical systems failure: a distributed situation awareness analysis of Air France 447." *Theoretical Issues in Ergonomics Science, 17*, 1 64–79. DOI: 10.1080/1463922X.2015.1106618.

Salmon, P. M., N. A. Stanton, G. H. Walker, D. Jenkins, C. Baber, and R McMaster. 2008. "Representing situation awareness in collaborative systems: A case study in the energy distribution domain." *Ergonomics, 51*, 3 367–384. DOI: 10.1080/00140130701636512.

Sarter, N. B., and D. D. Woods. 1995. "How in the world did we ever get into that mode? mode error and awareness in supervisory control." *Human Factors: The Journal of the Human Factors and Ergonomics Society, 37*, 1 5–19. DOI: 10.1518/001872095779049516.

Senders, J. W. 1997. "Distribution of visual attention in static and dynamic displays." *Human Vision and Electronic Imaging II, Volume 3016*. San Jose, CA: Society of Photo-Optical Instrumentation Engineers (SPIE). 186194. https://doi.org/10.1117/12.274513.

Sorensen, L. J., N. A. Stanton, and A. P. Banks. 2011. "Back to SA school: Contrasting three approaches to situation awareness in the cockpit." *Theoretical Issues in Ergonomics Science, 12*, 6 451–471. DOI: 10.1080/1463922X.2010.491874.

Stanton, N. A. 2002. "Developing and validating theory in ergonomics science." *Theoretical Issues in Ergonomics Science, 3*, 2 111–114. DOI: 10.1080/14639220210124085.

Stanton, N. A. 2006. "Hierarchical task analysis: Developments, applications, and extensions." *Applied Ergonomics, 37*, 1 55–79. DOI: 10.1016/j.apergo.2005.06.003.

Stanton, N. A. 2014. "Representing distributed cognition in complex systems: how a submarine returns to periscope depth." *Ergonomics, 57*, 3 403–418. DOI: 10.1080/00140139.2013.772244.

Stanton, N. A. 2016. "On the reliability and validity of, and training in, ergonomics methods: A challenge revisited." *Theoretical Issues in Ergonomics Science, 17*, 4 [Methodological Issues in Ergonomics Science I] 345–353. DOI: 10.1080/1463922X.2015.1117688.

Stanton, N. A., and S. V. Stevenage. 1998. "Learning to predict human error: Issues of acceptability, reliability and validity." *Ergonomics, 41*, 11 1737–1756. DOI: 10.1080/001401398186162.

Stanton, N. A., and M. S. Young. 1999a. "What price ergonomics?" *Nature, 399* 197–198. DOI: 10.1038/20298.

Stanton, N. A., and M. Young. 1999b. *Guide to Methodology in Ergonomics: Designing for Human Use*. Abingdon, UK: CRC Press.

Stanton, N. A., and C. Baber. 2002. "Error by design: Methods for predicting device usability." *Design Studies, 23*, 4 363–384. DOI: 10.1016/S0142-694X(01)00032-1.

Stanton, N. A., and M. S. Young. 2003. "Giving ergonomics away? The application of ergonomics methods by novices." *Applied Ergonomics, 34*, 5 479–490. DOI: 10.1016/S0003-6870(03)00067-X.

Stanton, N. A., and C. Baber. 2005. "Validating task analysis for error identification: reliability and validity of a human error prediction technique." *Ergonomics, 48*, 9 1097–1113. DOI: 10.1080/00140130500219726.

Stanton, N. A., and C. Baber. 2008. "Modelling of human alarm handling responses times: A case of the Ladbroke Grove rail accident in the UK." *Ergonomics, 51*, 4 423–440. DOI: 10.1080/00140130701695419.

Stanton, N. A., and C. Harvey. 2017. "Beyond human error taxonomies in assessment of risk in sociotechnical systems: A new paradigm with the EAST 'broken-links' approach." *Ergonomics, 60*, 2 221–233. DOI: 10.1080/00140139.2016.1232841.

Stanton, N. A., A. Dunoyer, and A. Leatherland. 2011. "Detection of new in-path targets by drivers using Stop & Go Adaptive Cruise Control." *Applied Ergonomics, 42*, 4 592–601. DOI: 10.1016/j.apergo.2010.08.016.

Stanton, N. A., M. S. Young, and C. Harvey. 2014. *Guide to Methodology in Ergonomics: Designing for Human Use*, 2nd edition. Boca Raton, FL: CRC Press.

Stanton, N. A., P. M. Salmon, and G. H. Walker. 2015. "Let the reader decide: A paradigm shift for situation awareness in sociotechnical systems." *Journal of Cognitive Engineering and Decision Making*, 9, 1 44–50. DOI: 10.1177/1555343414552297.

Stanton, N. A., P. M. Salmon, and G. H. Walker. 2017. "New paradigms in ergonomics." *Ergonomics*, 60, 2 151–156. DOI: 10.1080/00140139.2016.1240373.

Stanton, N. A., P. M. Salmon, D. Harris, A. Marshall, J. Demagalski, M. S. Young, T. Waldmann, and S. Dekker. 2009. "Predicting pilot error: Testing a new methodology and a multi-methods and analysts approach." *Applied Ergonomics*, 40, 3 464–471. DOI: 10.1016/j.apergo.2008.10.005.

Stanton, N. A., P. M. Salmon, L. A. Rafferty, G. H. Walker, C. Baber, and D. P. Jenkins. 2013. *Human Factors Methods: A Practical Guide for Engineering and Design*. London, UK: CRC Press. https://doi.org/10.1201/9781315587394.

Stanton, N. A., P. M. Salmon, G. H. Walker, E. Salas, and P. A. Hancock. 2017. "State-of-science: Situation awareness in individuals, teams and systems." *Ergonomics*, 60, 4 449–466. DOI: 10.1080/00140139.2017.1278796.

Stanton, N. A., P. M. Salmon, G. H. Walker, and M. Stanton. 2019. "Models and methods for collision analysis: A comparison study based on the Uber collision with a pedestrian." *Safety Science*, 120 117–128. DOI: 10.1016/j.ssci.2019.06.008.

Walker, G. H., N. A. Stanton, P. M. Salmon, and D. P. Jenkins. 2009. "A review of sociotechnical systems theory: A classic concept for new command and control paradigms." *Theoretical Issues in Ergonomics Science*, 9, 6 479–499. DOI: 10.1080/14639220701635470.

Walker, G. H., N. A. Stanton, C. Baber, L. Wells, H. Gibson, P. M. Salmon, and D. Jenkins. 2010. "From ethnography to the EAST method: A tractable approach for representing distributed cognition in Air Traffic Control." *Ergonomics*, 53, 2 184–197. DOI: 10.1080/00140130903171672.

16 Feedback in Highly Automated Vehicles

What Do Drivers Rely on in Simulated and Real-World Environments?

Joy Richardson, James W.H. Brown, and Jediah R. Clark
University of Southampton

Nermin Caber and Michael Bradley
University of Cambridge

Patrick Langdon
Edinburgh Napier University

Kirsten M. A. Revell and Neville A. Stanton
University of Southampton

CONTENTS

16.1　INTRODUCTION

As semi-autonomous SAE Level 2 vehicles are becoming increasingly popular and highly autonomous vehicles are on their way, in-vehicle alerts to convey driving mode or the potential need for a takeover are no longer a simple icon on the dashboard. They can include information on other in-vehicle information systems (IVIS) such as a head-up display (HUD), an infotainment screen including navigation and media outputs which is usually located in the centre of the dashboard or just below. Other options include haptic feedback in pedals, seats or steering wheel, ambient lighting, vocalisations, and other audio sounds. The increased number of modalities can be more efficient when communicating with the driver (Lylykangas et al. 2016; Politis, Brewster and Pollick 2017; Yoon, Kim and Ji 2019), and it has been shown that multi-modal alerts can significantly increase alertness (Politis, Brewster and Pollick 2017; Huang et al. 2019). However, too many modalities can become annoying (Politis, Brewster and Pollick 2015) and even be detrimental to the driving task, resulting in safety issues (Normark 2015; Gibson, Butterfield and Marzano 2016). There is no consensus in the literature on how best to provide takeover alerts (Huang et al. 2019); the challenge is therefore which of these modalities should be removed. Clark, Stanton, and Revell (2020) found that drivers of different skill levels required different stimuli to prompt them that a takeover was required. Gibson, Butterfield, and Marzano (2016) found participants had diverse and contradictory opinions regarding the elements that should be included on a dashboard display.

16.1.1　Challenges of Customisable Interfaces

A solution to this problem is to allow the driver to design their own space using flexible technologies. The ability to customise a suite of IVIS to suit themselves could create a more pleasant driving experience and reduce cognitive overload, increasing safety (Normark 2015). The challenge is to create both a safe interaction and a positive user experience (Gkouskos, Normark and Lundgren 2014). Drivers strive for the freedom of choice (Clark, Stanton and Revell 2020; Gibson, Butterfield and Marzano 2016; Normark 2015). But this must be carefully balanced with safety. When designing customisable interfaces, it is important to ensure it may only be customised when it is safe to do so, such as when the vehicle is parked, and that the safety-critical baseline information is always available (Normark 2015).

The default IVIS contains safety-critical and legal minimum information, but which of the other modalities should be included? Is it possible to create profiles for certain user groups, or should everyone receive the same default setting? In order to answer these questions, it is necessary to know which modalities drivers rely on when driving an automated vehicle. Despite the wide range of research conducted on the subject of takeover technologies, little is known about what types of feedback drivers are most reliant on when interacting with semi-autonomous vehicle interfaces, and if these preferences are the same during simulated and on-road driving.

16.1.2 What is Reliance?

The concept of reliance is used as a measure in a wide variety of fields, such as accountancy (Hampton 2005; Arel 2010), medicine (Janssens et al. 2004), philosophy (Budnik 2018), health information (Hall, Bernhardt and Dodd 2015), politics (Johnson and Kaye 2014), marketing (Mumuni et al. 2019), cell phone technology (Sato et al. 2013), and epistemology (Fantl, McGrath and Sosa 2019).

Within the field of philosophy, the difference between reliance and trust has been considered difficult to distinguish, causing much debate (Budnik 2018). Political studies have found that reliance is more complex than the amount of time spent using a particular technology, or how frequently it is used, as reliance is also dependent on its usefulness. Reliance therefore is a more robust measure, which can also include influence and confidence (Johnson and Kaye 2014).

Mumuni et al. (2019) started trying to define reliance by using definitions from the Cambridge Dictionary online and Dictionary.com. The Cambridge Dictionary defines reliance as: 'the state of depending on or trusting in something or someone' or, in business English, 'the state of needing or depending on something or someone in order to be able to do something'. Dictionary.com defines reliance as 'confident of trustful dependence' (Cambridge Dictionary 2019; Dictionary.com 2019). They continue to describe reliance graphically through models as the extent to which a person depends on one medium, relative to other mediums or technologies; this can include a tendency for a person to value and trust a particular technology (Mumuni et al. 2019). These models have been adapted in Figure 16.1 to apply to the automated driving human–machine interface (HMI).

16.1.3 Measuring Reliance

Similar to the disparity in definitions of reliability, there is no standard measure or scale to quantify reliance. Sato and colleagues employed a six-point scale, coding the response to five questions regarding mobile phone usage – 1, responding to feature less frequently used (ranging from 0–1 to 0–25 times per day), up to 6, responding to a feature more frequently used (ranging from >25 to >125 times per day) (Sato et al. 2013).

Likert responses are more frequently used, but with varying scales, some using an even figure scale, which only allows for positive or negative responses, while others use an odd figure scale allowing for a neutral response. Some use the lowest number

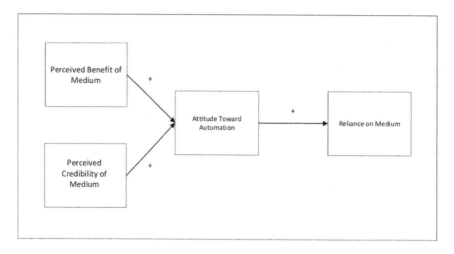

FIGURE 16.1 Model of reliance. (Adapted from Mumuni et al., 2019).

to mean positive and the highest to mean negative, while other researchers employ their scales inversely. Hall and colleagues employed a four-point Likert response scale with 1 meaning strongly agree and 4 strongly disagree (Hall, Bernhardt and Dodd 2015). Johnson and Kaye employ two identical five-point Likert scales with 1 meaning never rely and 5 heavily rely (Johnson and Kaye 2014). Mumuni and colleagues also employ a Likert response but with a scale of seven points – 1 meaning strongly disagree and 7 strongly agree (Mumuni et al. 2019). Hampton also employs a seven-point Likert-type scale with ranges from strongly disagree to strongly agree across four measures of reliance: user perception of agreement, level of confidence in output, user's confidence in their own judgement, and finally the reliance on their aid when forming a decision (Hampton 2005).

None of these existing scales are fit for the purpose of looking at reliance across a range of modalities such as within the takeover and for mode awareness in an automated vehicle.

16.1.4 Development of a New Reliance Scale

A novel reliance ranking scheme was created. A Likert scale was not adopted as multiple modalities may receive the same score, making analysis unclear. Therefore, the reliance ranking scale was developed using radio buttons; participants were able to rank in order (1–7) the elements that they most relied on for information on the automation system, with 1 being the least reliant and 7 being the most reliant (Figure 16.2). Each number could only be selected once. This had the benefit that the modalities could be clearly compared against each other to determine which were relied on most.

This scale was tested and evaluated in two studies to understand its strengths and weaknesses; it was presented as part of the post-task questionnaire in both cases. The two studies involved participants experiencing a range of modalities during manual and automated driving, including takeover requests, where the vehicle would inform the driver that they needed to resume manual control. The first study was simulator based, with 66 participants, and the second, smaller study was a follow-on, where 26 of the participants from the first study completed a very similar task on-road, on an active motorway. During both studies, the participants experienced default settings, but were then able to use a customisation matrix (Figure 16.3) on a touchscreen in the central cluster, to change the settings to their preference. They would then continue the study with their chosen settings.

Driving simulators are frequently used by researchers to study drivers' reactions to IVIS (Wynne, Beanland and Salmon 2019), and medium-fidelity, fixed-base simulators are considered an effective method for assessing such interactions (Eriksson, Banks

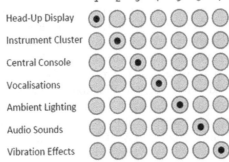

FIGURE 16.2 Reliance ranking scale question.

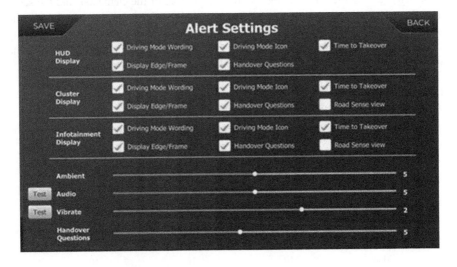

FIGURE 16.3 Customisation matrix screen.

and Stanton 2017). However, due to the lack of risk behaviours in a driving simulator, it may be more adventurous than those on-road (Eriksson, Banks and Stanton 2017; Wynne, Beanland and Salmon 2019). The opportunity presented by using this scale in a simulator study with a follow-on on-road study also allowed us to investigate if the trend for riskier behaviour in a simulator could be reflected in IVIS customisation.

The experimental design, procedure and results for each are discussed in turn.

16.2   EXPERIMENT 1 – SIMULATOR STUDY

16.2.1   METHOD

16.2.1.1   Participants

Ethical approval was gained via the University of Southampton Ethics and Research Governance Office (ERGO Number: 41761.A3). A total of 66 participants were recruited for this study. In order to be included, potential participants needed to have held a valid full UK driving license for more than two years and be generally healthy with no history of motion sickness. Corrected vision was acceptable so long as they met the minimum driving standards. Drivers who drove less than once per week, who had a conflict of interest (themselves or from close family or friends), and who were pregnant, were excluded from participating. The recruited participants were an equal split of female and male, and were divided into the following age groups: age group 1, 18–34; age group 2, 35–56; and age group 3, 57–82, as shown in Table 16.1. They included 40% premium vehicle owners (with vehicles less than 7 years old and from one of the following brands: Audi, BMW, JLR, Lexus, Mercedes, MINI, Porsche, Tesla, Volvo, and VW).

During the post-study debrief questionnaire, participants were asked to complete the new reliance ranking score, although it was explained to them that they were not monitored while completing the question. Unfortunately, 11 people answered the question incorrectly (giving the same number score to 2 or more of the elements), and the score from these 11 participants was removed from the analysis. These participants fell into the following demographic categories: females (age group 1 = 3, age group 2 = 1, age group 3 = 1) and males (age group 1 = 1, age group 2 = 3, age group 3 = 1). Table 16.2 shows the amended demographics of the participants who are represented in the reliance data.

16.2.1.2   Design

Participants were invited to take part in a study simulating manual and highly automated driving on a UK motorway in a hybrid vehicle. The focus of the study was the

TABLE 16.1
Demographics of Participants in Simulator Study

Gender	Age Group 1	Age Group 2	Age Group 3
Female	10	11	10
Male	10	12	10

TABLE 16.2
Amended Participant Demographics

Gender	Age Group 1	Age Group 2	Age Group 3
Female	7	10	9
Male	9	9	9

transition from the vehicle self-driving to manual control, assisted by a customisable HMI. The study consisted of four repeated-measures driving trials, each consisting of three handovers (driver to vehicle) and three takeovers (vehicle to driver). For each participant, half of the trials were defined as being a 'short' time out of the loop, consisting of one minute of automated control (three minutes in total for each short trial). The other half of the trials were defined as being a 'long' time out of the loop, consisting of 10 min of automated control (30 min in total for each long trial). The entire experiment took no longer than 2 h, 40 min to complete. The dependent variables collected were:

- A recording of where the participant was gazing, at a rate of 20 Hz, split into seven areas: road ahead, HUD, instrument cluster, (central infotainment) console, rear-view mirror, and right/left field of view.
- A record of behavioural data, including time, availability of automation, current HMI state, headway, lane position, longitudinal velocity, measures of control input, and information regarding what stage the driver is at within the takeover process.
- Demographical information: gender, age, annual mileage, years since passing test, any vehicles currently driven (including their own, family cars, work vehicles or any other regularly driven), any advanced driving qualifications, and what advanced driver-assistance systems drivers had tried, or used frequently.
- A record of any changes to the HMI that participants made at the end of each trial.

16.2.1.3 Apparatus

STISIM drive software was used to simulate a typical UK motorway environment. The simulator represented a hybrid Land Rover Discovery, equipped with a single front-view screen replicating the windscreen, with separate digital display wing mirrors and an augmented display for rear view. The simulator was equipped with a customisable digitalised instrument cluster, central infotainment console, and HUD. The vehicle was also equipped with vocal and audio information streams, ambient lighting to indicate driving mode (orange for manual, blue for automated), and a vibrating seat providing haptic feedback, which was initiated when a takeover was expected, and when control was safely transferred to the driver. All these streams of information were customisable by the participant via the customisation matrix displayed on the central console after each trial.

16.2.1.4 Procedure

On arrival to the laboratory, participants were given a brief verbal introduction to highly automated vehicles, the background and purpose of the study, and safety aspects. They then read the participant information sheet and signed a consent and attendance form. Following this, participants completed a demographic question-naire capturing age, gender, and driving experience both generally and with auto-mated features; at this point, they were able to ask questions, take a comfort break, or have refreshments. Once ready to proceed, participants were then guided into the driving simulator where they adjusted seat and steering wheel positioning. Drivers were then introduced to the controls and information displays, including the instru-ment cluster information, the HUD, the central console, audio and verbal interaction, vibrating seat, and ambient lighting. Drivers were then introduced to what would happen in the experiment, outlining how and when transitions were expected.

Drivers took part in a short out-of-the-loop condition (~7 min in total) to become familiar with the vehicle's controls and how to interact with the system. Drivers were then introduced to the customisation matrix and the settings that it entailed on the cen-tral console touchscreen (see Figure 16.3). They were then instructed that there were four trials in total, potentially taking up to 35 min to complete, and that breaks were encouraged prior to each trial. When the participant was satisfied that any remaining questions were addressed, the trials began.

For each trial, participants started in the hard shoulder of a motorway and were instructed to drive off into the middle lane and keep to the local speed limit (70 mph). After a minute, the system vocalised to the driver that automation was avail-able and presented the driver with a tone and visual wording/icon accordingly. Participants passed control to the vehicle by pressing two flashing green buttons with their thumbs on the steering wheel. Once activated, the black lighting of the displays and the orange ambient lighting transitioned to white displays/blue lighting. Vocal indications were given throughout, such as 'the car has control'.

During automated control, to simulate a secondary task, the driver played Tetris on a Window's tablet (they were told that score was being recorded). A visual indi-cator counting down the time left in automation (from 1 to 10 min depending on condition) was displayed on all three screens by default. At 5, 2, and 1 min before manual control was expected, an audio tone and a vocalised alert were given to the driver, notifying them of time remaining. When the countdown reached zero, the seat vibrated in co-occurrence with an audio and vocal alert. The takeover icon ani-mated the requirement to resume driving position. At this stage, the vehicle vocal-ised questions, which were also presented on each display. Questions were randomly generated from a list of 10 and asked the driver about vehicle status (remaining fuel, battery range, time displayed, or speed) or the driving environment (what colour is the vehicle in front, what kind of vehicle is in front, is there a bend in the road ahead, lane, or weather). Each answer was delivered vocally and was categorised as being either correct or incorrect by the researcher. Once the system was satisfied that more than half of questions were correctly answered, the vehicle indicated to the driver to take control of the vehicle by vocal and visual communication. Should questions come below the 50% threshold, an additional warning was given to the driver, but the takeover was still initiated. After pressing the two green buttons, the driver was in control, and audio, vocal, and visual alerts, and ambient lighting (now orange), were given and the vibrating seat pulsed one last time to confirm the takeover. This process represented one control cycle and was performed three times for each condition (12 for the entire experimental session) before being asked to pull over to the hard shoulder and bring the vehicle to a stop.

After each trial, the participant was asked to make changes to the HMI interaction by ticking/unticking boxes in the customisation matrix or using sliding scales. During this process, the researcher encouraged the participant to vocalise their reasoning behind the changes made, asking questions to ensure their decision-making process was clear. Once satisfied with their decisions, participants saved the matrix and the next trial used these settings. Once three trials had been completed, the driver left the vehicle and took part in a final debrief questionnaire which was in PDF form. This form included the reliance ranking scale. They were then thanked for their time, notified of payment, and advised not to drive for another 20 min.

16.2.2 Method of Analysis

The data, which was collected for the reliance ranking scale in the PDF form, was exported into a spreadsheet. As this scale had not been used before, the data was manipulated in several ways before selecting two methods of presentation. The following methods were chosen as they were most effective at illustrating the diversity of scores given by the participants. The first is a bar chart (Figure 16.4), which presents how many of each score was awarded to each modality – the lowest score of 1 in the left of each group and the highest score of 7 in the right of each group.

The second method of presentation is the radar graph. The innermost points of the shape represent a score of 1, and the outermost points of the shape represent a score of 7. All of the individual radars have been uniquely coloured and laid on top of each other to produce a composite graph (Figure 16.5). Here, all of the ranking scores from each participant have been converted into a radar, with each point of the

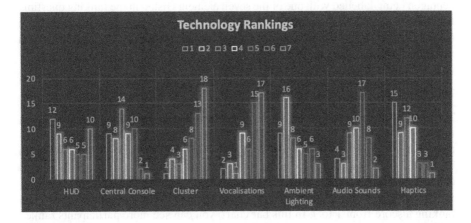

FIGURE 16.4 Bar chart showing the results of ranking question across all participants.

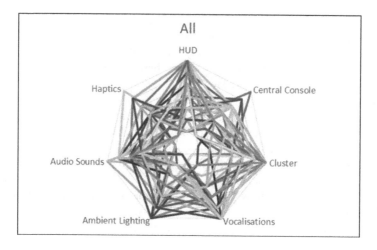

FIGURE 16.5　Radar graph showing the results of ranking question across all participants.

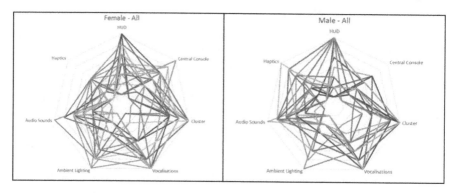

FIGURE 16.6　Radar graphs showing the results of ranking question divided by gender.

heptagon being labelled with one of the seven elements listed in the ranking question. It is constructed from seven concentric shapes, the points of each of which respond to the ranking scores. The individual radars were sorted further by demographics such as age (Figure 16.6) and age and gender (Figure 16.7).

16.2.3　RESULTS

The results from the reliance questionnaire have been represented graphically in two ways. Figure 16.4 is a bar chart showing the cumulative frequency and distribution of the reliance ranking scores from all participants, as previously identified in Table 16.2. Each of the different elements is listed along the *x*-axis: HUD, central console, (instrument) cluster, vocalisations, ambient lighting, audio sounds, and haptics. Each bar then represents how many participants ranked that element, with each ranking a score from 1 to 7. In this bar chart, you can see more participants ranked ambient lighting (a total of 25 participants) and haptics (a total of 24 participants) as

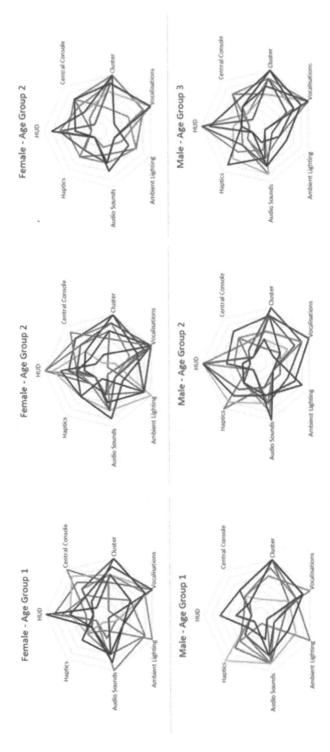

FIGURE 16.7 Radar graphs showing the results of ranking question divided by gender and age.

the elements scoring the lowest two scores, of 1 or 2, meaning they were least reliant on these methods for information on the automation system. This reflects the design of the system as these elements are designed to reinforce mode awareness via the ambient lighting, or that a takeover is either required or has been completed via the haptic feedback seat; they were not designed to be a primary element of information about the automation system.

The bar chart also shows that more participants ranked the cluster (a total of 31 participants) and vocalisations (a total of 32 participants) as the elements scoring one of the highest two scores, 6 or 7, meaning they were most reliant on these methods for information on the automation system. However, there is a wide distribution of ranking scores across all of the elements.

Figure 16.5 shows the results from the same ranking question displayed as a radar graph. Again, this graph shows there was a wide variety between the ranking scores across the participant group. Both graphs also show how every element was scored the highest ranking of 7, therefore relied upon the most, by at least one participant. This composite graph was divided by gender to create two new graphs, as can be seen in Figure 16.6. The scores from the female participants were generally well distributed across the graph; however, there is a noticeable absence of scores of 5, 6, or 7, relating to the most reliance, for haptics. The scores for the male participants were also generally well distributed; however in this case, no males scored the central console as either 6 or 7, relating to the most reliance.

The individual scores were divided again into both age and gender groups to create six new graphs as seen in Figure 16.7. The lower reliance on haptics for the female group and the lower reliance on the central console for the male group are reflected in these new graphs, but it also appears that the younger male age group are inclined to rely less on the HUD, with none of the younger male age group scoring the HUD the highest score of 6 or 7 for reliance. No two participants share the same radar profile, meaning that no two participants scored the seven elements by which information regarding the automation system was in the same rank order.

16.3 EXPERIMENT 2 – ON-ROAD STUDY

16.3.1 METHOD

16.3.1.1 Participants

Ethical approval was gained via the University of Southampton Ethics and Research Governance Office (ERGO Number: 49792.A2). A total of 24 participants were recruited from the pool that had previously experienced the simulator study featuring an almost identical HMI. Due to insurance requirements, they were now required to be a minimum of 28 years old, but all other requirements remained the same. The recruited participants were split across gender and age as shown in Table 16.3.

16.3.1.2 Design

For this second experiment, the participants were invited to take part in a replication of the previous study, but on road, in an electric vehicle on the motorway. The focus remained on the transition from the vehicle in self-driving mode to manual control,

TABLE 16.3
Demographics of On-Road Study Participants

Gender	Age Group 1	Age Group 2	Age Group 3
Female	3	3	3
Male	4	6	4

assisted by the same customisable HMI. A repeated-measures design was employed for the experiment. Participants experienced two differing customisations of vehicle HMI, the HMI customisation representing the independent variable. Two trials took place: trial 1 using default HMI settings and trial 2 using participant selected settings. A final customisation of the HMI took place after the completion of trial 2. The entire experiment lasted four hours.

16.3.1.3 Apparatus

The experimental vehicle was a 2017 Jaguar i-PACE EV400 AWD pre-production model. The automation system consisted of a combination of factory standard, lane keep assist, and adaptive cruise control. When utilising these systems, the standard car would issue a frequent warning to maintain hands on the wheel; these warnings were specifically removed from the system in order for it to simulate Level 3 automation. The car was fitted with a PC (In-CarPC 2020) running SBG software (SBG 2020) for data and video logging. A dSPACE MicroAutoBox II system was employed to interface with the vehicle's systems (dSPACE 2020), together with three controller area network (CAN) interfaces (Connective Peripherals 2020) and a vector CAN box (Vector 2020). Power was supplied via an inverter, connected to the 12 V battery circuit. An Ethernet switch and USB hub provided communications between systems. The visual aspects of the HMI consisted of a 14″ thin-film transistor (TFT) panel fitted to the centre console, a 10″ TFT fitted in place of the original manufacturer's cluster, and a HUD comprising a small 1000×250 px TFT and reflector screen (Bysameyee 2020). The car's original cluster was moved into the passenger footwell to allow monitoring by the safety driver. The interactive element of the HMI constituted two illuminating green buttons (iYeal 2020) fitted in the thumb positions of the steering wheel. The button functionality was integrated using an Arduino (Arduino 2020). Ambient lighting was supplied via LED lighting strips (Elfeland 2020). Haptics were provided in the seat base via Leggett and Platt motors (Legett & Platt Automotive 2020) controlled through an Arduino Micro and motor control board (Arduino 2020). Five cameras (KT&C 2020) were installed within the I-PACE to provide footage from the forward view, driver-facing, over-the-driver-shoulder, footwell, and rear views. A dashcam was also fitted in the safety car to collect footage of the participants' vehicle from an external viewpoint. A cognitive loading task (*n*-back) was controlled from a mobile phone using MIT's AgeLab delayed digit recall task app (AgeLab 2019), which was connected to a generic Bluetooth speaker. Additional speakers were also placed within the car and connected to the PC to relay the vocalised and audio cue aspects of the HMI. The HMI was controlled via a

Microsoft Surface Pro tablet providing 'Wizard of Oz' control of the system. Custom software was developed for the HMI using C++.

16.3.1.4 Procedure

On arrival at the test track, participants were welcomed by experimenter 1 and asked to provide their driving license, which was checked, and a visitor pass provided. They were given a refresher on the background of the study and how it related to the simulator study they had completed the previous year, including being reacquainted with the HMI, how to engage/disengage the automation, and the customisation matrix. The procedure for the day was talked through, including the test track training, route, HMI customisation, and questionnaire completion. They were then shown an example of the n-back task and performed a test of this to ensure it was understood. It was stressed that if completing this secondary task was impairing their ability to monitor the road, they should stop and maintain monitoring as their priority. The participant was given the participant information sheet and privacy statement to read, and once any questions had been answered, they were asked to sign some consent forms. Once confident, the participant was taken to the vehicle.

On completion of all forms, they were led from the reception area to the car park and asked to sit in the passenger side of the car, while both experimenters took their places in the rear seats. The participant was introduced to a safety driver who explained the basic controls on the car, and the elements of the interface. The safety driver then drove the car through security to the proving ground and demonstrated the vehicle's performance before running through the transition to automated mode and back to manual control. The automation system was operated by one of the experimenters using a Wizard of Oz system from a Windows tablet in the back of the car. To offer automation, an experimenter would press a start button on a custom control panel app running on the tablet; the HMI then indicated automation as being available via the three graphical interfaces and a vocal alert. To enable automation, the driver simultaneously pressed the two green buttons mounted on the steering wheel and released all of the controls, including the accelerator. The system would then engage automation and the HMI would indicate that the automated mode was active. During automation, the cognitive load task was controlled by the other experimenter via a mobile telephone-based app linked to a Bluetooth speaker. The safety driver demonstrated the automation system multiple times, including the n-back task, and requested the safety car overtake and brake in front of the vehicle, when in automated mode, to illustrate how it reacts to maintain a gap to the car in front. The safety driver then stopped in a safe area and swapped places with the participant. The participant was allowed some time to drive the car on the proving ground to become familiar with the controls. The automation was then made available to them, and they experienced multiple handovers and takeovers, including the use of the n-back task while in automation.

The participant was asked to drive manually to the start point of the experiment at the Southbound Warwick services on the M40. Two miles prior to the services, while on the M40, a road-based practice handover was conducted. The automation was offered to the participant; once activated, a short 30-s period of automation followed, which included the participant carrying out the n-back task. The participant then experienced the takeover protocol and resumed manual control before stopping at the services. After confirming that the participant was happy to continue, the on-board systems were checked and configured for the first trial, data logging was started, and the video and audio were synchronised using a clapper board.

The participant was instructed to drive from the services onto the M40, proceeding in the left lane at approximately 58 mph. After one minute of manual driving, automation was offered to the participant. Once activated, after a further period of 30 s, the *n*-back cognitive loading task was started. This task was run in 2-min intervals, separated by 30 s. Following 10 min of automation, the HMI started the takeover protocol, and once completed, the participant pressed both steering wheel buttons and resumed manual control. After one minute of manual control, automation was again offered to the participant and the process repeated. On completion of the second takeover, the manual driving period was extended to approximately 7 min in order to pass a section of motorway (M40 J9 Southbound) that would have adversely affected the automation due to the lane becoming a slip road. Due to proximity to the end point of trial 1, handover 3's automation period was reduced to 8 min. The *n*-back task was started simultaneously with the automation for 30 s, before reverting to 30 s off and 2 min on until the protocol started. When the participant completed the protocol and assumed manual control, they continued in manual mode the short distance remaining to the motorway services and parked the car. Data logging was stopped as this constituted the end point of trial 1. The participant was presented with a set of PDF-based forms on a laptop and asked to complete them based upon their experience, specifically of the takeover protocol, process, and HMI elements. When the forms were complete, the customisation screen was displayed on the vehicle's centre console and the participant was asked to make any changes that they would like to the HMI. During this customisation process, they were asked to vocalise their opinions, considerations, and rationales for customisations. This information was recorded for future analysis. This concluded the data collection for trial 1; the participant at this point had access to facilities and refreshments. Trial 2 was a repeat of the processes and data collection carried out in trial 1; however, the route was Northbound to the start point for trial 1. On completion of trial 2, the final debrief questionnaire, used in the simulator study, was completed which included the reliance ranking question. Due to some participants having some difficulty completing the reliance ranking question in the simulator study, they were supervised while completing it for this study. The safety driver returned to the driving seat from this point and drove the vehicle back to the Jaguar Land Rover facility. The participant was thanked for their time and signed out at reception.

16.3.2 Method of Analysis

Analysis of the data from this study followed the same procedure as the previous simulator study, producing a bar chart detailing the scores awarded to each modality (Figure 16.8) and a radar graph showing the scoring of all participants (Figure 16.9). Due to the smaller numbers in the on-road study, it was not possible to identify themes within age and gender. However, as a follow-on study, there was now the possibility of comparing a participant's reliance ranking score from the simulator

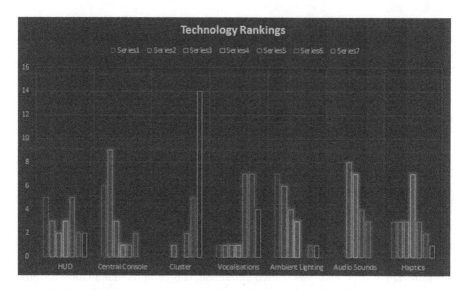

FIGURE 16.8   Bar chart showing the results of ranking question across all modalities.

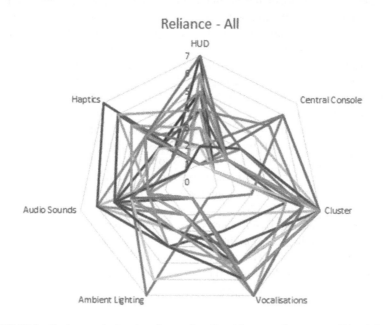

FIGURE 16.9   Radar graph showing the results of ranking question across all participants.

study with that of the on-road study. It was possible to create these new graphs with the data from 20 participants who had completed the on-road questionnaire correctly and who also took part in the on-road study. The individual results from the simulator study and on-road study reliance scores were compared. It was discovered that no participant awarded the same scores in both studies. However, eight participants did not change which modality they scored as 7 (Figure 16.10).

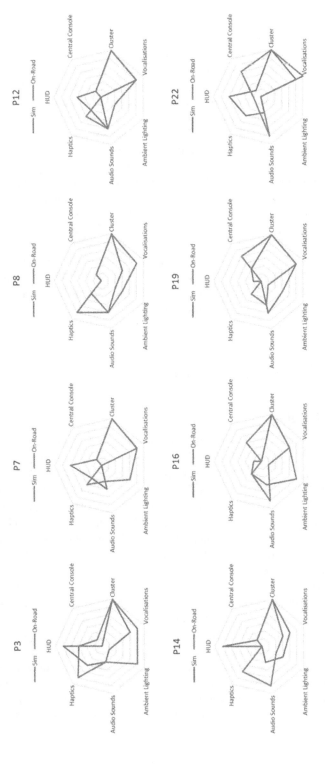

FIGURE 16.10 Radar graphs showing participant reliance ranking scores from simulator and on-road studies.

16.3.3 RESULTS

The results from the reliance questionnaire have been represented graphically in the same two ways as for the previous study, with Figure 16.8, a bar chart showing the cumulative frequency and distribution of the scores, and Figure 16.9, a radar graph. In the bar chart, you can see again how more participants ranked ambient lighting (a total of 13 participants) as one of the elements scoring the lowest two scores, 1 or 2, similar to the simulator study. However, the mode receiving the most scores of 1 and 2 (with 15 participants) was the central console. This mode also received no scores of 7, meaning that no participant relied on this the most.

The bar chart also shows more participants ranked the cluster (a total of 19 participants) and vocalisations (a total of 11 participants) as the elements scoring the highest scores of 6 or 7, meaning that they were most reliant on these methods for information on the automation system. In this second study, not all scores were received by all modalities; however, it was a much smaller sample size. Nevertheless, there still remains a wide distribution of ranking scores across all the elements.

Figure 16.9 illustrates the results from the same ranking questions as a radar graph. Again, each coloured line represents the scores from an individual participant. It shows, in a different way, the wide variety of ranking scores across the group and also how each participants' score profile was unique, as in the previous study, no two participants scored all of the modalities the same. Figure 16.10 shows a selection of the new participant layered radars where they retained the same most relied on modality. This was a total of 40% of the participants. None of the participants scored all the modalities in the on-road study the same as they had done in the simulator study.

16.4 DISCUSSION

During the simulator study, evaluating all the modes the participants had been exposed to and been able to adapt to suit their preference throughout all trials, highlights those which participants found themselves relying on most and enabled the analysis of reliance based on age and gender. This analysis highlighted some differences in the results reported between the male and female participants. Notable is the female participants' seeming choice to not rely on the haptic feedback from the seat. Some previous studies have indicated that females are more sensitive than males in detecting haptic feedback or feel it more intensively (Goff et al. 1965; Verrillo 1979; Neely and Burström 2006; Forta 2009). Despite this, it has been observed that female participants scored haptic feedback a lower score for satisfaction than their male counterparts when using this as an alert to indicate the need to perform a takeover task in an automated driving context (Duthoit et al. 2018). This could suggest that while the female participants were more sensitive to the haptic alert, they did not find this an adequate alert for indicating a takeover was required. Further division of the radars into age and gender groups indicated the younger males relied much more on the simple dashboard and audio. Van Der Laan, Heino, and De Waard (1997) reported that older drivers found more information while driving was useful and satisfying, whereas younger drivers stated the opposite. It could be possible the reliance ranking results reflect this.

During the on-road study, the reliance ranking score was useful to see the wide range of scores, but due to the lower number of participants, it was not able to show visible trends relating to age and/or gender. However, it was possible to compare participants' scores from the simulator study with their scores from the on road study. This showed that every participant changed their reliance ranking scores between the two studies, which could be due to increasing familiarity with the system or feeling the need to rely on the elements differently in real life compared to simulation. For the highest-ranking modality, 33% of participants reported relying on the traditional dashboard cluster the most during the simulator study, but this increased hugely to 86% in the on-road study. Conversely, the HUD, a modality not experienced by any participant in a vehicle prior to this study, had the highest score from 18% of participants in the simulator and just 10% on-road. There does appear to be a correlation with the trend previously reported by Eriksson, Banks, and Stanton (2017) and Wynne, Beanland, and Salmon (2019) to be more venturesome in a risk-free, simulated environment.

16.5   CONCLUSION

In order to determine which modalities a driver relied on most when driving a highly automated vehicle, a ranking scale was developed and used in a simulator and an on-road study. The trust and acceptance scales employed during the course of these studies were effective at demonstrating the opinion of a participant on the overall automation system at different points during the study, and can be used to evaluate how the participants' opinions may change as they progress through the multiple trials in the study. However, the reliance ranking scale can provide a novel insight into the participants' use and opinion of the individual modes of communication used during the study by isolating them and comparing them against one another. The ability to seek out trends visually can enable the researcher to pick up on areas for further research easily – in this case, the reasoning for female drivers to score the haptic seat lowly and for younger males to rely on the dashboard and audio, rather than a more high-tech HUD. With a smaller number of participants, this scale can still be a useful tool to compare individual participants' scores over a period of time.

It is clear from the wide range of scores in both studies that there are no distinct profiles for different user groups, but rather drivers value a freedom of choice when selecting their preferred modalities. Design of customisable interfaces should therefore always be safety led, starting with a default which provides all essential details, a baseline of which is always visible (Normark 2015). This could be in the most traditional forms of IVIS, as in this on-road study it is shown that drivers rely on this the most for information relating to the automation system. From this base, it is safe to customise alerts to one's own preference, which can both increase safety (Politis, Brewster and Pollick 2017; Huang et al. 2019) and the positive diving experience (Gkouskos, Normark and Lundgren 2014; Normark 2015; Gibson, Butterfield and Marzano 2016).

Future use of this tool with large and longitudinal study groups would further validate its usefulness as an interesting method of analysing reliance in multimodal studies.

ACKNOWLEDGEMENTS

This work was supported by Jaguar Land Rover and the UK-EPSRC Grant EP/N011899/1 as part of the jointly funded 'Towards Autonomy: Smart and Connected Control (TASCC) Programme' HI:DAVe Project.

REFERENCES

AgeLab. 2019. *MIT AgeLab Study Tools*. Accessed February 17, 2020. https://agelab.mit.edu/study-tools.

Arduino. 2020. Accessed May 7, 2020. https://www.arduino.cc/.

Arel, B. 2010. "The influence of litigation risk and internal audit source on reliance decisions." *Advances in Accounting*, 26, 2 170–176. DOI: 10.1016/j.adiac.2010.05.002.

Budnik, C. 2018. "Trust, reliance, and democracy." *International Journal of Philosophical Studies*, 26, 2 221–239. DOI: 10.1080/09672559.2018.1450082.

Bysameyee. 2020. *Head-up Display GPS Navigation*. Accessed February 17, 2020. https://www.amazon.co.uk/Head-up-Display-GPS-Navigation-Reflective/dp/B075T2DH12/ref=sr_1_10?keywords=car+head+up+display&qid=1573804882&sr=8-10.

Cambridge Dictionary. 2019. *Reliance*. Accessed June 19, 2019. https://dictionary.cambridge.org/dictionary/english/reliance.

Clark, J. R., N. A. Stanton, and K. M.A. Revell. 2020. "Automated vehicle handover interface design: Focus groups with learner, intermediate and advanced drivers." *Automotive Innovation*, 3 14–29. DOI: 10.1007/s42154-019-00085-x.

Connective Peripherals. 2020. *USB2-F-7101*. Accessed May 7, 2020. https://www.connectiveperipherals.com/products/canbus-solutions/usb2-f-7101.html.

Dictionary.com. 2019. *Reliance*. Accessed June 19, 2019. https://www.dictionary.com/browse/reliance.

dSPACE. 2020. *MicroAutoBox II*. Accessed May 7, 2020. https://www.dspace.com/en/ltd/home/products/hw/micautob/microautobox2.cfm.

Duthoit, V., J.M. Sieffermann, E. Enrègle, C. Michon, and D. Blumenthal. 2018. "Evaluation and optimization of a vibrotactile signal in an autonomous driving context." *Sensory Studies*, 33, 1 e12308. DOI: 10.1111/joss.12308.

Elfeland. 2020. *LED Strip Lights Kits*. Accessed May 7, 2020. https://www.amazon.co.uk/Elfeland-Waterproof-Controller-Self-Adhesive-Decoration/dp/B07H2DR5DY/ref=sr_1_6?keywords=led+rgb+strip&qid=1559907897&s=gateway&sr=8-6.

Eriksson, A., V. A. Banks, and N. A. Stanton. 2017. "Transition to manual: Comparing simulator with on-road control transitions." *Accident Analysis & Prevention*, 102 227–234. DOI: 10.1016/j.aap.2017.03.011.

Fantl, J., M. McGrath, and E. Sosa. 2019. "Reliance." In *Contemporary Epistemology: An Anthology*, by E. Sosa, J. Fantl and M. McGrath, 75–146. New Jersey: John Wiley & Sons.

Forta, N. G. 2009. *Vibration intensity difference thresholds*. Doctoral Thesis, University of Southampton, Institute of Sound and Vibration Research.

Gibson, Z., J. Butterfield, and A. Marzano. 2016. "User-centered design criteria in next generation vehicle consoles." *Procedia CIRP*, 55 260–265. DOI: 10.1016/j.procir.2016.07.024.

Gkouskos, D., C. J. Normark, and S. Lundgren. 2014. "What drivers really want: Investigating dimensions in automobile user needs." *International Journal of Design*, 8, 1 59–71.

Goff, G. D., B. S. Rosner, T. Detre, and D. Kennard. 1965. "Vibration perception in normal man and medical patients." *Journal of Neurology, Neurosurgery, and Psychiatry*, 28, 6 503–509. DOI: 10.1136/jnnp.28.6.503.

Hall, A. K., J. M. Bernhardt, and V. Dodd. 2015. "Older adults' use of online and offline sources of health information and constructs of reliance and self-efficacy for medical decision making." *Journal of Health Communication*, 20, 7 751–758. DOI: 10.1080/10810730.2015.1018603.

Hampton, C. 2005. "Determinants of reliance: An empirical test of the theory of technology dominance." *International Journal of Accounting Information Systems*, 6, 4 217–240. DOI: 10.1016/j.accinf.2005.10.001.

Huang, G., C. Steele, X. Zhang, and B. J. Pitts. 2019. "Multimodal cue combinations: A possible approach to designing in-vehicle takeover requests for semi-autonomous driving." *Proceedings of the Human Factors and Ergonomics Society Annual Meeting*, 63, 1 1739–1743. DOI: 10.1177/1071181319631053.

In-CarPC. 2020. *Vehicle and Car Computer Systems | In-CarPC*. Accessed May 7, 2020. https://www.in-carpc.co.uk/.

iYeal. 2020. *24mm LED Illuminated Push Button*. Accessed February 17, 2020. https://www. amazon.co.uk/iYeal-Illuminated-Buttons-Joystick-Raspberry/dp/B081D4W33K/ref=sr _1_151?keywords=green+led+push+button&qid=1574084349&sr=8-151.

Janssens, J.P., A. Héritier-Praz, M. Carone, L. Burdet, J.W. Fitting, C. Uldry, J.M. Tschopp, and T. Rochat. 2004. "Validity and reliability of a French version of the MRF-28 health-related quality of life questionnaire." *Respiration*, 71, 6 567–574. DOI: 10.1159/000081756.

Johnson, T. J., and B. K. Kaye. 2014. "Site effects: How reliance on social media influences confidence in the government and news media." *Social Science Computer Review*, 33, 2 127–144. DOI: 10.1177/0894439314537029.

KT&C. 2020. *KNC-HDi47*. Accessed February 25, 2020. http://www.ktncusa.com/ip/ip_miniature/knc-hdi47.

Legett & Platt Automotive. 2020. *Vehicle Seating and Lumbar Support*. Accessed February 25, 2020. https://leggett-automotive.com//products.

Lylykangas, J., V. Surakka, K. Salminen, A. Farooq, and R. Raisamo. 2016. "Responses to visual, tactile and visual–tactile forward collision warnings while gaze on and off the road." *Transportation Research Part F: Traffic Psychology and Behaviour*, 40 68–77. DOI: 10.1016/j.trf.2016.04.010.

Mumuni, A. G., K. M. Lancendorfer, K. A. O'Reilly, and A. MacMillan. 2019. "Antecedents of consumers' reliance on online product reviews." *Journal of Research in Interactive Marketing*, 13, 1 26–46. DOI: 10.1108/JRIM-11-2017-0096.

Neely, G., and L. Burström. 2006. "Gender differences in subjective responses to hand–arm vibration." *International Journal of Industrial Ergonomics*, 36, 2 135–140. DOI: 10.1016/j.ergon.2005.09.003.

Normark, C. J. 2015. "Design and evaluation of a touch-based personalizable in-vehicle user interface." *International Journal of Human-Computer Interaction*, 31, 11 731–745. DOI: 10.1080/10447318.2015.1045240.

Politis, I., S. Brewster, and F. Pollick. 2015. "Language-based multimodal displays for the handover of control in autonomous cars." *AutomotiveUI - Proceedings of the 7th International Conference on Automotive User Interfaces and Interactive Vehicular Applications*. Nottingham, UK: Association for Computing Machinery. 3–10. https://doi.org/10.1145/2799250.2799262.

Politis, I., S. Brewster, and F. Pollick. 2017. "Using multimodal displays to signify critical handovers of control to distracted autonomous car drivers." *International Journal of Mobile Human-Computer Interaction*, 9, 3 1–16. DOI: 10.4018/ijmhci.2017070101.

Sato, T., B. A. Harman, L. T. Adams, J. V. Evans, and M. K. Coolsen. 2013. "The cell phone reliance scale: Validity and reliability." *Individual Differences Research*, 11, 3 121–132.

SBG. 2020. *Fusion. SBG Sports Software*. Accessed May 7, 2020. https://sbgsportssoftware.com/product/fusion/.

Van Der Laan, J. D., A. Heino, and D. De Waard. 1997. "A simple procedure for the assessment of acceptance of advanced transport telematics." *Transportation Research Part C: Emerging Technologies, 5*, 1 1–10. DOI: 10.1016/S0968-090X(96)00025-3.

Vector. 2020. *Know-How & Solutions for CAN/CAN FD.* Accessed May 7, 2020. https://www.vector.com/us/en-us/know-how/technologies/networks/can/.

Verrillo, R. T. 1979. "Comparison of vibrotactile threshold and suprathreshold responses in men and women." *Perception & Psychophysics, 26* 20–24. DOI: 10.3758/BF03199857.

Wynne, R. A., V. Beanland, and P. M. Salmon. 2019. "Systematic review of driving simulator validation studies." *Safety Science, 117* 138–151. DOI: 10.1016/j.ssci.2019.04.004.

Yoon, S. H., Y. W. Kim, and Y. G. Ji. 2019. "The effects of takeover request modalities on highly automated car control transitions." *Accident Analysis & Prevention, 123* 150–158. DOI: 10.1016/j.aap.2018.11.018.

Part V

On-Road and Design Guidelines

17 Can Allowing Interface Customisation Increase Driver Confidence and Safety Levels in Automated Vehicle TORs?

James W.H. Brown, Kirsten M. A. Revell, Joy Richardson, and Jisun Kim
University of Southampton

Nermin Caber
University of Cambridge

Patrick Langdon
Edinburgh Napier University

Michael Bradley
University of Cambridge

Simon Thompson and Lee Skrypchuk
Jaguar Land Rover

Neville A. Stanton
University of Southampton

CONTENTS

17.1 INTRODUCTION

With automated vehicles on the horizon with their plethora of potential benefits, there remains the issue of how to ensure drivers safely take back control from automation when necessary. Although much research has been carried out into takeover request (TOR) design, and resultant post-takeover performance, such designs are generally a one-fits-all approach. Providing a customisable TOR interface enables users to adapt their TOR to their requirements; however, the effects on takeover performance, and associated confidence and safety when TORs are customised are largely unknown.

There is currently a strong push from multiple governments worldwide to introduce driverless vehicles to public roads. The benefits for this were said by Chris Patton of Fujitsu's EMEIA (Europe, Middle East, India, and Africa) Transport Team, to be the potential to render transport safer and cheaper while reducing environmental harm and congestion (Billington 2019). Jeffrey Zients, the Director of the US National Economic Council, stated that automated vehicles would *'...save time, money and lives'* (Kang 2016). In the UK, development is being aided by the government's Centre for Connected and Autonomous Vehicles production of a code of practice for automated vehicle trialling (Department for Transport 2015), and the introduction of new legislation aimed at encouraging testing. In the US, a report produced by the National Science and Technology Council and the US Department of Transportation (United States Department of Transport 2020) outlines a total of ten areas of interest relating to the development and use of automated vehicles. The first is the prioritisation of safety, and within this area, the Federal Motor Carrier Safety Administration is currently carrying out human factors-based research into aspects, including human–machine interface (HMI) design, communication, and driver readiness, a factor that is strongly related to confidence.

Indeed, with safety clearly the primary concern, much research has been focused on the driver readiness aspect, in particular, how to bring the driver back into the loop safely via the TOR (Gold et al. 2013; Walch et al. 2015; Eriksson and Stanton 2017a, 2017b). TORs occur in vehicles that, while they feature automation capabilities, cannot cope sufficiently with all circumstances that might include adverse weather conditions, badly marked highways, or sensor issues. In such cases, the vehicle submits a TOR. In order to frame the requirements for TORs, they should be viewed in the context of automation capability. The Society of Automotive Engineers outline six distinct levels of automation (SAE International 2018), as described in Table 17.1.

It can be seen from Table 17.1 that TORs will occur primarily at automation Levels 2 and 3. TORs can be presented to drivers using different modalities involving a range of methods, such as audible tones, vocalisations, graphical icons, textual information, and haptics. It is possible to present a single modality of information; however, findings have suggested that this is not optimal (Yoon, Kim and Ji 2019; Politis, Brewster and Pollick 2015). A considerable amount of literature has been published on the use of multimodal TORs (Melcher et al. 2015; Petermeijer et al. 2017; Yoon, Kim and Ji 2019). This research has frequently focused on identifying the most effective combinations of modalities. While effective combinations may be identified, each of these modalities can have multiple attributes. For example, audio cues and vocalisations can vary in amplitude and frequency. Visual modalities such as icons and text-based alerts could be placed in differing locations such as on a cluster or centre console; they could vary by size, colour, and either be static or animated in some way. The haptic modality has multiple parameters, such as duration

---

TABLE 17.1

The Six Levels of Automation as Specified by the SAE (2018)

Automation Level	Definition	Supervision
0	Fully manual driving experience, no advanced driving assistance systems (ADAS)	All times
1	No automation but some ADAS, such as cruise control	All times
2	Limited automation, the driver must supervise at all times	All times
3	Conditional automation, the driver must take control when prompted	Only to take control when prompted
4	Highly automated, no driver intervention required; however, the vehicle cannot cope with all conditions	No
5	Fully autonomous, vehicle can complete journeys entirely independently in all conditions	No

of pulse, pulse pattern, and amplitude. With these variations, it becomes clear that there are many permutations of potential settings. However, selecting a single permutation of settings across all modalities, even with an effective set of modalities, is unlikely to represent an ideal configuration for all users due to wide variations in requirements and desires (Normark 2015). Stanton and Baber (1992) describe user characteristics as one of the multiple factors in defining usability – these encompass aspects such as skills, knowledge, and motivation. There is also the necessity for inclusive design (Hickson et al. 2010); driver's experiencing TORs will have varying levels of vision, hearing (Hickson et al. 2010), cognitive function (Herbert et al. 2016), and haptic sensitivity. These abilities will likely not only vary by demographic criteria such as age and gender, but there may well be a subjective aspect in addition, where drivers simply prefer one setting to another (Tractinsky et al. 2011). Stanton and Baber (1992) also described task match; this specific usability factor concerns how well a system fulfils the requirements and needs of the user. It therefore follows that a user given the opportunity to adapt, or customise, a TOR interface to their preferences may conceive an interface that closely matches their requirements and needs. An additional benefit is that a customisable interface can be adapted as the users' needs and requirements change (Normark 2015) and as the driving circumstances change. Customisation is the term employed when a user selects their own preferences; this contrasts with personalisation where the system behind the interface adapts the interface to the user (Babich 2017). It should be noted that some sources use these terms interchangeably. Gibson, Butterfield, and Marzano (2016) describe the design challenge of ensuring dashboards include all necessary information while remaining simple, non-distracting, efficient, and subjectively acceptable. Indeed, visual clutter can result in erroneous decision-making (Baldassi, Megna and Burr 2006). As in-car technologies have evolved from analogue to digital, systems such as LCD screens provide a significant scope for customisation (Walker, Stanton and Young 2001). Little research has been conducted on driver interface customisation, and the use of customisable TORs may be a novel research path. Gibson, Butterfield, and Marzano (2016) carried out an experiment involving interface customisation and subsequent subjective performance of 35 participants assessing the usability of three distinct vehicle consoles using questionnaires. Among a range of findings, the participants indicated that the displayed information should be relevant and there should not be an excess. Responding to questions regarding whether they would welcome the opportunity to have an input in dashboard design, over 54% agreed. When asked if they would like to redesign their instrument panel, 43% agreed. 66% also reported that they would not consider an in-vehicle information system to be useful. Additional questions revealed that there were highly variable opinions on what aspects should constitute a dashboard. Gibson, Butterfield, and Marzano (2016) concluded that drivers do not have a clear idea of their needs or requirements with regard to vehicle information. This is highly suggestive that the inclusion of the ability to customise vehicle interfaces would be welcomed by drivers and would enable them to try various configurations until they find one that they like. Gibson, Butterfield, and Marzano's (2016) questionnaire-based experiment revealed that when asked, participants were less concerned about safety and reliability and more about the drivers' experience and how they feel about the vehicle. A customisable interface would allow the information presented to the driver to be tailored to individual preferences, from both quantitatively practical and subjectively aesthetic viewpoints. Normark (2015) employed a research-through-design approach, evaluating a prototype vehicle–user interface using a range of methods, including subjective measures and simulator-based trials. Four research questions were presented, relating to the effect of personalisation on traffic safety, usefulness in traffic, definition of the user experience, and any usability issues that could affect traffic safety. The prototype consisted of a simulator featuring a customisable head-up display (HUD), cluster, central console, and upper console. The default configuration included consisted of a speedometer on the cluster. Apps featuring information such as fuel level and GPS were added using an 'app-store'-style interface. Driver performance was measured using lane position standard deviation and lane departures. Normark (2015) measured driving performance of participants while customising, as opposed to post-customisation, and compared against a control condition where participants simply completed a drive. He found no statistically significant effects on standard deviation of speed, standard deviation of lane position, or lane exceedances, although 44% of participants considered that it may be distracting. It was stated by all participants that there were no negative effects associated with driving using their customised interface and all participants agreed that they would like the system on their own vehicles. Crucially, participants considered that the customisations they made to the interface could possibly improve safety levels due to matching their requirements and desires.

With safety as the primary requirement, it is necessary to examine the potential effects of customisation on driver performance post-takeover, with a particular focus on driver readiness and confidence. A range of metrics can be employed to measure performance (Louw et al. 2017) – of particular interest are the primary control inputs, steering, and throttle (braking was a very rare occurrence post-takeover), as they dictate lateral and longitudinal behaviours, respectively. Unexpected or large amplitude control inputs resulting in significant changes in lateral or longitudinal accelerations could be an indication that a driver was not ready to take control, or they may have been surprised by the TOR. When employing multimodal information, including alerts, the avoidance of a startle or surprise response is crucial. Startle is defined as *'a response to an intense and surprising stimulus'* (Grillon 2002). A startle or surprise response can occur due to high audio volumes (Fagerlönn 2011), high-amplitude haptic signals (Väänänen-Vainio-Mattila et al. 2014), and visual stimuli (Biondi et al. 2014). It is important in terms of safety, as the effects of startle or surprise in drivers can lead to slower reaction times (Fagerlönn 2010) and even mild surprise can result in unintentional motor reflexes (Biondi et al. 2014). Both of these are clearly detrimental in terms of driving performance and could have safety implications; indeed, Biondi et al. (2014) found that when exposed to a sufficient acoustic stimulus, drivers were distracted and exhibited an involuntary reduction in throttle and applied similarly involuntary steering inputs. Individuals have differing thresholds for startle. Kofler et al. (2001a, 2001b) found a significant effect of both gender and age on auditory startle influences. This strengthens the argument for a customisable interface as drivers are able to adjust settings to their specifications, and thus remove any surprise element and the associated detrimental effects. Steering inputs and the resultant lateral accelerations should also be considered against expected behaviour or previously defined parameters. For example, Lee et al. (2011) defined a standard lane change scenario as a resultant steering input of 12° and a resultant lateral acceleration of $1 \, \text{m/s}^2$. Post-takeover values would be expected to be well under these levels as no lane changes occurred. Equally, throttle and longitudinal acceleration can be compared with expected values. Deceleration is particularly relevant in terms of safety as, in motorway conditions, it can lead to congestion (Kerner 2002; Findley et al. 2015) and the potential for accidents (Kim and Zhang 2008). From a safety perspective, ideally a TOR should be a seamless transition from automation to manual with no perceivable change in speed or lateral position (Clark and Feng 2017). However, in the real world, vehicles are exposed to multiple influences such as side winds, road surface imperfections (Miura 2014), and camber that can result in a requirement for minor control inputs. Some control inputs therefore would be expected.

Much research has employed takeover times (TOTs) as a metric representative of readiness and HMI efficacy (Zhang et al. 2019; Gold, Happee and Bengler 2018; Eriksson and Stanton 2017b). Politis, Brewster, and Pollick (2017) found that in critical takeover scenarios, multimodal TORs result in lower TOTs than unimodal visual TORs. TORs should be of an appropriate length, and they need to be presented at appropriate times, doing so can reduce TOTs (McDonald et al. 2019).

While the benefits of multimodal alerts have been highlighted in multiple studies (Geitner et al 2019; Yoon, Kim and Ji 2019; Eriksson et al. 2019), it is important to consider not only the placement of visual elements, but also magnitudes of non-visual elements. Lee et al. (2006) carried out a study into the performance effects of alert modality and adaptive cruise control. Their findings suggested that a multimodal alert could result in slower reaction times than a dual modality alert. They concluded that this may be due to participants perceiving their alert in the form of multiple cues, opposed to their cognitively forming a single cue. Citing Driver and Spence (1998) and Spence (2002), they further suggest that it may be necessary to identify and adjust the parameters of multimodal signals in order to achieve this single cohesive cue. The ability to customise may therefore provide users with the ability to achieve this forming of a single cue independently, and therefore adapting the interface to maximise usability.

This paper builds upon research into multimodal TOR interfaces by presenting a study specifically on the post-takeover effects of user-customised interfaces. The advantages of customisation have been well established (Normark 2015; Gibson, Butterfield and Marzano 2016); however, it is critical to ensure that safety is not compromised in any way, by examining performance post-takeover. The hypothesis is that customisation will not have a detrimental effect upon post-takeover performance with regard to safety or driver confidence.

After introducing and discussing the subject domain literature and presenting the main arguments and hypotheses, this paper outlines the method, including participant demographics, experimental design, equipment, and the procedure undertaken.

The analysis then describes in detail the statistical methods employed in order to obtain the final results. The results are presented by data channel with the relevant statistical tests and graphs; TOTs are also analysed. The findings are discussed with reference to the literature outlined in the introduction and the hypotheses, leading to a synthesis of conclusions and future work.

17.2 METHOD

17.2.1 PARTICIPANTS

A total of 24 participants were recruited from a pool that had previously experienced a simulator-based study featuring an almost identical HMI. All were required to be a minimum of 28 years old and hold full UK driving licenses. The gender split was 15 males/9 females, across three age ranges: 28–34, 35–56, and 57–82 (Table 17.2). Participants had to provide signed consent prior to involvement in the study which was approved by the University of Southampton's Ethics and Research Governance Office (ERGO Number: 49792.A2).

17.2.2 EXPERIMENTAL DESIGN

A repeated-measures design was employed for the experiment. Participants experienced two differing customisations of vehicle HMI, the HMI customisation representing the independent variable. Two trials took place: trial 1 using default HMI settings and trial 2 using participant's selected settings. A final customisation of the HMI took place after the completion of trial 2. Customisations were carried out using a matrix presented to the participant on the experimental vehicle's central console screen. The graphical elements of the HMI were displayed on a HUD, the vehicles' cluster, and infotainment display. Additional elements of the HMI included a haptic seat, ambient lighting, vocalisations, and audio cues. The customisable elements are listed in Table 17.3.

Figures 17.1 and 17.2 illustrate the customisation matrix and the drivers' HMI, and Table 17.4 lists the ten questions used in various orders during the takeover protocol. The driving mode wording is situated below the driving mode icon; this is centrally placed in the infotainment display and cluster and placed on the left side in the HUD. The colour acts as a mode indicator – blue indicating automated and orange for manual. Time to takeover is shown in the bottom-left corner of the cluster and infotainment display, and the top-right of the HUD. The edge frame mode indicator

TABLE 17.2
Participant Demographics

No. of Part's	Age Bracket	F	M
7	28–34 (Mean 29.3) (SD 2.2)	3	4
10	35–56 (Mean 42.6) (SD 5.5)	3	7
7	57–82 (Mean 63.4) (SD 4.5)	3	4

TABLE 17.3

HMI Customisable Elements and Parameters

HUD Display	Cluster Display	Infotainment Display	Ambient Lighting	Audio Volume	Haptic Amplitude	Takeover Questions
Driving mode wording (ON/OFF)	Driving mode wording (ON/OFF)	Driving mode wording (ON/OFF)	(1–11) 1 calibrated to be visible at all times	(1–11) 1 calibrated to be audible at all times	0–3	1–10
Driving mode icon (ON/OFF)	Driving mode icon (ON/OFF)	Driving mode icon (ON/OFF)	-	-	-	-
Time to takeover (ON/OFF)	Time to takeover (ON/OFF)	Time to takeover (ON/OFF)	-	-	-	-
Display edge frame (ON/OFF)	Display edge frame (ON/OFF)	Display edge frame (ON/OFF)	-	-	-	-
Takeover questions (ON/OFF)	Takeover questions (ON/OFF)	Takeover questions (ON/OFF)	-	-	-	-
-	Road sense view (ON/OFF)	Road sense view (ON/OFF)	-	-	-	-

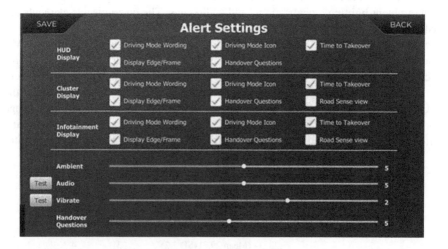

FIGURE 17.1 The customisation matrix displayed at the end of each trial on the vehicle's infotainment display.

FIGURE 17.2 The three visual aspects of the HMI: on the left is the central console, bottom-right is the cluster, and the HUD is shown on the top-right.

TABLE 17.4

Ten Questions That Were Employed to Raise Situational Awareness as Part of the Takeover Protocol

Question

What lane are you in?

Can you see any motorcycles?

Is the road ahead straight?

Is your right blind spot warning light on?

Are there any vehicles alongside?

What is your speed?

What traffic is behind you?

How many car lengths is the vehicle in front away from you?

Is the traffic busy ahead?

What is the weather like?

is only visible on the cluster and infotainment display. Takeover questions and associated icons appear in the centre of each of the HMI display elements.

Dependent variables in terms of performance data included a range of variables taken from the vehicle's controller area network (CAN), and outputs from the HMI system. The categories of performance data that was gathered can be seen in Table 17.5.

17.2.3 EQUIPMENT

The experimental vehicle was a 2017 Jaguar I-PACE EV400 AWD pre-production model (Figure 17.3). A full overview of the system can be seen in Figure 17.4. The automation system consisted of a combination of factory standard lane keep assist and adaptive cruise control. When utilising these systems, the standard car would issue frequent warnings to maintain hands on the wheel; these warnings were

TABLE 17.5
The Recorded Performance Data Variables

Variable	Units
Speed	mph
Steering angle	degrees
Throttle	percentage
Brake	percentage
Lateral acceleration	m/s^2
Longitudinal acceleration	m/s^2
Steering wheel angle speed	deg/s

FIGURE 17.3   The test vehicle, a 2017 Jaguar I-PACE EV400 AWD.

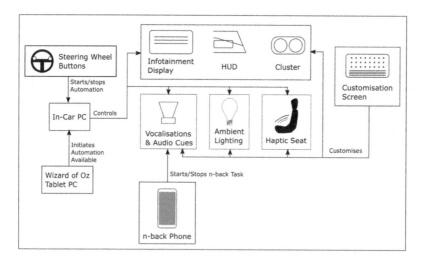

FIGURE 17.4   The configuration of the vehicle's systems.

specifically removed from the system in order for it to simulate Level 3 automation. The car was fitted with a PC (In-CarPC 2020) (Figure 17.5) running SBG software (SBG 2020) for data and video logging. A dSPACE MicroAutoBox II system was employed to interface with the vehicle's systems (dSPACE 2020), together with three CAN interfaces (Connective Peripherals 2020) and a Vector CAN box (Vector 2020). Power was supplied via an inverter, connected to the 12 V battery circuit. An Ethernet switch and USB hub provided communications between systems. The visual aspects of the HMI consisted of a 14″ thin-film transistor (TFT) panel fitted to the centre console, a 10″ TFT fitted in place of the original equipment manufacturer's cluster, and a HUD comprising a small 1000×250 px TFT and reflector screen (Bysameyee 2020) (Figures 17.6 and 17.7). The car's original cluster was moved into the passenger footwell to allow monitoring by the safety driver. The interactive element of the HMI constituted two illuminating green buttons (iYeal 2020) fitted in the thumb positions of the steering wheel. The button functionality was integrated using an Arduino (Arduino 2020). Ambient lighting was supplied via LED lighting strips (Elfeland 2020). Haptics were provided in the seat base via Leggett and Platt motors (Leggett & Platt Automotive 2020) controlled through an Arduino Micro and motor control board (Arduino 2020). Five cameras (KT&C 2020) were installed within the I-PACE to provide footage from the forward view, driver-facing, over-the-driver-shoulder, footwell, and rear views. A dashcam was also fitted in the

FIGURE 17.5   The in-car PC fitted in the boot of the I-PACE, the dSPACE system, Ethernet switch, USB hub, and switch panel for system initialisation.

FIGURE 17.6   The car in manual mode with the HUD and cluster clearly visible.

FIGURE 17.7   The view from the over-the-shoulder camera, the green buttons on the steering wheel are visible, as are the three visual elements of the HMI – the HUD, cluster, and centre display. The ambient blue lighting is also apparent, indicating the vehicle is in automated mode. The driver-facing camera can be seen fitted on the dashboard.

safety car to collect footage of the participants' vehicle from an external viewpoint. A cognitive loading task (*n*-back) was controlled from a mobile phone using MIT's AgeLab delayed digit recall task app (AgeLab 2019); this was connected to a generic Bluetooth speaker. Additional speakers were also placed within the car and connected to the PC to relay the vocalised and audio cue aspects of the HMI. The HMI was controlled via a Microsoft Surface Pro tablet providing Wizard of Oz (WOZ) control of the system. Custom software was developed for the HMI using C++.

17.2.4 Procedure

On arrival at the Jaguar Land Rover (JLR) facility at Fen End, the participant was welcomed and asked to sign in, presenting their driving license and receiving a visitor pass. They were then briefed on the aim of the project and how the on-road study follows on from the simulator study that they completed the previous year. A brief explanation of the sequence of events for the study was then given. This included the route (Figure 17.8), the takeover procedure, form filling on trial completion, customisation of the HMI, and the *n*-back cognitive load task. It was stressed that they were required to maintain the same level of attention as if they were driving and should be ready to take control at all times. They were also advised that if the cognitive load task was detrimentally affecting their ability to retain a safe level of attention, they should stop the *n*-back

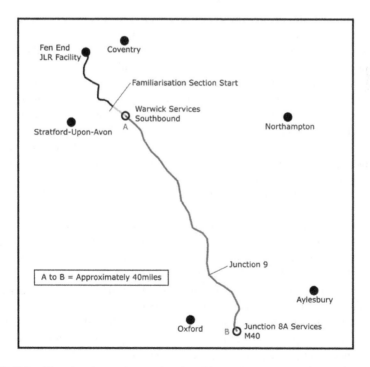

FIGURE 17.8   The experimental route showing the location of JLR's Fen End facility.

task. They were presented with a copy of the participant information sheet; this same sheet was made available to them when they were recruited. A reminder sheet showing screenshots of the HMI they would be using was provided, including modal variation of the cluster, HUD and in-vehicle interface, the green buttons, and the customisation matrix screen. They were shown a route map that highlighted where the trials would take place, but also areas of caution and one junction that required traversal in manual mode. A privacy policy was presented, and a sheet explaining the cognitive load n-back task was provided along with a brief verbal explanation. The participant was asked if they had any questions, and if they were happy to continue, they were provided with two consent forms – an attendance form and an events team form – to sign. On completion of all forms, they were led from the reception area to the car park and asked to sit in the passenger side of the car; both experimenters took their places in the rear seats. The participant was introduced to a safety driver who explained the basic controls of the car, and the elements of the interface. The safety driver then drove the car through security to the proving ground and demonstrated the vehicle's performance before running through the transition to automated mode and back to manual control. The automation system was operated by one of the experimenters using a WOZ system, from a Windows tablet in the back of the car. To offer automation, an experimenter would press a start button on a custom control panel app running on the tablet; the HMI then indicated automation as being available via the three graphical interfaces and a vocal alert. In order to enable automation, the driver simultaneously pressed the two green buttons mounted on the steering wheel and released all of the controls, including the accelerator. The system would then engage automation and the HMI would indicate that the automated mode was active. During automation, the cognitive load task was controlled by the other experimenter via a mobile telephone-based app linked to a Bluetooth speaker. The safety driver demonstrated the automation system multiple times, including the n-back task, and requested the safety car overtake and brake in front of the vehicle, when in automated mode, to illustrate how it reacts to maintain a gap to the car in front. The safety driver then stopped in a safe area and swapped places with the participant. The participant then was allowed some time to drive the car on the proving ground to become familiar with the controls. The automation was then made available to them, and they experienced multiple handovers and takeovers, including the use of the n-back task while in automation.

In Figure 17.8, the 2-mile on-road familiarisation section is shown just prior to point A. The main trial route between Warwick Services Southbound (A) and Junction 8A Services M40 (B) was approximately 40 miles long and required just over 40 min to complete. Trial 1 was conducted Southbound from A to B, whereas trial 2 used the reversed route, Northbound from B to A.

The participant was asked to drive manually to the start point of the experiment at the Southbound Warwick services on the M40. Two miles prior to the services, while on the M40, a road-based practice handover was conducted. The automation was offered to the participant; once activated, a short 30-s period of automation followed which included the participant carrying out the n-back task. The participant then experienced the takeover protocol and resumed manual control before stopping at the

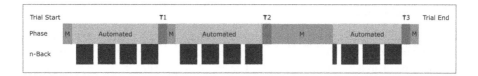

FIGURE 17.9 A timeline showing phases of one trial illustrating manual mode (M), automated mode, and takeovers (T1/T2/T3). *N*-back application periods are shown by the grey bars.

services. After confirming that the participant was happy to continue, the on-board systems were checked and configured for the first trial, data logging was started, and the video and audio were synchronised using a clapper board.

The participant was instructed to drive from the services onto the M40, proceeding in the left lane at approximately 58 mph. Figure 17.9 illustrates the following procedure. After 1 min of manual driving, automation was offered to the participant. Once activated, after a further period of 30 s, the *n*-back cognitive loading task was started. This task was run in 2-min intervals, separated by 30 s. Following 10 min of automation, the HMI started the takeover protocol; once completed, the participant pressed both steering wheel buttons and resumed manual control. After 1 min of manual control, automation was again offered to the participant and the process repeated. On completion of the second takeover, the manual driving period was extended to approximately 7 min in order to pass a section of motorway (M40 J9 Southbound) that would have adversely affected the automation due to the lane becoming a slip road. Due to proximity to the end point of trial 1, takeover 3-s automation period was reduced to 8 min. The *n*-back task was started simultaneously with the automation for 30 s, before reverting to 30 s off and 2 min on until the protocol started. When the participant completed the protocol and assumed manual control, they continued in manual mode the short distance remaining to the motorway services and parked the car. Data logging was stopped as this constituted the end point of trial 1. The participant was presented with a set of PDF-based forms on a laptop (including the NASA-Task Load Index, system usability scale, and system acceptance scale) and asked to complete them based upon their experience specifically of the takeover protocol, process, and HMI elements. When the forms were complete, the customisation screen was displayed on the vehicle's centre console and the participant was asked to make any changes that they would like to the HMI. During this customisation process, they were asked to vocalise their opinions, considerations, and rationales for customisations. This information was recorded for future analysis. This concluded the data collection for trial 1; the participant at this point had access to facilities and refreshments. Trial 2 was a repeat of the processes and data collection carried out in trial 1; however, the route was Northbound to the start point for trial 1. On completion of trial 2, one additional form was provided to collect subjective post-task data (post-trial interview). The safety driver swapped places with the participant from this point and drove the vehicle back to the JLR facility at Fen End. The participant was thanked for their time and signed out at reception.

17.3 ANALYSIS

Performance analysis involved six channels of data, falling broadly into two categories, namely lateral and longitudinal. The lateral category comprised steering angle, steering speed, and lateral acceleration. The longitudinal category comprised throttle input, longitudinal acceleration, and speed. Although braking data was recorded, it was omitted due to the low frequency of usage. Data was initially examined to determine quality; the majority of data sets from participants were complete; however due to technical issues, in one case, data for trial 2 was missing, and in a second case, the system failed to initiate a takeover. These participant sets were therefore rejected. The focus of the research question was on performance post-takeover; therefore, the 60 s of data immediately post-takeover was isolated and analysed for each participant. This revealed that in many cases, participants initiated lane changes after approximately 20 s. This was considered to constitute a confounding variable, and therefore, the period 20 s post-takeover was selected for analysis. Initial analysis consisted of time-series graphs for each trial to study trends for all six data channels. Graphs of standard deviations were also generated for each channel. TOTs were calculated for each trial, both in terms of time to complete the full takeover protocol and in terms of the reaction time to the request to resume manual control at the end of the protocol. Paired t-tests were carried out on a per-time increment basis (20 Hz/0.05 s), to generate graphs for each data channel indicating significance based on trial.

To test for a significant effect of customisation, a mean value of each data channel was calculated for each participant, one for each trial. Paired t-tests were then applied to the data sets of 21 values per trial, resulting in a single value for significance for each data channel. The collated p-values can be seen in Table 17.6. These data sets were also graphed using boxplots to reveal distributions.

17.4 RESULTS

17.4.1 THROTTLE

Mean throttle usage values in both trials (Figure 17.10) start at very low values at the point of takeover; for the default takeover settings (DTS) (trial 1), this is 5% and slightly higher at approximately 7% for customised takeover settings (CTS) (trial 2).

TABLE 17.6

Significance of Effect of Trials on Data Channels

Data Channel	Paired t-Test Results (Participant Means by Trial)	Significant
Throttle	$t=-1.453$, df $=20$, p-value $=0.1617$	No
Speed	$t=-2.1206$, df $=20$, p-value $=0.04665$	Yes
Longitudinal acceleration	$t=-2.1446$, df $=20$, p-value $=0.04445$	Yes
Steering angle	$t=0.85367$, df $=20$, p-value $=0.4034$	No
Steering angle speed	$t=1.1025$, df $=20$, p-value $=0.2833$	No
Lateral acceleration	$t=2.5007$, df $=20$, p-value $=0.0212$	Yes

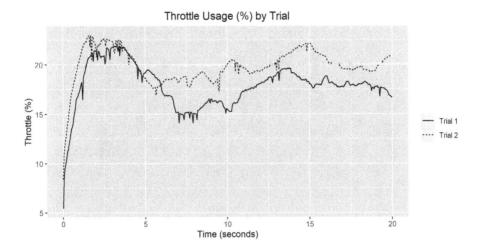

FIGURE 17.10   Mean throttle usage over time by trial. Trial 1 is default settings, whereas trial 4 is customised settings.

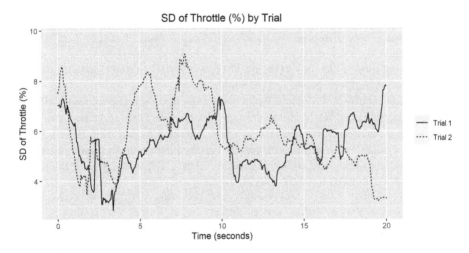

FIGURE 17.11   SD of mean throttle usage over time, by trial. Trial 1 is default settings, while trial 4 is customised settings.

In both trials, drivers initially increased throttle to slightly over 20%; in the case of CTS, the rate of increase was slightly higher, peaking after approximately 1.5 s, compared to approximately 3 s in DTS. Both trials then showed a decrease in throttle from around 3 s; in CTS, this then stabilised at around 18%; however in DTS, it continued to decrease to 15%. From 7 s post-takeover, the percentage of applied throttle in CTS remains consistently higher than in DTS, with both increasing to secondary lower peaks between 12.5 and 15 s.

Examining the standard deviation of throttle values (Figure 17.11) reveals a fairly wider spread of throttle application levels in the initial 2.5 s and between 4

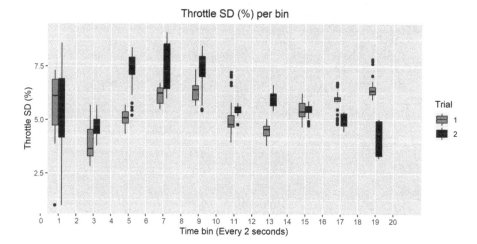

FIGURE 17.12 A boxplot showing the standard deviation of throttle percentage in 2-s time bins. Trial 1 represents the default settings, and trial 2 is the customised settings.

and 10 s. In both cases, CTS exhibits a wider spread of values. There also appears to be a considerable difference in standard deviation values between trials in the final second.

Examining binned SD values (Figure 17.12), the variations and similarities in SD values between trials can be viewed with more clarity.

No significant effect of trial was found comparing mean throttle values for all participants by trial ($p = 0.1617$). However, a boxplot (Figure 17.13) indicated a slightly lower median value of just over 18% in DTS compared to just over 19% in CTS. Participants in DTS also exhibited a greater variance between minimum and maximum values (approximately 10% compared to 7% in CTS), and a greater interquartile range (4% compared to 2% in trial 2).

17.4.2 SPEED

Mean initial vehicle speeds in both trials post-takeover (Figure 17.14) were between 52.5 and 55 mph; this was defined by the requirement to engage automation at approximately 55 mph. In the case of both trials, speed values initially decrease over the first 1 or 2 s, and then increase fairly linearly by around 2.5 mph until approximately 6 s post-takeover. Participants in DTS then appeared to largely maintain their speed between 57.5 and 58.75 mph. Participants in CTS, however, continued to accelerate almost linearly to over 61 mph at 20 s.

Examining the standard deviations of speed values (Figure 17.15), there appears to be little difference between trials, with CTS values consistently slightly higher. In both trials, the SD values remained at between 2 and 3 mph for the initial 5 s post-takeover. Subsequently, SD values increased linearly to 20 s, with values for CTS generally slightly higher.

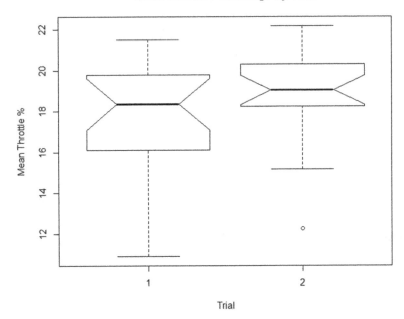

FIGURE 17.13 A boxplot indicates the variance in mean throttle percentage by trial. Trial 1 is default settings, and trial 2 is customised settings.

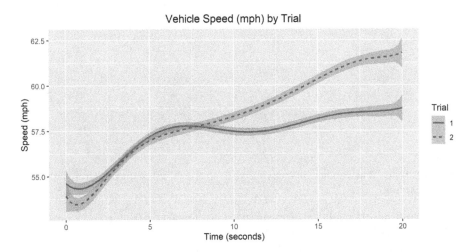

FIGURE 17.14 Smoothed mean vehicle speed over time. Trial 1 is default settings, and trial 2 is customised settings.

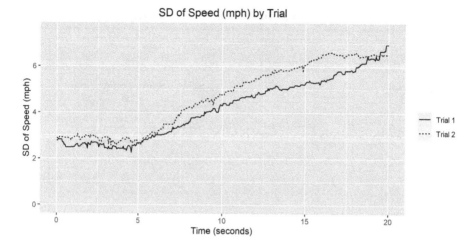

FIGURE 17.15   SD of mean speed by trial.

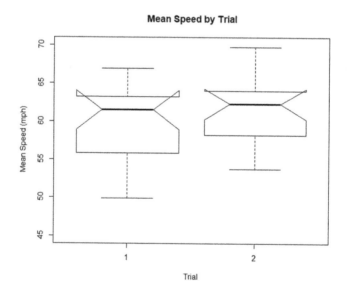

FIGURE 17.16   The variance in mean speed. Trial 1 is default settings, and trial 2 is customised settings.

A significant effect of trial on mean speed was found ($p = 0.04665$) (Figure 17.16). Median values for speed were broadly similar with CTS being slightly higher at 63 mph; the interquartile range in DTS was slightly larger, approximately 8 mph, compared to 6 mph for CTS. Both 25th and 75th percentile values, and minimum and maximum values were all higher for CTS. The 25th percentile value for DTS was 56 mph, compared to 58 mph for CTS. The minimum value for DTS was 50 mph compared to 54 mph for CTS.

17.4.3 Longitudinal Acceleration

Values for longitudinal acceleration (Figure 17.17) naturally mirror the throttle inputs; this is likely enhanced further due to the characteristics of the electric test vehicle. In both trials, participants accelerated over the initial 2.5 s post-takeover; in CTS, the acceleration level peaked at just over 0.5 m^{-2}, whereas in DTS, the peak level was 0.35 m/s^2. After peaking at 2.5 s, acceleration levels dropped to between 0.1 and 0.2 m/s^2 in CTS; however in DTS, a slight deceleration of just under 0.1 m/s^2 was recorded between 6.5 and 11 s. In both trials, there is some variation in acceleration between 10 and 20 s, in both cases, fluctuating by approximately 0.2 m/s^2.

Standard deviation values (Figure 17.18) for longitudinal acceleration indicate some minor fluctuations largely constrained between 0.25 and 0.5 m/s^2. However, CTS appeared to show higher levels of standard deviation particularly at approximately 5 s and 7.5 s post-takeover.

A significant effect of trial on mean longitudinal acceleration was found ($p = 0.04445$) (Figure 17.19). In general, values for CTS were higher than for DTS; it should also be noted that negative values signify a mean deceleration. The DTS median value was between 0 and 0.05 m/s^2 compared to a median value of 0.1 m/s^2 in CTS. The interquartile ranges were similar between trials; however, in DTS, the upper quartile value was approximately 0.08 m/s^2, whereas the lower value was negative (-0.05 m/s^2). CTS's interquartile range was all positive (0.03 m/s^2 to 0.15 m/s^2). The maximum values for trials 1 and 2 were 0.26 and 0.15 m/s^2, respectively. The minimum values were -0.08 and -0.22 m/s^2, respectively.

17.4.4 Steering Angle

Mean steering angles by trial (Figure 17.20) showed considerably more fluctuation in DTS compared to CTS. In CTS, the highest mean value of steering angle was

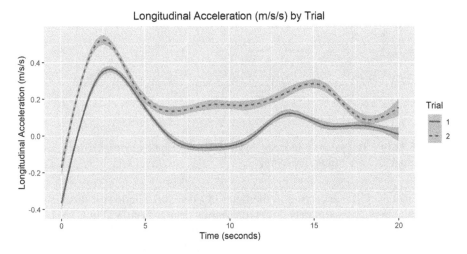

FIGURE 17.17 Smoothed mean longitudinal acceleration over time. Trial 1 is default settings, and trial 4 is customised settings.

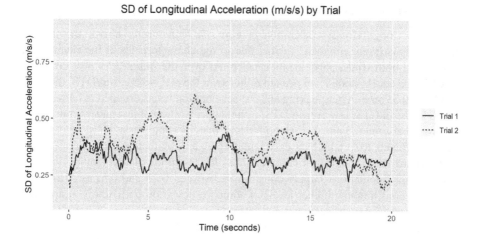

FIGURE 17.18 The longitudinal acceleration SD over time. Trial 1 is default settings, and trial 2 is customised settings.

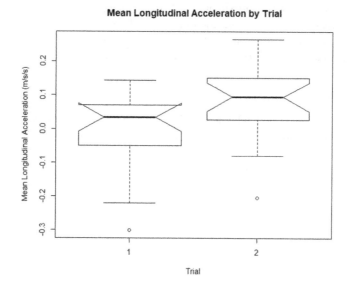

FIGURE 17.19 The variance in mean longitudinal acceleration by trial. Trial 1 is default settings, and trial 2 is customised settings.

approximately 1.75°, occurring at the point of takeover. The lowest mean steering angle value was approximately 0.8°, representing a spread of 0.95° in CTS. In DTS, the highest mean steering angle was 2.6° and the lowest was 0.5°, resulting in a spread of 2.1°. In DTS, three peaks of steering input occurred at approximately similar frequency but varying amplitude. The initial steering input in DTS initiated with a large positive steering input; however in CTS, the steering input remained stable at

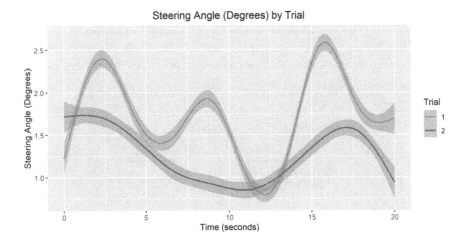

FIGURE 17.20 Smoothed mean steering angles by time. Trial 1 is default settings, and trial 2 is customised settings.

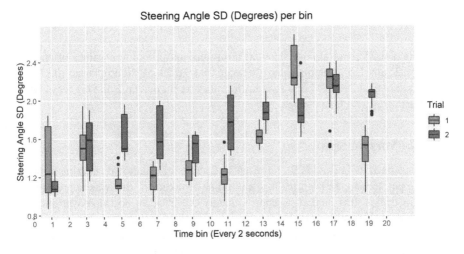

FIGURE 17.21 Variance in steering angle SD in 2-s bins. Trial 1 is default settings, and trial 2 is customised settings.

around 1.7°, prior to a mild, low-frequency fluctuation. The grey areas represent the confidence interval of 95%.

Examining the standard deviations of steering angle values per bin (Figure 17.21), it appears that steering angles had a wider spread of variance in trial 2 in the period between 3 and 13 s post-takeover.

No significant effect of trial on mean steering angle was found ($p = 0.4034$) (Figure 17.22). Indeed, the data for both trials was very similar with almost identical

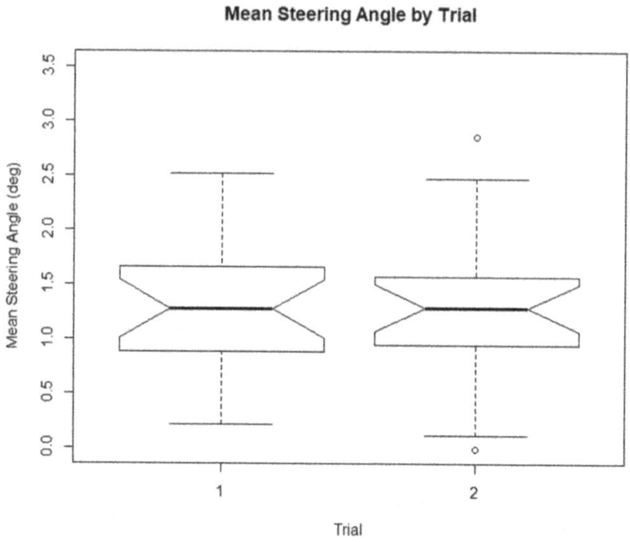

Mean Steering Angle by Trial

FIGURE 17.22 Variance in mean steering angle by trial. Trial 1 is default settings, and trial 2 is customised settings.

medians of 1.25°, and almost identical minimum and maximum values of approximately 0.2° and 2.5°. The interquartile range for trial 1 was slightly larger (0.8°–1.7°) compared to trial 2 (0.9°–1.6°). Two outliers are present in trial 2, the higher of which may be a result of a lane change.

17.4.5 STEERING SPEED

Steering speed (Figure 17.23) represents the rate of rotational steering input in degrees per second. In both trials at the point of takeover, this was approximately 0.1°/s; in trial 1, this peaked at approximately 0.375°/s, 3 s post-takeover and then reduced to 0.23°/s at 6 s, prior to increasing back to 0.375°/s by 20 s with some minor fluctuations. Trial 2 showed a less steep increase in steering speed over the initial 3 s post-takeover; this then stabilised with minor fluctuations around 0.25°/s.

Steering wheel speed standard deviation values (Figure 17.24) showed multiple high-frequency fluctuations of similar amplitudes in both trials; there were two minor exceptions at around 3 s where trial 1 values peaked and trial 2 values exhibited a similar peak at 4 s.

No significant effect of trial on mean steering angle speed was found ($p = 0.2833$) (Figure 17.25). Participants in trial 2 appeared to exhibit slightly lower mean steering angle speeds in general. Median values were 0.24°/s compared to 0.29°/s in trial 1. Interquartile ranges were similar in both trials (approximately 0.19°/s); however, the upper quartile was 0.4°/s for trial 1 and 0.37°/s for trial 2. The maximum and minimum values for trial 1 were both higher than for trial 2, at 0.61°/s compared to 0.54°/s, respectively, and 0.14°/sand 0.08°/s, respectively.

FIGURE 17.23 Smoothed mean steering angle speed by time. Trial 1 is default settings, and trial 2 is customised settings.

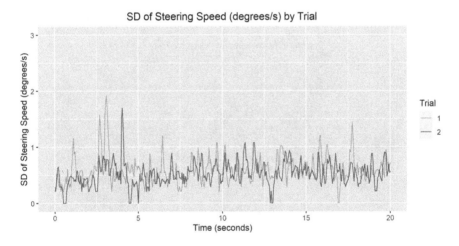

FIGURE 17.24 The SD of steering speed by time. Trial 1 is default settings, and trial 2 is customised settings.

17.4.6 LATERAL ACCELERATION

Lateral accelerations (Figure 17.26) appeared to be more stable in trial 2 than in trial 1 with only minor fluctuations between 0 and $-0.05\,\text{m/s}^2$ in the initial 15 s post-takeover. A slight increase in magnitude occurs from 15 s to $-0.1\,\text{m/s}^2$ before reverting to $-0.025\,\text{m/s}^2$ at 20 s. Trial 1 lateral accelerations fluctuate considerably at a frequency of approximately 0.15 Hz. The greatest magnitude change occurs

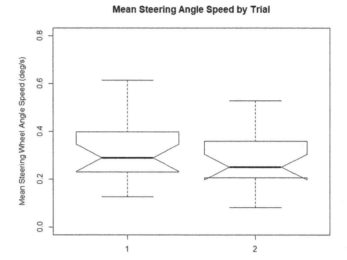

FIGURE 17.25   The variance in mean steering angle speed by trial.

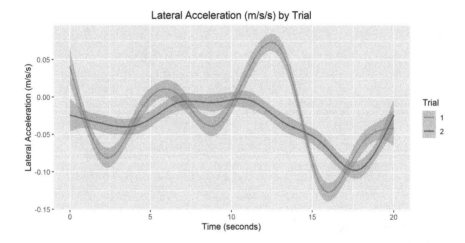

FIGURE 17.26   Smoothed mean lateral acceleration over time. Trial 1 is default settings, and trial 2 is customised settings.

approximately from 10 s, increasing to 0.075 m/s² at 12.5 s and then decreasing to 0.125 m/s² at 16 s. It then appears to stabilise at a similar level to trial 2 at 20 s.

Examining standard deviation values (Figure 17.27) for mean lateral accelerations by trial reveals that both trials exhibit broadly similar results. There appeared to be a slightly wider spread of values in trial 1 between 1 and 3 s post-takeover, and again between 15 and 16 s.

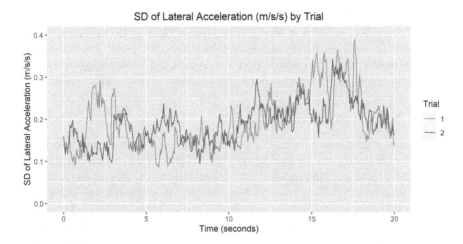

FIGURE 17.27 The SD of lateral acceleration by time. Trial 1 is default settings, and trial 2 is customised settings.

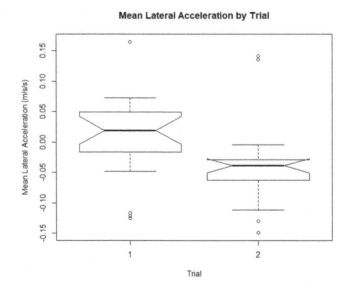

FIGURE 17.28 The variance in mean lateral acceleration by trial. Trial 1 is default settings, and trial 2 is customised settings.

A significant effect of trial on mean lateral acceleration was found ($p = 0.0212$) (Figure 17.28). Median values were approximately 0.02 and $-0.04\,\text{m/s}^2$ for trials 1 and 2, respectively. This suggests that in trial 2, a greater magnitude of lateral acceleration is occurring; indeed, the upper and lower quartile values in trial 2 are both negative (-0.025 and $-0.06\,\text{m/s}^2$, respectively). In trial 1, the interquartile range is larger with a lower quartile value of $-0.02\,\text{m/s}^2$ and an upper quartile value of $0.05\,\text{m/s}^2$. This suggests that in trial 1, a greater variation in lateral acceleration is occurring. The minimum and maximum values for trials 1 and 2 were -0.05 to $0.075\,\text{m/s}^2$ (trial 1) and 0.01 to $0\,\text{m/s}^2$ (trial 2), respectively. Some outliers are evident in both trials; these may have been due to lane changes or reactions to traffic.

17.4.7 TAKEOVER PROTOCOL TIME

The takeover protocol time represents the time required for participants to complete the takeover protocol; this duration starts with the HMI requesting the participant to 'get ready for handover', and ends with the driver pressing the buttons on the steering wheel to resume manual control. The box-and-whisker graph (Figure 17.29) shows that takeover protocol times appear to be lower for trial 2 than for trial 1. The median value for trial 1 is approximately 29 s, and the median value for trial 2 is approximately 26 s. Interquartile ranges are similar at approximately 5 s; however, the first quartile value for trial 2 is 24 s, approximately 2 s lower than that for trial 1. The minimum and maximum values showed similar trends, with a minimum value for trial 2 of 20 s, 5 s lower than the minimum value for trial 1. Three outliers occurred – the lowest in trial 1 may be attributed to a participant misunderstanding the protocol and taking control too early, while the lowest in trial 2 may be a result of a participant who selected the minimum number of questions and answered quickly.

Examining the protocol TOTs using bins of 1-s duration (Figure 17.30), slightly more detail regarding the distribution of bin frequencies is revealed.

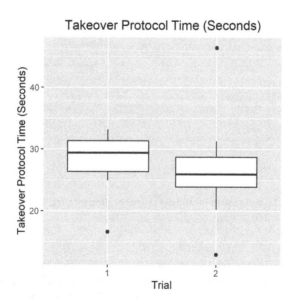

FIGURE 17.29 The variance in the time required for participants to complete the full takeover protocol. Trial 1 is default settings, and trial 2 is customised settings.

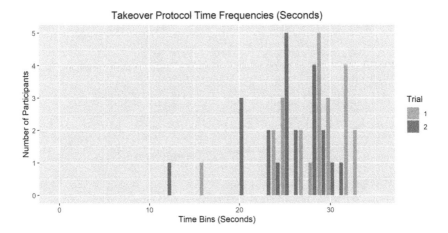

FIGURE 17.30 Distribution of protocol takeover times. Trial 1 is default settings, and trial 2 is customised settings.

17.4.8 TAKEOVER REACTION TIME

Takeover reaction times represent the time taken for the participant to press the steering wheel buttons to resume manual control after completion of the takeover protocol and takeover prompt. The box-and-whisker plots (Figure 17.31) indicate very similar results from both trials. Median values are both around 1.8 s, while the interquartile ranges are both approximately 0.7 s with the similar first and third quartile values. The minimum and maximum values are also similar at approximately 0.7 and 3.4 s, respectively.

Plotting the takeover reaction times in bins of 1-s duration (Figure 17.32) reveals the distribution. Only one participant took control in under 1 s; in the 1- to 2-s bin, there were approximately four more participants from trial 1 than from trial 2, indicating that takeovers were potentially slightly faster in trial 1. Similarly, in the 2- to 3-s bin, there were approximately six more participants from trial 2 than from trial 1, again suggesting that participants in trial 2 were taking slightly more time to react.

17.5 DISCUSSION

The research question was to identify if a customised TOR interface can improve driver confidence and safety; the hypothesis stated that customisation will not have a detrimental effect upon post-takeover performance with regard to safety or driver confidence. A wide range of conditions can affect takeover performance, including fatigue, density of traffic, and driver feedback (Merat and de Waard 2014). The driver feedback aspect was of interest in this experiment as participants had the ability to adjust multiple modalities to their preferences.

Two trials took place: one using default settings and a second with customised settings; the performance data for multiple channels was analysed and a test for a significant effect of customisation carried out for each. The data channels fell broadly

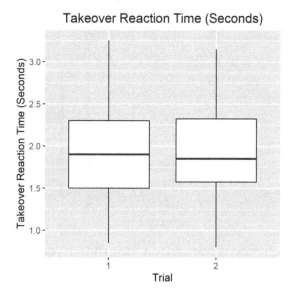

FIGURE 17.31 Variance in the time required for participants to react to the prompt to take control. Trial 1 is default settings, and trial 2 is customised settings.

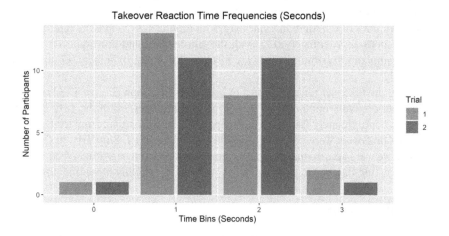

FIGURE 17.32 Distribution of takeover reaction times. Trial 1 is default settings, and trial 2 is customised settings.

into two groups: longitudinal (speed, throttle, longitudinal acceleration) and lateral (steering angle, steering speed, lateral acceleration). Within groups, they are clearly closely associated with each other. Within the longitudinal group, the data for mean throttle percentage application post-takeover showed no significant effect; however, it was interesting to note that more throttle was applied at the moment of takeover when using the customised interface (CTS). This might be suggestive that participants exhibited a higher level of readiness whereby they had already anticipated that throttle would be applied at the point of takeover and pre-applied it. This has multiple potential benefits with regard to safety; by reducing a period where too little throttle is applied immediately post-takeover, any resultant deceleration is decreased. This is particularly relevant for electric vehicles such as the Jaguar I-PACE that feature regenerative braking, as dependent upon a setting, zero throttle can result in a high level of deceleration (Jaguar Land Rover 2020). In a highway setting, these decelerations are exacerbated by drag, which increases with the square of speed (NASA 2015). Decelerations on the highway can be problematic due to the required reactions by the following drivers. In certain traffic flow conditions, even small decelerations can result in congestion (Kerner 2002) and associated accidents (Kim and Zhang 2008). It is possible therefore that customisation could improve safety by improving driver readiness and avoiding unnecessary decelerations that could cause collisions. Examining the mean throttle values over time, it is also apparent that there is less variation when using the customised settings (CTS). Less variation in throttle will result in fewer decelerations again, potentially improving safety with regard to reducing disturbances to traffic flow (Kerner 2002; Kim and Zhang 2008) and responses required from following drivers. Quddus (2013) states that accident rates are increased by 0.3% following a 1% increase in speed variation. Therefore, by minimising small decelerations on takeover, traffic flow speed variations are decreased and safety increased. There was a significant effect of customisation on mean speed; speeds in the trial using the customised interfaces were higher; the speed/time graph shows that 7.5 s post-takeover, participants using the customised interfaces continued to accelerate smoothly. However, the speed limit was not exceeded, and the rate of acceleration was not above what might be expected on a motorway. This data suggests that customising an interface appears to have no detrimental effects on performance in terms of safety. A significant effect of customisation on longitudinal acceleration was evident; acceleration levels were consistently higher in the trial featuring the customised interface. Longitudinal acceleration is closely coupled with throttle input, especially in electric vehicles where maximum torque is available from zero rpm. Therefore, the rationale for this finding is also associated with a higher level of readiness. Initial decelerations are visible for both trials immediately post-takeover; however, participants using the CTS had less initial deceleration for a shorter duration, which indicates a potential safety benefit as previously discussed with regard to reducing congestion (Kerner 2002) and accident rates (Quddus 2013). Where a higher level of readiness is exhibited, it might perhaps be due to the information being presented to participants in a perceivably less cluttered way. Baldassi, Megna, and Burr (2006) examined the effect of visual clutter in a perceptual decision-based experiment. They found that visual clutter can cause participants to make erroneous decisions with high confidence levels. Relating this to the context of TOR design, this is an important finding as TOR information can be considered safety-critical. Customised interfaces may therefore be able to minimise this issue as users can remove items that they do not consider useful. The effect of this on post-takeover performance is difficult to predict; it might be reasonably expected that a higher confidence level should relate to a higher level of readiness and a smoother takeover.

In addition to this, participants may be used to a particular vehicle's information layout, and by broadly replicating it, they may feel more comfortable and therefore more prepared.

Within the lateral data group, mean steering angle data exhibited no significant effect of customisation. The angles recorded in both trials due to the nature of motorway driving were minimal (peak of 2.6°). This is approximately 25% of the steering input that might be expected for a non-emergency lane change (Lee et al. 2011). Miura (2014) states that large steering inputs can result in lane excursions; however, the recorded inputs are low and therefore are likely to represent an adjustment within lane. Examining the plot of steering angle against time, it is apparent, however, that there was a greater fluctuation in steering angles with the default interface; this could indicate a lack of readiness, resulting in the initial large angle input which then required correction. Green (1995) citing Zwahlen, Adams, and DeBals (1988) and Noy (1989, 1990) outlines how lateral deviations can occur due to the attentional demand of an HMI, but also due to fatigue. The default HMI might therefore be considered to be more demanding, e.g. if the participant is searching for information, or if they consider they are lacking some information at the point of taking control. It could also be frustrating the participant and inducing fatigue if the participant wants to take control, but the protocol is taking longer than they would like. Conversely, if a participant considers that they were not prompted by sufficient questions, they may not feel that they have achieved a sufficient level of readiness and may be unprepared. In both cases, it is plausible that this could potentially have contributed to a less smooth takeover in terms of steering and throttle. Participants were also not made specifically aware of the number of protocol questions in the default profile. In this case, participants may not have been able to anticipate when they were nearing the end of the protocol and would be requested to take manual control. With the customised number of questions, they would likely be more able to anticipate how close they were to resuming control. The fluctuations could potentially be viewed as a mild form of pilot-induced oscillation (Mitchell and Klyde 2012); however, the frequency is low (approximately 0.2 Hz), and the magnitude is also low. Low-amplitude fluctuations occurring with high frequency are described by Miura (2014) as unlikely to cause large changes in lane position. The recorded fluctuations occurred with low frequency, but experimenters within the car witnessed no lane excursions in any trials. Miura (2014) describes how minor fluctuations in steering inputs can be a result of wind shear and imperfections in the road surface – these and other variables would require driver input to maintain the vehicle in a straight line. However, crucially, participants using their customised interfaces appeared to initially maintain a fixed steering angle, which is suggestive that they were confident of their trajectory, potentially due to effectively receiving the information that they required to attain readiness. It may also be the case that using a customised interface removed any potential startle effect of the TOR as the amplitudes and levels of haptics, and visual audio alerts, were reduced to what were considered to be comfortable levels, thus removing any involuntary muscular inputs (Biondi et al. 2014; Fagerlönn 2011; Väänänen-Vainio-Mattila et al. 2014). This is a considerable benefit, particularly taking into account that sensitivity to startle is both age and gender dependent (Kofler et al., 2001a, 2001b), and the potentially serious side effects of startle.

Mean steering angle speed showed no significant effect of customisation; however, there was a higher mean rate recorded in the first 5 s when using the default interface settings. This indicates that participants were making faster inputs soon after taking manual control. One possible rationale for this is a lack of readiness; participants who are aware of their surroundings both internally and externally might be considered more likely to achieve a smoother transition to manual control. Additionally, this could be a result of the customised interface being more predictable in duration and reducing startle, as participants would exhibit less involuntary inputs immediately post-takeover (Biondi et al. 2014). Mean lateral acceleration values showed a significant effect of customisation ($p = 0.0212$); however, they were generally very low-level values with a peak of approximately $0.17 \, \text{m/s}^2$.

Protocol TOTs appeared to reduce slightly with customisation; this may partly be an effect of participants selecting less protocol questions. It does, however, represent a positive finding as participants are likely to have reduced the number of questions to their preference, thus reducing the length of the TOR as recommended by McDonald et al. (2019). Protocol reaction times appeared to be largely unaffected by customisation; this also represents a positive outcome as it suggests that there is no negative effect associated with customisation.

In terms of limitations, the nature of the study, whereby participants experienced a default interface in the first trial, followed by their customised interface in the second trial, may have introduced some learning effects. However, all participants experienced multiple laps and TORs on a test track using the default interface, followed by a manual drive and a short TOR trial on road prior to trial 1. This will have enabled them to acclimatise to the vehicle and the TOR interface. If a learning effect had occurred, it might be presumed that greater prior experience with the default settings would result in superior performance post-takeover to the custom interface; however, this was not evident. Although there were multiple uncontrolled variables within this experiment, they was a side effect of running a study in the public domain. The advantages of gathering highly ecologically valid, real-world data were considered to be an acceptable compromise. The length of the experiment route necessitated running the first trial in one direction and the second in the opposite direction. While the same motorway was used, there were some variations, such as the order in which corners were encountered. The locations at which participants resumed control were also entirely dependent upon the rate at which they completed the takeover protocols. Customised interfaces allowed the number of protocol questions to be changed, thus affecting the TOT and the position on the motorway at which this occurred. Both light and weather represented uncontrollable variables; the trials were not run in conditions that were deemed to represent any risk; however, in some circumstances, in the first trials (Southbound), bright, low sunlight caused the automation to fail occasionally. In these scenarios, the participant took immediate manual control. However, post-takeover data was only utilised after a fully completed WOZ initiated TOR. Traffic levels and public vehicle behaviour also represented uncontrollable variables; however, traffic levels were relatively consistent due to the consistent timings of the trials; the behaviour of other drivers was also consistent with what might be expected.

Ultimately, the findings suggest that customisation is beneficial in terms of usability and has no perceivable detrimental effects on post-takeover performance. This represents an important finding as the benefits of customisation can be substantial, particularly in terms of subjective considerations (Normark 2015). In the context a TOR HMI, within the motor industry much effort is put into optimising subjective aspects and many options exist in selecting colours, materials, and technologies within vehicles. However, dashboard displays generally only exhibit a degree of personalisation. Customisation also provides more inclusivity – the ability to adjust individual aspects could be considered advantageous in a range of scenarios. For example, for those with hearing impairment, TOR volume alone may be increased to make it more audible and lessen the cognitive workload associated with interpreting a quiet vocalisation (Martini et al. 2014). Hickson et al. (2010) discovered that hearing impairment in older drivers can detrimentally affect driving performance, as indicated by a range of metrics. However, when controlling for age, Herbert et al. (2016) found no effect of hearing loss on driving performance in terms of lateral deviation and gaze dispersal. This suggests that it is possible that it is primarily the elderly with hearing impairment who will benefit from the ability to customise interface volumes and thus improve their driving performance. Equally, haptics levels, if too high, could potentially have a detrimental effect if it surprises the participant (Väänänen-Vainio-Mattila et al. 2014). These examples highlight the potential advantages of customisation over personalisation. Customisation enables the user to match the HMI to their task and to their individual characteristics, thus attending to two important aspects of usability (Stanton and Baber 1992).

17.6 CONCLUSIONS

It appears that customisation allows users to increase the usability of their TOR interface. This may be positively influencing their readiness and confidence levels as indicated by the statistically significant increases in speed, and longitudinal and lateral acceleration. In terms of safety, these increases remain within expected and small bounds; the increase in longitudinal acceleration may be considered a positive outcome in terms of congestion reduction. Where results were insignificant, it conveys the message that customisation has no negative effects on performance. Therefore, it can be concluded that customisation provides the ability to increase usability with no obvious safety impacts in terms of post-takeover performance. There does exist the possibility that participants might consider that they are capable of gaining sufficient situation awareness with a sparse TOR HMI and customise their interface to reduce questions, visual elements, and modality magnitudes to a minimum. However, uni-modality should be avoided; therefore, customisations should have imposed minimum settings. Further work might examine the possibility of dynamic customisation where an adaptive gaze-based system identifies frequently utilised interface elements and prioritises their display while hiding non-essential elements that are not utilised.

ACKNOWLEDGEMENTS

This work was supported by Jaguar Land Rover and the UK-EPSRC Grant EP/N011899/1 as part of the jointly funded 'Towards Autonomy: Smart and Connected Control (TASCC) Programme' HI:DAVe Project.

REFERENCES

AgeLab. 2019. *MIT AgeLab Study Tools*. Accessed February 17, 2020. https://agelab.mit. edu/study-tools.

Arduino. 2020. Accessed February 25, 2020. https://www.arduino.cc/.

Babich, N. 2017. *The Difference between Customization and Personalization*. 18 April. Accessed May 4, 2020. https://uxplanet.org/the-difference-between-customization-and-personalization-624ddd70b163.

Baldassi, S., N. Megna, and D. C. Burr. 2006. "Visual clutter causes high-magnitude errors." *PLoS Biology, 4, 3* e56. doi: 10.1371/journal.pbio.0040056.

Billington, J. 2019. *UK Government Pledges to Get Driverless Cars on the Road by 2021*. 8 February. Accessed April 30, 2020. https://www.autonomousvehicleinternational.com/news/legislation/uk-government-pledges-to-get-driverless-cars-on-the-road-by-2021. html.

Biondi, F., R. Rossi, M. Gastaldi, and C. Mulatti. 2014. "Beeping ADAS: reflexive effect on drivers' behavior." *Transportation Research Part F: Traffic Psychology and Behaviour, 25, Part A* 27–33. https://doi.org/10.1016/j.trf.2014.04.020.

Bysameyee. 2020. *Head-Up Display GPS Navigation*. Accessed February 17, 2020. https://www.amazon.co.uk/Head-up-Display-GPS-Navigation-Reflective/dp/B075T2DH12/ref=sr_1_10?keywords=car+head+up+display&qid=1573804882&sr=8-10.

Clark, H., and J. Feng. 2017. "Age differences in the takeover of vehicle control and engagement in non-driving-related activities in simulated driving with conditional automation." *Accident Analysis & Prevention, 106* 468–479. https://doi.org/10.1016/j. aap.2016.08.027.

Connective Peripherals. 2020. *USB2-F-7101*. Accessed February 25, 2020. https://www.connectiveperipherals.com

Department for Transport. 2015. *The Pathway To Driverless Cars: A Code Of Practice For Testing*. Accessed April 30, 2020. https://assets.publishing.service.gov.uk/government/uploads/system/uploads/attachment_data/file/446316/pathway-driverless-cars.pdf.

Driver, J., and C. Spence. 1998. "Cross-modal links in spatial attention." *Philosophical Transactions of the Royal Society of London. Series B: Biological Sciences, 353, 1373* 1319–1331. doi: 10.1098/rstb.1998.0286.

dSPACE. 2020. *MicroAutoBox II*. Accessed February 17, 2020. https://www.dspace.com/en/ltd/home/products/hw/micautob/microautobox2.cfm.

Elfeland. 2020. *LED Strip Lights Kits*. Accessed February 17, 2020. https://www.amazon.co.uk/Elfeland-Waterproof-Controller-Self-Adhesive-Decoration/dp/B07H2DR5DY/ref=sr_1_6?keywords=led+rgb+strip&qid=1559907897&s=gateway&sr=8-6.

Eriksson, A., and N. A. Stanton. 2017a. "Driving performance after self-regulated control transitions in highly automated vehicles." *Human Factors, 59, 8* 1233–1248. doi: 10.1177/0018720817728774.

Eriksson, A., and N. A. Stanton. 2017b. "Takeover time in highly automated vehicles: noncritical transitions to and from manual control." *Human Factors: The Journal of the Human Factors and Ergonomics Society, 59, 4* 689–705. https://doi.org/10.1177/0018720816685832.

Eriksson, A., S. M. Petermeijer, M. Zimmermann, J.C.F. de Winter, K. J. Bengler, and N. A. Stanton. 2019. "Rolling out the red (and green) carpet: supporting driver decision making in automation-to-manual transitions." *IEEE Transactions on Human-Machine Systems, 49, 1* 20–31. doi: 10.1109/THMS.2018.2883862.

Fagerlönn, J. 2010. "Distracting effects of auditory warnings on experienced drivers." *The 16th International Conference on Auditory Display*. Washington DC: ICAD 2010. 127–132.

Fagerlönn, J. 2011. "Urgent alarms in trucks: effects on annoyance and subsequent driving performance." *IET Intelligent Transport Systems, 5, 4* 252–258. doi: 10.1049/iet-its.2010.0165.

Findley, D. J., B. J. Schroeder, C. M. Cunningham, and T. H. Brown. 2015. *Highway Engineering: Planning, Design, and Operations.* Oxford, UK: Butterworth-Heinemann.

Geitner, C., F. Biondi, L. Skrypchuk, P. Jennings, and S. Birrell. 2019. "The comparison of auditory, tactile, and multimodal warnings for the effective communication of unexpected events during an automated driving scenario." *Transportation Research Part F: Traffic Psychology and Behaviour, 65* 23–33. https://doi.org/10.1016/j.trf.2019.06.011.

Gibson, Z., J. Butterfield, and A. Marzano. 2016. "User-centered design criteria in next generation vehicle consoles." *Procedia CIRP, 55* 260–265. https://doi.org/10.1016/j.procir.2016.07.024.

Gold, C., D. Damböck, L. Lorenz, and K. Bengler. 2013. ""Take over!" How long does it take to get the driver back into the loop?" *Proceedings of the Human Factors and Ergonomics Society Annual Meeting, 57, 1* 1938–1942. https://doi.org/10.1177/1541931213571433.

Gold, C., R. Happee, and K. Bengler. 2018. "Modeling take-over performance in level 3 conditionally automated vehicles." *Accident Analysis & Prevention, 116* 3–13. doi: 10.1016/j.aap.2017.11.009.

Green, P. 1995. *Measures and Methods Used to Assess the Safety and Usability of Driver Information Systems.* Technical Report FHWA-RD-94-088, McLean: Federal Highway Administration.

Grillon, C. 2002. "Startle reactivity and anxiety disorders: aversive conditioning, context, and neurobiology." *Biological Psychiatry, 52, 10* 958–975. doi: 10.1016/s0006-3223(02)01665-7.

Herbert, N., N. Thyer, S. Isherwood, and N. Merat. 2016. "The effect of a simulated hearing loss on performance of an auditory memory task in driving." *Transportation Research Part F: Traffic Psychology and Behaviour, 43* 122–130. https://doi.org/10.1016/j.trf.2016.10.011.

Hickson, L., J. Wood, A. Chaparro, P. Lacherez, and R. Marszalek. 2010. "Hearing impairment affects older people's ability to drive in the presence of distracters." *Journal of the American Geriatrics Society, 58, 6* 1097–1103. doi: 10.1111/j.1532-5415.2010.02880.x.

In-CarPC. 2020. *Vehicle and Car Computer Systems | In-CarPC.* Accessed February 17, 2020. https://www.in-carpc.co.uk/.

iYeal. 2020. *24mm LED Illuminated Push Button.* Accessed February 17, 2020. https://www.amazon.co.uk/iYeal-Illuminated-Buttons-Joystick-Raspberry/dp/B081D4W33K/ref=sr_1_151?keywords=green+led+push+button&qid=1574084349&sr=8-151.

JaguarLandRover. 2020. *2019I-Pace Regenerative Braking.* Accessed May 7, 2020. https://www.ownerinfo.jaguar.com/document/4K/2019/T32090/29962_en_GBR/proc/G2174091.

Kang, C. 2016. *Self-Driving Cars Gain Powerful Ally: The Government.* 19 September. Accessed April 30, 2020. https://www.nytimes.com/2016/09/20/technology/self-driving-cars-guidelines.html.

Kerner, B. S. 2002. "Synchronized flow as a new traffic phase and related problems for traffic flow modelling." *Mathematical and Computer Modelling, 35, 5* 481–508. https://doi.org/10.1016/S0895-7177(02)80017-6.

Kim, T., and H. M. Zhang. 2008. "A stochastic wave propagation model." *Transportation Research Part B: Methodological, 42, 7* 619–634. https://doi.org/10.1016/j.trb.2007.12.002.

Kofler, M., J. Müller, L. Reggiani, and J. Valls-Solé. 2001a. "Influence of age on auditory startle responses in humans." *Neuroscience Letters, 307, 2* 65–68. doi: 10.1016/s0304-3940(01)01908-5.

Kofler, M., J. Müller, L. Reggiani, and J. Valls-Solé. 2001b. "Influence of gender on auditory startle responses." *Brain Research, 921, 1* 206–210. https://doi.org/10.1016/S0006-8993(01)03120-1.

KT&C. 2020. *KNC-HDi47*. Accessed February 25, 2020. http://www.ktncusa.com/ip/ip_miniature/knc-hdi47.

Lee, J. D., D. V. McGehee, T. L. Brown, and D. Marshall. 2006. "Effects of adaptive cruise control and alert modality on driver performance." *Transportation Research Record: Journal of the Transportation Research Board, 1980, 1* 49–56. https://doi.org/10.1177/0361198106198000108.

Lee, T., B. Kim, K. Yi, and C. Jeong. 2011. "Development of lane change driver model for closed-loop simulation of the active safety system." *2011 14th International IEEE Conference on Intelligent Transportation Systems*. Washington DC: IEEE. 56–61. doi: 10.1109/ITSC.2011.6083039.

Leggett & Platt Automotive. 2020. *Vehicle Seating and Lumbar Support*. Accessed May 7, 2020. https://leggett-automotive.com//products.

Louw, T., G. Markkula, E. Boer, R. Madigan, O. Carsten, and N. Merat. 2017. "Coming back into the loop: drivers' perceptual-motor performance in critical events after automated driving." *Accident Analysis & Prevention, 108* 9–18. https://doi.org/10.1016/j.aap.2017.08.011.

Martini, A., A. Castiglione, R. Bovo, A. Vallesi, and C. Gabelli. 2014. "Aging, cognitive load, dementia and hearing loss." *Audiology and Neurotology, 19, 1* 2–5. doi: 10.1159/000371593.

McDonald, A. D., H. Alambeigi, J. Engström, G. Markkula, T. Vogelpohl, J. Dunne, and N. Yuma. 2019. "Toward computational simulations of behavior during automated driving takeovers: a review of the empirical and modeling literatures." *Human Factors: The Journal of the Human Factors and Ergonomics Society, 61, 4* 642–688. https://doi.org/10.1177/0018720819829572.

Melcher, V., S. Rauh, F. Diederichs, H. Widlroither, and W. Bauer. 2015. "Take-over requests for automated driving." *Procedia Manufacturing, 3* 2867–2873. https://doi.org/10.1016/j.promfg.2015.07.788.

Merat, N., and D. de Waard. 2014. "Human factors implications of vehicle automation: current understanding and future directions." *Transportation Research Part F: Traffic Psychology and Behaviour, 27, Part B* 193–195. https://doi.org/10.1016/j.trf.2014.11.002.

Mitchell, D. G., and D. H. Klyde. 2012. "Identifying a pilot-induced oscillation signature: new techniques applied to old problems." *Journal of Guidance, Control, and Dynamics, 31, 1*. https://doi.org/10.2514/1.31470.

Miura, S. 2014. "Direct adaptive steering – independent control of steering force and wheel angles to improve straight line stability." *5th International Munich Chassis Symposium 2014*. Wiesbaden: Springer. 75–89. https://doi.org/10.1007/978-3-658-05978-1_8.

NASA. 2015. *The Drag Equation*. Accessed May 7, 2020. https://www.grc.nasa.gov/www/k-12/airplane/drageq.html.

Normark, C. J. 2015. "Design and evaluation of a touch-based personalizable in-vehicle user interface." *International Journal of Human-Computer Interaction, 31, 11* 731–745. https://doi.org/10.1080/10447318.2015.1045240.

Noy, Y. I. 1989. "Intelligent route guidance: will the new horse be as good as the old?" *First Vehicle Navigation and Information Systems Conference*. Toronto: IEEE. 49–55. doi: 10.1109/VNIS.1989.98739.

Noy, Y. I. 1990. *Attention and Performance While Driving with Auxiliary In-vehicle Displays*. Downsview: Ontario Ministry of Transportation.

Petermeijer, S. M., P. Bazilinskyy, K. Bengler, and J. de Winter. 2017. "Take-over again: investigating multimodal and directional TORs to get the driver back into the loop." *Applied Ergonomics, 62* 204–215. https://doi.org/10.1016/j.apergo.2017.02.023.

Politis, I., S. Brewster, and F. Pollick. 2015. "Language-based multimodal displays for the handover of control in autonomous cars." *AutomotiveUI - Proceedings of the 7th International Conference on Automotive User Interfaces and Interactive Vehicular Applications*. Nottingham: Association for Computing Machinery. 3–10. https://doi.org/10.1145/2799250.2799262.

Politis, I., S. Brewster, and F. Pollick. 2017. "Using multimodal displays to signify critical handovers of control to distracted autonomous car drivers." *International Journal of Mobile Human-Computer Interaction, 9, 3* 1–16. https://doi.org/10.4018/ijmhci.2017070101.

Quddus, M. 2013. "Exploring the relationship between average speed, speed variation, and accident rates using spatial statistical models and GIS." *Journal of Transportation Safety & Security, 5, 1* 27–45. https://doi.org/10.1080/19439962.2012.705232.

SAE International. 2018. *SAE International Releases Updated Visual Chart for Its "Levels of Driving Automation" Standard for Self-Driving Vehicles.*. Accessed November 3, 2020. https://www.sae.org/standards/content/j3016_201806/.

SBG. 2020. *Fusion. SBG Sports Software.* Accessed February 25, 2020. https://sbgsportssoftware.com/product/fusion/.

Spence, C. 2002. "Multisensory attention and tactile information-processing." *Behavioural Brain Research, 135, 1* 57–64. doi: 10.1016/s0166-4328(02)00155-9.

Stanton, N. A., and C. Baber. 1992. "Usability and EC Directive 90270." *Displays, 13, 3* 151–160. https://doi.org/10.1016/0141-9382(92)90017-L.

Tractinsky, N., R. Abdu, J. Forlizzi, and T. Seder. 2011. "Towards personalisation of the driver environment: investigating responses to instrument cluster design." *International Journal of Vehicle Design, 55, 2* 208–236. https://doi.org/10.1504/IJVD.2011.040584.

United States Department of Transport. 2020. *Ensuring American Leadership in Automated Vehicle Technologies - Automated Vehicles 4.0.* January. Accessed April 30, 2020. https://www.transportation.gov/sites/dot.gov/files/2020-02/EnsuringAmericanLeadershipAVTech4.pdf.

Väänänen-Vainio-Mattila, K., J. Heikkinen, A. Farooq, G. Evreinov, E. Mäkinen, and R. Raisamo. 2014. "User experience and expectations of haptic feedback in in-car interaction." *Proceedings of the 13th International Conference on Mobile and Ubiquitous Multimedia.* Melbourne: Association for Computing Machinery. 248–251. https://doi.org/10.1145/2677972.2677996.

Vector. 2020. *Know-How & Solutions for CAN/CAN FD.* Accessed February 17, 2020. https://www.vector.com/us/en-us/know-how/technologies/networks/can/.

Walch, M., K. Lange, M. Baumann, and M. Weber. 2015. "Autonomous driving: investigating the feasibility of car-driver handover assistance." *AutomotiveUI - Proceedings of the 7th International Conference on Automotive User Interfaces and Interactive Vehicular Applications.* Nottingham: Association for Computing Machinery. 11–18. https://doi.org/10.1145/2799250.2799268.

Walker, G. H., N. A. Stanton, and M. S. Young. 2001. "Where is computing driving cars?" *International Journal of Human-Computer Interaction, 13, 2* 203–229. https://doi.org/10.1207/S15327590IJHC1302_7.

Yoon, S. H., Y. W. Kim, and Y. G. Ji. 2019. "The effects of takeover request modalities on highly automated car control transitions." *Accident Analysis & Prevention, 123* 150–158. https://doi.org/10.1016/j.aap.2018.11.018.

Zhang, B., J. de Winter, S. Varotto, R. Happee, and M. Martens. 2019. "Determinants of take-over time from automated driving: a meta-analysis of 129 studies." *Transportation Research Part F: Traffic Psychology and Behaviour, 64* 285–307. https://doi.org/10.1016/j.trf.2019.04.020.

Zwahlen, H. T., C. C. Adams, and D. P. DeBals. 1988. "Safety aspects of CRT touch panel controls in automobiles." *Vision in Vehicles II. Second International Conference on Vision in Vehicles.* Nottingham: Elsevier. 335–344.

18 Effects of Customisable HMI on Subjective Evaluation of Takeover Experience on the Road

Jisun Kim, Kirsten M. A. Revell,
James W.H. Brown, and Joy Richardson
University of Southampton

Nermin Caber and Michael Bradley
University of Cambridge

Patrick Langdon
Edinburgh Napier University

Simon Thompson and Lee Skrypchuk
Jaguar Land Rover

Neville A. Stanton
University of Southampton

CONTENTS

18.1 INTRODUCTION

Highly automated driving technology allows users to stop focusing on driving physically and mentally, and to engage in non-driving-related tasks when conditions are met (SAE International 2018). This means they can take their hands and feet off the wheel and the pedals, and do other tasks, e.g. reading, working, or resting (Bazilinskyy et al. 2018). In this circumstance, drivers can be taken out of the loop due to a lack of interaction with the vehicle and monitoring (Lu, Coster and de Winter 2017). Nevertheless, they are required to drive when takeover requests (TORs) are issued when the automation reaches its limits (SAE International 2018). Therefore, reconstruction of a sufficient level of situation awareness is one of the most significant factors of timely and safe takeover (Lu et al. 2016; Gold, Happee and Bengler 2018).

Relevant studies regarding drivers' takeover have shown that the performance and their subjective experience can exhibit differences among individuals (Eriksson and Stanton 2017; Li et al. 2019; Walch et al. 2016). This is not surprising as the process of situation awareness development, and decision-making, could be influenced by individual factors, such as long-term memory stores, information processing mechanism, experience, and training (Endsley 1995).

In this sense, being able to customise the interface could be a means to help drivers create the automated driving environment to better suit their needs. It is because human–machine interface (HMI) in autonomous vehicles plays a significant role as a communication channel between the driver and the vehicle. It guides the driver's attention, provides appropriate feedback and feedforward about system status, and helps elicit the driver's rapid and effective interventions (van den Beukel, van der Voort and Eger 2016). Given this, this study posits that a customised interface can lead to an enhanced takeover experience.

Drivers' subjective experience is of importance to predict potential usage of autonomous vehicles (Choi and Ji 2015; Xu et al. 2018). It is widely accepted that the actual use of technology is explained by usefulness and satisfaction of the experience

(Davis 1989; Venkatesh et al. 2003; Xu et al. 2018). Workload could negatively impact drivers' willingness to accept the technology (Nees 2016). In addition, trust has also been identified as an antecedent of the acceptance (Panagiotopoulos and Dimitrakopoulos 2018). Therefore, how drivers perceive the automated driving and takeover experience is worthy of investigation to promote the use of the technology. As described previously, customisation could facilitate the use across a wider range of users – by helping ease difficulties, complexity, discomfort, and mental barriers – and strengthen confidence and effectiveness.

Workload is defined as 'the cost of accomplishing mission requirements for the human operator' (Hart 2006). It is the amount of effort incurred by the operator to interact with the system (Moller et al. 2009), and to achieve a certain level or performance. Thus, it might differ from person to person as task parameters are perceived differently (Hart and Staveland 1988). A variety of aspects could influence interaction that imposes workload, such as the operator's behaviour, and skills, the situation in which a task is performed, and the requirements (Hart and Staveland 1988). Autonomous driving could lead to an increase or a decrease in workload in relation to manual driving. The increase could result from a constant monitoring and vigilance (Stanton, Dunoyer and Leatherland 2011; Banks and Stanton 2016), and the decrease could be attributed to reduction in mental activity related to manual driving, and physical activities such as operating the pedals and steering wheel (de Winter et al. 2014).

Trust is defined as 'willingness of a party to be vulnerable to the actions of another party based on the expectation that the other will perform a particular action important to the trustor, irrespective of the ability to monitor and control that other party' (Mayer, Davis and Schoorman 1995). It is considered as a multidimensional construct embracing ability (trustee's skill and knowledge), benevolence (trustee's specific attachment to the trustor), and integrity (trustee's adherence to a set of acceptable principles) (Mayer, Davis and Schoorman 1995). In the domain of automation, dimensions of trust are identified as performance, process, purpose, and foundation, which enables other aspects of trust to be established. Performance is based on consistency, stability, and desirability of performance of automation. Process represents users' knowledge of the underlying qualities, or algorithms governing behaviour of the system. Purpose reflects on the producers' intention in developing the system (Lee and Moray 1992). Trust is a significant factor for drivers of autonomous vehicles as its usage embraces uncertain and vulnerable situations when the driver's well-being is entrusted to the system (Körber, Baseler and Bengler 2018). Further, interpersonal relationships are mediated by trust, and those between human automation may also be mediated by it (Choi and Ji 2015). Trust could have an influence on drivers' strategies about using the automation (Lee and Moray 1994). Moreover, it is a key antecedent of adoption and use of autonomous vehicles (Choi and Ji 2015; Kaur and Rampersad 2018).

Usability is one of the key criteria to assess driver–vehicle interaction (Forster et al. 2018). Usability is defined as 'the extent to which a product can be used by specified users to achieve specified goals with effectiveness, efficiency and satisfaction in a specific context of use' (ISO 1998). It is further clarified that it represents an attribute of the 'interaction between a product, a user and the task, or set of tasks, that he or she is trying to complete'. Hence, usability can be affected by various properties of the user that affect usability. Consequently, it is vital to understand about the user, and his or her characteristics (Jordan 1998).

Acceptance of autonomous vehicle technology accounts for how willing individuals are to accept, use, and intend to use the technology (Liu et al. 2019). According to the technology acceptance model, intention to use technology could be determined by perceived usefulness and ease of use (Davis, Bagozzi and Warshaw 1989; Choi and Ji 2015). Perceived usefulness is explained as a belief that a user has toward the system that using it would improve their job performance. Perceived ease of use is represented by the perception of a user that using it would require little effort (Davis 1989; Choi and Ji 2015). Acceptance has importance for the success of implementation of intelligent transport systems (Brookhuis et al. 2009).

Driver subjective experience in takeover situation has been explored in related studies. Yoon, Kim, and Ji's (2019) study results demonstrated that a higher level of usefulness, satisfaction, and trust were induced by TORs with multiple modal alerts than those with unimodal (visual only) alerts. Lu et al. (2019) presented that reduced workload and enhanced acceptance were recorded in the situation which issued a TOR preceded by a monitoring request (MR: which encouraged the drivers to monitor the environment before takeover). The situation resulted in better subjective experience than the TOR-only situation (MR absence). The reason could be that the MR might have helped improve the drivers' situation awareness along with cognitive and mental preparedness. Eriksson et al. (2019) explored the effects of HMI with conditions with different visual effects, including augmented representation during takeover situation. The participants found visual overlays highlighting areas in the lane (or lane markings) significantly more satisfying than indicating a slow-moving vehicle. Furthermore, they rated lower workload in augmented conditions than in the condition with no visual effects. Li et al. (2019) measured the influences of TOR with or without information about the reasons to takeover, and the vehicle status. The findings suggested that the condition with the information produced better attitudes, lower workload, and better performance. The results were interpreted that the participants' increased situation awareness led to better outcomes.

As described above, TORs preferably with multiple modalities, and with additional information about the reasons, seem to help enhance drivers' perceptions. One of the reasons is multimodal, and informative conditions for TOR could enhance drivers' situation awareness and preparedness. However, the development of situation awareness and the decision-making process of drivers could vary from driver to driver (Endsley 1995). Thus, in this study, customisation in the multimodal HMI is proposed as a method for further improving drivers' subjective perceptions and narrowing the differences across drivers. Accordingly, this study aims to understand drivers' takeover experience in customisable multimodal interface in highly automated driving situation on road.

18.2 METHOD

18.2.1 PARTICIPANTS

A total of 24 participants were recruited from a pool that had previously experienced a simulator-based study featuring an almost identical HMI. All were required to be

TABLE 18.1

Participant Demographics

No. of Parts	Age Bracket	F	M
7	28–34 (Mean 29.3) (SD 2.2)	3	4
10	35–56 (Mean 42.6) (SD 5.5)	3	7
7	57–82 (Mean 63.4) (SD 4.5)	3	4

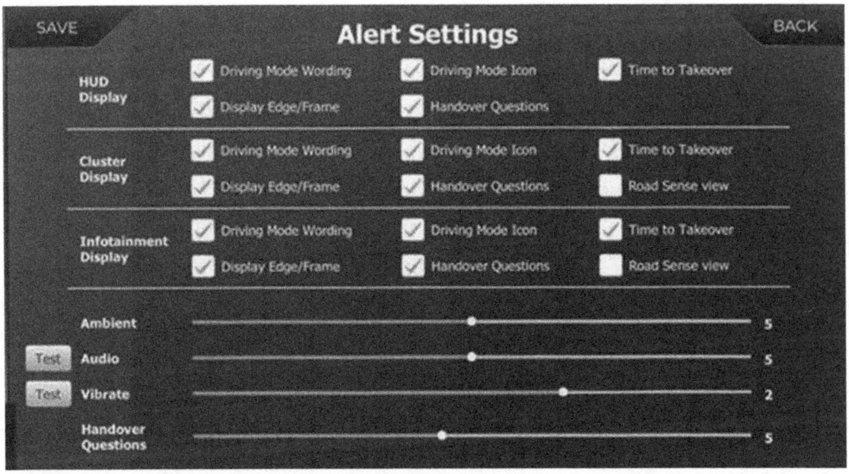

FIGURE 18.1 The customisation matrix displayed at the end of each trial on the vehicle's infotainment display.

a minimum of 28 years old and hold full UK driving licenses. The gender split was 15 males/9 females, across three age ranges: 28–34, 35–56, and 57–82 (Table 18.1). Participants had to provide a signed consent prior to involvement in the study, which was approved by the University of Southampton's Ethics and Research Governance Office (ERGO Number: 49792.A2).

18.2.2 EXPERIMENTAL DESIGN

A repeated-measures design was employed for the experiment. Participants experienced two differing customisations of vehicle HMI, the HMI customisation representing the independent variable. Two trials took place: trial 1 using default HMI settings and trial 2 using participant selected settings. A final customisation of the HMI took place after the completion of trial 2. Customisations were carried out using a matrix presented to the participant on the experimental vehicle's central console screen. The graphical elements of the HMI were displayed on a head-up display (HUD), the vehicles' cluster, and infotainment display. Additional elements of the HMI included a haptic seat, ambient lighting, vocalisations, and audio cues.

TABLE 18.2

HMI Customisable Elements and Parameters

HUD Display	Cluster Display	Infotainment Display	Ambient Lighting	Audio Volume	Haptic Amplitude	Takeover Questions
Driving mode wording (ON/OFF)	Driving mode wording (ON/OFF)	Driving mode wording (ON/OFF)	(1–11) 1 calibrated to be visible at all times	(1–11) 1 calibrated to be audible at all times	0–3	1–10
Driving mode icon (ON/OFF)	Driving mode icon (ON/OFF)	Driving mode icon (ON/OFF)	-	-	-	-
Time to takeover (ON/OFF)	Time to takeover (ON/OFF)	Time to Takeover (ON/OFF)	-	-	-	-
Display edge frame (ON/OFF)	Display edge frame (ON/OFF)	Display edge frame (ON/OFF)	-	-	-	-
Takeover questions (ON/OFF)	Takeover questions (ON/OFF)	Takeover questions (ON/OFF)	-	-	-	-
-	Road sense view (ON/OFF)	Road sense view (ON/OFF)	-	-	-	-

FIGURE 18.2 The three visual aspects of the HMI: on the left is the central console, bottom-right is the cluster, and the HUD is shown on the top-right.

The customisable elements are listed in Table 18.2. The customisation matrix and the drivers' HMI are illustrated in Figures 18.1 and 18.2.

The driving mode wording is situated below the driving mode icon, which is centrally placed in the infotainment display and cluster and placed on the left side in the HUD. The colour acts as a mode indicator – blue indicating automated and orange for manual. Time to takeover is shown in the bottom-left corner of the cluster and infotainment display, and the top-right of the HUD. The edge frame mode indicator is only visible on the cluster and infotainment display. Takeover questions and associated icons appear in the centre of each of the HMI display elements.

Dependent variables in terms of performance data included a range of variables taken from the vehicle's controller area network (CAN), and outputs from the HMI system but these will not be discussed in this paper.

Dependent variables in the form of questionnaire survey data comprised work-load, usability, acceptance, and trust scores, which were collected to answer the fol-lowing research questions:

- How do participants assess takeover experience in terms of workload, usability, acceptance, and trust before and after customising the interface?
- Will subjective evaluations of takeover experience vary across user groups?

Workload, usability, acceptance, and trust were measured by standard scales as fol-lows. The questionnaires were administered soon after each trial finished in the vehicle. The NASA-Task Load Index (NASA-TLX) was selected to assess perceived workload of takeover experience (Hart and Staveland 1988) since it is one of the most commonly used tools to assess subjective workload in experimental conditions (Ko and Ji 2018). It evaluates six independent constructs of workload, based on an assumption that it is represented by different aspects of experience while perform-ing tasks. It consists of the following six items: mental demand, physical demand, temporal demand, effort, frustration, and performance (Hart and Staveland 1988). For analysis, unweighted mean scores (raw TLX) were used since raw scores are simple to employ, and they have been used in a wide range of studies. Either weighted or unweighted scores can be used as unweighted ones were more, equally, or less sensitive to the weighted scores (Hart 2006). The system usability scale (SUS) was adopted to measure participants' subjective usability regarding the interface for takeover experience. It covers various aspects of usability which include complexity, training, and the need for support, which show a high degree of face validity. It adopts a Likert scale consisting of 10 items regarding subjective usability and learnability. Overall scores were calculated according to the scale scoring method that allows for the generation of a single figure between 0 and 100, representing a composite score (Brooke 1996). The system acceptance scale (SAS) developed by Van Der Laan, Heino, and De Waard (1997) was chosen to measure acceptance. It consists of two subscales that evaluate usefulness (five items) and satisfaction (four items) of accep-tance of new technology. The scale reacted sensitively to differences in view between groups of drivers. Before- and after-measurement scores can be compared to identify how respondents' experience with the system resulted in changes in their opinion (Van Der Laan, Heino and De Waard 1997). The scale of trust in automated systems was used to gauge drivers' trust (Jian, Bisantz and Drury, 1998). It is comprised of 12 unidimensional items whose scores can be averaged into a representative score indicating trust. An overall trust score was generated using the mean of the scores of the 12 items (Blömacher, Nöcker and Huff 2018).

18.2.3 EQUIPMENT

The experimental vehicle was a 2017 Jaguar I-PACE EV400 AWD pre-production model (Figure 18.3). The automation system consisted of a combination of factory

FIGURE 18.3   The test vehicle, a 2017 Jaguar I-PACE EV400 AWD.

FIGURE 18.4   View of in-car PC fitted in the boot of the I-PACE, the dSPACE system, Ethernet switch, USB hub, and switch panel for system initialisation.

standard lane keep assist and adaptive cruise control. When utilising these systems, the standard car would issue frequent warnings to maintain hands on the wheel; these warnings were specifically removed from the system in order for it to simulate Level 3 automation. The car was fitted with a PC (In-CarPC 2020) (Figure 18.4) running SBG software (SBG 2020) for data and video logging. A dSPACE MicroAutoBox II system was employed to interface with the vehicle's systems (dSPACE 2020), together with three CAN interfaces (Connective Peripherals 2020) and a Vector CAN box (Vector 2020). Power was supplied via an inverter, connected to the 12 V battery

FIGURE 18.5   The car in manual mode with the HUD and cluster clearly visible.

circuit. An Ethernet switch and USB hub provided communications between systems. The visual aspects of the HMI consisted of a 14" thin-film transistor (TFT) panel fitted to the centre console, a 10" TFT fitted in place of the original equipment manufacturer's cluster, and a HUD comprising a small 1000×250 px TFT and reflector screen (Bysameyee 2020) (Figures 18.5 and 18.6). The car's original cluster was moved into the passenger footwell to allow monitoring by the safety driver. The interactive element of the HMI constituted two illuminating green buttons (iYeal 2020) fitted in the thumb positions of the steering wheel. The button functionality was integrated using an Arduino (Arduino 2020). Ambient lighting was supplied via LED lighting strips (Elfeland 2020). Haptics were provided in the seat base via Leggett and Platt motors (Leggett & Platt Automotive 2020) controlled through an Arduino Micro and motor control board (Arduino 2020). Five cameras (KT&C 2020) were installed within the I-PACE to provide footage from the forward view, driver-facing, over-the-driver-shoulder, footwell, and rear views. A dashcam was also fitted in the safety car to collect footage of the participants' vehicle from an external viewpoint. A cognitive loading task (*n*-back) was controlled from a mobile phone using MIT's AgeLab delayed digit recall task app (AgeLab 2019), which was connected to a generic Bluetooth speaker. Additional speakers were also placed within the car and connected to the PC to relay the vocalised and audio cue aspects of the HMI. The HMI was controlled via a Microsoft Surface Pro tablet providing Wizard of Oz control of the system. Custom software was developed for the HMI using C++.

FIGURE 18.6 View from the over-the-shoulder camera.

Figure 18.4 pictures the in-car PC fitted in the boot of the I-PACE, the dSPACE system, Ethernet switch, USB hub, and switch panel for system initialisation. Figure 18.6 shows the view from the over-the-shoulder camera, the green buttons on the steering wheel are visible, as are the three visual elements of the HMI: the HUD, cluster, and centre display. The ambient blue lighting is also apparent, indicating the vehicle is in automated mode. The driver-facing camera can be seen fitted on the dashboard.

18.2.4 PROCEDURE

On arrival at the Jaguar Land Rover (JLR) facility at Fen End, the participant was welcomed and asked to sign in, presenting their driving license and receiving a visitor pass. They were then briefed on the aim of the project and how the on-road study follows on from the simulator study that they completed the previous year (see Chapter 13). A brief explanation of the sequence of events for the study was then given. This included the route (Figure 18.7), the takeover procedure, form filling on trial completion, customisation of the HMI, and the *n*-back cognitive load task. It was stressed that they were required to maintain the same level of attention as if they were driving and should be ready to take control at any time. They were also advised that if the cognitive load task was detrimentally affecting their ability to retain a safe level of attention, they should stop the *n*-back task. They were presented with a copy of the participant information sheet; this same sheet was made available to them when they were recruited. A reminder sheet showing screenshots of the HMI they would be using was provided, including modal variation of the cluster, HUD and other in-vehicle information, the green buttons, and the customisation matrix screen. They were shown a route map that highlighted where the trials would take place, but also areas of caution and one junction that required traversal in manual mode. A privacy policy was presented, and a sheet explaining the cognitive load n-back task was provided along with a brief verbal explanation. The participant was asked if they had any questions, and if they were happy to continue, they were provided with two consent forms: an attendance form and an events team form to sign. On completion of all forms, they were led from the reception area to the car park and asked to sit in the passenger side of the car; both experimenters took their places in the rear seats. The participant was introduced to a safety driver who explained the basic controls on the car, and the elements of the interface. The safety driver then drove the car through security to the proving ground and demonstrated the vehicle's performance before running through the transition to automated mode and back to manual control. The automation system was operated by one of the experimenters using a Wizard of Oz system, from a Windows tablet in the back of the car. To offer automation, an experimenter would press a start button on a custom control panel app running on the tablet; the HMI then indicated automation as being available via the three graphical interfaces and a vocal alert. To enable automation, the driver simultaneously pressed the two green buttons mounted on the steering wheel and released all of the controls, including the accelerator. The system would then engage automation, and the HMI would indicate that the automated mode was active. During automation, the cognitive load task was controlled by the other experimenter via a mobile telephone-based app linked to a Bluetooth speaker. The safety driver demonstrated the automation system multiple times including the n-back task, and requested the safety car overtake and brake in front of the vehicle, when in automated mode, to illustrate how it reacts to maintain a gap to the car in front. The safety driver then stopped in a safe area and swapped places with the participant. The participant then was allowed some time to drive the car on the proving ground to become familiar with the controls. The automation was then made available to them, and they experienced multiple handovers and takeovers, including the use of the n-back task while in automation.

Figure 18.7 shows the experimental route and indicates the location of JLR's Fen End facility. The 2-mile on-road familiarisation section is shown just prior to point A. The main trial route between Warwick Services Southbound (A) and Junction 8A Services M40 (B) was approximately 40 miles long and required just over 40 min to complete. Trial 1 was conducted Southbound from A to B, while trial 2 used the reversed route, Northbound from B to A.

The participant was asked to drive manually to the start point of the experiment at the Southbound Warwick services on the M40. Two miles prior to the services, while on the M40, a road-based practice handover was conducted. The automation was offered to the participant; once activated, a short 30-s period of automation followed which included the participant carrying out the n-back task. The participant then experienced the takeover protocol and resumed manual control before stopping at the services. After confirming that the participant was happy to continue, the on-board systems were checked and configured for the first trial, data logging was started, and the video and audio were synchronised using a clapper board.

The participant was instructed to drive from the services onto the M40 (Figure 18.7), proceeding in the left lane at approximately 58 mph. The following process is illustrated in Figure 18.8. After 1 min of manual driving, automation was offered to the participant. Once activated, after a further period of 30 s, the n-back

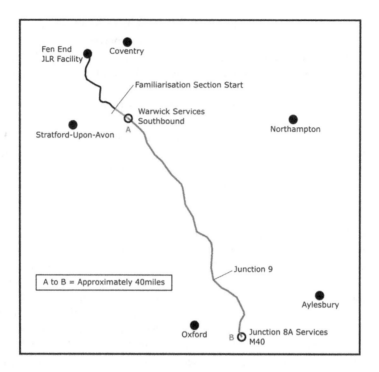

FIGURE 18.7   The experimental route showing the location of JLR's Fen End facility.

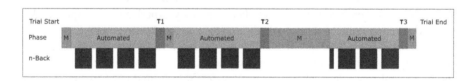

FIGURE 18.8   A timeline showing phases of one trial illustrating manual mode (M), auto-mated mode, and takeovers (T1/T2/T3). *N*-back application periods are shown by the grey bars.

cognitive loading task was started. This task was run in 2-min intervals, separated by 30 s. Following 10 min of automation, the HMI started the takeover protocol; once completed, the participant pressed both steering wheel buttons and resumed manual control. After 1 min of manual control, automation was again offered to the participant and the process repeated. On completion of the second takeover, the manual driving period was extended to approximately 7 min in order to pass a sec-tion of motorway (M40 J9 Southbound) which would have adversely affected the automation due to the lane becoming a slip road. Due to proximity to the end point of trial 1, takeover 3-s automation period was reduced to 8 min. The *n*-back task was started simultaneously with the automation for 30 s, before reverting to 30 s off and 2 min on until the protocol started. When the participant completed the proto-col and resumed manual control, they continued in manual mode the short distance remaining to the motorway services, where they parked the car. Data logging was stopped as this constituted the end point of trial 1. The participant was presented with a set of PDF-based forms on a laptop (NASA-TLX, SUS, SAS, Trust) and asked to complete them based upon their experience specifically of the takeover protocol, process, and HMI elements. When the forms were complete, the customisation screen was displayed on the vehicle's centre console and the participant was asked to make any changes that they would like to the HMI. During this customisation process, they were asked to vocalise their opinions, considerations, and rationales for customisations. This information was recorded for future analysis. This concluded the data collection for trial 1; the participant at this point had access to facilities and refreshments. Trial 2 was a repeat of the processes and data collection carried out in trial 1; however, the route was Northbound to the start point for trial 1. On completion of trial 2, one additional form was provided to collect subjective post-task data (post-trial interview). The safety driver swapped places with the participant from this point and drove the vehicle back to the JLR facility at Fen End. The participant was thanked for their time and signed out at reception.

18.2.5 SAMPLE AND DATA SCREENING

Questionnaire data collected from 23 participants was taken into analysis because one of the participants (participant 24) could not complete the second trial due to technical issues. The participants comprised 14 males and 9 females. They were aged between 28 and 75, classified into three age groups (younger: 28–34 ($N = 6$); middle: 35–56 ($N = 10$); and older: 57–82 ($N = 7$)). Missing responses were replaced with the median value. They were comprised of one missing response in trial 1 and two missing responses in trial 2.

18.2.6 DATA ANALYSIS

Descriptive and inferential data analysis techniques were applied. Box plots were created to identify general trends of data showing the range of scores, such as minimum, maximum, median presented with quartiles, and outliers (Field 2017; Mcgill, Tukey and Larsen 1978). In order to assess the differences between trials, the paired sample t-test or the Wilcoxon signed-rank test was used. The former was applied for the data which presented normal distribution of the differences between scores (Field 2017). The latter was employed to examine the data that did not meet the assumption since the test is a nonparametric technique that enables analysis of data with no assumption of normality (Woolson 2008; Field 2017). The Mann–Whitney U-test was applied to examine differences between genders. It was selected because it allows to test for differences between two independently sampled groups with no assumption of normality (Mcknight and Najab 2010). The Kruskal–Wallis test was conducted to compare differences among the three age groups. It was chosen because it enables the examination of differences between two or more independent groups of an ordinal or continuous independent variable as a nonparametric test (Corder and Foreman 2009).

For acceptance subscores (usefulness and satisfying), reliability of each score for each user group, and trial, was tested (Van Der Laan, Heino and De Waard 1997).

In general, Cronbach's alpha values for each case were rated above 0.65, which the measurement proposes as the sufficiently high level of reliability. However, the cases of (1) female group, trial 1 satisfying score; trial 2 usefulness score; (2) middle age group, trial 1 usefulness; trial 2 usefulness; (3) older age group, trial 1 usefulness, generated Cronbach's alpha values ranging between 0.498 and 0.633. The values could be accepted because the Cronbach's alpha values ranging between 0.45 and 0.89 are suggested as 'acceptable' (Taber 2018). However, even after considering the more generous criterion, there were two cases whose internal consistency was not satisfactory. Firstly, the Cronbach's alpha for the female group, trial 1 usefulness was not acceptable, and it seemed to result from item nine (sleep inducing – raising alertness). Once it was removed, the alpha value was marked 0.563. The inconsistency resulting from the response to the item might demonstrate that the participants' perception may have been affected by the status during the automation period. It was when longitudinal and lateral control was conducted by the vehicle without feeling the need to intervene, which was followed by takeover. This means that the responses' relatively lower ratings (closer to sleep inducing which was supposed to be negative reaction) did not seem to be consistent to responses to the other items, which were closer to the positive end in the usefulness subscale. Secondly, the Cronbach's alpha for the older age group, trial 1 satisfying, was below the acceptable range. It resulted from item eight (undesirable – desirable). One of the participants' score (−1) was lower than other participants, and this had an impact on the general interrelatedness. Nevertheless, both cases did not seem to affect the usefulness score considerably. Therefore, an overall score was generated as proposed by the guidelines (Van Der Laan, Heino and De Waard 1997).

18.3 RESULTS AND DISCUSSIONS

Findings from subjective judgement of takeover experience will be described. In order to explore the potential influence of customisation, comparisons between trial 1 and trial 2 (conducted before and after the participants customised the interface) will be made. Additionally, the effects will be investigated by examining the differences between demographic subgroups, such as gender and age.

18.3.1 COMPARISON BETWEEN TRIALS

Difference in perceived workload, usability, acceptance, and trust between trial 1 and trial 2 will be analysed to identify potential effects of customisation of the interface on participants' takeover experience. In general, workload was scored lower; usability, acceptance, and trust were higher in trial 2 than in trial 1. The differences in workload and trust between the trials showed statistical significance.

18.3.1.1 Workload

The data considered for workload can be viewed in Figure 18.9 and Table 18.3. Workload was lower in trial 2 compared to trial 1 considering maximum and minimum scores and the quartiles. The difference was statistically significant ($p = 0.002$, $t = 3.430$, paired samples t-test; skewness $= -0.258$, kurtosis $= 0.272$; $p = 0.200$,

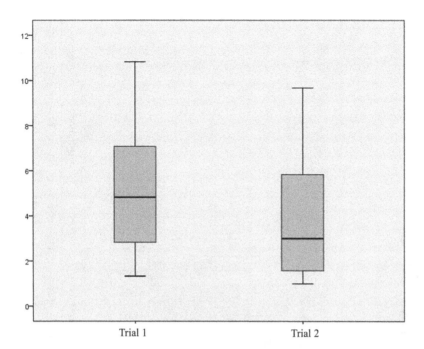

FIGURE 18.9 Perceived workload (unweighted average) in trial 1 and trial 2.

TABLE 18.3
Perceived Workload Scores between Trials

Trial	N	Mean	Std. Deviation	Minimum	Maximum	Wilcoxon Signed-Rank Test sig. (2-Tailed)
1	23	4.913	2.769	1.330	10.830	$Z = -2.893$
2	23	3.993	2.920	1.000	9.670	$P = 0.004$

Kolmogorov–Smirnov test of normality). The unweighted averages of workload for trials 1 and 2 were 24.56% and 19.96%, respectively. The scores were well below the 43.5% for manual driving and 38.6% for adaptive cruise control driving reported in De Winter et al.'s review. The workload for trial 2 was also rated lower than 22.7% for highly automated driving proposed in their study (de Winter et al. 2014). A 4.6% decrease in trial 2 was found in comparison to trial 1.

When looking into the individual components, reduction was observed in all items. Among them, decreases in temporal demand (5.43%) and effort (11.52%) were statistically significant ($p = 0.022$, $Z = -2.296$; $Z = -3.236$, $p = 0.001$, respectively). However, the amount of reduction in mental and physical workload was smaller than other components. This information is given in Table 18.4.

A decrease in workload in trial 2 could be interpreted that conducting takeovers in trial 2 was perceived less demanding than in trial 1. The customised interface

TABLE 18.4
Unweighted Subjective Workload of Each Element and the Averages for Trial 1 and Trial 2

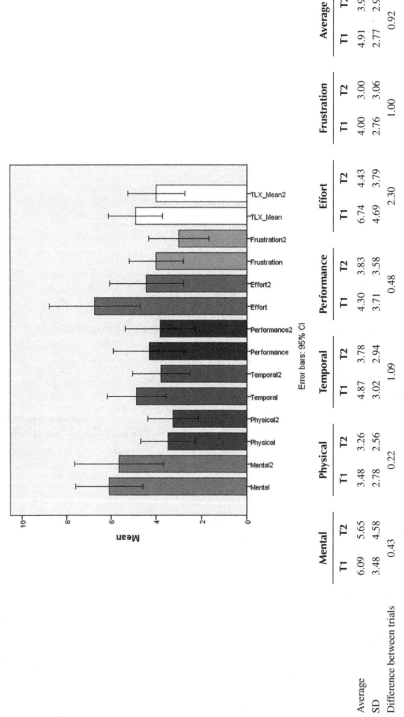

	Mental		Physical		Temporal		Performance		Effort		Frustration		Average	
	T1	T2	T1	T2	T1	T2	T1	T2	T1	T2	T1	T2	T1	T2
Average	6.09	5.65	3.48	3.26	4.87	3.78	4.30	3.83	6.74	4.43	4.00	3.00	4.91	3.99
SD	3.48	4.58	2.78	2.56	3.02	2.94	3.71	3.58	4.69	3.79	2.76	3.06	2.77	2.92
Difference between trials	0.43		0.22		1.09		0.48		2.30		1.00		0.92	

may have helped reduce the drivers' commitment of physical and mental resources to achieve the takeover tasks in trial 2 (Hart and Staveland 1988). This may have been achieved because the participants were able to find and retrieve the information needed from the display screens, or from vocal warnings. Therefore, they may have made less physical movement (e.g. to move their gaze) to find the information they needed or carefully look at the screens to identify the information they wanted. On the basis of statistical differences of individual elements between trials, the participants felt less time pressure during takeover in trial 2 than in trial 1. The statistically significant reduction in temporal demand might have resulted from the customised system helping them to gather necessary information quickly that could shorten the decision-making process, and spare more time for monitoring the environment to ensure safe takeover. Further, they did not feel that the burden of demand in trial 2 was as high as trial 1 as the effort score exhibited a statistically significant decrease in trial 2. The minor decrease in mental demand may be accounted for by monitoring and supervising roles that the drivers needed to fulfil which could not be replaced by automation (Stapel, Mullakkal-Babu and Happee 2019). The demand level could have not been decreased significantly as a result of the customised HMI because the experiment was conducted in actual traffic. In the environment, unpredictability and uncertainty in traffic are high, and the risk is perceived at a greater level than in a simulation environment (Eriksson, Banks and Stanton 2017). The small reduction in physical demand may be explained by the general characteristics of the takeover situation, which requires drivers to take steering and pedal control back. The actions were required regardless of the settings for the HMI.

18.3.1.2   Usability

Usability was higher in trial 2 in comparison to trial 1 considering maximum and minimum scores and the quartiles (seen in Figure 18.10 and Table 18.5). The mean scores were 86.52 in trial 2 and 81.96 in trial 1. Although the difference was not significant, positive changes were identified. The standard deviation for trial 2 was smaller than for trial 1, and the usability score represented an 'excellent' level of usability (85+) according to Brooke (2013). This shows that the HMI was perceived as highly usable in both trials and the customisation may help strengthen the positive perception.

The improvement in the score could be interpreted that the participants perceived the system as easier to use, and they felt that less support was required to complete the takeover. This could be understood that the HMI adjusted to the drivers' needs and preferences was found to be more usable to operate. To assess the usability, the driving environment needed to be incorporated, which required task switching from the secondary task to regaining situation awareness, accompanied by monitoring the traffic and the system. This was followed by takeover control. The series of process carried out with the assistance of the customised HMI seemed to support the drivers better in trial 2 than in trial 1.

18.3.1.3   Acceptance

Both the usefulness and satisfying scores increased in trial 2 in comparison to trial 1 considering positions of minimum, median, and maximum scores, and quartiles

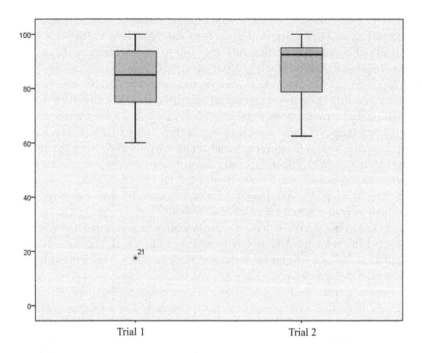

FIGURE 18.10 Usability in trial 1 and trial 2.

TABLE 18.5
Usability Scores between Trials

Trial	N	Mean	Std. Deviation	Minimum	Maximum	Wilcoxon Signed-Rank Test sig. (2-Tailed)
1	23	81.957	18.090	17.50	100	$Z=-1.562$
2	23	86.522	12.007	62.50	100	$p=0.118$

(seen in Figure 18.11). Statistical differences between the scores for both trials were not identified.

Differences in scores between pre- and post-experience of the takeover with customised interface were calculated because they represent whether, and in which direction, their opinion was changed (Van Der Laan, Heino and De Waard 1997). Changes in the participants' attitudes (usefulness and satisfying) are shown in Table 18.6. A greater proportion of participants' opinions about takeover experience changed more positively than negatively.

Mean values for each subscore in each trial (seen in Table 18.7) were as follows: 1.3565 and 1.4348 for usefulness, and 1.4565 and 1.53 for satisfying, in trials 1 and 2, respectively. The differences in the mean subscores between the trials (0.0783 and 0.0735, respectively) demonstrate that the participants' positive opinions were reinforced after the second trial.

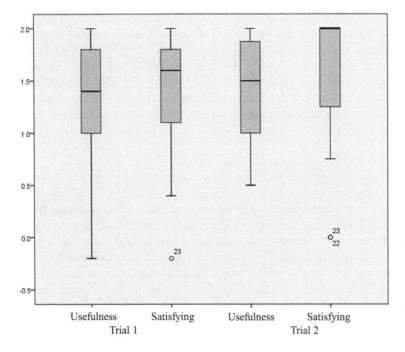

FIGURE 18.11 Acceptance scores in trial 1 and trial 2.

TABLE 18.6

Changes in Acceptance in Trial 2 Compared to Trial 1

	Usefulness		Satisfying	
	Number	%	Number	%
Positive	8	34.78	10	43.48
Unchanged	11	47.82	9	39.13
Negative	4	17.39	4	17.39

TABLE 18.7

Acceptance Scores between Trials

Subscale	Trial	N	Cronbach's Alpha	Mean	Std. Deviation	Min.	Max.	Wilcoxon Signed-Rank Test sig. (2-Tailed)
Usefulness	1	23	0.757	1.357	0.549	−0.200	2.000	$Z=-1.565$
	2	23	0.822	1.435	0.587	−0.200	2.000	$P=0.118$
Satisfying	1	23	0.727	1.457	0.463	0.500	2.000	$Z=-1.192$
	2	23	0.910	1.533	0.632	0.000	2.000	$P=0.233$

The increase in acceptance could represent that the participants perceived the takeover experience through the support of customised HMI in trial to be more useful and favourable. This is meaningful because relevant theories explain that a positive attitude may precede actual usage and purchase (Davis 1989; Venkatesh et al. 2003). The results may have been attributed to the design of alerts offering more assistance to the driver and the structure of setting options, leading to strengthening the density and intensity of necessary information, and reducing redundant information (Shedroff 1999).

18.3.1.4 Trust

In terms of trust, the level increased in trial 2 in comparison to trial 1 with consideration of the positions of minimum, median, maximum, lower, and upper quartiles, as seen in Figure 18.12. The level of trust in trial 2 differed significantly from that of trial 1 (see Table 18.8). This shows the degree of participants' trust during the takeover experience significantly increased after customising the interface.

The improvement in trust could be construed that the customised HMI enabled the participants to identify the behaviour, and the underlying intention of the autonomous system more easily. This may have resulted from the participants' status in which they were informed more effectively by the alert settings in the manner through the options they preferred. For example, the ones who preferred textual information regarding actions to take (e.g. hands on wheel) could remove the animated icon which could otherwise distract their vision. Consequently, the likelihood that the HMI could lead to distrust may have been reduced.

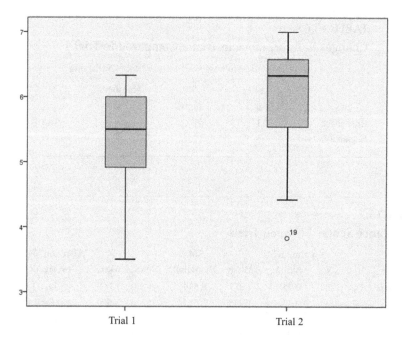

FIGURE 18.12   Trust scores in trial 1 and trial 2.

TABLE 18.8
Trust Scores between Trials

Trial	N	Mean	Std. Deviation	Minimum	Maximum	Wilcoxon Signed-Rank Test sig. (2-Tailed)
1	23	5.3183	0.818	3.50	6.33	Z=−3.552
2	23	5.7387	0.839	3.83	7.000	p=0.000

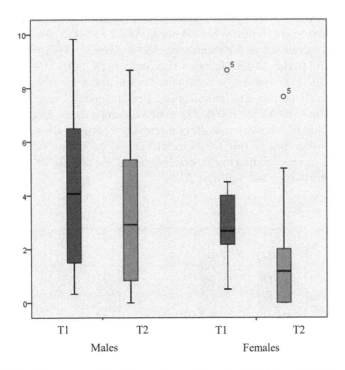

FIGURE 18.13   Perceived workload for males and females in trial 1 and trial 2.

18.3.2   COMPARISON BETWEEN GENDERS

Overall, there were no statistical differences in workload, usability, acceptance, and trust between genders in both trials. However, within each group, significant differences between trials were found. This could be understood that subjective experience of the takeover did not differ significantly across user groups with differing abilities and characteristics.

18.3.2.1   Workload

In general, workload was lower in trial 2 in comparison to trial 1 both in males' and in females' cases (see Figure 18.13). Unweighted average workload scores for males were 5.24 (SD = 3.06): 26.2% in trial 1, and 4.56 (SD = 3.05): 22.8% in trial 2; and those for females were 4.41 (SD = 2.33): 22.05% in trial 1, and 3.09 (SD = 2.61): 15.45%. The unweighted workload of the female participants was lower than that of male participants in both trials, but the differences were not statistically significant. The non-significant result parallels Ko and Ji's (2018) study findings that showed no difference between genders in workload and cognitive demand.

18.3.2.2 Usability

Increase in usability was shown in trial 2 in comparison with trial 1 for both male and female data (see Figure 18.14). Gender difference was detected neither in trial 1 ($p = 0.43$, the Mann–Whitney U-test) nor in trial 2 ($p = 0.98$, the Mann–Whitney U-test). The overall scores for males were 85.54 (SD = 12.37) in trial 1 and 86.07 (SD = 13.36) in trial 2, and for females, they were 76.38 (SD = 24.37) in trial 1 and 87.22 (SD = 10.27) in trial 2. The position of the females' score was altered from the 'good and excellent' range to 'excellent and best imaginable' range (Brooke 2013; Bangor, Kortum and Miller 2009). The positive change in the females' attitude in trial 2 is meaningful because the slight discrepancy between the genders was narrowed down after customising the interface. This is contradictory to expectations since it has been reported that female drivers showed less positive attitudes compared to male drivers (König and Neumayr 2017).

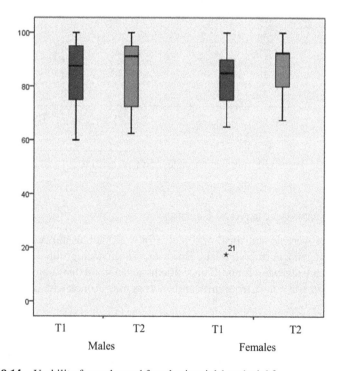

FIGURE 18.14   Usability for males and females in trial 1 and trial 2.

18.3.2.3 Acceptance

Both usefulness and satisfying scores were similar between trials for males. However, the scores seemed to be greater in trial 2 than in trial 1 for females (see Figure 18.15). Overall usefulness scores for the males were 1.33 (SD=0.67) in trial 1, and 1.33 (SD=0.69) in trial 2, and those for the females were 1.40 (SD=0.32) in trial 1, and 1.60 (SD=0.35) in trial 2. Overall satisfying scores for the males were 1.38 (SD=0.49) in trial 1 and 1.36 (SD=0.72) in trial 2, and those for the females were 1.58 (SD=0.41) in trial 1 and 1.81 (SD=0.35) in trial 2. The females' usefulness and satisfying scores showed an increase, and the difference in the satisfying score was statistically significant ($p=0.038$, $Z=-2.070$: the Wilcoxon signed-rank test). Considering differences between trials, female participants' acceptance changed more positively than males (see Table 18.9). The females' satisfying score showed a significant increase in trial 2 compared to trial 1.

The results imply that acceptance about takeover experience may not differ between genders before and after they customised the interface settings. However, acceptance of the female participants showed greater improvement after customising the settings. This seems encouraging because male drivers are more willing to relinquish driving control, and less worried about the technology of autonomous vehicles (Charness et al. 2018). In a similar vein, female drivers expressed a lower degree of comfort about certain driving functions and tasks to be undertaken by an

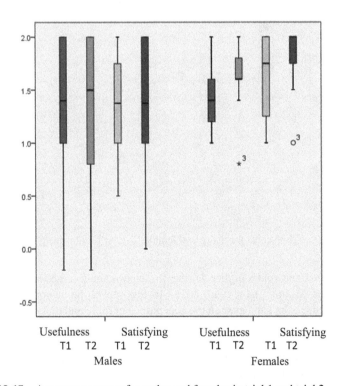

FIGURE 18.15 Acceptance scores for males and females in trial 1 and trial 2.

TABLE 18.9
Changes in Participants' Attitudes of Acceptance

		Usefulness		Satisfying	
		Number	%	Number	%
Male	Positive	3	21.43	5	35.71
	Unchanged	3	57.14	4	35.71
	Negative	8	57.14	5	28.57
Female	Positive	5	55.56	5	55.56
	Unchanged	3	33.33	4	44.44
	Negative	1	11.11	0	0.00

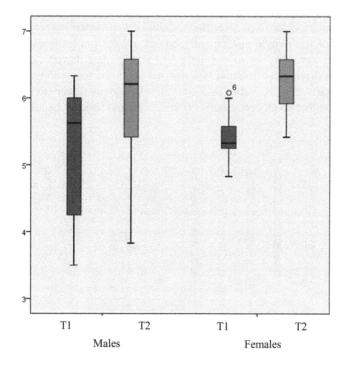

FIGURE 18.16   Trust scores for males and females in trial 1 and trial 2.

autonomous vehicle, and a higher degree of concern about issues regarding autonomous vehicles (Cunningham et al. 2019). In this sense, the customisable interface might help reduce the gap between the acceptance of male and that of female drivers.

18.3.2.4   Trust

Trust scores seemed to be increased in trial 2 compared to trial 1 for both males' and females' cases (see Figure 18.16). The levels of trust for males in trials 1 and 2 were 5.3 (SD = 1.00) and 5.87 (SD = 0.99). Those for females in trials 1 and 2 were 5.42

(SD = 0.42) and 6.22 (SD = 0.51). Significant differences between genders for both trials were not found. For both males' and females' cases, increases in trust in trial 2 compared to trial 1 were observed, presenting statistical significance (males' case: $p = 0.017$, $Z = -2.392$; females' case: $p = 0.008$, $Z = -2.670$, the Wilcoxon signed-rank test). The amount of increase seemed greater in the females' case.

Non-significant differences between males and females might represent that a similar level of trust was elicited by the assistance from HMI during the takeover for the males and females in both trials. This result seems contradictory to previous findings. For example, male drivers showed higher trust in self-driving technology, including reliability, functionality, and helpfulness (Tussyadiah, Zach and Wang 2017). Moreover, male drivers expressed a higher level of trust regarding the autonomous vehicle's decision-making based on detected information (Deb et al. 2016). Nonetheless, the HMI adopted in the present study was able to induce similar levels of trust from both genders and helped improve the females' trust to a greater extent.

18.3.3 Comparisons between Age Groups

Comparison of subjective evaluation of takeover experience among three age groups, between the first and second trials, was made. Statistically significant differences among the groups were not found. Furthermore, overall, perceived workload was lower, and usability, acceptance, and trust were higher in the second trial in comparison to the first trial.

18.3.3.1 Workload

Perceived workload tended to show a decrease in younger and middle age groups in trial 2 compared with trial 1 except for the maximum score for the younger age group. Although the maximum and the third quartile were lower in trial 2, the median value was higher for the older group (see Figure 18.17). Unweighted average workload scores for younger, middle, and older groups in trials 1 and 2 are presented in Table 18.10. In general, differences among the three age groups were not statistically significant in trial 1 nor in trial 2 ($p = 0.53$, and $p = 0.24$, respectively, the Kruskal–Wallis test result).

Participants in the older age group showed an increase in mental and physical demands in trial 2 in comparison to trial 1, while other elements showed decreasing trends (see Figure 18.18). This might be explained by age-related degradation in cognitive capacity (Rogers and Fisk 2001), and decline in speed of information processing and decision-making (Shaheen and Niemeier 2001). Further, driving leads to a greater mental workload for older drivers than for younger drivers, and this effect was exacerbated by the more complex driving context (Cantin et al. 2009). In each trial, the participants were required to drive for approximately 40 min. Moreover, older drivers tended to trust the automated system less than the younger groups, and they were not familiar with the system; therefore, they were more prone to monitor the system carefully (Favarò et al. 2019). Older drivers also experienced greater cognitive workload than younger drivers, and complexity in driving exacerbated the effect (Cantin et al. 2009). Hence, it could be judged that the older participants may have experienced a higher level of workload during the takeover in trial 2 which took place

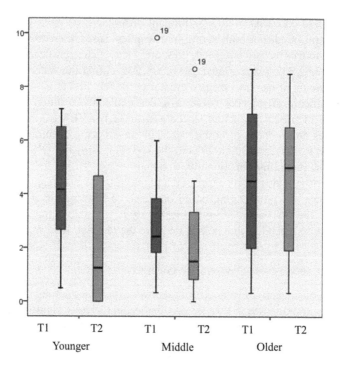

FIGURE 18.17   Perceived workload for three age groups in trial 1 and trial 2.

TABLE 18.10
Unweighted Average Workload in Trial 1 and Trial 2

Age Group	Unweighted Average Workload in Trial 1	Unweighted Average Workload in Trial 2
Younger	5.20 (SD = 2.46)	3.45 (SD = 3.11)
Middle	4.22 (SD = 2.80)	3.35 (SD = 2.63)
Older	5.50 (SD = 3.21)	5.38 (SD = 3.10)

after a long period of driving tasks, with reduced capability. This is also supported with previous studies that discussed the advanced driver assistance systems which provide useful assistance to older drivers. This can be fulfilled by alleviating the difficulties resulting from limitations in motion perception, peripheral vision, selective attention, and decreased speed of processing information and decision-making (Shaheen and Niemeier 2001).

18.3.3.2   Usability
Usability was higher in trial 2 than in trial 1 in all cases apart from the outlier for the younger group, and the minimum score for the middle group (see Figure 18.19).

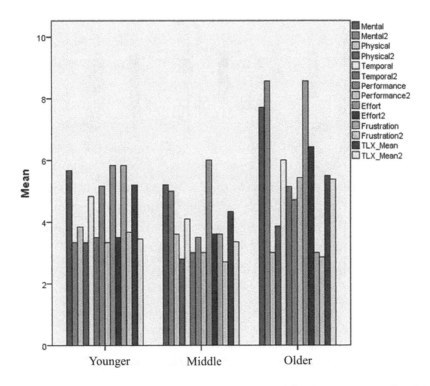

FIGURE 18.18 Individual elements of perceived workload for three age groups in trial 1 and trial 2.

Average scores for younger and middle age groups were within the 'excellent' level in both trials. Older participants' scores were within the 'good' range; however, the score for trial 2 was just below the 'excellent' level (Brooke 2013; Bangor, Kortum and Miller 2009; Pyae et al. 2016) (see Table 18.11).

There was no significant difference in workload among the three groups in both trials ($p = 0.799$ for trial 1, $p = 0.720$ for trial 2, the Kruskal–Wallis test results). Further, although there was no statistically significant difference in scores between trials within the three age groups, they all exhibited an increase in scores. Although the usability of the older group was lower than that of the other two groups, the highest increase in trial 2 compared to trial 1 was seen in the older age group. This is important because older adults are known to be more concerned about autonomous driving technology (Hulse, Xie and Galea 2017), and less capable to overcome difficulties with new technology (Venkatesh, Thong and Xu 2012). However, this result presents that the older drivers' opinion was changed positively more than the younger groups in terms of needing support, training, and perceived complexity of the system (Brooke 1996).

18.3.3.3 Acceptance

There were no statistical differences among the three age groups in usefulness (U) and satisfying (S) scores neither in trial 1 (U: $p = 0.204$; S: $p = 0.086$) nor in trial 2

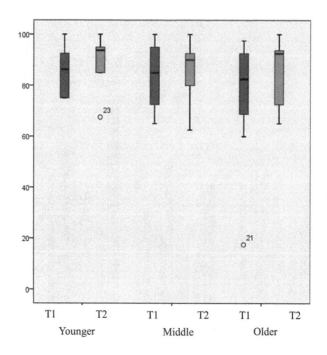

FIGURE 18.19   Usability scores for three age groups in trial 1 and trial 2.

TABLE 18.11

Usability Score Range for Younger, Middle, and Older Age Groups in Trial 1 and Trial 2

Age Group	Usability in Trial 1	Range	Usability in Trial 2	Range
Younger	85.83 (SD = 10.56)	Excellent	89.17 (SD = 11.69)	Excellent
Middle	85.00 (SD = 12.13)	Excellent	86.50 (SD = 11.62)	Excellent
Older	74.29 (SD = 28.05)	Good	84.29 (SD = 14.12)	Good

(U: $p = 0.262$; S: $p = 0.253$). In a previous study, older drivers reported a more positive attitude towards using HAD despite their lower self-assessed self-efficacy and environmental conditions facilitating HAD usage (technical support) compared to younger drivers (Hartwich et al. 2019). Non-significant differences presented in this study may show that the HMI was designed in an inclusive manner which could be accepted across all three age groups. For instance, as described by Hartwich et al.'s (2019) study, the interface could have facilitated takeover actions to compensate perceived lack of self-efficacy.

18.3.3.4   Trust

Generally, trust showed increasing trends in all the age groups although it presented widespread in the first quartile of the older age group in trial 2 measured against

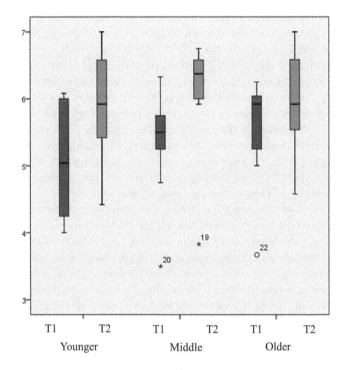

FIGURE 18.20 Trust scores for three age groups in trial 1 and trial 2.

trial 1 (see Figure 18.20). A statistically significant difference between age groups was not identified in either trial 1 or trial 2 ($p=0.578$ and $p=0.820$). This seems to be in line as young male drivers' trust was not different from young and old females and older males (Zoellick et al. 2019). Comparisons between trials in each age group were made. The results demonstrated significant increases in younger ($p=0.028$, $Z=-2.201$: the Wilcoxon signed-rank test), middle ($p=0.025$, $Z=-2.247$: the Wilcoxon signed-rank test), and older age group ($p=0.041$, $Z=-2.043$: the Wilcoxon signed-rank test). This can be interpreted that the levels of trust in the system during the takeover did not differ across the age groups in both trials, while a significant improvement was achieved in trial 2 across all age groups. This delivers an important message that customisation conducted by the participants based on their own decision according to their needs did not have a negative impact on reliability, security, and confidence; nor did it increase the level of distrust which were defined as factors of trust in Jian, Bisantz, and Drury (1998).

18.3.4 BENEFITS AND EFFECTS OF CUSTOMISATION

The customisable interface allowed the drivers to optimise the driving information system to better suit their needs. In the default settings, all the options for alert settings were selected apart from road sense view which visualised a vehicle in front on the cluster and infotainment displays. The settings were used for the first trial, and the participants had an opportunity to change the settings after the run. Overall, they reduced the amount of information by deselecting information that they perceived as not necessary on the particular screen, or redundant. This may be supported by the research findings presented that satisfaction with digital interface could be attributed to the simplified interface design (Choi and Lee 2012). Consequently, on the customised interfaces, the participants may have been able to find necessary information more easily. It was because they could make the information they needed clearly visible in places where they felt comfortable to look, and in the appropriate modalities that they preferred. Simplified information settings could facilitate drivers' decision-making during takeover events especially to rapidly respond to short TORs by helping shorten the time for information acquisition and decision-making. This, in turn, seemed to help improve the participants' subjective experience of takeover. Considering the advantages of multimodal information provision during takeover situation, all the participants received information through all the three modalities (visual, auditory, and vibrotactile) in both trials. They did not exclude any of the modalities but adjusted the intensities. This may have been connected to the enhancement of satisfaction; e.g. female drivers, who are generally more sensitive to vibrotactile stimulus (Ji, Lee and Hwang 2010), could reduce the intensity for tactile alerts. Furthermore, using all the modalities in trial 2 which were better adjusted to the participants seems consistent to the Yoon, Kim, and Ji's (2019) study findings that showed enhanced usefulness, satisfaction, and trust in multimodal TOR situation than in unimodal TOR. Furthermore, the participants kept more than one protocol question with appreciating the utility of the questions which were helpful for them to divert their attention back to manual driving (e.g. by looking at the road ahead, recognising the speed, checking blind spot), and get ready to drive manually. This in turn helped improve positive responses, such as acceptance and trust towards the system. This may be similar to the Lu et al.'s (2019) findings which issued MRs prior to TORs and led to enhanced acceptance and trust, and relaxed workload. Although the HMI adopted in this study did not offer reasons to takeover, by asking the protocol questions, the participants were able to have a better understanding about the surroundings and the vehicle status. Given the participants were able to increase or decrease the number of the questions, it could be judged that the number of questions was more optimal to each participant in trial 2. The established situation awareness gained through answering the questions in trial 2 (not too few: too little information, or too many: not necessary information), and resultant positive subjective experience could be supported by the Li et al.'s (2019) study result. In their experiment, issuing TOR together with offering reasons for takeover helped the drivers have better attitudes, enhance performance, and reduce workload.

Overall, customisation may have been useful for them to tailor the driving environment for them to conduct the driving tasks, including takeover more effectively (functional aspects) and feel more pleasant (emotional aspects). Given that drivers have differing abilities, preferences, and interests, the customisable HMI offers a possible alternative to meet them as an inclusive approach.

In terms of group differences, for both trials 1 and 2, there was no statistical difference between genders or age groups. In trial 2, a significant decrease in workload and a significant increase in trust were found compared to trial 1. Although the differences were not statistically significant, increases in usability and trust were seen in trial 2 based on descriptive analysis results. Furthermore, a significant decrease in workload and a significant increase in acceptance and trust of female participants in trial 2 might demonstrate that customisation could help meet female participants' needs effectively. This could imply that enabling users to customise the interface according to their own preference and needs could lead to enhancement of the experience of female drivers, identified as a group of users who do not perceive autonomous driving as useful or acceptable as much as male drivers. Further work in this area with a larger sample size is warranted.

18.3.5 INFORMATION SETTINGS FOR SAFE AND TIMELY TAKEOVER

Participants kept all the essential information (text-based visual information) regarding driving mode wording, time to takeover, either on the cluster or on the HUD. This might be interpreted that they did not customise the interface settings in the manner that may have a negative effect on safety, such as in the way that they could miss important warning. In other words, they made a reasonable decision to modify the interface settings within the range that would not lead to any detrimental effects of their performance.

18.4 CONCLUSION

This study investigated the effects of customisable HMI on drivers' subjective assessment of takeover experience in highly automated driving in an on-road setting. It was motivated by a lack of available information about the impact of driver-initiated customisation in the field of autonomous driving systems. Twenty-three drivers, comprised 14 males and 9 females, classified in three age groups, participated in the study. They conducted two automated driving trials which included three takeovers in each trial in a highly autonomous vehicle on the road. The participants were able to customise the HMI between the trials. The results presented that their subjective experience was enhanced in trial 2, demonstrating positive influence of customisation. Statistical significance was identified in perceived workload lowered, and trust increased in trial 2. This may represent that customised HMI was able to better support the participants' varying cognitive and physical abilities during the takeover situation. This in turn could elicit more positive attitudes towards the system and lead to an enhanced subjective experience.

This research has limitations because the other studies used within-subjects design. Learning effects may have contributed to the formation of participants' perceptions towards takeover experience. However, the participants who took part in this study were familiar with the HMI adopted in this experiment because they participated in the precedent simulation study which used almost identical HMI. Moreover, a training session was provided on the test track prior to the on-road trial, which could be completed only when they were confident enough to proceed to the on-road trial. Therefore, the learning effects should have been minimised through this process.

ACKNOWLEDGEMENTS

This work was supported by Jaguar Land Rover and the UK-EPSRC Grant EP/N011899/1 as part of the jointly funded 'Towards Autonomy: Smart and Connected Control (TASCC) Programme' HI:DAVe Project.

REFERENCES

AgeLab. 2019. *MIT AgeLab Study Tools.* Accessed February 17, 2020. https://agelab.mit.edu/study-tools.

Arduino. 2020. Accessed February 25, 2020. https://www.arduino.cc/.

Bangor, A., P. Kortum, and J. Miller. 2009. "Determining what individual SUS scores mean: adding an adjective rating scale." *Journal of Usability Studies, 4, 3* 114–123.

Banks, V. A., and N. A. Stanton. 2016. "Keep the driver in control: Automating automobiles of the future." *Applied Ergonomics, 53, Part B* 389–395. https://doi.org/10.1016/j.apergo.2015.06.020.

Bazilinskyy, P., S. M. Petermeijer, V. Petrovych, D. Dodou, and J. C. F. de Winter. 2018. "Take-over requests in highly automated driving: a crowdsourcing survey on auditory, vibrotactile, and visual displays." *Transportation Research Part F: Traffic Psychology and Behaviour, 56* 82–98. https://doi.org/10.1016/j.trf.2018.04.001.

Blömacher, K., G. Nöcker, and M. Huff. 2018. "The role of system description for conditionally automated vehicles." *Transportation Research Part F: Traffic Psychology and Behaviour, 54* 159–170. https://doi.org/10.1016/j.trf.2018.01.010.

Brooke, J. 1996. "SUS: a 'quick and dirty' usability scale." In *Usability Evaluation In Industry,* by P. W. Jordan, B. Thomas, I. L. McClelland, and B. Weerdmeester, 189–194. London: Taylor & Francis Group. https://doi.org/10.1201/9781498710411.

Brooke, J. 2013. "SUS: a retrospective." *Journal of Usability Studies, 8, 2* 29–40.

Brookhuis, K. A., C. J. G. van Driel, T. Hof, B. van Arem, and M. Hoedemaeker. 2009. "Driving with a congestion assistant; mental workload and acceptance." *Applied Ergonomics, 40, 6* 1019–1025. https://doi.org/10.1016/j.apergo.2008.06.010.

Bysameyee. 2020. *Head-up Display GPS Navigation.* Accessed February 17, 2020. https://www.amazon.co.uk/Head-up-Display-GPS-Navigation-Reflective/dp/B075T2DH12/ref=sr_1_10?keywords=car+head+up+display&qid=1573804882&sr=8-10.

Cantin, V., M. Lavallière, M. Simoneau, and N. Teasdale. 2009. "Mental workload when driving in a simulator: effects of age and driving complexity." *Accident Analysis & Prevention, 41, 4* 763–771. doi: 10.1016/j.aap.2009.03.019.

Charness, N., J. S. Yoon, D. Souders, C. Stothart, and C. Yehnert. 2018. "Predictors of attitudes toward autonomous vehicles: the roles of age, gender, prior knowledge, and personality." *Frontiers in Psychology, 9* 2589. doi: 10.3389/fpsyg.2018.02589.

Choi, J. H., and H. J. Lee. 2012. "Facets of simplicity for the smartphone interface: a structural model." *International Journal of Human-Computer Studies, 70, 2* 129–142. https://doi.org/10.1016/j.ijhcs.2011.09.002.

Choi, J. K., and Y. G. Ji. 2015. "Investigating the importance of trust on adopting an autonomous vehicle." *International Journal of Human–Computer Interaction, 31, 10* 692–702. https://doi.org/10.1080/10447318.2015.1070549.

Connective Peripherals. 2020. *USB2-F-7101.* Accessed February 25, 2020. https://www.connectiveperipherals.com

Corder, G. W., and D. I. Foreman. 2009. *Nonparametric Statistics for Non-Statisticians: A Step-by-Step Approach.* New Jersey: John Wiley & Sons, Inc.

Cunningham, M. L., M. A. Regan, T. Horberry, K. Weeratunga, and V. Dixit. 2019. "Public opinion about automated vehicles in Australia: results from a large-scale national survey."

Transportation Research Part A: Policy and Practice, 129 1–18. https://doi.org/10.1016/j.tra.2019.08.002.

Davis, F. D. 1989. "Perceived usefulness, perceived ease of use, and user acceptance of information technology." *MIS Quarterly, 13, 3* 319–340. doi: 10.2307/249008.

Davis, F. D., R. P. Bagozzi, and P. R. Warshaw. 1989. "User acceptance of computer technology: a comparison of two theoretical models." *Management Science, 35, 8* 903–1028. https://doi.org/10.1287/mnsc.35.8.982.

de Winter, J. C. F., R. Happee, M. H. Martens, and N. A. Stanton. 2014. "Effects of adaptive cruise control and highly automated driving on workload and situation awareness: a review of the empirical evidence." *Transportation Research Part F: Traffic Psychology and Behaviour, 27, Part B* 196–217. https://doi.org/10.1016/j.trf.2014.06.016.

Deb, S., B. Warner, S. R. Poudel, and S. Bhandari. 2016. "Identification of external design preferences in autonomous vehicles." *Proceedings of the 2016 Industrial and Systems Engineering Research Conference.* Anaheim. 69.

dSPACE. 2020. *MicroAutoBox II.* Accessed February 17, 2020. https://www.dspace.com/en/ltd/home/products/hw/micautob/microautobox2.cfm.

Elfeland. 2020. *LED Strip Lights Kits.* Accessed February 17, 2020. https://www.amazon.co.uk/Elfeland-Waterproof-Controller-Self-Adhesive-Decoration/dp/B07H2DR5DY/ref=sr_1_6?keywords=led+rgb+strip&qid=1559907897&s=gateway&sr=8-6.

Endsley, M. R. 1995. "Toward a theory of situation awareness in dynamic systems." *Human Factors: The Journal of the Human Factors and Ergonomics Society, 37, 1* 32–64. http://doi.org/10.1518/001872095779049543.

Eriksson, A., and N. A. Stanton. 2017. "Takeover time in highly automated vehicles: non-critical transitions to and from manual control." *Human Factors: The Journal of the Human Factors and Ergonomics Society, 59, 4* 689–705. https://doi.org/10.1177/0018720816685832.

Eriksson, A., V. A. Banks, and N. A. Stanton. 2017. "Transition to manual: comparing simulator with on-road control transitions." *Accident Analysis & Prevention, 102* 227–234. https://doi.org/10.1016/j.aap.2017.03.011.

Eriksson, A., S. M. Petermeijer, M. Zimmermann, J. C. F. de Winter, K. J. Bengler, and N. A. Stanton. 2019. "Rolling out the red (and green) carpet: supporting driver decision making in automation-to-manual transitions." *IEEE Transactions on Human-Machine Systems, 49, 1* 20–31. doi: 10.1109/THMS.2018.2883862.

Favarò, F. M., P. Seewald, M. Scholtes, and S. Eurich. 2019. "Quality of control takeover following disengagements in semi-automated vehicles." *Transportation Research Part F: Traffic Psychology and Behaviour, 64* 196–212. https://doi.org/10.1016/j.trf.2019.05.004.

Field, A. 2017. *Discovering Statistics Using IBM SPSS Statistics*, 5th edition. Thousand Oaks, CA: SAGE Publications.

Forster, Y., S. Hergeth, F. Naujoks, and J. F. Krems. 2018. "How usability can save the day - methodological considerations for making automated driving a success story." *AutomotiveUI - Proceedings of the 10th International Conference on Automotive User Interfaces and Interactive Vehicular Applications.* Toronto: Association for Computing Machinery. 278–290. https://doi.org/10.1145/3239060.3239076.

Gold, C., R. Happee, and K. Bengler. 2018. "Modeling take-over performance in level 3 conditionally automated vehicles." *Accident Analysis & Prevention, 116* 3–13. doi: 10.1016/j.aap.2017.11.009.

Hart, S. G. 2006. "Nasa-task load index (NASA-TLX); 20 years later." *Proceedings of the Human Factors and Ergonomics Society Annual Meeting*, 904–908. https://doi.org/10.1177/154193120605000909.

Hart, S. G., and L. E. Staveland. 1988. "Development of NASA-TLX (task load index): results of empirical and theoretical research." *Advances in Psychology, 52* 139–183. http://doi.org/10.1016/S0166-4115(08)62386-9.

Hartwich, F., C. Witzlack, M. Beggiato, and J. F. Krems. 2019. "The first impression counts – a combined driving simulator and test track study on the development of trust and acceptance of highly automated driving." *Transportation Research Part F: Traffic Psychology and Behaviour, 65* 522–535. https://doi.org/10.1016/j.trf.2018.05.012.

Hulse, L. M., H. Xie, and E. R. Galea. 2017. "Perceptions of autonomous vehicles: relationships with road users, risk, gender and age." *Safety Science, 102* 1–13. https://doi.org/10.1016/j.ssci.2017.10.001.

In-CarPC. 2020. *Vehicle and Car Computer Systems | In-CarPC.* Accessed February 17, 2020. https://www.in-carpc.co.uk/.

ISO. 1998. "ISO 9241-11: ergonomic requirements for office work with visual display terminals (VDTs) — Part 11: guidance on usability." Accessed November 3, 2020. https://www.iso.org/standard/63500.html.

iYeal. 2020. *24mm LED Illuminated Push Button.* Accessed February 17, 2020. https://www.amazon.co.uk/iYeal-Illuminated-Buttons-Joystick-Raspberry/dp/B081D4W33K/ref=sr_1_151?keywords=green+led+push+button&qid=1574084349&sr=8-151.

Ji, Y. G., K. Lee, and W. Hwang. 2010. "Haptic perceptions in the vehicle seat." *Human Factors and Ergonomics in Manufacturing & Service Industries, 21, 3* 305–325. https://doi.org/10.1002/hfm.20235.

Jian, J. Y., A. M. Bisantz, and C. G. Drury. 1998. "Towards an empirically determined scale of trust in computerized systems: distinguishing concepts and types of trust." *Proceedings of the Human Factors and Ergonomics Society Annual Meeting, 42, 5* 501–505. https://doi.org/10.1177/154193129804200512.

Jordan, P. W. 1998. *An Introduction To Usability.* Boca Raton, FL: CRC Press.

Kaur, K., and G. Rampersad. 2018. "Trust in driverless cars: investigating key factors influencing the adoption of driverless cars." *Journal of Engineering and Technology Management, 48* 87–96. https://doi.org/10.1016/j.jengtecman.2018.04.006.

Ko, S. M., and Y. G. Ji. 2018. "How we can measure the non-driving-task engagement in automated driving: comparing flow experience and workload." *Applied Ergonomics, 67* 237–245. https://doi.org/10.1016/j.apergo.2017.10.009.

König, M., and L. Neumayr. 2017. "Users' resistance towards radical innovations: the case of the self-driving car." *Transportation Research Part F: Traffic Psychology and Behaviour, 44* 42–52. https://doi.org/10.1016/j.trf.2016.10.013.

Körber, M., E. Baseler, and K. Bengler. 2018. "Introduction matters: manipulating trust in automation and reliance in automated driving." *Applied Ergonomics, 66* 18–31. https://doi.org/10.1016/j.apergo.2017.07.006.

KT&C. 2020. *KNC-HDi47.* Accessed February 25, 2020. http://www.ktncusa.com/ip/ip_miniature/knc-hdi47.

Lee, J., and N. Moray. 1992. "Trust, control strategies and allocation of function in human-machine systems." *Ergonomics, 35, 10* 1243–1270. doi: 10.1080/00140139208967392.

Lee, J. D., and N. Moray. 1994. "Trust, self-confidence, and operators' adaptation to automation." *International Journal of Human-Computer Studies, 40, 1* 153–184. https://doi.org/10.1006/ijhc.1994.1007.

Leggett & Platt Automotive. 2020. *Vehicle Seating and Lumbar Support.* Accessed February 25, 2020. https://leggett-automotive.com//products.

Li, S., P. Blythe, W. Guo, A. Namdeo, S. Edwards, P. Goodman, and G. Hill. 2019. "Evaluation of the effects of age-friendly human-machine interfaces on the driver's takeover performance in highly automated vehicles." *Transportation Research Part F: Traffic Psychology and Behaviour, 67* 78–100. https://doi.org/10.1016/j.trf.2019.10.009.

Liu, H., R. Yang, L. Wang, and P. Liu. 2019. "Evaluating initial public acceptance of highly and fully autonomous vehicles." *International Journal of Human-Computer Interaction, 35, 11* 919–931. https://doi.org/10.1080/10447318.2018.1561791.

Lu, Z., X. Coster, and J. C. F. de Winter. 2017. "How much time do drivers need to obtain situation awareness? A laboratory-based study of automated driving." *Applied Ergonomics,* 60 293–304. https://doi.org/10.1016/j.apergo.2016.12.003.

Lu, Z., R. Happee, C. D. D. Cabrall, M. Kyriakidis, and J. C. F. de Winter. 2016. "Human factors of transitions in automated driving: a general framework and literature survey." *Transportation Research Part F: Traffic Psychology and Behaviour, 43* 183–198. https://doi.org/10.1016/j.trf.2016.10.007.

Lu, Z., B. Zhang, A. Feldhütter, R. Happee, M. Martens, and J. C. F. de Winter. 2019. "Beyond mere take-over requests: the effects of monitoring requests on driver attention, take-over performance, and acceptance." *Transportation Research Part F: Traffic Psychology and Behaviour, 63* 22–37. https://doi.org/10.1016/j.trf.2019.03.018.

Mayer, R. C., J. H. Davis, and F. D. Schoorman. 1995. "An integrative model of organizational trust." *The Academy of Management Review, 20, 3* 709–734. doi: 10.2307/258792.

Mcgill, R., J. W. Tukey, and W. A. Larsen. 1978. "Variations of box plots." *The American Statistician, 32, 1* 12–16. doi: 10.1080/00031305.1978.10479236.

Mcknight, P., and J. Najab. 2010. "Mann-Whitney U test." In *The Corsini Encyclopedia of Psychology,* by I. B. Weiner, and W. E. Craighead, 960–961. Hoboken, NJ: John Wiley & Sons, Inc. doi: 10.1002/9780470479216.corpsy0524.

Moller, S., K. Engelbrecht, C. Kuhnel, I. Wechsung, and B. Weiss. 2009. "A taxonomy of quality of service and quality of experience of multimodal human-machine interaction." *2009 International Workshop on Quality of Multimedia Experience.* San Diego. 7–12. doi: 10.1109/QOMEX.2009.5246986.

Nees, M. A. 2016. "Acceptance of self-driving cars: an examination of idealized versus realistic portrayals with a self- driving car acceptance scale." *Proceedings of the Human Factors and Ergonomics Society Annual Meeting, 60, 1* 1449–1453. https://doi.org/10.1177/1541931213601332.

Panagiotopoulos, I., and G. Dimitrakopoulos. 2018. "An empirical investigation on consumers' intentions towards autonomous driving." *Transportation Research Part C: Emerging Technologies, 95* 773–784. https://doi.org/10.1016/j.trc.2018.08.013.

Pyae, A., T. N. Liukkonen, T. Saarenpää, M. Luimula, P. Granholm, and J. Smed. 2016. "When Japanese elderly people play a Finnish physical exercise game: a usability study." *Journal of Usability Studies, 11, 4* 131–152.

Rogers, W. A., and A. D. Fisk. 2001. "Understanding the role of attention in cognitive aging research." In *Handbook of the Psychology of Aging,* by J. E. Birren, and K. W. Schaie, 267–287. San Diego, CA: Academic Press.

SAE International. 2018. *SAE International Releases Updated Visual Chart for Its "Levels of Driving Automation" Standard for Self-Driving Vehicles.* 11 December. Accessed June 11, 2019. https://www.sae.org/news/press-room/2018/12/sae-international-releases-updated-visual-chart-for-its-%E2%80%9Clevels-of-driving-automation%E2%80%9D-standard-for-self-driving-vehicles.

SBG. 2020. *Fusion. SBG Sports Software.* Accessed February 25, 2020. https://sbgsportssoftware.com/product/fusion/.

Shaheen, S. A., and D. A. Niemeier. 2001. "Integrating vehicle design and human factors: minimizing elderly driving constraints." *Transportation Research Part C: Emerging Technologies, 9, 3* 155–174. https://doi.org/10.1016/S0968-090X(99)00027-3.

Shedroff, N. 1999. "Information interaction design: a unified field theory of design." *Information Design* 267–292.

Stanton, N. A., A. Dunoyer, and A. Leatherland. 2011. "Detection of new in-path targets by drivers using stop & go adaptive cruise control." *Applied Ergonomics, 42, 4* 592–601. https://doi.org/10.1016/j.apergo.2010.08.016.

Stapel, J., F. A. Mullakkal-Babu, and R. Happee. 2019. "Automated driving reduces perceived workload, but monitoring causes higher cognitive load than manual driving."

Transportation Research Part F: Traffic Psychology and Behaviour, 60 590–605. doi: 10.1016/j.trf.2018.11.006.

Taber, K. S. 2018. "The use of Cronbach's alpha when developing and reporting research instruments in science education." *Research in Science Education, 48* 1273–1296. https://doi.org/10.1007/s11165-016-9602-2.

Tussyadiah, I. P., F. J. Zach, and J. Wang. 2017. "Attitudes toward autonomous on demand mobility system: the case of self-driving taxi." *Information and Communication Technologies in Tourism 2017.* Cham: Springer. 755–766. https://doi.org/10.1007/978-3-319-51168-9_54.

van den Beukel, A. P., M. C. van der Voort, and A. O. Eger. 2016. "Supporting the changing driver's task: exploration of interface designs for supervision and intervention in automated driving." *Transportation Research Part F: Traffic Psychology and Behaviour, 43* 279–301. https://doi.org/10.1016/j.trf.2016.09.009.

Van Der Laan, J. D., A. Heino, and D. De Waard. 1997. "A simple procedure for the assessment of acceptance of advanced transport telematics." *Transportation Research Part C: Emerging Technologies, 5, 1* 1–10. http://doi.org/10.1016/S0968-090X(96)00025-3.

Vector. 2020. *Know-How & Solutions for CAN/CAN FD.* Accessed February 17, 2020. https://www.vector.com/us/en-us/know-how/technologies/networks/can/.

Venkatesh, V., M. G. Morris, G. B. Davis, and F. D. Davis. 2003. "User acceptance of information technology: toward a unified view." *MIS Quarterly, 27, 3* 425–478. doi: 10.2307/30036540.

Venkatesh, V., J. Y. L. Thong, and X. Xu. 2012. "Consumer acceptance and use of information technology: extending the unified theory of acceptance and use of technology." *MIS Quarterly, 36, 1* 157–178. doi: 10.2307/41410412.

Walch, M., T. Sieber, P. Hock, M. Baumann, and M. Weber. 2016. "Towards cooperative driving: involving the driver in an autonomous vehicle's decision making." *AutomotiveUI - Proceedings of the 8th International Conference on Automotive User Interfaces and Interactive Vehicular Applications.* Ann Arbor: Association for Computing Machinery. 261–268. https://doi.org/10.1145/3003715.3005458.

Woolson, R. F. 2008. "Wilcoxon signed-rank test." *Wiley Encyclopedia of Clinical Trials* 1–3. https://doi.org/10.1002/9780471462422.eoct979.

Xu, Z., K. Zhang, H. Min, Z. Wang, X. Zhao, and P. Liu. 2018. "What drives people to accept automated vehicles? Findings from a field experiment." *Transportation Research Part C: Emerging Technologies, 95* 320–334. https://doi.org/10.1016/j.trc.2018.07.024.

Yoon, S. H., Y. W. Kim, and Y. G. Ji. 2019. "The effects of takeover request modalities on highly automated car control transitions." *Accident Analysis & Prevention, 123* 150–158. https://doi.org/10.1016/j.aap.2018.11.018.

Zoellick, J. C., A. Kuhlmey, L. Schenk, D. Schindel, and S. Blüher. 2019. "Amused, accepted, and used? Attitudes and emotions towards automated vehicles, their relationships, and predictive value for usage intention." *Transportation Research Part F: Traffic Psychology and Behaviour, 65* 68–78. https://doi.org/10.1016/j.trf.2019.07.009.

19 Accommodating Drivers' Preferences Using a Customised Takeover Interface on UK Motorways

Nermin Caber
University of Cambridge

Patrick Langdon
Edinburgh Napier University

Michael Bradley
University of Cambridge

James W.H. Brown
University of Southampton

Simon Thompson
Jaguar Land Rover

Joy Richardson and Jisun Kim
University of Southampton

Lee Skrypchuk
Jaguar Land Rover

Kirsten M. A. Revell
University of Southampton

P. John Clarkson
University of Cambridge

Neville A. Stanton
University of Southampton

CONTENTS

19.1   INTRODUCTION

Autonomous driving systems promise many benefits ranging from improved road safety to reduced congestion (Fagnant and Kockelman 2015). Once regulatory issues are resolved (Bishop 2019), the first systems to enter the market will be semi-autonomous in accordance with SAE Level 3 (McNamara 2017; SAE J3016 2016). Hands- and feet-free driving will thereby become legal and is likely to be adopted by many drivers (Payre, Cestac and Delhomme 2014). Although SAE Level 3 systems do not require drivers to permanently monitor the dynamic driving task, they do require them to intervene when requested – this is called a 'takeover request' (TOR). These driver interventions are required when the system limits are reached – e.g. leaving operational design domain, or a malfunction occurs – e.g. sensor failure

(Eriksson and Stanton 2017; Gold et al. 2018; Merat et al. 2012; Stanton and Marsden 1996; Strand et al. 2014).

The challenges these TORs pose are worsened through potentially exacerbating effects of prolonged automation on the human operator (Bainbridge 1983), such as heightened fatigue (Desmond, Hancock and Monette 1998; Neubauer, Matthews and Saxby 2014; Schömig et al. 2015) and reduced situation awareness (Endsley 2010; Endsley and Kiris 1995; Strand et al. 2014). This duration of automation is called 'out-of-the-loop' (OOTL) time and leads in connection with the aforementioned effects to longer reaction times (Young and Stanton 2007). An extensive research body looking into the interface design of TORs (Borojeni et al. 2016; Eriksson et al. 2019; Naujoks, Mai and Neukum 2014; Petermeijer et al. 2017; Petermeijer, Cieler and de Winter 2017; Politis, Brewster and Pollick, 2015; 2017) and the time needed for safe control transitions from autonomous to manual driving (Bourrelly et al. 2019; Eriksson and Stanton 2017; Gold et al. 2013; Körber et al. 2016; Melcher et al. 2015; Merat et al. 2014) has evolved. A comprehensive literature review on TORs and takeover time has shown that drivers vary starkly and take up to 25s to resume manual control from automation (Eriksson and Stanton 2017). For example, older drivers tend to take longer to resume control from automation than younger drivers (Chapter 14). Considering these results and the known, wide range of capabilities among drivers (Green 2008), it is essential to design the TOR inclusively and with the driver in mind (Eriksson and Stanton 2017).

User-tailorable interfaces are a potential approach to accommodate the stark variance among drivers by optimising the takeover interface design to the respective driver. Two different techniques are widely used to tailor interfaces to a specific driver and environment: customisation and adaptation. Personalisation can be considered as a subtype of adaptation and is therefore not listed as a separate technique (García-barrios, Mödritscher and Gütl 2005).

Customisation requires an active involvement of the user to make decisions about and changes to the system and its design (Arora et al. 2008). In the automotive field, customisation dates back to the 1920s when the implementation of the seat-sliding mechanism enabled drivers of varying sizes to tailor their distance to the driving controls (Akamatsu, Green and Bengler 2013), increasing safety and comfort. Over the years, customisation has become ubiquitous, ranging from mirror and steering wheel position adjustments to customisable lumbar support and side bolster for the seats, to accommodate drivers of different body sizes and shapes. This focus on anthropometry has shifted more towards individual preferences and human cognition, such as coping with the driving task and the in-vehicle information system simultaneously (Akamatsu, Green and Bengler 2013). Recent developments therefore enable drivers to, e.g., change gearbox, steering, engine, and suspension characteristics (Caber, Langdon and Clarkson 2018); switch between different fragrances in the car interior (Mercedes-Benz 2020); and alter the level of support from specific driver assistance systems (Caber, Langdon and Clarkson 2019), achieving even greater individuality. These settings can be saved in driver profiles and therefore be recalled anytime by the user (Caber, Langdon and Clarkson 2019; Pollard 2018).

Adaptation is the method applied to tackle the increased cognitive demand (Akamatsu, Green and Bengler 2013), to overcome drivers' tendency not to use available customisation (Reagan et al. 2018), and to improve road safety (Blaschke et al. 2009) as it does not require active involvement of the driver. In contrast to customisation, adaptation alters the system based on some gathered stimuli (García-barrios, Mödritscher and Gütl 2005; Supulniece 2012). Examples for such stimuli are the driver's state and behaviour (Bellet and Manzano 2007; Dong et al. 2011; Gonçalves and Bengler 2015; Ma et al. 2017; Miyajima and Takeda 2016) as well as environmental and contextual interactions (Amditis et al. 2010; Bellotti et al. 2005; Heigemeyr and Harrer 2014; Michon 1993; Wright et al. 2017). Some of these functions cannot be reproduced by customisation as the driver is actively driving while the system is altered; delaying a phone call due to a demanding driving environment, e.g. a busy roundabout, is a good example (Bellotti et al. 2005). An adaptive system is called 'personalised' when it alters system characteristics to a specific user.

Gathering relevant data and modelling drivers along with their behaviour and state are major challenges in the step from customisation towards adaptation (Miyajima and Takeda 2016; Terzi, Sagiroglu and Demirezen 2018). The large inter- and intra-driver variability poses a significant problem to driver modelling (Miyajima and Takeda 2016). Different approaches have been taken to model drivers: 'average', 'personalised', and 'clustering-aided'. The so-called average system takes all drivers together without discriminating between them. A 'personalised system', in contrast, learns from individuals' historical data and, consequently, has proved as more effective (Yi et al. 2019a). However, a main limitation of a 'personalised system' is the usually missing historical data for a specific individual. This has motivated 'clustering-aided systems' which try to exploit similarities among drivers and thereby create clusters which then can be used to improve driver modelling (Ping et al. 2018; Yi et al. 2019b). Research in psychology demonstrating consistent human traits and preferences supports this approach (Roberts and DelVecchio 2000).

Although huge data sets have been collected on driver behaviour and driver state, using vehicle and physiological measures, driver feedback, along with facial and body expressions (Aghaei et al. 2016; Barnard et al. 2016; National Academies of Sciences, Engineering, and Medicine 2014; Fridman et al. 2019; Miyajima and Takeda 2016), there is little historical data on drivers' preferences regarding takeover interface design. The initial data collection in Chapter 14 is extended by an on-road study featuring an almost identical multimodal, user-tailorable takeover interface. In total, 24 participants from the previous simulator study were invited to take part in this on-road study. This provided further data on drivers' preferences regarding takeover interface design along with higher ecological validity, being one of the few studies to research TORs in a naturalistic driving environment (Naujoks et al. 2019).

Since we assume clustering-aided systems to be the mid- to long-term solution for improving performance in driver modelling (Ping et al. 2018; Yi et al. 2019b), we performed a hierarchical agglomerative cluster analysis on the newly gathered on-road data and compared it with the previously collected simulator data and clusters. This allowed us to explore inter-driver and intra-driver variability and contribute to future driver modelling of drivers' preferences regarding multimodal, user-tailorable takeover interfaces.

19.2 METHODS

19.2.1 System Description

19.2.1.1 Study Vehicle

A semi-autonomous driving system (SAE Level 3) was simulated by using an SAE Level 2 semi-autonomous vehicle, namely a pre-production Jaguar I-PACE EV400 AWD. The car was equipped with adaptive cruise control and lane centring whose hands-on detection system was deactivated. This allowed hands-free, semi-autonomous SAE Level 3 driving on a UK motorway. A safety driver accompanied the participant throughout the experiment for safety reasons. Their job was to intervene in case the semi-autonomous system dropped out and ceased operations. For this purpose, the parking brake switch was relocated to the centre hub, enabling the safety driver to perform an emergency stop; the original equipment manufacturer cluster was placed in the footwell of the passenger seat, allowing the safety driver to monitor system operation.

To make the car fit for purpose, it was equipped with a PC (In-CarPC 2020), a dSPACE MicroAutoBox II system (dSPACE 2020), three controller area network (CAN) interfaces (Connective Peripherals 2020), and a Vector CAN box (Vector 2020) (see Figure 19.1). The PC ran SBG software (SBG 2020) and was connected to five cameras – forward view, rear view, footwell, driver face, and over-the-driver-shoulder, and a microphone. Three screens were added to the car's interior: a 14″ thin-film transistor (TFT) display was mounted on the centre console; a 10″ TFT display replaced the original instrument cluster; and a head-up display

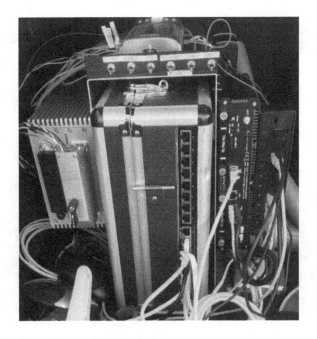

FIGURE 19.1 Car equipment placed in the boot.

(HUD) dash mount in combination with a 1000×250 px TFT display was placed on top of the instrument cluster.

LED lighting strips (Elfeland 2020) were placed around the display on the centre console, underneath the driver's dashboard, and on top of the windshield to provide ambient lighting. A motor (Legett & Platt Automotive 2020) was fitted to the seat base to enable tactile feedback. Two illuminating green buttons were mounted in the thumb positions of the steering wheel and integrated via Arduino (Arduino 2020) to activate and deactivate the semi-autonomous driving system when pressed. Speakers were placed within the car and connected to the PC and a smartphone.

A Microsoft Surface Pro tablet, connected to the PC via WiFi, was used to provide a user interface for the experimenter. The smartphone provided the cognitive load task, delayed digit recall task (1-back), via the MIT AgeLab app (MIT AgeLab 2020). The power supply for all the equipment was provided via the 12 V battery and an interposed inverter.

This set-up enabled the experimenters to interact with the vehicle's assistance systems; to collect CAN Bus, customisation, audio, and video data; and to manipulate the human–machine interface (HMI).

19.2.1.2 Human–Machine Interface

The HMI for the takeover consisted of three visual displays: HUD, infotainment display (ID), and instrument cluster; auditory outputs, audio cues, and vocalisation; seat vibration; ambient lighting; and green buttons mounted to the steering wheel for activation and deactivation of the semi-autonomous system (see Figure 19.2).

FIGURE 19.2 Study vehicle HMI.

The HMI presented in this on-road study was almost identical to the simulator study's one presented in Chapter 14. The current driving mode, manual or autonomous mode, was conveyed via icons and ambient lighting, both colour-coded; amber indicated the manual mode, and blue represented the autonomous mode (see Figure 19.3).

The system informed the driver about the availability of automation applying the manual mode icon with pulsing green buttons, an audio cue, and the message and vocalisation 'Automation Available'. This message was issued by the experimenter via the Wizard of Oz HMI. Once the participant pressed the green buttons, the autonomous mode was confirmed via the message and vocalisation 'Automation Activated', the ambient lighting, and the autonomous mode icon (see Figure 19.4).

After a set period of time, the participant had to take control of the car. This involved three steps. Firstly, the HMI released the message and vocalisation 'Get Ready For Handover' along with seat vibration and an animated icon showing amber hands moving and arrows pointing towards the blue steering wheel. This icon became static after a few seconds, showing amber hands on and amber arrows pointing to the blue steering wheel. Secondly, the takeover protocol started and required the participant to answer driving-related questions which were evaluated by the experimenter controlling the Wizard of Oz HMI. An overview of the questions and accepted, exemplary answers is given in Table 19.1. Finally, the driver was prompted to take control by the message and vocalisation 'Press Green Buttons To Take Control Now'. After the participant pressed the green buttons, the transition to manual mode was confirmed by seat vibration, ambient lighting, the manual mode icon, and the message and vocalisation 'Automation Deactivated. You Are In Control'.

FIGURE 19.3   Driving mode icons.

FIGURE 19.4   Interfaces for 'Automation Activated' message.

TABLE 19.1

Exemplary Answers to the Questions Asked in the Takeover Protocol

Question	Exemplary Answers
What lane are you in?	'Left lane'; 'slow lane'; 'centre lane'; 'fast lane'
Can you see any motorcycles?	'Yes'; 'one'; 'two'; 'no'
Is the road ahead straight?	'Yes'; 'no'; 'there is a bend coming up'
Is your right blind spot warning light on?	'Yes'; 'no'; 'there is a car in the blind spot'
Are there any vehicles alongside?	'Yes'; 'no'; 'there are a lorry and a car'
What is your speed?	'60'; '60 mph'; '60 miles per hour'
What traffic is behind you?	'It is a blue car'; 'It looks like a white Range Rover'
How many car lengths is the vehicle in front away from you?	'3'; 'more than 20'
Is the traffic busy ahead?	'Yes'; 'no'; 'only a few cars'
What is the weather like?	'Nice'; 'rainy'; 'sunny'

19.2.1.3 Customisation Settings

An overview of selectable customisation settings is shown in Figure 19.5. The offered settings were almost identical to the one presented in the previous simulator study (see Chapter 14) and consisted of six binary settings and four ordinal settings. The binary settings changed the amount of information presented in the three screens: ID, HUD, and instrument cluster.

The ordinal settings influenced the intensity of the specific characteristic. The only change to the simulator study settings was that the 'Audio' setting could not be turned to zero; i.e., all audio cues and vocalisations were muted. This decision was made for safety reasons, ensuring mode awareness. All other ordinal settings were calibrated to be noticeable at all times from Levels 1 to 10; to match the intensity experienced in the simulator study; and, therefore, to ensure comparability. Since the questions were an essential part of the takeover interface, the minimum setting for the number of questions was 1, matching the simulator study.

19.2.2 STUDY DESIGN

A within-subject design with a single independent variable – takeover interface – with two levels, default and customised takeover interface, was employed for the on-road study. When looking at the whole project, the previous simulator experiment and the on-road study featured almost identical experimental designs that can be seen as an additional independent variable, namely study type. Although the subsequent paragraphs and sections focus on the on-road study, the data of the two studies will be compared in Section 19.4.

Two trials were conducted successively for the on-road study. In the first trial, participants experienced a default takeover interface which they had experienced before in the previous simulator study. At the end of the first trial, participants customised the settings by changing the customisation matrix (Figure 19.5) via the ID.

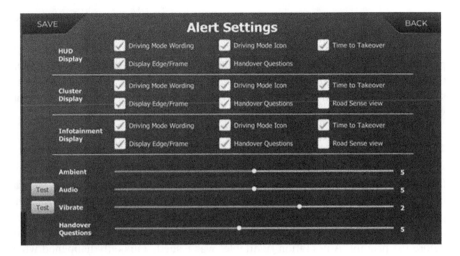

FIGURE 19.5   Customisation settings.

The resulting customised takeover interface was experienced in the second trial. Participants were allowed to make a final adjustment to their customisation profile after the second trial.

The dependent variable – customisation profile – was gathered after each trial. For the binary settings, we collected whether the specific setting was selected or deselected. In regard to the ordinal settings, we gathered the level of the specific setting. Overall, two customisation profiles were collected for each participant during the on-road study.

19.2.3   Procedure

19.2.3.1   Pre-trial

The experiment started at the test track facility of Jaguar Land Rover in Fen End, where participants were welcomed and briefed on the project goal, the connection between the previous simulator study and the upcoming on-road study, and the procedure of the latter one. Subsequently, participants completed the necessary pre-trial documentation, including consent form and attendance list. Before getting into the car, participants were reminded of the HMI design, the available customisation settings, and the takeover procedure, and were introduced to the route and the cognitive load task 1-back.

Once in the car, participants took place in the passenger seat and were introduced to the specifications of the car and its interface by the safety driver. Following the introduction, the safety driver drove onto the test track, explained the essential car controls, and demonstrated semi-autonomous driving, control transitions from and to the car, and the cognitive load task. After the safety driver's demonstration, the participant took place in the driver seat and experienced the previously mentioned situations on the test track. When the participant felt safe to operate the car on an open road, the participant left the test track facility and drove towards a service station from where the study started.

After joining the UK motorway and before the service station, the participant was prompted by the HMI to activate automation. This autonomous drive had a short OOTL period of 1 min that allowed the participant to experience the semi-autonomous system along with control transitions and the cognitive load task on the UK motorway before the start of the actual trial.

19.2.3.2 Trial

After stopping at the service station, the participant was asked whether they felt safe to commence the study. All participants agreed to go ahead. At that point, the experimenter started the data logging and the safety driver told the participant to drive onto the motorway, stay in the left lane, and keep a speed of approximately 60 mph. A short period of 1 min of manual driving followed, before the prompt 'Automation Available' was issued. This only happened when the driving environment was deemed safe by the safety driver. After 30 s, the 1-back cognitive load task was started and ran in 2-min intervals with 30-s breaks in-between.

The OOTL time was set to 10 min for the first automation periods and to 8 min for the third. While the manual drive between the first and the second automation period was 1 min before the semi-autonomous system became available again, it was 7 min between the second and the third. An overview of the trial design is given in Figure 19.6. All of this was necessary due to a lane becoming a slip road, giving a situation the semi-autonomous system would not have been able to handle, and the short distance between this section and the turning point at the upcoming service station.

After the participant parked the car at the service station, the customisation matrix was called on the ID. The participant was then asked whether they would like to make any changes to the HMI. The resulting customisation profile was logged for analysis. This constituted the end of the first trial, and participants had a break during which they were able to access facilities and refreshments.

The second trial involved the same procedures as the first one but incorporated the reversed route ending at the service station where the study started. After completion of the trial, the safety driver swapped with the participant and drove back to the test track facility.

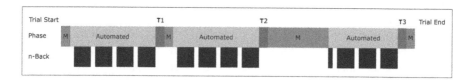

FIGURE 19.6 A timeline showing phases of one trial illustrating manual mode (M), automated mode, and takeovers (T1/T2/T3). *N*-back application periods are shown by the grey bars.

19.2.3.3 Post-trial

At the test track facility, participants were thanked for their participation and accompanied to the exit. In total, the study lasted approximately 240 min.

19.2.4 PARTICIPANTS

In total, 24 (9 female) UK drivers, aged between 28 and 75 ($M = 46$, SD $= 14$), were recruited by an external agency for participation in the on-road study. All of them had participated in a previous simulator study (Chapter 14), which was almost identical to this study in regard to study design. Participants were divided into three age bands in adherence with inclusive design guidelines (Langdon and Thimbleby 2010): (1) 18–34 years ($N = 7$, 3 females), (2) 35–56 years ($N = 10$, 3 females), and (3) 57–82 years ($N = 7$, 3 females). These age bands were defined on a principal using statistical analysis (Langdon and Stanton, personal communication, 2017). Participants had years of driving experience from 9 up to 48 ($M = 24$, SD $= 13$) and covered yearly between 4300 and 30,000 miles ($M = 11,317$, SD $= 5104$).

The study complied with the American Psychological Association Code of Ethics and was approved by the Department of Engineering Research Ethics Committee at the University of Cambridge. All participants gave their informed consent to take part in the study.

19.2.5 DATA ANALYSIS

The data analysis was performed using 'IBM SPSS® Statistics 25' and 'RStudio', and the significance threshold was set at 0.05.

19.2.5.1 Customisation Settings

In order to calculate the overall interface score, the values were normalised into the range [0,1] and added up, giving a minimum of 0 and a maximum of 20. Dividing the values by the maximum score of 20 results in the interface density.

To analyse the significance of change in interface density and the ordinal settings from the simulator study to the on-road study, we applied the nonparametric Wilcoxon signed-rank test. The non-normal distribution, detected with the Kolmogorov–Smirnov test, was the main motivation for using a nonparametric test.

19.2.5.2 Cluster Analysis

As in the simulator study (Chapter 14), a hierarchical agglomerative cluster analysis was performed to cluster participants and interfaces, due to its wide application in psychology-related domains (Everitt et al. 2011). For the clustering of binary interfaces, the squared Euclidean distance was used to measure similarity between cases.

Two different approaches, namely Euclidean distance and Gower distance, were explored to measure similarity between participants. The motivation to use Gower distance was that it allows to calculate distances for attributes with the mixed categorical and numerical values. For the approach using Euclidean distance, all values were normalised into the range [0,1] to ensure a common scaling.

After testing different linkage methods, we decided to apply Ward's linkage as the inter-group proximity measure because it revealed the clearest distinction for all approaches. To decide on the cut-off distance and determine the number of clusters, we used the R package 'NbClust', which provides 30 indices (Charrad et al. 2014), along with the R package 'dynamicTreeCut', which provides 'novel dynamic branch cutting methods' (Langfelder, Zhang and Horvath 2008). These packages allow us to take a methodological approach for determining the number of clusters.

19.3   HYPOTHESES

Hypothesis 1: Because of non-normally distributed results in driving performance (Eriksson and Stanton 2017), clusters found within drivers' data (Yi et al. 2019b), and high inter-driver variability (Miyajima and Takeda 2016), we expected to find distinct customisation setting clusters among participants, as seen in Chapter 14.

Hypothesis 2: Based on the results of the simulator study which showed a clear distinction in drivers' perception of the different visual displays (Chapter 14), we expected to see clusters for the visual display settings again.

Hypothesis 3: Assuming consistent human preferences and traits (Roberts and DelVecchio 2000), we expected to observe a high similarity between participants' simulator and on-road customisation settings despite anticipating some adaptation due to the changed environment.

19.4   RESULTS

The results are structured by theme for overview reasons. Their implications for the hypotheses will be inferred in the discussion.

19.4.1   Overview Customisation Settings

In total, 20 participants (83%) deviated from the default settings and all of them were unique in their combination. The mean interface density was $M = 68\%$ with a standard deviation of $SD = 15\%$ and a range of $R = 41\%$. To measure the dispersion of drivers' preferences, the coefficient of variation was calculated ($CV = 22\%$).

19.4.1.1   Binary Customisation Settings

An overview of the selected and unselected binary settings is shown in Figure 19.7. It shows that the majority of participants (79%) decided to keep the road view option turned off on the cluster as well as the central infotainment screen, indicating low appreciation for it. This is supported by the fact that in the case of the cluster road view setting, 5 out of 10 participants who experienced it in trial 2 decided to deselect it afterwards.

When comparing the different interfaces, HUD, cluster, and ID, we can observe that in total 28 HUD settings, 7 cluster settings, and 35 ID settings were turned off. This means that participants decided to keep the cluster the densest interface.

19.4.1.2   Ordinal Customisation Settings

The descriptive statistics of mean, standard deviation, and the number of participants who turned the setting up or down, along with the number of levels, are shown in

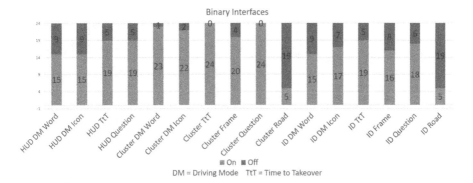

FIGURE 19.7   Selected binary customisation settings in final customisation.

TABLE 19.2
Descriptive Statistics for Ordinal Customisation Settings

Ordinal Setting	Mean	SD	Change in Number of Participants and (Levels)
Ambient	5.83	1.76	↑ 5 (20) — 19 (0) ↓ 0 (0)
Audio	5.29	1.10	↑ 4 (9) — 19 (0) ↓ 1 (2)
Haptic	2.04	0.69	↑ 5 (5) — 16 (0) ↓ 3 (4)
Questions	3.83	1.37	↑ 1 (2) — 8 (0) ↓ 15 (30)

Table 19.2. The number of questions was the ordinal setting which decreased the most with regard to participants as well as levels. This means participants considered the default number of five questions to be too high.

19.4.2   Cluster Analysis of Customisation Settings

19.4.2.1   Clustering Participants

The first hierarchical agglomerative cluster analysis grouped participants based on their customisation profile. The separate analyses of the two distance measures Euclidean and Gower resulted in similar clusters, shown as a tanglegram in Figure 19.8, indicating that the scaling of the ordinal settings mitigated the issue of mixed categorical and numerical variables for the Euclidean distance approach.

Due to its wide application in psychology (Clatworthy et al. 2005; Nosofsky 1985), the Euclidean distance was used in the further analysis to determine the number of clusters with NbClust. Because of the specific characteristics of our data, partly dichotomous and mixed types, we were able to apply 20 out of the 30 indices included in the NbClust package. Based on the majority rule, the best number of clusters is two when limiting the maximum plausible number of clusters to six. However, the margin was very small with three indices indicating two clusters, and two indices indicating five clusters. The subsequent analysis with 'dynamicTreeCut' resulted in

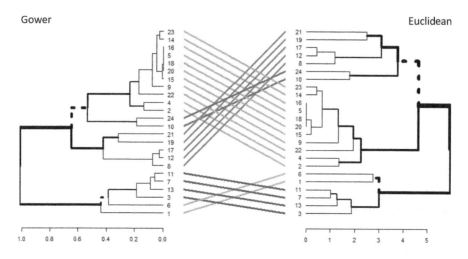

FIGURE 19.8　Clustering participants: Tanglegram Gower and Euclidean distance.

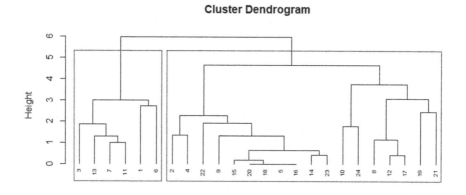

FIGURE 19.9　Clustering participants: Dendrogram using Euclidean distance.

no observable clusters. Figure 19.9 shows the corresponding dendrogram with red rectangles indicating the two possible clusters.

While analysing the interface density and other possible explanations for the two clusters, we discovered that binary interface density can explain the clusters (see Table 19.3), more specifically the selected binary options for the ID. In the smaller cluster, nearly all binary options for the ID were turned off by participants.

19.4.2.2　Clustering Binary Interfaces

The second hierarchical agglomerative cluster analysis grouped binary interface settings based on their psychological similarity perceived by participants. Out of the 20 performance indices calculated using NbClust, seven supported two clusters (see Figure 19.10), five favoured four clusters (see Figure 19.11), and five advocated six

TABLE 19.3

Interface Densities of Identified Clusters

Interface	N	Mean Interface Density		
		Binary Settings (%)	Ordinal Settings (%)	Overall Settings (%)
Heavy	18	79	50	73
Light	6	50	55	51

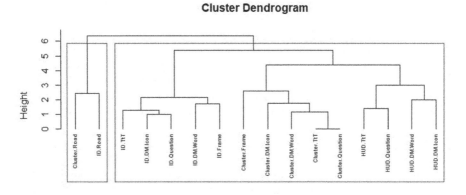

FIGURE 19.10 Clustering binary interfaces: Dendrogram with two clusters.

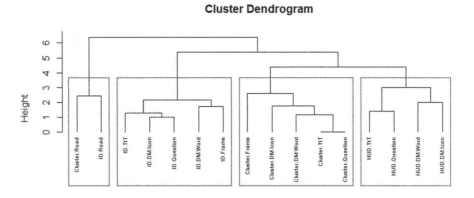

FIGURE 19.11 Clustering binary interfaces: Dendrogram with four clusters.

clusters. The dynamicTreeCut package indicated no observable clusters. According to the majority rule, the optimal number of clusters is two.

19.4.2.3   Comparing Simulator and On-Road Study

Comparing the simulator to the on-road customisation settings, we can observe that average interface density was non-normally distributed (simulator: $D(24)=0.178$, $p=0.049$; on-road: $D(24)=0.251$, $p<0.001$) and increased by 13.12%. However, a Wilcoxon signed-rank test revealed that the difference is insignificant ($Z=-1.543$, $p=0.123$). When analysing the average number of questions, we can also observe an increase from 3.5 in the simulator study to 3.83 in the on-road study. This difference was not significant though ($Z=-0.772$, $p=0.44$). A significant increase from simulator to on-road study was found for the ambient setting ($Z=-2.895$, $p=0.004$), indicating that participants preferred a stronger ambient lighting during the on-road study. All this shows that there are indications for participants' desire for more salient interfaces for on-road driving.

To compare the extent to which participants were grouped similarly in the simulator and on-road study, a tanglegram, as shown in Figure 19.12, was created. We can observe that nearly all participants, with the exception of two, were not grouped in similar clusters. This indicates that the consistency of clusters was low over the course of the two studies and that intra-driver variability was high.

19.5   DISCUSSION

This study collected drivers' customisation data for a user-tailorable takeover interface of an SAE Level 3 semi-autonomous car driving on the UK motorway. The results show a high inter-driver and intra-driver variability with many unique customisation profiles. Subsequently, we discuss them with regard to our hypotheses and study limitations.

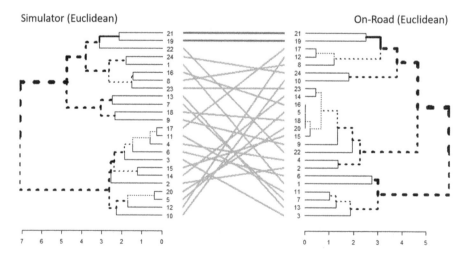

FIGURE 19.12   Comparison simulator and on-road study: Tanglegram.

19.5.1 HYPOTHESES

Hypothesis 1 is accepted since the observed participant clusters showed two differences: usage of the ID and the overall interface density. The light group, in contrast to the heavy, deselected most of the settings on the ID, indicating they are not making use of the interface for takeover-specific information. This understandably resulted in a lower interface density in comparison to the other group as both factors correlate with each other. The identified clusters for drivers' customisation data extend the current research which found and used clusters in performance and behaviour data (Ping et al. 2018; Yi et al. 2019b).

The number of unique customisation profiles (83%) demonstrates clearly the individuality of drivers and their interface preferences, and is thereby in agreement with research showing large inter-driver variability (Miyajima and Takeda 2016). The growth of customisation in many areas of the automotive sector (Akamatsu, Green and Bengler 2013; Caber, Langdon and Clarkson, 2018, 2019; Mercedes-Benz 2020; Pollard 2018) is worthwhile being pursued to meet individuals' preferences and, as this study indicates, should be carried over to takeover interfaces.

The naturalistic data sets on driver behaviour and state (Aghaei et al. 2016; Barnard et al. 2016; National Academies of Sciences, Engineering, and Medicine 2014; Fridman et al. 2019; Miyajima and Takeda 2016) have been extended by the newly collected data on drivers' preferences regarding user-tailorable takeover interfaces for non-critical transitions. However, the data collection needs to be continued in order to provide sufficiently large data sets enabling the application of further, more data-intensive machine-learning techniques.

Hypothesis 2 is accepted since at least two clusters can be observed for the binary interfaces based on the analysis performed. Although the majority rule supports two clusters, the narrow result along with qualitative analysis of the settings being clustered favours four clusters. As hierarchical agglomerative clustering is an explorative technique, there is no clear answer to this question. Therefore, both solutions and their implications are discussed.

For the two-cluster solution, the clear distinction between them is that one group accommodates default and the other one non-default settings. Drivers seem to perceive a higher psychological inter-group than intra-group distance for default and non-default settings. This result matches with the previous simulator study (Chapter 14) and research showing drivers' tendency to keep the factory settings (Reagan et al. 2018). A possible explanation for this observation could be that drivers assume the default settings to be optimised. In summary, drivers tend not to use customisation to its full extent and consequently miss out on potential benefits, such as comfort and safety (Akamatsu, Green and Bengler 2013; Blaschke et al. 2009; Caber, Langdon and Clarkson, 2018, 2019).

The four-cluster solution separates the different visual displays, instrument cluster, HUD, and ID, indicating that participants perceived them as independent entities. This differs from the results in Chapter 14 where HUD and ID were seen as one entity independent from the instrument cluster. In fact, this time HUD and cluster merge first, showing higher similarity to each other than to the ID. Considering the spatial distance and the fact that they are in the driver's field of view when driving supports this observation.

Despite the higher psychological similarity between HUD and cluster along with increased availability of HUDs in many cars, participants' customisation profiles suggest that participants, by keeping the cluster the densest display, relied on it the most when searching for specific information. This implies that the cluster is still the most important source of information and should not be replaced for now. This reliance on the cluster display conforms with previous research (Chapter 14).

Hypothesis 3 is rejected since the tanglegram comparing the clusters from the simulator and the on-road study (Figure 19.12) suggests high intra-driver variability. Although we anticipated some variability due to the changed environment, we nevertheless expected drivers within the same cluster to adapt similarly and, as a consequence, to stay in the same cluster. The significant increase in ambient setting along with the indications of overall higher interface density illustrates that drivers indeed adapted to the environment, supporting the first part of our hypothesis. The tanglegram, on the other hand, disproves the second part of the hypothesis. Our results seem to disagree with research in consistency of personality traits (Roberts and DelVecchio 2000). However, it is possible to argue that simulator and naturalistic on-road situations evoke different personality traits, supporting the argument for more naturalistic studies and data (Miyajima and Takeda 2016; Naujoks et al. 2019).

Nevertheless, the data and results from both studies feed into data sets on drivers' interaction with a user-tailorable takeover interface and can be used as input into further, more-advanced machine-learning techniques. Adaptive, context- and driver-aware systems, altering the in-car HMI to some stimuli (Amditis et al. 2010; Bellotti et al. 2005; Heigemeyr and Harrer 2014; Ma et al. 2017; Michon 1993; Wright et al. 2017), illustrate the long-term goal. This study lays the foundation to extend these adaptive systems to a tailorable takeover interface by providing participant clusters and their preferences. In future, the effect of external factors, such as weather and journey length, on participants should be addressed to get a more comprehensive behaviour model.

19.5.2 STUDY LIMITATIONS

The proactive approach to ask participants whether they would like to make changes to the takeover interface along with the high rate of occurrences is a clear study limitation as it may have encouraged participants to make changes albeit not seeing the necessity. In this sense, we can expect a stark deviation in the number of unique settings between this study and the real world where drivers may stick to factory settings.

Another limitation is the restricted scale of the collected data as this study looked into a very specific scenario, the non-critical transition, and a very specific interface design. Consequently, the observed clusters and driver preferences may not be fully transferable to other areas and interfaces. Even in close fields of application, such as critical control transitions, we must anticipate totally different driver responses. The transferability of clusters and preferences needs to be addressed in future work.

As an exploratory method, hierarchical agglomerative clustering cannot provide a ground truth and always requires qualitative, human analysis to determine proximity measure, clustering method, and the number of clusters. As a consequence, different decisions lead to different results. The use of standardised validation methods aimed at minimising bias.

19.6 CONCLUSION

The takeover interface will become a crucial part of the in-car HMI, when semi-autonomous systems enter the market, and drivers were hypothesised to show different preferences concerning its exact layout due to inter-driver variability. The study results reveal distinguishable clusters showing different preferences for a user-tailorable takeover interface in a naturalistic on-road driving environment. However, not only inter-driver variability but also intra-driver variability seems to be high as drivers were not grouped similarly when comparing the results of the previous simulator to the current on-road study.

ACKNOWLEDGEMENTS

This work was supported by Jaguar Land Rover and the UK-EPSRC Grant EP/N011899/1 as part of the jointly funded 'Towards Autonomy: Smart and Connected Control (TASCC) Programme'.

REFERENCES

Aghaei, A. S., B. Donmez, C. C. Liu, D. He, G. Liu, K. N. Plataniotis, H. Y. W. Chen, and Z. Sojoudi. 2016. "Smart driver monitoring: when signal processing meets human factors: in the driver's seat." *IEEE Signal Processing Magazine, 33, 6* 35–48. https://doi.org/10.1109/MSP.2016.2602379.

Akamatsu, M., P. Green, and K. Bengler. 2013. "Automotive technology and human factors research: past, present, and future." *International Journal of Vehicular Technology, 2013* 1–27. https://doi.org/10.1155/2013/526180.

Amditis, A., L. Andreone, K. Pagle, G. Markkula, E. Deregibus, M. Romera Rue, F. Bellotti, and et al. 2010. "Towards the Automotive HMI of the Future: Overview of the AIDE-Integrated Project Results." *IEEE Transactions on Intelligent Transportation Systems, 11, 3* 567–578. https://doi.org/10.1109/TITS.2010.2048751.

Arduino. 2020. Accessed May 7, 2020. https://www.arduino.cc/.

Arora, N., X. Dreze, A. Ghose, J. D. Hess, R. Iyengar, B. Jing, Y. Joshi, and et al. 2008. "Putting one-to-one marketing to work: personalization, customization, and choice." *Marketing Letters, 19* 305–321. https://doi.org/10.1007/s11002-008-9056-z.

Bainbridge, L. 1983. "Ironies of automation." *Analysis, Design and Evaluation of Man–Machine Systems. Proceedings of the IFAC/IFIP/IFORS/IEA Conference.* Baden-Baden: Elsevier Ltd. 129–135. https://doi.org/10.1016/B978-0-08-029348-6.50026-9.

Barnard, Y., F. Utesch, N. van Nes, R. Eenink, and M. Baumann. 2016. "The study design of UDRIVE: the naturalistic driving study across Europe for cars, trucks and scooters." *European Transport Research Review, 8, 14.* https://doi.org/10.1007/s12544-016-0202-z.

Bellet, T., and J. Manzano. 2007. "To be available or not, that is the question: a pragmatic approach to avoid drivers overload and manage in-vehicle information." *Proceedings of the 20th International Technical Conference on the Enhanced Safety of Vehicles (ESV).* Lyon.

Bellotti, F., A. De Gloria, R. Montanari, N. Dosio, and D. Morreale. 2005. "COMUNICAR: designing a multimedia, context-aware human-machine interface for cars." *Cognition, Technology & Work, 7* 36–45. https://doi.org/10.1007/s10111-004-0168-9.

Bishop, R. 2019. *Is 2020 The Year For Eyes-Off Automated Driving?* 1 October. Accessed April 21, 2020. https://www.forbes.com/sites/richardbishop1/2019/10/01/is-2020-the-year-for-eyes-off-automated-driving/.

Blaschke, C., F. Breyer, B. Färber, J. Freyer, and R. Limbacher. 2009. "Driver distraction based lane-keeping assistance." *Transportation Research Part F: Traffic Psychology and Behaviour, 12* 288–299. https://doi.org/10.1016/j.trf.2009.02.002.

Borojeni, S. S., L. Chuang, W. Heuten, and S. Boll. 2016. "Assisting drivers with ambient take-over requests in highly automated driving." *AutomotiveUI - Proceedings of the 8th International Conference on Automotive User Interfaces and Interactive Vehicular Applications.* Ann Arbor: Association for Computing Machinery. 237–244. https://doi.org/10.1145/3003715.3005409.

Bourrelly, A., C. Jacobé de Naurois, A. Zran, F. Rampillon, J. Vercher, and C. Bourdin. 2019. "Long automated driving phase affects take-over performance." *IEEE Intelligent Transport Systems, 13, 8* 1249–1255. doi: 10.1049/iet-its.2019.0018.

Caber, N., P. Langdon, and P. J. Clarkson. 2018. "Intelligent driver profiling system for cars – a basic concept." *International Conference on Universal Access in Human-Computer Interaction. Virtual, Augmented, and Intelligent Environments.* Cham: Springer. 201–213. https://doi.org/10.1007/978-3-319-92052-8_16.

Caber, N., P. Langdon, and P. J. Clarkson. 2019. "Designing adaptation in cars: an exploratory survey on drivers' usage of ADAS and car adaptations." *International Conference on Applied Human Factors and Ergonomics - Advances in Human Factors of Transportation.* Cham: Springer. 95–106. https://doi.org/10.1007/978-3-030-20503-4_9.

Charrad, M., N. Ghazzali, V. Boiteau, and A. Niknafs. 2014. "NbClust: an R package for determining the relevant number of clusters in a data set." *Journal of Statistical Software, 61, 6* 1–36. doi: 10.18637/jss.v061.i06.

Clatworthy, J., D. Buick, M. Hankins, J. Weinman, and R. Horne. 2005. "The use and reporting of cluster analysis in health psychology: a review." *British Journal of Health Psychology, 10* 329–358. https://doi.org/10.1348/135910705X25697.

Connective Peripherals. 2020. *USB2-F-7101.* Accessed May 7, 2020. https://connectiveperipherals.com/

Desmond, P. A., P. A. Hancock, and J. L. Monette. 1998. "Fatigue and automation-induced impairments in simulated driving performance." *Transportation Research Record: Journal of the Transportation Research Board, 1628, 1* 8–14. https://doi.org/10.3141/1628-02.

Dong, Y., Z. Hu, K. Uchimura, and N. Murayama. 2011. "Driver inattention monitoring system for intelligent vehicles: a review." *IEEE Transactions on Intelligent Transportation Systems, 12, 2* 596–614. https://doi.org/10.1109/TITS.2010.2092770.

dSPACE. 2020. *MicroAutoBox II.* Accessed May 7, 2020. https://www.dspace.com/en/ltd/home/products/hw/micautob/microautobox2.cfm.

Elfeland. 2020. *LED Strip Lights Kits.* Accessed May 7, 2020. https://www.amazon.co.uk/Elfeland-Waterproof-Controller-Self-Adhesive-Decoration/dp/B07H2DR5DY/ref=sr_1_6?keywords=led+rgb+strip&qid=1559907897&s=gateway&sr=8-6.

Endsley, M. R. 2010. "Level of automation effects on performance, situation awareness and workload in a dynamic control task." *Ergonomics, 42, 3* 462–492. https://doi.org/10.1080/001401399185595.

Endsley, M. R., and E. O. Kiris. 1995. "The out-of-the-loop performance problem and level of control in automation." *Human Factors, 37, 2* 381–394. https://doi.org/10.1518/001872095779064555.

Eriksson, A., and N. A. Stanton. 2017. "Takeover time in highly automated vehicles: noncritical transitions to and from manual control." *Human Factors: The Journal of the Human Factors and Ergonomics Society, 59, 4* 689–705. https://doi.org/10.1177/0018720816685832.

Eriksson, A., S. M. Petermeijer, M. Zimmermann, J. C. F. de Winter, K. J. Bengler, and N. A. Stanton. 2019. "Rolling out the red (and green) carpet: supporting driver decision making in automation-to-manual transitions." *IEEE Transactions on Human-Machine Systems, 49, 1* 20–31. doi: 10.1109/THMS.2018.2883862.

Everitt, B. S., S. Landau, M. Leese, and D. Stahl. 2011. *Cluster Analysis*, 5th Edition. Chichester: John Wiley & Sons, Ltd. https://doi.org/10.1002/9780470977811.

Fagnant, D. J., and K. Kockelman. 2015. "Preparing a nation for autonomous vehicles: opportunities, barriers and policy recommendations." *Transportation Research Part A: Policy and Practice, 77* 167–181. https://doi.org/10.1016/j.tra.2015.04.003.

Fridman, L., D. E. Brown, M. Glazer, W. Angell, S. Dodd, B. Jenik, J. Terwilliger, and et al. 2019. "MIT advanced vehicle technology study: large-scale naturalistic driving study of driver behavior and interaction with automation." *IEEE Access, 7* 102021–102038. https://doi.org/10.1109/ACCESS.2019.2926040.

García-barrios, V. M., F. Mödritscher, and C. Gütl. 2005. "Personalisation versus adaptation? A user-centred model approach and its application." *Proceedings of the International Conference on Knowledge Management (I-KNOW).* 120–127.

Gold, C., D. Damböck, L. Lorenz, and K. Bengler. 2013. ""Take over!" How long does it take to get the driver back into the loop?" *Proceedings of the Human Factors and Ergonomics Society Annual Meeting, 57, 1* 1938–1942. https://doi.org/10.1177/1541931213571433.

Gold, C., F. Naujoks, J. Radlmayr, H. Bellem, and O. Jarosch. 2018. "Testing scenarios for human factors research in level 3 automated vehicles." In: N. Stanton (eds) *Advances in Human Aspects of Transportation. AHFE 2017. Advances in Intelligent Systems and Computing*, vol 597. Cham: Springer. 551–559. https://doi.org/10.1007/978-3-319-60441-1_54.

Gonçalves, J., and K. Bengler. 2015. "Driver state monitoring systems– transferable knowledge manual driving to HAD." *Procedia Manufacturing, 3* 3011–3016. https://doi.org/10.1016/j.promfg.2015.07.845.

Green, P. A. 2008. "Motor vehicle-driver interfaces." In A. Sears, and J. A. Jacko (eds) *Human Computer Interaction Handbook: Fundamentals, Evolving Technologies, and Emerging Applications, Human Factors and Ergonomics.* New York: Lawrence Erlbaum Association. 701–719.

Heigemeyr, A., and A. Harrer. 2014. "Information management for adaptive automotive human machine interfaces." *AutomotiveUI - Proceedings of the 6th International Conference on Automotive User Interfaces and Interactive Vehicular Applications.* Seattle: Association for Computing Machinery. 1–8. https://doi.org/10.1145/2667317.2667341.

In-CarPC. 2020. *Vehicle and Car Computer Systems | In-CarPC.* Accessed May 7, 2020. https://www.in-carpc.co.uk/.

Körber, M., C. Gold, D. Lechner, and K. Bengler. 2016. "The influence of age on the takeover of vehicle control in highly automated driving." *Transportation Research Part F: Traffic Psychology and Behaviour, 39* 19–32. https://doi.org/10.1016/j.trf.2016.03.002.

Langdon, P., and H. Thimbleby. 2010. "Inclusion and interaction: designing interaction for inclusive populations." *Interacting with Computers, 22* 439–448. https://doi.org/10.1016/j.intcom.2010.08.007.

Langfelder, P., B. Zhang, and S. Horvath. 2008. "Defining clusters from a hierarchical cluster tree: the dynamic tree cut package for R." *Bioinformatics, 24, 5* 719–720. https://doi.org/10.1093/bioinformatics/btm563.

Leggett & Platt Automotive. 2020. *Vehicle Seating and Lumbar Support.* Accessed May 7, 2020. https://leggett-automotive.com//products.

Ma, Z., M. Mahmoud, P. Robinson, E. Dias, and L. Skrypchuk. 2017. "Automatic detection of a driver's complex mental states." In: O. Gervasi, B. Murgante, S. Misra, G. Borruso, C. M. Torre, A. M. A. C. Rocha, D. Taniar, and et al. (eds) *Computational Science and Its Applications – ICCSA 2017. ICCSA 2017. Lecture Notes in Computer Science*, vol 10406. Cham: Springer. 678–691. https://doi.org/10.1007/978-3-319-62398-6_48.

McNamara, P. 2017. *How Did Audi Make the First Car with Level 3 Autonomy?* 12 July. Accessed May 2, 2018. https://www.carmagazine.co.uk/car-news/tech/audi-a3-level-3-autonomy-how-did-they-get-it-to-market/.

Melcher, V., S. Rauh, F. Diederichs, H. Widlroither, and W. Bauer. 2015. "Take-over requests for automated driving." *Procedia Manufacturing, 3* 2867–2873. https://doi.org/10.1016/j.promfg.2015.07.788.

Merat, N., A. H. Jamson, F. C. H. Lai, and O. Carsten. 2012. "Highly automated driving, secondary task performance, and driver state." *Human Factors: The Journal of the Human Factors and Ergonomics Society, 54* 762–771. https://doi.org/10.1177/0018720812442087.

Merat, N., A. H. Jamson, F. C. H. Lai, M. Daly, and O. M. J. Carsten. 2014. "Transition to manual: driver behaviour when resuming control from a highly automated vehicle." *Transportation Research Part F: Traffic Psychology and Behaviour, 27* 274–282. https://doi.org/10.1016/j.trf.2014.09.005.

Mercedes-Benz. 2020. *A Fragrance for the New S-Class.* Accessed May 12, 2020. https://www.mercedes-benz.com/en/innovation/vehicle-development/a-fragrance-for-the-new-s-class/.

Michon, J. A. 1993. *Generic Intelligent Driver Support: a Comprehensive Report on GIDS.* London: Taylor & Francis.

MIT AgeLab. 2020. *Delayed Digit Recall (n-back) Task.* Accessed May 7, 2020. https://agelab.mit.edu/delayed-digit-recall-n-back-task.

Miyajima, C., and K. Takeda. 2016. "Driver-behavior modeling using on-road driving data: a new application for behavior signal processing." *IEEE Signal Processing Magazine, 33, 6* 14–21. https://doi.org/10.1109/MSP.2016.2602377.

National Academies of Sciences, Engineering, and Medicine. 2014. *Naturalistic Driving Study: Technical Coordination and Quality Control.* Washington, DC: The National Academies Press. https://doi.org/10.17226/22362.

Naujoks, F., C. Mai, and A. Neukum. 2014. "The effect of urgency of take-over requests during highly automated driving under distraction conditions." *5th International Conference on Applied Human Factors and Ergonomics.* Krakow: AHFE. 431–438.

Naujoks, F., C. Purucker, K. Wiedemann, and C. Marberger. 2019. "Noncritical state transitions during conditionally automated driving on german freeways: effects of non-driving related tasks on takeover time and takeover quality." *Human Factors: The Journal of the Human Factors and Ergonomics Society, 61, 4* 596–613. https://doi.org/10.1177/0018720818824002.

Neubauer, C., G. Matthews, and D. Saxby. 2014. "Fatigue in the automated vehicle: do games and conversation distract or energize the driver?" *Proceedings of the Human Factors and Ergonomics Society Annual Meeting, 58, 1* 2053–2057. https://doi.org/10.1177/1541931214581432.

Nosofsky, R. 1985. "Overall similarity and the identification of separable-dimension stimuli: a choice model analysis." *Perception & Psychophysics, 38* 415–432. https://doi.org/10.3758/BF03207172.

Payre, W., J. Cestac, and P. Delhomme. 2014. "Intention to use a fully automated car: attitudes and a priori acceptability." *Transportation Research Part F: Traffic Psychology and Behaviour, 27* 252–263. https://doi.org/10.1016/j.trf.2014.04.009.

Petermeijer, S. M., P. Bazilinskyy, K. Bengler, and J. de Winter. 2017. "Take-over again: investigating multimodal and directional TORs to get the driver back into the loop." *Applied Ergonomics, 62* 204–215. https://doi.org/10.1016/j.apergo.2017.02.023.

Petermeijer, S. M., S. Cieler, and J. C. F. de Winter. 2017. "Comparing spatially static and dynamic vibrotactile take-over requests in the driver seat." *Accident Analysis & Prevention, 99* 218–227. https://doi.org/10.1016/j.aap.2016.12.001.

Ping, P., W. Qin, Y. Xu, C. Miyajima, and T. Kazuya. 2018. "Spectral clustering based approach for evaluating the effect of driving behavior on fuel economy." *2018 IEEE International Instrumentation and Measurement Technology Conference (I2MTC).* Houston: IEEE. 1–6. https://doi.org/10.1109/I2MTC.2018.8409675.

Politis, I., S. Brewster, and F. Pollick. 2015. "Language-based multimodal displays for the handover of control in autonomous cars." *AutomotiveUI - Proceedings of the 7th International Conference on Automotive User Interfaces and Interactive Vehicular Applications.* Nottingham: Association for Computing Machinery. 3–10. https://doi.org/10.1145/2799250.2799262.

Politis, I., S. Brewster, and F. Pollick. 2017. "Using multimodal displays to signify critical handovers of control to distracted autonomous car drivers." *International Journal of Mobile Human-Computer Interaction, 9, 3* 1–16. https://doi.org/10.4018/ijmhci.2017070101.

Pollard, T. 2018. *BMW's New iDrive OS 7.0: Hands-On Test.* 17 April. Accessed May 2, 2018. https://www.carmagazine.co.uk/car-news/tech/bmw-idrive-os-70-what-you-need-to-know-about-the-new-2018-idrive/.

Reagan, I. J., J. B. Cicchino, L. B. Kerfoot, and R. A. Weast. 2018. "Crash avoidance and driver assistance technologies – are they used?" *Transportation Research Part F: Traffic Psychology and Behaviour, 52* 176–190. https://doi.org/10.1016/j.trf.2017.11.015.

Roberts, B. W., and W. F. DelVecchio. 2000. "The rank-order consistency of personality traits from childhood to old age: a quantitative review of longitudinal studies." *Psychological Bulletin, 126* 3–25. https://doi.org/10.1037/0033-2909.126.1.3.

SAE J3016. 2016. *Taxonomy and Definitions for Terms Related to Driving Automation Systems for On-Road Motor Vehicles.* Accessed November 3, 2020. https://www.sae.org/standards/content/j3016_201806/.

SBG. 2020. *Fusion. SBG Sports Software.* Accessed May 7, 2020. https://sbgsportssoftware.com/product/fusion/.

Schömig, N., V. Hargutt, A. Neukum, I. Petermann-Stock, and I. Othersen. 2015. "The interaction between highly automated driving and the development of drowsiness." *Procedia Manufacturing, 3* 6652–6659. https://doi.org/10.1016/j.promfg.2015.11.005.

Stanton, A. A., and P. Marsden. 1996. "From fly-by-wire to drive-by-wire: safety implications of automation in vehicles." *Safety Science, 24* 35–49. https://doi.org/10.1016/S0925-7535(96)00067-7.

Strand, N., J. Nilsson, I. C. M. Karlsson, and L. Nilsson. 2014. "Semi-automated versus highly automated driving in critical situations caused by automation failures." *Transportation Research Part F: Traffic Psychology and Behaviour, 27* 218–228. https://doi.org/10.1016/j.trf.2014.04.005.

Supulniece, I. 2012. "Conceptual Aspects of User-Oriented Adaptive Systems." *ICIS 2012 Proceedings. Presented at the International Conference on Information Systems (ICIS).* 116–124.

Terzi, R., S. Sagiroglu, and M. U. Demirezen. 2018. "Big data perspective for driver/driving behavior." *IEEE Intelligent Transportation Systems Magazine, 1* 20–35. https://doi.org/10.1109/MITS.2018.2879220.

Vector. 2020. *Know-How & Solutions for CAN/CAN FD.* Accessed May 7, 2020. https://www.vector.com/us/en-us/know-how/technologies/networks/can/.

Wright, J., Q. Stafford-Fraser, M. Mahmoud, P. Robinson, E. Dias, and L. Skrypchuk. 2017. "Intelligent scheduling for in-car notifications." *2017 IEEE 3rd International Forum on Research and Technologies for Society and Industry (RTSI).* Modena: IEEE. 1–6. https://doi.org/10.1109/RTSI.2017.8065957.

Yi, D., J. Su, C. Liu, and W. H. Chen. 2019a. "Personalized driver workload inference by learning from vehicle related measurements." *IEEE Transactions on Systems, Man, and Cybernetics: Systems, 49* 159–168. https://doi.org/10.1109/TSMC.2017.2764263.

Yi, D., J. Su, C. Liu, and W. H. Chen. 2019b. "New driver workload prediction using clustering-aided approaches." *IEEE Transactions on Systems, Man, and Cybernetics: Systems, 49* 64–70. https://doi.org/10.1109/TSMC.2018.2871416.

Young, M. S., and N. A. Stanton. 2007. "Back to the future: brake reaction times for manual and automated vehicles." *Ergonomics, 50, 1* 46–58. https://doi.org/10.1080/00140130600980789.

20 Validating OESDs in an On-Road Study of Semi-Automated Vehicle-to-Human Driver Takeovers

Neville A. Stanton, James W. H. Brown,
Kirsten M. A. Revell, Jisun Kim, and Joy Richardson
University of Southampton

Patrick Langdon
Edinburgh Napier University

Michael Bradley and Nermin Caber
University of Cambridge

Lee Skrypchuk and Simon Thompson
Jaguar Land Rover

CONTENTS

20.1 INTRODUCTION

The design of the interface between human driver and vehicle automation will be critical for successful takeovers, especially when control of the vehicle is being handed back to the human driver (Eriksson et al. 2019; Eriksson and Stanton 2017a, 2017b; Clark, Stanton and Revell 2019a). There are many examples of the failure to hand vehicle control back successfully, from both simulator studies (Banks, Stanton and Harvey 2014; Stanton, Young and McCaulder 1997) and real on-road collisions that have ended in fatalities (Banks, Plant and Stanton 2018; Stanton et al. 2019). The problem of handing back control from an automated system to a human operator is not restricted to the road domain, as examples from aviation have shown (Stanton and Marsden 1996; Salmon, Walker and Stanton 2016). In aviation, mode confusion in a perfectly functioning aircraft has led to situations whereby pilots have crashed. Air France flight AF447 is a case in point. The hand-back of control of the aircraft was misunderstood by the pilots, who seemed to think that the aircraft was in an over-speed condition. This is a classic example of mode confusion (Sarter and Woods 1995). The confusion led the pilot flying to pull back on the control stick, which induced a wing stall in the aircraft, causing the aircraft to drop into the ocean, killing all on board (Salmon, Walker and Stanton 2016). Mode confusion is also of great concern in automated road vehicles (Stanton, Dunoyer and Leatherland 2011). When vehicle control is handed back to a human driver, it is particularly important that the driver is aware of the vehicle status, road environment, and pertinent road infrastructure as well as other road users (Stanton et al. 2017). Therefore, the design of the takeover is critical for success (Banks et al. 2018; Clark, Stanton and Revell 2019b; Eriksson et al. 2019).

A structured approach to design is required for developing the takeover requirements. Operator event sequence diagrams (OESDs) are one such approach among others (Stanton et al. 2013). OESDs were selected for this project because they had been used previously in simulator studies of driver takeovers (Stanton et al. 2020). OESDs have been used in the design of human–machine interaction and interfaces for over 60 years (Kurke 1961). Since that time, they have been used in a wide variety of applications, including analysis of aircraft landing procedures (Sorensen, Stanton and Banks 2011), evaluation of single pilot operations (Harris, Stanton and Starr 2015), analysis of the relationship between air traffic control and flight decks on civilian airliners (Walker et al. 2010), analysis of the activities between maintenance operator in the field and operations in the central control room of an electrical energy distribution company (Salmon et al. 2008), comparison of traditional and new procedures for managing collision avoidance in the maritime domain (Kurke 1961), and comparing approaches to automatic emergency braking systems in road vehicles (Banks, Stanton and Harvey 2014). OESDs represent the different aspects of systems (including the human operator, interfaces, and technical elements) in separate columns (colloquially called 'swim-lanes' in OESD parlance), against time.

Previous studies into the design of takeover protocols for vehicle automation have found that the OESDs have good predictive validity, with a median Phi of over 0.8 (Stanton et al. 2020). This means that the OESDs were able to predict the majority of the behaviours drivers engaged in during the vehicle control takeover process.

In both of these studies, OESDs were constructed in workshops with experts in vehicle engineering, computing, and human factors. The behaviour of the driver during that takeover process was described in the driver column together with any interactions with other aspects of the vehicle (such as hearing or reading any information as well as making any verbal or physical responses). The drivers' behaviours during the hand-back of vehicle control in the OESD were then compared to those observed in the video recordings from the driving simulators. In the first study, there were over 100 drivers in two separate driving simulators: one desktop ($N=49$) and one full vehicle ($N=60$). In both studies, the predictive validity of the OESDs was very good (Stanton et al. 2020). In the second study, there were 65 drivers undertaking four trials and the correlational data showed there was also a good predictive validity for each trial (Stanton et al. 2020b). As both of these previous studies report on data collected in driver simulators, so it is an important to see if these findings generalise to on-road studies. While it may seem reasonable to expect the predictive validity of OESDs for the hand-back of vehicle control to human drivers in on-road studies to mirror those of driving simulators, it is important to test validity rather than to assume it (Stanton and Young 1999; Stanton 2016). To that end, the OESDs were constructed prior to the on-road studies, as described in the next section.

20.2   CONSTRUCTION OF OESDs

Further guidance on the development of OESDs may be found in Kirwan and Ainsworth (1992) and Stanton et al. (2013). The analysis presented in this paper is based on a use-case of vehicle automation takeover scenario on a UK motorway with a SAE Level 3 vehicle (SAE J3016 2016). It is assumed that drivers will drive manually onto the motorway and hand the driving task over to vehicle automation when it becomes available. While vehicle automation is engaged, the driver is free to engage in non-driving tasks (such as reading, emailing, working on a tablet computer). The vehicle would alert the driver of the need to take back control of the vehicle in a planned, non-emergency takeover in a timely manner before the exit junction. These takeovers are described using the task elements from OESDs, as shown in Table 20.1.

The OESDs as shown in Figures 20.1–20.5 were developed in workshops with Human Factors and Automotive Engineering experts.

The nine swim-lanes show the different 'actors' under consideration in the design of the handovers to and takeovers from the vehicle automation by the human driver (via the instrument cluster – instruments viewed through the steering wheel), head-up display (HUD – viewed in the windscreen or windshield), centre console (the upper part of the centre of the dashboard), ambient (lighting around the dashboard and vehicle interior), and haptic (vibration through the driver's seat) displays. The arrows are connectors that show the links between the events in the swim-lanes. The takeover protocol presented in Figures 20.1–20.5 was designed to raise the situation awareness of drivers, by presenting them with contextually relevant questions about the vehicle status, other road users as well as the surrounding environment and infrastructure. This was based on the research evidence that degraded performance of drivers of automated vehicles is, in part, due to poor situation awareness (Stanton et al. 2017). For example, the collisions in the Tesla and Uber vehicles report that

TABLE 20.1

Key for the Operator Event Sequence Diagrams

OESD Task Elements   Description

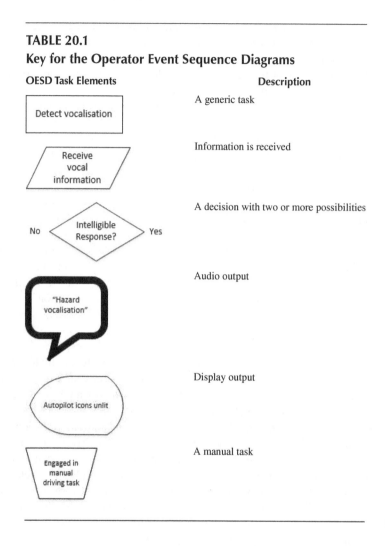

A generic task

Information is received

A decision with two or more possibilities

Audio output

Display output

A manual task

the driver was not aware of the environment outside the vehicle (Banks et al. 2018; Stanton et al. 2019). As can be seen in Figure 20.1, it is assumed that the vehicle is under manual control until the system detects that the road is suitable for automation to operate. Then, the system prompts the driver with the message that automation is available, via the four interfaces (cluster, HUD, centre console, and ambient display), should they wish to use it.

If the driver chooses to engage vehicle automation, then they would press two buttons on the steering wheel with their thumbs simultaneously (assuming that their hands are in the ten-to-two clock position on the steering wheel). At this point, the interfaces would display 'Automation Activated' followed by 'The car is in control' (see Figure 20.2). At the same time, the ambient lighting in the car would change from orange (indicating manual driving mode) to blue (indicating automated driving mode). Then, the driver is able to engage in non-driving tasks (on a tablet computer in

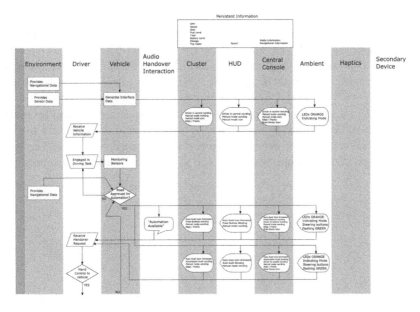

FIGURE 20.1 Vehicle in manual mode with automation available.

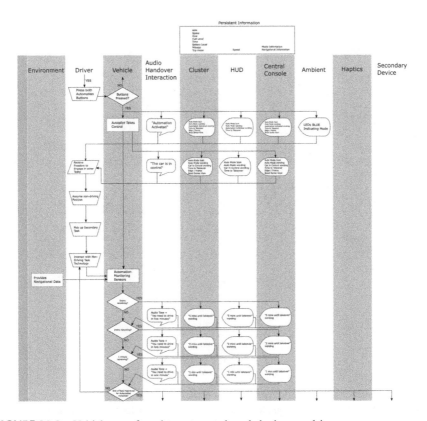

FIGURE 20.2 Vehicle transferred to automated mode by human driver.

this scenario, as SAE (2016) Level 3 is assumed, so there is no need for the driver to monitor the automated driving system). The scenario assumes that there is a planned takeover of driving from automation back to the human driver (such as when their exit from the motorway is coming up, which would have been pre-programmed into the satellite navigation system). The driver is given a 5-, then 2-, then 1-min notice that the takeover process will begin (see Figure 20.2).

Upon the prompt from the automated system that the driver needs to get ready to drive, it is assumed that the driver ceases the non-driving task, puts down the tablet computer, and resumes the driving position (as shown in Figure 20.3). The system then presents a series of questions designed to raise the situation awareness of the

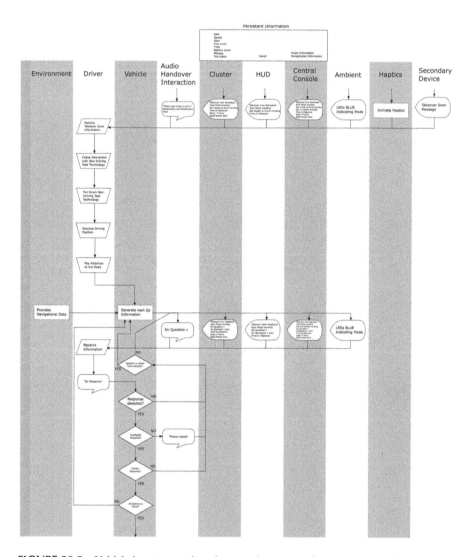

FIGURE 20.3   Vehicle in automated mode preparing to transfer to manual mode.

driver (such as What speed is the vehicle currently travelling at? What lane are you currently in? What colour is the vehicle in front of you? What is your remaining fuel range? Can you see a bend in the road ahead?). The driver is expected to respond to these questions (which are presented auditorily as well as on all of the visual interfaces). If the answer is correct, then the next question is presented until all questions have been answered. If the answer is incorrect, then the question is repeated for a maximum of two additional times before moving onto the next question. When all questions have been presented, the takeover interaction moves on to that presented in Figure 20.3.

As Figure 20.4 shows, the driver is then requested to take manual control of the vehicle, which will mean placing both hands on the steering wheel and positioning their foot on the accelerator pedal. To transfer control from the automated system to the driver, they need to press two buttons, mounted on the steering wheel at the ten-to-two clock position, with their thumbs (in the same manner as they do for handing control over to the vehicle automation system).

When control of the vehicle is passed back to the human driver, the ambient lighting changes back from blue to orange (indicating the vehicle is now in manual driving mode) and the words 'Automation deactivated' are presented auditorily as well as on the visual displays. This is followed by the words 'You are in control', which are also presented auditorily and on the visual displays (as shown in Figure 20.5). The human driver is now driving the vehicle.

The next section describes the study in which the video data was collected from human drivers of an automated vehicle on the road. These data were used to validate the takeover from vehicle automation by human driver, as shown in Figures 20.3–20.5.

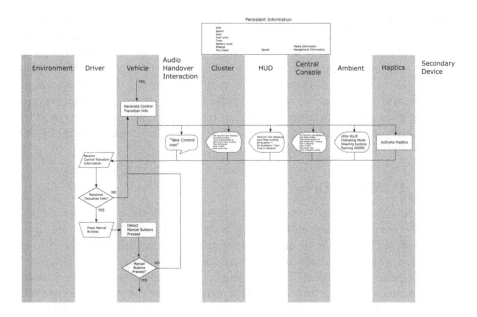

FIGURE 20.4   Vehicle transferring from automated mode to manual mode.

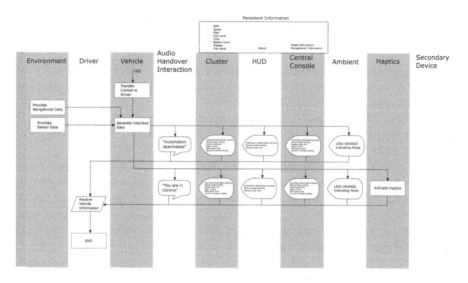

FIGURE 20.5 Vehicle back in manual mode.

20.3 METHOD

The experimental methods used in this study cover the participants, experimental design, equipment, procedure, data reduction, and analysis.

20.3.1 Participants

Although 24 participants were recruited from a pool that had previously taken part in a simulator-based study featuring an almost identical human–machine interface (HMI) (Stanton et al. 2020b), data from only 16 was good enough to be used in this study, due to equipment and recording failures. All drivers held full UK driving licenses. There were ten males and six females. The age range of drivers was 29–67 years (mean = 46.3 years, SD = 11.7 years). Participants provided a signed consent prior to involvement in the study, which was approved by the University of Southampton's Ethics and Research Governance Office (ERGO Number: 49792.A2).

20.3.2 Experimental Design

A repeated-measures design was employed for the experiment, covering three take-over events using the default HMI settings, as shown in Figure 20.6.

As seen in Figure 20.7, the driving mode wording is situated below the driving mode icon, which is centrally placed in the infotainment display and cluster and placed on the left side in the HUD. The colour acts as a mode indicator – blue indicating automated and orange for manual. Time to takeover is shown in the bottom-left corner of the cluster and infotainment display, and the top-right of the HUD. The edge frame mode indicator is only visible on the cluster and infotainment display.

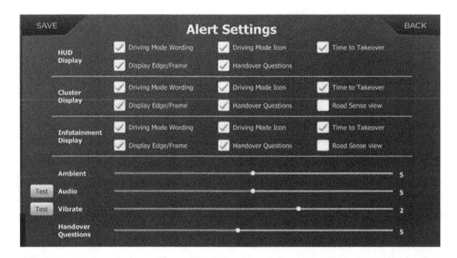

FIGURE 20.6   The customisation matrix displayed at the end of each trial on the vehicle's infotainment display.

FIGURE 20.7   The interfaces in automated mode, showing centre infotainment display (left), head-up display (top-right), and instrument cluster (bottom-right).

Takeover questions and associated icons appear in the centre of each of the HMI display elements.

Video cameras were positioned around the interior cabin of the vehicle to record the driver interaction with vehicle automation so that it could be compared with the OESDs.

20.3.3   EQUIPMENT

The experimental vehicle was a 2017 Jaguar I-PACE EV400 AWD pre-production model, pictured in Figure 20.8. The automation system consisted of a combination of factory standard lane keep assist and adaptive cruise control. When utilising these systems, the standard car would issue a frequent warning to maintain hands on the wheel; these warnings were specifically removed from the system in order for it to

FIGURE 20.8   The test vehicle, a 2017 Jaguar I-PACE EV400 AWD.

simulate SAE Level 3 automation. The visual aspects of the HMI consisted of a 14″ thin-film transistor (TFT) panel fitted to the centre console, a 10″ TFT fitted in place of the original equipment manufacturer cluster, and a HUD comprising a small 1000×250 px TFT and reflector screen (Bysameyee 2020) (see Figure 20.9). The car's original cluster was moved into the passenger footwell to be monitored by the safety driver. The interactive element of the HMI constituted two illuminating green buttons fitted in the thumb positions of the steering wheel (approximately at the ten-to-two clock position), as seen in Figure 20.10. Ambient lighting was supplied via LED lighting strips. Haptics were provided in the seat base via Leggett and Platt motors, controlled through an Arduino Micro and motor control board. Five KT&C cameras were installed within the I-PACE to provide footage from the forward view, driver-facing, over-the-driver-shoulder, footwell, and rear views. A dashcam was also fitted in the safety car to collect footage of the participants' vehicle from an external viewpoint.

20.3.4 PROCEDURE

On arrival at the Jaguar Land Rover facility at Fen End, the participant was welcomed and asked to sign in, presenting their driving license and receiving a visitor pass. They were then briefed on the aim of the project and how the on-road study follows on from the simulator study that they completed the previous year. A brief explanation of the sequence of events for the study was then given. This included the route, the takeover procedure, form filling on trial completion, customisation of the HMI, and the *n*-back cognitive load task (to simulate the driver undertaking another (verbal) task while the vehicle was being driven by the automation). It was stressed that they were required to maintain the same level of attention as if they were driving and be ready to take control at all times. They were also advised that if the cognitive load task was detrimentally affecting their ability to retain a safe level of attention, they should stop the *n*-back task. They were presented with a copy of the participant information sheet; this same sheet was made available to them when they were recruited. A reminder sheet showing screenshots of the HMI they would be using was provided, including modal variation of the cluster, HUD and in-vehicle information systems, the green buttons, and the

FIGURE 20.9   The car in manual mode with the HUD and instrument cluster.

FIGURE 20.10   The green buttons on the steering wheel and the infotainment display (ambient blue lighting indicates that the vehicle is in automated mode). The driver-facing camera can be seen fitted on the dashboard.

customisation matrix screen. They were shown a route map that highlighted where the trials would take place, but also areas of caution and one junction that required traversal in manual mode. A privacy policy was presented, and a sheet explaining the cognitive load n-back task was provided, along with a brief verbal explanation. The participant was asked if they had any questions, and if they were happy to continue, they were provided with two consent forms – an attendance form and an events team form to sign. On completion of all forms, they were led from the reception area to the car park and asked to sit in the passenger side of the car, while both experimenters took their places in the rear seats. The participant was introduced to a safety driver who explained the basic controls on the car, and the elements of the interface. The safety driver then drove the car through security to the proving ground and demonstrated the vehicle's performance before running through the transition to automated mode and back to manual control. The automation system was operated by one of the experimenters using a Wizard of Oz system, from a Windows tablet in the back of the car. To offer automation, an experimenter would press a start button on a custom control panel app running on the tablet; the HMI then indicated automation as being available via the three graphical interfaces and a vocal alert. In order to enable automation, the driver simultaneously pressed the two green buttons mounted on the steering wheel and released all of the controls, including the accelerator. The system would then engage automation and the HMI would indicate that the automated mode was active. During automation, the cognitive load task was controlled by the other experimenter via a mobile telephone-based app, linked to a Bluetooth speaker. The safety driver demonstrated the automation system multiple times, including the n-back task, and requested the safety car overtake and brake in front of the vehicle, when in automated mode, to illustrate how it reacts to maintain a gap to the car in front. The safety driver then stopped in a safe area and swapped places with the participant. The participant was then allowed some time to drive the car on the proving ground to become familiar with the controls. The automation was then made available to them, and they experienced multiple handovers and takeovers, including the use of the n-back task while in automated mode.

The participant was asked to drive manually to the start point of the experiment at the Southbound Warwick services on the M40. Two miles prior to the services, while on the M40, a road-based practice handover was conducted. The automation was offered to the participant; once activated, a short 30-s period of automation followed, which included the participant carrying out the n-back task. The participant then experienced the takeover protocol and resumed manual control before stopping at the services. After confirming that the participant was happy to continue, the on-board systems were checked and configured for the first trial, data logging was started, and the video and audio were synchronised using a clapper board.

The participant was instructed to drive from the services onto the M40, proceeding in the left lane at approximately 58 mph. After 1 min of manual driving, automation was offered to the participant. Once activated, after a further period of 30 s, the n-back cognitive loading task was started. This task was run in 2-min intervals, separated by 30 s. Following 10 min of automation, the HMI started the takeover protocol, and once completed, the participant pressed both steering wheel buttons and resumed manual control. After 1 min of manual control, automation was again offered to the participant and the process was repeated. On completion of the second takeover, the manual driving period was extended to approximately 7 min in order to pass a section of motorway (M40 J9 Southbound) that would have adversely affected the automation due to the lane becoming a slip road. Due to the proximity to the end point of the trial, takeover 3-s automation period was reduced to 8 min. The *n*-back task was started simultaneously with the automation for 30 s, before reverting to 30 s off and 2 min on until the protocol started. When the participant completed the protocol and assumed manual control, they continued in manual mode the short distance remaining to the motorway services and parked the car. This concluded the data collection for the trial and the safety driver swapped places with the participant from this point and drove the vehicle back to the Jaguar Land Rover facility at Fen End. The participant was thanked for their time and signed out at reception.

20.3.5 DATA REDUCTION AND ANALYSIS

The validation of the OESDs was assessed by comparing the video data collected of the driver on the road during the automation takeover process with those tasks identified in the OESDs. In the OESDs, 15 tasks were defined as follows:

1. Receive 'Get Ready to Take Over' Information
2. Resume Driving Position
3. Receive SA Question 1
4. Answer Question 1
5. Receive SA Question 2
6. Answer Question 2
7. Receive SA Question 3
8. Answer Question 3
9. Receive SA Question 4
10. Answer Question 4
11. Receive SA Question 5
12. Answer Question 5
13. Receive 'Take Control Now' Information
14. Received Transition Information
15. Press Manual Buttons.

These data were processed using the signal detection paradigm (Green and Swets 1966), which discerns between four events: hits, misses, false alarms, and correct rejections, as illustrated in Figure 20.11. In the context of this experiment, it provided a method by which to compare the predicted driver behaviour, illustrated on OESDs, with driver behaviour observed during the on-road trials of the vehicle control takeover process.

The four events in Figure 20.11 are defined as follows:

- **Hits**: Present in OESD and present in the video of automation–driver takeover.
- **Misses**: Not present in the OESD but present in the video.
- **False alarms**: Present in the OESD but not present in video.
- **Correct rejections**: Not present in OESD nor in the video.

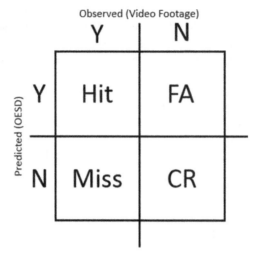

FIGURE 20.11   Signal detection theory matrix.

The latter category can be difficult to calculate as it could be infinity, but for the purposes of this investigation, it should be based on the total number of unique misses generated by all of participants, minus the number of misses for each individual participant. Additionally, the hit rate and the false alarm rate were calculated by Equations 20.1 and 20.2, respectively.

$$Hit\ rate = \frac{Hits}{Hits + Misses} \tag{20.1}$$

$$False\ alarm\ rate = \frac{False\ alarms}{False\ alarms + Correct\ rejections} \tag{20.2}$$

Finally, Matthews (1975) correlation coefficient (Phi – a correlation coefficient for dichotomous data) was applied to the data generated by the signal detection theory analysis. This quantified the correlation between the expected and the observed behaviour, as a means to validate the OESDs. The Matthews correlation coefficient formula is given by Equation 20.3.

$$\varphi = \frac{Hit \times CR - FA \times Miss}{\sqrt{(Hit + FA)(Hit + Miss)(CR + FA)(CR + Miss)}} \tag{20.3}$$

20.4   RESULTS

The data from the three takeovers, from the automated vehicle to the human driver, are summarised in Table 20.2 for each participant. As there were 15 tasks per takeover, 45 hits represent a perfect score, meaning all of the tasks in the OESD were observed in the video. One participant had 13 false alarms because they cut the

TABLE 20.2

Presentation of Data by Participant, with Hits, Misses, False Alarms (FA), and Correct Rejections (CR)

Participant	Hits	Misses	FA	CR	Hit Rate	FA Rate	Phi
1	44	1	1	17	0.977778	0.055556	0.922222
2	44	8	1	10	0.846154	0.090909	0.634663
3	45	2	0	16	0.957447	0	0.922531
4	45	3	0	15	0.9375	0	0.883883
7	45	0	0	18	1	0	1
8	45	1	0	17	0.978261	0	0.961204
9	45	3	0	15	0.9375	0	0.883883
12	45	1	0	17	0.978261	0	0.961204
13	45	0	0	18	1	0	1
17	45	1	0	17	0.978261	0	0.961204
19	45	3	0	15	0.9375	0	0.883883
20	45	3	0	15	0.9375	0	0.883883
21	45	0	0	18	1	0	1
22	32	3	13	15	0.914286	0.464286	0.494975
23	42	0	3	18	1	0.142857	0.894427
24	45	0	0	18	1	0	1

TABLE 20.3

Misses in the Three Takeovers from Vehicle Automation by Human Driver

Misses	Takeover 1	Takeover 2	Takeover 3	Total
Failed to assume driving position	1	0	0	1
Covered buttons with thumbs	6	4	5	15
Took control early	1	0	0	1
Assumed driving position at end	1	1	4	6
Removed hands during protocol	1	1	3	5
Assumed driving position early	0	1	0	1
Total	10	7	12	29

takeover short by assuming control of the vehicle without passing through all of the preceding tasks (participant 22). The misses are shown in Table 20.3 with the frequencies by takeover. As there were a total of six unique misses per takeover (observed from the video data), the total correct rejections were 18 if none were observed. Hit rate, false alarm rate, and Phi were calculated as described in the data reduction section.

The misses are an interesting category of events, as they describe activities that were not predicted by the OESD but were present in the video of the driver.

Over half of the misses (15/29) were related to the driver covering the green buttons on the steering wheel with their thumbs (the ten-to-two position), early on in the takeover process. This is not a safety concern however, as it shows that the drivers were readying themselves to resume manual control. On one occasion, the driver failed to assume the driving position at the right point in the process (task 2), as they remained with their hands on their lap and their feet away from the pedals. The only consequence of this is that it increases vehicle control takeover time, but as this time is entirely driver-paced, there is no adverse consequence for safety. On one occasion, a driver (participant 22) took control over the vehicle early in the process, which shortened the takeover protocol considerably. While any takeover protocol should not hamper drivers, or be overly intrusive, there is a fine balance to ensure that the driver's awareness has been raised sufficiently for them to resume control of the vehicle (Stanton et al. 2017). Six drivers did not assume driving position until right at the end of the takeover protocol, whereas it had been anticipated that they would be in this position from the start. Again, there were no safety implications from this behaviour. Five participants removed their hands from the steering wheel during the course of the takeover, having previously placed them on the wheel. Finally, just one driver assumed the driving position before the protocol had started. In summary, none of these misses presented any safety concerns to the study, particularly as the takeover was at the pace of the driver, rather than the vehicle automation.

As Figure 20.12 shows, most of the hits were almost at the maximum level ($n = 45$), except for two participants, most notably participant 22, who resumed manual control early from the automation system for one of the three takeovers, which also generated the most false alarms. Participant 2 is less of an outlier, being just below the mean for the hits and just above the mean for the false alarms in the box-and-whisker plots. Participant 2 also had more misses, and consequentially fewer correct rejections, than the other participants. Misses were generally quite low and correct rejections were high for most participants.

Given the findings in Table 20.2, it is no surprise that the hit rate is high and the false alarm rate is low, as shown in Figure 20.13. Again, there are the outliers of participants 2 and 22 for false alarm rate, with the latter being more pronounced. This means that missing task steps out in the return of control to the human driver is more likely to affect the predictive validity of the OESDs.

Finally, Phi (the correlation coefficient for dichotomous data), as shown in Figure 20.14, reveals good predictive validity for OESDs, above 0.8 (which is generally considered to be the criterion for correlational data). The two outliers are again participants 2 and 22. The latter represents the lower Phi coefficient.

In summary, this means that OESDs have been shown to be good predictors of the observed takeover behaviours when going from automated control to manual control in 14 of the 16 cases.

20.5 DISCUSSION

The main finding from the on-road study of takeovers from vehicle automation to human drivers is that the OESDs predicted most of the observed activities.

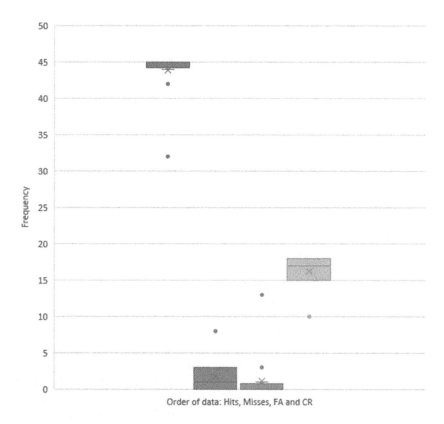

FIGURE 20.12 Box-and-whisker plots for hits, misses, false alarms, and correct rejections.

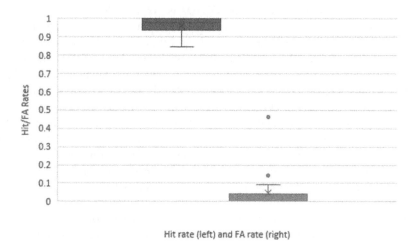

FIGURE 20.13 Box-and-whisker plots for hit rate and false alarm rate.

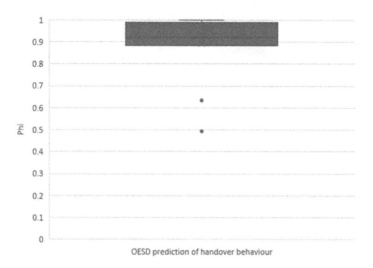

FIGURE 20.14 Box-and-whisker plot for Phi.

To that end, the findings of the study mirrored those from simulator studies undertaken previously (i.e. median Phi > 0.8), albeit with a much smaller sample (Stanton et al. 2020). Also, all takeovers from vehicle automation to human driver were successful, in that they largely followed the protocol designed in the OESD, with the exception of two participants (numbers 2 and 22) who, for one of their three takeovers, took control of the vehicle back early. Nevertheless, there was no evidence of mode confusion in any of the takeovers (Sarter and Woods 1995; Stanton, Dunoyer and Leatherland 2011). So, as an exercise in design, we may conclude that OESD was a useful human factors method (Kurke 1961; Kirwan and Ainsworth 1992; Stanton et al. 2013) and was able to predict automation–driver takeover behaviour.

The validation performance of the OESDs is comparable with the best of the human factors methods (Stanton and Young 1999; Stanton et al. 2013). Some of the best methods in the field include those associated with the prediction of task performance time (Card, Newell and Moran 1983; Baber and Mellor 2001; Stanton and Baber 2008; Harvey and Stanton 2013; Moray, Groeger and Stanton 2017) and prediction of human error (Baber and Stanton 1996; Harris et al. 2005; Stanton and Baber 2002, 2005; Stanton et al. 2009). While a wide range in the predictive performance of human factors methods has been observed (Stanton and Young 1999), those better-performing methods typically achieve validity statistics in excess of 0.8 (which in correlational terms is generally considered to be of strong predictive value). OESDs appear to be performing as well as could be expected, for the limited range of automation takeover tasks, both in driving simulators and in a real vehicle on the road.

While the validation evidence for the practical application of OESD is growing, we cannot assume validity generalisation until more extensive studies have been undertaken. This means that it is an important goal for researchers to undertake studies with larger numbers of participants and in a broader range of applications and domains. So far, studies have been limited to automation–driver takeover tasks and, with the exception of this latest on-road study, mainly in driving simulators (Stanton et al. 2020). As an engineering discipline, Human Factors needs to be able to provide evidence that its methods actually work (Stanton and Young 2003; Stanton et al. 2013; Stanton 2016). This means providing evidence for the validity and reliability of the methods to support their continued use.

There is little by way of studies of reliability for OESDs, which is a shortcoming in the research. Reliability has two facets (Stanton and Young 1999; Stanton 2016): stability over time (called 'inter-analyst reliability') and stability between people conducting the analysis (called 'intra-analyst reliability'). Establishing both forms of reliability is an important future goal for the methodological research into OESD. The repeatability of a method, both within and between analysts, would give confidence in its continued use, although it should be noted that OESDs have been in continuous use for over 60 years without this evidence.

Also, little has been written on training people to use human factors methods (including OESDs), as noted by Stanton and Stevenage (1998). Stanton and Young (1999, 2003) report on an evaluation of the ease with which novices were able to acquire expertise in human factors methods. They showed quite a large spread of training times, which is, to some extent at least, associated with the complexity of the methods tested. No such study has been reported on the ease or difficulty in acquiring expertise in developing OESDs. This represents a gap in knowledge about human factors methods, although there are estimates reported in Stanton et al. (2013). Typically, the most popular human factors methods are those that are quick to learn and apply (Stanton et al. 2013). More research could be undertaken to streamline methods and improve their ease of use.

There were some limitations to this study, which are worth noting in terms of full disclosure. Firstly, the sample size was very limited for claims about the predictive validity of a method, although the Matthews (1975) correlation coefficient was chosen as an acknowledgement of the sampling problem. Secondly, the scope of the study was limited to the takeover from vehicle automation by human driver. A wider range of task and domains is required to determine validity generalisation. Finally, as discussed previously, the reliability of the method should be established over time as well as between analysts.

20.6 CONCLUSIONS

OESDs are able to predict takeover behaviour from vehicle automation by human driver in on-road trials. Although the sample size for this study was quite small compared to the previous driving simulator studies, the findings are encouraging and a large sample size for future work is to be encouraged. For now, the use of OESDs for developing design of the interfaces between vehicle automation and human drivers seems to be reasonable. An important goal for future research is to establish the reliability of OESD generation, both within and between analysts, as well as validity generalisation across a broader range of domains and application.

ACKNOWLEDGEMENTS

This work was supported by Jaguar Land Rover and the UK-EPSRC Grant EP/N011899/1 as part of the jointly funded 'Towards Autonomy: Smart and Connected Control (TASCC) Programme'.

REFERENCES

Baber, C., and B. Mellor. 2001. "Using critical path analysis to model multimodal human–computer interaction." *International Journal of Human-Computer Studies, 54, 4* 613–636. https://doi.org/10.1006/ijhc.2000.0452.

Baber, C., and N. A. Stanton. 1996. "Human error identification techniques applied to public technology: predictions compared with observed use." *Applied Ergonomics, 27, 2* 119–131. https://doi.org/10.1016/0003-6870(95)00067-4.

Banks, V. A., K. L. Plant, and N. A. Stanton. 2018. "Driver error or designer error: using the perceptual cycle model to explore the circumstances surrounding the fatal tesla crash on 7th May 2016." *Safety Science, 108* 278–285. https://doi.org/10.1016/j.ssci.2017.12.023.

Banks, V. A., N. A. Stanton, and C. Harvey. 2014. "Sub-systems on the road to vehicle automation: Hands and feet free but not "mind" free driving." *Safety Science, 62* 505–514. http://doi.org/10.1016/j.ssci.2013.10.014.

Banks, V. A., N. A. Stanton, G. Burnett, and S. Hermawati. 2018. "Distributed cognition on the road: Using EAST to explore future road transportation systems." *Applied Ergonomics, 68* 258–266. https://doi.org/10.1016/j.apergo.2017.11.013.

Bysameyee. 2020. *Head-up Display GPS Navigation.* Accessed February 17, 2020. https://www.amazon.co.uk/Head-up-Display-GPS-Navigation-Reflective/dp/B075T2DH12/ref=sr_1_10?keywords=car+head+up+display&qid=1573804882&sr=8-10.

Card, S. K., A. Newell, and T. P. Moran. 1983. *The Psychology of Human-Computer Interaction.* Hillsdale, MI: L. Erlbaum Associates Inc.

Clark, J. R., N. A. Stanton, and K. Revell. 2019a. "Directability, eye-gaze, and the usage of visual displays during an automated vehicle handover task." *Transportation Research Part F: Traffic Psychology and Behaviour, 67* 29–42. https://doi.org/10.1016/j.trf.2019.10.005.

Clark, J. R., N. A. Stanton, and K. M. Revell. 2019b. "Identified handover tools and techniques in high-risk domains: using distributed situation awareness theory to inform current practices." *Safety Science, 118* 915–924. https://doi.org/10.1016/j.ssci.2019.06.033.

Eriksson, A., and N. A. Stanton. 2017a. "Takeover time in highly automated vehicles: noncritical transitions to and from manual control." *Human Factors: The Journal of the Human Factors and Ergonomics Society, 59, 4* 689–705. https://doi.org/10.1177/0018720816685832.

Eriksson, A., and N. A. Stanton. 2017b. "Driving performance after self-regulated control transitions in highly automated vehicles." *Human Factors, 59, 8* 1233–1248. doi: 10.1177/0018720817728774.

Eriksson, A., S. M. Petermeijer, M. Zimmermann, J. C. F. de Winter, K. J. Bengler, and N. A. Stanton. 2019. "Rolling out the red (and green) carpet: supporting driver decision making in automation-to-manual transitions." *IEEE Transactions on Human-Machine Systems, 49, 1* 20–31. doi: 10.1109/THMS.2018.2883862.

Green, D. M., and J. A. Swets. 1966. *Signal Detection Theory and Psychophysics* (Volume 1). New York: John Wiley.

Harris, D., N. A. Stanton, A. Marshall, M. S. Young, J. Demagalski, and P. Salmon. 2005. "Using SHERPA to predict design-induced error on the flight deck." *Aerospace Science and Technology, 9, 6* 525–532. https://doi.org/10.1016/j.ast.2005.04.002.

Harris, D., N. A. Stanton, and A. Starr. 2015. "Spot the difference: operational event sequence diagrams as a formal method for work allocation in the development of single-pilot operations for commercial aircraft." *Ergonomics, 58, 11* 1773–1791. https://doi.org/10.1080/00140139.2015.1044574.

Harvey, C., and N. A. Stanton. 2013. "Modelling the hare and the tortoise: predicting the range of in-vehicle task times using critical path analysis." *Ergonomics, 56, 1* 16–33. doi: 10.1080/00140139.2012.733031.

Kirwan, B., and L. K. Ainsworth. 1992. *A Guide To Task Analysis: The Task Analysis Working Group.* London: CRC Press. https://doi.org/10.1201/b16826.

Kurke, M. I. 1961. "Operational sequence diagrams in system design." *Human Factors: The Journal of the Human Factors and Ergonomics Society, 3, 1* 66–73. https://doi.org/10.1177/001872086100300107.

Matthews, B. W. 1975. "Comparison of the predicted and observed secondary structure of T4 phage lysozyme." *Biochimica et Biophysica Acta (BBA) - Protein Structure, 405, 2* 442–451. https://doi.org/10.1016/0005-2795(75)90109-9.

Moray, N., J. Groeger, and N. A. Stanton. 2017. "Quantitative modelling in cognitive ergonomics: predicting signals passed at danger." *Ergonomics, 60, 2* 206–220. https://doi.org/10.1080/00140139.2016.1159735.

SAE J3016. 2016. *Taxonomy and Definitions for Terms Related to Driving Automation Systems for On-Road Motor Vehicles.* Accessed November 3, 2020. https://www.sae.org/standards/content/j3016_201806/.

Salmon, P. M., N. A. Stanton, G. H. Walker, D. Jenkins, C. Baber, and R. McMaster. 2008. "Representing situation awareness in collaborative systems: a case study in the energy distribution domain." *Ergonomics, 51, 3* 367–384. https://doi.org/10.1080/00140130701636512.

Salmon, P. M., G. H. Walker, and N. A. Stanton. 2016. "Pilot error versus sociotechnical systems failure: a distributed situation awareness analysis of Air France 447." *Theoretical Issues in Ergonomics Science, 17, 1* 64–79. https://doi.org/10.1080/1463922X.2015.1106618.

Sarter, N. B., and D. D. Woods. 1995. "How in the world did we ever get into that mode? Mode error and awareness in supervisory control." *Human Factors: The Journal of the Human Factors and Ergonomics Society, 37, 1* 5–19. https://doi.org/10.1518/001872095779049516.

Sorensen, L. J., N. A. Stanton, and A. P. Banks. 2011. "Back to SA school: contrasting three approaches to situation awareness in the cockpit." *Theoretical Issues in Ergonomics Science, 12, 6* 451–471. https://doi.org/10.1080/1463922X.2010.491874.

Stanton, N. A. 2016. "On the reliability and validity of, and training in, ergonomics methods: a challenge revisited." *Theoretical Issues in Ergonomics Science, 17, 4 [Methodological Issues in Ergonomics Science I]* 345–353. https://doi.org/10.1080/1463922X.2015.1117688.

Stanton, N. A., and C. Baber. 2002. "Error by design: methods for predicting device usability." *Design Studies, 23, 4* 363–384. https://doi.org/10.1016/S0142-694X(01)00032-1.

Stanton, N. A., and C. Baber. 2005. "Validating task analysis for error identification: reliability and validity of a human error prediction technique." *Ergonomics, 48, 9* 1097–1113. doi: 10.1080/00140130500219726.

Stanton, N. A., and C. Baber. 2008. "Modelling of human alarm handling responses times: a case of the Ladbroke Grove rail accident in the UK." *Ergonomics, 51, 4* 423–440. doi: 10.1080/00140130701695419.

Stanton, N. A., and P. Marsden. 1996. "From fly-by-wire to drive-by-wire: safety implications of automation in vehicles." *Safety Science, 24* 35–49. https://doi.org/10.1016/S0925-7535(96)00067-7.

Stanton, N. A., and S. V. Stevenage. 1998. "Learning to predict human error: issues of acceptability, reliability and validity." *Ergonomics, 41, 11* 1737–1756. https://doi.org/10.1080/001401398186162.

Stanton, N. A., and M. S. Young. 1999. "What price ergonomics?" *Nature, 399* 197–198. https://doi.org/10.1038/20298.

Stanton, N. A., and M. S. Young. 2003. "Giving ergonomics away? The application of ergonomics methods by novices." *Applied Ergonomics, 34, 5* 479–490. https://doi.org/10.1016/S0003-6870(03)00067-X.

Stanton, N. A., A. Dunoyer, and A. Leatherland. 2011. "Detection of new in-path targets by drivers using Stop & Go Adaptive Cruise Control." *Applied Ergonomics, 42, 4* 592–601. https://doi.org/10.1016/j.apergo.2010.08.016.

Stanton, N. A., J. W. H. Brown, K. M. Revell, P. Langdon, M. Bradley, I. Politis, L. Skrypchuk, S. Thompson, and M. Mouzakitis. 2020. "Validating operator event sequence diagrams: the case of automated vehicle to human driver takeovers." *Submitted to the International Journal of Human Factors and Ergonomics in the Manufacturing and Service Industries (under review).*

Stanton, N. A., M. Young, and B. McCaulder. 1997. "Drive-by-wire: the case of driver workload and reclaiming control with adaptive cruise control." *Safety Science, 27, 2* 149–159. https://doi.org/10.1016/S0925-7535(97)00054-4.

Stanton, N. A., P. M. Salmon, D. Harris, A. Marshall, J. Demagalski, M. S. Young, T. Waldmann, and S. Dekker. 2009. "Predicting pilot error: testing a new methodology and a multi-methods and analysts approach." *Applied Ergonomics, 40, 3* 464–471. https://doi.org/10.1016/j.apergo.2008.10.005.

Stanton, N. A., P. M. Salmon, G. H. Walker, and M. Stanton. 2019. "Models and methods for collision analysis: a comparison study based on the Uber collision with a pedestrian." *Safety Science, 120* 117–128. https://doi.org/10.1016/j.ssci.2019.06.008.

Stanton, N. A., P. M. Salmon, G. H. Walker, E. Salas, and P. A. Hancock. 2017. "State-of-science: situation awareness in individuals, teams and systems." *Ergonomics, 60, 4* 449–466. doi: 10.1080/00140139.2017.1278796.

Stanton, N. A., P. M. Salmon, L. A. Rafferty, G. H. Walker, C. Baber, and D. P. Jenkins. 2013. *Human Factors Methods: A Practical Guide for Engineering and Design.* London: CRC Press. https://doi.org/10.1201/9781315587394.

Walker, G. H., N. A. Stanton, C. Baber, L. Wells, H. Gibson, P. M. Salmon, and D. Jenkins. 2010. "From ethnography to the EAST method: a tractable approach for representing distributed cognition in air traffic control." *Ergonomics, 53, 2* 184–197. https://doi.org/10.1080/00140130903171672.

21 Design Constraints and Guidelines for the Automation– Human Interface

Neville A. Stanton, Kirsten M. A. Revell,
James W. H. Brown, Jisun Kim, and Joy Richardson
University of Southampton

Nermin Caber and Michael Bradley
University of Cambridge

Patrick Langdon
Edinburgh Napier University

CONTENTS

21.1 DESIGN CONSTRAINTS

The design constraints were developed by the HI:DAVe team to guide our studies and design of the automation–human interfaces. As such, they represent some of our early thinking on the project. Thirty-one, high-level, design constraints were identified as follows.

21.1.1 ALLOW DRIVER TO TAKE CONTROL AT ANY POINT DURING TAKEOVER, BE SURE HANDS ON WHEEL AND FEET ON PEDALS

System needs to prompt driver to have hands on wheel and feet on pedals prior to takeover. Pedals and wheel need to move in response to vehicle inputs, such as braking and accelerating. If driver inputs are detected, then automation should drop out.

21.1.2 PERSONALISE TAKEOVER BASED ON DRIVER PREFERENCES (AND SITUATION)

Takeover customisation menu allowing driver to personalise the takeover protocol – above the minimum requirements (for both short and long takeover protocols).

21.1.3 ALLOW OPTION TO COMPLETE NON-DRIVING TASK (EVEN IF IT MEANS MISSED TAKEOVER FOR JUNCTION/EXIT)

If no inputs are detected from the driver by the automation in response to the take-over protocol (after detecting that driver is awake and active in the vehicle), then carry on to next junction (may have to exit and turn back if the next junction does not fit in with planned route).

21.1.4 ALLOW SUFFICIENT TIME FOR TAKEOVER (BIG INDIVIDUAL DIFFERENCES IN OUR STUDIES)

Default takeover needs to occur with sufficient time to exit junction (such as 5 min before junction – this is an arbitrary figure). Allow the driver to customise this time in the takeover set-up menu.

21.1.5 CUSTOMISE TAKEOVER BASED ON DURATION OF BEING OUTSIDE OF THE CONTROL LOOP AND FREQUENCY OF TAKEOVER (AND CONTEXT: ROAD, WEATHER, OTHER ROAD USERS, INFRASTRUCTURE, SIGNAGE) – MULTIMODAL HUMAN–MACHINE INTERFACE (HMI)

See design constraint number 2. This may be too complex to undertake. Remember KISS design principle: Keep It Simple Stupid!

21.1.6 QUERYING SITUATION AWARENESS OF DRIVER BY 'VEHICLE AVATAR'

The chatty co-driver concept: natural language interaction with vehicle automation on aspects related to the driving task.

21.1.7 MAKE EXPLICIT WHO IS IN CONTROL OF VEHICLE – MODE AWARENESS HMI (LIGHT-UP STEERING WHEEL)

Need colour coding for light-up wheel to show who is driving. Not light would mean driver is in control. Could be traffic light colours. Red = driver must take control; Amber = driver prepare to take control; Green = automation is in control; Blue = auto-mation can take control if driver releases wheel. Trouble with colour coding is colour blindness. This coding needs to be supplemented with information about mode of control on the head-up display (HUD).

21.1.8   Recommended Settings Based on Customer Profiles for Customisation

Need to take customer preferences and profiles into account. For example, the driver may be deaf.

21.1.9   Pre-set Defaults for Takeover

See design constraint number 4.

21.1.10   Graduated Alert to Takeover Visual, Audio, Haptic (Escalating)

Takeover prompt begins with visual information. If this is not responded to within 1 min (arbitrary figure), it is backed up with auditory information. If the auditory information is not responded to with 30 s (arbitrary figure), it is backed up with haptic information. If the haptic information is not responded to within 15 s (arbitrary figure), the driver is informed that the car will progress to the next junction (via visual and auditory information). If there are no more junctions, then the vehicle needs to pull over on the hard shoulder and make a safe stop.

21.1.11   Cue Driver to 'Grab' Steering Wheel

Light up steering wheel to indicate that driver needs to take control.

21.1.12   Make 'Takeover Button' Easy to Access (e.g. Put on Gear Stick)

Not sure we need a 'takeover' button, as the driver just needs to grab the steering wheel and/or place feet on pedals. Suggest that if the driver only grabs the steering wheel, then longitudinal control is still automated (like adaptive cruise control).

21.1.13   'Repeat' Button and 'OK' Button?

Place 'repeat' button on steering wheel, so that the last verbal information or instruction can be repeated back to the driver. Also, place an 'OK' button on steering wheel so that driver can acknowledge system without having to give verbal acknowledgement (if they do not wish to).

21.1.14   Encourage (Facilitate) Visual Checks in Environment and Controls of Vehicle

Where possible, show information on the HUD, such as the identification of other vehicles, hazards, road signage, path of vehicle, vehicle status, and vehicle intentions.

21.1.15   Display the Vehicle Status and Intention

Use HUD on windscreen to display status of the vehicle (such as the current mode of the vehicle and the status of vehicle sensors) as well as the intended path of the vehicle (arrow on road) and also inform driver of any planned manoeuvre (such as a lane change, braking, and accelerating).

21.1.16   Driver's HMI Actions Need to Be Clearly Fed Back (Link to 1 – Volvo Hands on Wheel to Flip Both Paddles)

See design constraint number 7.

21.1.17   Eyes Out

Place information on HUD where possible. Use the windscreen as the interface between the driver and the automation.

21.1.18   Use System to Aid Manual Driving

Option to have HUD switched on even when driving manually, to show path of vehicle and pick out relevant information in road environment (such as the detection of other vehicles and road signage).

21.1.19   Some Level of Personalisation and Setting of Levels

Basic minimum takeover protocol includes instruction to grab vehicle controls and motorway exit distance (if appropriate). Additional information is available that can be supplied if desired, which can be added in a personalisation takeover menu.

21.1.20   Longer Automated Vehicle-to-Human Driver Takeover in Urban Environment (Compared to Motorway)

This could be set up in the defaults and personalisation menu. We are focusing on motorway takeover in our use-case, so wonder if we should tackle urban yet as it seems to be too complex.

21.1.21   Takeover Strategy That Guides Visual Search

Use HUD to pick out information of relevance to takeover, such as position of other vehicles that pose potential hazards, signage for exit (if relevant), path of vehicle, lane position, and so on.

21.1.22   Feedback to Every Driver Action (Process Needs Adapting to Driver and Situation)

This has already been addressed in other design constraints.

21.1.23   Checklist

This constraint conveys the need to have a systematic takeover process, from vehicle by driver as well as handover from driver to vehicle.

21.1.24 OPTION TO REQUEST SPECIFIC INFORMATION OF IMPORTANCE TO DRIVER (IF NOT IN PROTOCOL)

Design a question and answer protocol with an 'Alexa-like' interaction.

21.1.25 EDUCATION OF DRIVERS IN RATIONALE AND TECHNIQUE

Education of the driver in the takeover and limitations of vehicle automation could be provided in the manual, on-line training (You Tube?) and within the vehicle via head-down display and HUD (when the vehicle is stationary).

21.1.26 TRAINING (VIDEO) BEFORE BEING ABLE TO USE AUTOPILOT ON ROADS

Training – similar media to design constraint number 25.

21.1.27 OLDER DRIVERS DO NOT LIKE TO CONSTANTLY MONITOR AUTOMATION FOR TAKEOVER (TIMER ONLY) TREND ONLY

Provide information of automation status and intention via HUD where possible.

21.1.28 DIFFERENCES BETWEEN USER PREFERENCE AND RANKINGS OF USEFULNESS

Performance is more important than preference, so need to make such that the take-over protocol elicits the best performance from the driver. Should not be irritating and frustrating though.

21.1.29 CHARACTERISTICS OF MODALITY

Capitalise on multimodality, such as presenting important information visually, auditorily, and haptically.

21.1.30 SYNCHRONISE MULTIMODAL CUES – COMBINING OR SINGLE MODALITY

Need to decide if multimodal cues are presented together or in an escalating manner. Suggest escalating for non-emergency situations and together for emergency situations.

21.1.31 LONGITUDINAL STUDIES

Drivers should be brought back into the simulator daily over the course of a week/ month to see how their behaviour adapts to the system. The average commuter drive lasts about 22 min on a motorway.

21.2 DESIGN GUIDELINES

In this section, the HI:DAVe team have summarised their main guidance and advice from the work undertaken into three classes of guideline: design methods, interface and interaction design, and user trials. These are presented in the three following subsections. Where appropriate, the chapters in this book are referenced for further information. Interpretation and use of the design guidelines need to be undertaken with care, as some of the guidelines might appear, at face value, to be contradictory or in conflict with each other. In these instances, it is strongly recommended that the reader go back to the source material in the relevant chapter.

21.2.1 DESIGN METHODOLOGICAL GUIDELINES (DMG)

Six guidelines were identified for design methodology, as follows.

> *DMG1: Always triangulate data from multiple sources when addressing the research question. Do not allow yourself to become over-reliant on one data source such as a survey, interview, or focus group.*
>
> **Supporting advice and evidence**: Both qualitative and quantitative sources of evidence are valuable. However, they are subject to different inherent biases and degrees of reliability. This makes them complimentary and allows meaningful comparison of data. For example, a survey addressed the public opinion of autonomous vehicles in the UCEID (user-centred ecological interface design) approach, but this can only represent a snapshot in time for a limited and non-stratified sample. Hence, focus groups were used to look at detailed individual and group responses to the same issues, and carrying out two allowed a technical focus in one and a general approach in another, with the same exemplar materials. These are weak samples but allow a detailed analysis of response to presented materials. Thematic analyses reflected the original attitudes of all these data sources and later allowed a detailed summation in the formative design process. Finally, the summation across sources, combined with an iterative design process and appreciation of the strengths and weaknesses of each method, allowed a methodologically sound convergence on the detailed concepts for quantitative hypothesis testing (Chapters 1–3, 5).
>
> *DMG2: Use a design workshop format which presents a divergent idea generation phase with a convergent summation and summarising phase in order to effectively traverse the design space.*
>
> **Supporting advice and evidence**: It has been well established that many formal and informal design processes fail to adequately sample the design solution space for a parameterised problem. Although some recommend a depth-first search that exhaustively pursues variants on one concept or a breadth-first search that aims to exhaust all concepts at a low-fidelity level before examining their variants, the result is a biased search. This may lead to non-optimal solutions or even design failures. In qualitative design, one mitigating approach is to initiate an initial prompt or encouragement to produce multiple solutions quickly, such as brainstorming. This is then followed by an elimination and pruning phase, using convergent techniques such as Pugh design matrices to filter the number of concepts to a useful set for prototyping. The chapters in this book dealing with the UCEID stages of

design show how this approach was successful in developing detailed HMI concepts for simulation and road testing (Chapters 1–3, 5).

DMG3: Preserve an iterative but rigorous design approach throughout multiple stages of design even when methods vary in timing and the nature of their outputs.

Supporting advice and evidence: It is inevitable during a multidisciplinary engineering project that a mixture of quantitative and qualitative methods will be used, and in some cases multiple different methods of both types. Good design requires a sequential progress through stages from initial design and requirements phases to concept generation to late prototypes. Although this sequential approach is both traditional and attractive for project planning, it is now well established that cascade design approaches lack flexibility and can lead to expensive prototype faults. It is generally accepted that iteration is required at each stage of design in order to capture the effect and improvements resulting from feedback from the evaluation of design alternatives. However, different methods require different time spans to complete, and quantitative methods are often prolonged and resource intensive. Hence, the results may be unavailable when project management initiates the next scheduled method. It is vital that the integration of different methods is managed in itself, and this should ensure that both concurrent and parallel information streams ensure effective continuity and review of the design by stakeholders (Chapters 1–3, 5–7).

DMG4: Document the outcomes of different stages in detail and store in an organised way with an index. This should record method outputs without bias of post hoc knowledge and prior expectations so that the progress of the design process can be accurately understood.

Supporting advice and evidence: The duration of a large project, such as this one, may span many years. During that time, a large amount of material in a variety of formats will accumulate, often on different sites. It is easy for information to become mislaid in conventional computer and physical filing systems, especially when stored on multiple sites. Method outputs can become incoherent and often disorderly as a result. It is therefore important to establish a master repository at an early stage with a coherent organisation that reflects project stages and specific methods, as well as key meetings and design decisions. Ultimately, experimental data should be stored in compliance with data protection regulation. The storage index should be easy to use, and a graphical diagram of storage organisation, including flowcharts and dates, is recommended (Chapters 1–3 and 5).

DMG5: It is essential to model the interactions between drivers and vehicle takeover technology as the prototype interfaces are being developed.

Supporting advice and evidence: Operators event sequence diagrams (OESDs) offer a valid way of modelling and predicting the interactions between the driver and the takeover technology in the vehicle. OESDs provide a framework for identifying all the agents (human and non-human) in a swim-lane format, together with their activities and interactions against a timeline. The OESDs are also a useful way to undertake co-evolution of the sociotechnical system (developing both the social (tasks of the driver) and technical subsystems at the same time – rather than the driver's task being defined by the technical system alone). Together, Chapters 8, 15, and 20 show that there was a good correspondence between the OESDs and the behaviour of the driver (above 80%). Modelling is also useful in the design of the interaction and interfaces, as it helps in the identification of design requirements provided that the modelling work is undertaken along-side the design phase. The modelling work informed the prototype development phases as it progressed from the lower-fidelity simulators through the higher-fidelity simulators and finally to the on-road prototypes. In this way, modelling and design with OESDs supported the agile development of takeover technology in the vehicle.

DMG6: The reliability and validity of methods used to model and predict automation–driver interaction needs to be formally tested.

Supporting advice and evidence: While the evidence based for using OESD is good (see Chapters 8, 15, and 20), this cannot always be assumed across tasks and domains of application. Reliability takes two main forms: stability over time (called 'inter-analyst reliability') and stability between people conducting the analysis (called 'intra-analyst reliability'). There are four types of validity for ergonomics methods: construct, content, concurrent, and predictive. Construct validity addressed the underlying theoretical basis of a method, whereas content validity addressed with the credibility that a method is likely to gain among its users. Concurrent and predictive validity address the extent to which an analysed performance is representative of the performance that might have been analysed. It is important that the methods possess a level of concurrent or predictive validity appropriate for their application.

21.2.2 Interface and Interaction Design Guidelines (IDG)

The following 36 guidelines on interface and interaction design were identified.

IDG1: Try to design for users of least skill, motivation, and capability. A good proxy for these users is to design for, and test with, the eldest adults, representative of the extremes of the user population to ensure acceptability to as many users as possible.

Supporting advice and evidence: This can be difficult as older adults beyond retirement age are usually not present within the easily accessible pool of volunteers available in the workplace. Older adults who tend to volunteer to take part in product evaluations tend to be more capable than the average, and older adults who are less capable than the average can be extremely difficult to recruit. Since the divergence between the most capable and the least capable in older adults is much wider than in younger adults, this problem is particularly acute. Recruiting participants through building long-term relationships with specific groups and individuals can help overcome this barrier.

IDG2: If designing interfaces for the untrained general population, then make the interfaces require as little prior knowledge and understanding as possible to reduce the exclusion caused by the cognitive demands of learning, recall, and operation.

Supporting advice and evidence: Do not hide task steps and functionality in obscurely labelled controls, nor provide status feedback in symbols which are difficult to see, unclear, or ambiguous in meaning. This can range from controls not being labelled for the functions they control, to byzantine symbols wholly inadequate at providing sufficient clue as to their function (an example of a highly exclusive interface being the Tesla Model S autopilot controls and displays).

IDG3: If designing interfaces for the untrained general population, or interfaces that are used infrequently, under unpredictable circumstances, then design them to require as little learning as possible.

Supporting advice and evidence: As a general principle, designers and engineers have substantially greater capability in most areas than the general population and are more interested and motivated to engage with technology. Consequently, many technologies require motivation, capability, and an engagement level that is not congruent with many potential users in the population. Do not require users to have a detailed understanding of mode-specific behaviours that are not easily predictable. For example, steering assist systems cease to function under random conditions, without warning and without clear mode status feedback.

IDG4: Design the takeover to accommodate a large range of individual differences in performance rather than the 'average' performance (such as mean, median, or mode performer).

Supporting advice and evidence: Takeover times have shown that there are large differences in the performance of drivers, ranging from a few seconds to nearly a minute (Chapters 5, 7, 12, and 17). This could be even longer in non-experimental situations, especially if the driver has been out of the driving control loop for some time and focusing on non-driving tasks. Allowing the driver to be in control of the pace of the takeover to suit their temperament and situation is generally preferred by drivers. Designing the takeover so that it is paced by the driver, rather than paced by the automated system, also seems to result in greater manual driving stability immediately following the takeover.

IDG5: Avoid mode confusion by limiting the possible number of states the system can be in and making each state unambiguous to the driver at all times.

Supporting advice and evidence: Mode confusion has been implicated in a number of collisions and fatalities associated with automated vehicles. Simplification of the number of modes is important to help the drivers' understanding of which mode the system is in. Ideally, there should be just two modes: human driver in control and vehicle automation in control. In the Jaguar Land Rover (JLR) simulation study and the on-road study, only these two modes were present. The displays (centre console, instrument cluster, and HUD) contained the information of either 'The Car is in

Control' or 'The Driver is in Control' together with an accompanying icon and blue or orange ambient lighting (where blue indicated that the vehicle was being controlled by automation and orange indicated that the car was being controlled manually by the driver). A clear indication was given to the driver that a mode change could occur. The handover protocol from manual to automated driving and takeover from automated to manual driving is described further in Chapters 15 and 20.

IDG6: Provide highly explicit mode awareness capability.

Supporting advice and evidence: Drivers who are using SAE Level 3 or above autonomous vehicles are going to be attending to other activities, some of which will be highly engaging and distracting. Enabling them to be reassured and check the mode that the vehicle is in, without undue effort will be important to minimising mode error, but also will help them to take advantage of the autonomy benefit – being able to focus on other tasks and activities.

IDG7: Before handing the control of the vehicle to a human driver, the automated system needs to raise the awareness of the driver to relevant situational and contextual features of the road environment, infrastructure, and other road users.

Supporting advice and evidence: Bringing the human driver back into the vehicle control loop is not simply a matter of handing over the controls, but also one of raising their situation awareness to prepare them for the driving task. In Level 3 vehicles, the driver could be undertaking non-driving tasks, so they need to be brought back into the driving task in a gradual manner. Driving comprises many tasks, including, but not limited to, hazard detection and avoidance, and vehicle guidance and navigation. The driver will need to be appraised of the situation by the automated vehicle, so the information transfer prior to driving is likely to include the road environment, weather conditions, behaviour of other vehicles, hazards on the road, status of their own vehicle, as well as immediate actions required, such as vehicle navigation and guidance.

IDG8: Allow driver to take over manual control of the vehicle at any time during the takeover process.

Supporting advice and evidence: In all of the takeover studies reported during the simulator and on-road trials, some drivers occasionally took manual control over the vehicle before the takeover process was complete (Chapters 8, 15, and 20). It is important that the human driver is in control of the takeover, both in terms of the pacing of the takeover and in terms of making the decision to cut the takeover process short if they are ready to take control early. This may be due to many factors – such as they were already monitoring the road prior to the takeover process or the road environment is relatively simple without any obvious hazards. The important design point is to put the driver in charge of the takeover process and not to unnecessarily impede their progress through it.

IDG9: Provide the ability for the driver to take control from a vehicle in automated mode, quickly, and in a manner that is familiar to them.

Supporting advice and evidence: Many drivers will have years of experience driving non-automated vehicles. In a stressful situation, when they perceive they need to take control, they are likely to revert to manual driving behaviours to do so. Therefore, steering wheel inputs should be able to overcome the lateral automation, and brake pedal inputs to overcome longitudinal automation. There is an argument that in this scenario, the vehicle should evaluate the impact of the intervention and moderate the inputs to the systems to avoid dangerous manoeuvres, but this has not been tested in this project.

IDG10: When designing new icons for in-vehicle displays (of which there is no ISO standard example), it is important that design is user-led, user-tested, and iterative.

Supporting advice and evidence: It is shown in Chapter 11 that when new icons are designed, original equipment manufacturers (OEMs) may come to different design decisions, creating inconsistency between vehicles which can cause confusion or safety issues. Therefore, new icons should be designed using the existing human factors methodologies. Where possible, they should be of the pictorial or concrete type and avoid the more abstract which are hard to learn. Designers should start by using the target population to help develop candidate icons and, once refined, these should be evaluated using the ISO 9186 test for graphical symbols by potential users from varying countries and cultures. If the icons fail the test, they should be redesigned, and the process repeated until they pass. This will ensure they are simple clear, concise, and universally understood.

IDG11: Customisation of multimodal interface settings is worthy of consideration as a method to help enhance drivers' subjective experience of takeover.

Supporting advice and evidence: Interface settings customised by drivers led to enhancement in subjective evaluation of takeover experience, as described in Chapters 13 and 18. Drivers need to take control back whenever requested by the automation, when the system reaches its limits in highly automated driving. The performance could be influenced by the drivers' condition, as well as the environment. Considering the individual differences and situational variations, a one-size-fits-all approach cannot meet their varying needs. In order to strengthen the drivers' positive experience, offering customisable interfaces seemed to be beneficial, as verified in the both sets of studies. Enabling drivers to tailor the interface settings (such as places, amounts, and intensities of information) is advisable. This is supported by the results of the studies that presented a reduction in workload, and improvement in usability, acceptance, and trust.

IDG12: Simplicity and clarity for presentation and delivery of information are recommendable to better assist drivers during the takeover process.

Supporting advice and evidence: In highly automated driving situations, takeovers can take place when drivers are engaged in non-driving-related tasks, which could induce a slower reaction to a takeover request. Thus, a rapid reconstruction of situation awareness is required for safe and timely takeover. Therefore, vehicle interfaces need to facilitate information seeking, retrieval, and processing procedure effectively. As presented in Chapter 18, a clear and concise presentation of information regarding takeover requests (such as automation mode, actions to take, and remaining time to takeover), located in appropriate places, was seen beneficial to enhance user experience since it may have helped drivers to find essential information for their decision-making process quickly and easily. This in turn seemed to be able to lead to a decrease in workload, and an increase in usability and acceptance, which is associated with intention to use the system.

IDG13: Multimodal alerts should be customisable but always with a safety-driven baseline.

Supporting advice and evidence: Multimodal alerts are more effective than unimodal but too many modalities can be annoying or even cause cognitive overload, resulting in a detrimental driving effect. Chapter 16 showed that the drivers in our studies each had unique profiles of what modalities they relied on most in the vehicle; therefore, it is difficult to predict which set-up would be most useful to any user group. It was also shown that during the on-road study, the majority of drivers relied on the traditional dashboard the most for information relating to the automation system. In order to increase safety and the positive user experience, in-vehicle information systems should be customisable by the driver. However, there should remain a safety-driven baseline which is always visible, and it would seem reasonable for this information to remain on the dashboard.

IDG14: Design for conflicting intentions between driver and automation.

For example, when the driver begins or ends a manual manoeuvre, there needs to be a means for this to be communicated to the existing assistive technologies so alternate models of autonomous behaviour can be activated, or where necessary, autonomous assistance is disabled until the manual manoeuvre is complete (Chapter 9 – case study 1).

IDG15: Design for reassurance.

Supporting advice and evidence: Such as when an atypical change in longitudinal or lateral locomotion is initiated by the automation, the action and its reason needs to be successfully communicated to the driver to avoid unnecessary or potentially risky manual intervention by the driver. Visual dashboard displays may be insufficient if the driver is focused 'eyes out' of the windscreen, observing potential hazards, so HUDs or alerts may be more appropriate. Visual and vocal displays should be used to convey semantics to the driver to support cognitive processing and decision selection (Chapter 9 – case study 1).

IDG16: Design for appropriate mental models.

By ensuring the function of any assistive automation is effectively understood by the driver and does not conflict directly with their existing mental models, to promote a positive transfer of knowledge. Without an appropriate schema or mental model of automation function, the predictability of vehicle behaviour and appropriate human response is diminished. At worst, safety is compromised, and at best, user trust and commercial adoption is diminished (Chapter 9 – case studies 1 and 2).

IDG17: Design for context.

Supporting advice and evidence: For example, recognise that the expected behaviour by a car when following a large vehicle will differ from that when following a standard vehicle. Interaction design should either appropriately set the drivers' expectations in a range of contexts, or the automation should be programmed to adjust to differing expectations based on context (Chapter 9 – case study 1).

IDG18: Design for the variety of human capabilities and the variety of environmental conditions of use.

Supporting advice and evidence: The interface elements should be designed to accommodate the visual, hearing, cognitive, reach, and dexterity capabilities of the potential user population. Where automation sensors (radar etc.) experience limitations in sensing the road, drivers also have limitations in sensing the low contrast or diminutive display icons in non-ideal weather conditions (e.g. bright sunshine). Digital displays provide the flexibility to offer redundancy in presentation, and traditional ergonomic guidelines should be adhered to and tested against the variety of relevant capabilities in the user population (Chapter 9 – case study 3).

IDG19: Design an adequately prominent mode state.

Supporting advice and evidence: While occurring from different causes in different contexts and relating primarily to different triads of the Perceptual Cycle Model, each case study ultimately resulted in mode errors. This supports the extensive body of literature investigating human interaction with automated systems. The importance of an explicit mode state, and feedback which is detectable and interpretable by the user population's capabilities when the modes changes, cannot be understated (Chapter 9 – case studies 1–3).

IDG20: Disambiguate controls of different functions.

Supporting advice and evidence: Disambiguate two very different actions by changing the design of controls – e.g. by using different shape and activation of controls, or by positioning controls in distinct locations, or requiring different hands or methods of activation (Chapter 9 – case study 3).

IDG21: Provide situation awareness guidance in interaction design to support transfer of control between automation and user.

Supporting advice and evidence: This provides a better user experience in terms of workload and usability for users of all ages, than interfaces that omit this guidance, by pacing the takeover task and guiding attention and information seeking to ensure the user is 'in the loop' and ready for safe manual driving (Chapter 6).

IDG22: Levels of feed-forward takeover request information can be increased without negatively affecting performance post-takeover.

Supporting advice and evidence: Chapter 7 examined the performance effects of increasing the levels of feed-forward information to participants using multiple modalities. The findings suggested that increasing information levels resulted in a minimal impact on post-takeover performance. This additional information may, however, be advantageous in raising situation awareness, readiness, and confidence prior to takeover, with potential safety benefits. It should be noted, however, that there will be an upper bound in terms of information density, beyond which distraction and overload may occur. The experiment outlined in Chapter 7 examined HMIs of varying levels of information – the levels assessed showed no detrimental effects that might be associated with distraction or increased workload.

IDG23: Multimodal HMIs tend to be associated with improved post-takeover performance when compared to unimodal HMIs.

Supporting advice and evidence: Examination of four HMIs of varying levels of information in Chapter 7 showed that the use of a unimodal HMI results in a lower level of performance compared with multimodal HMIs. Participants' performance appeared to exhibit a less smooth and considered takeover when using a unimodal HMI; this could be due to a lower level of situation awareness, or a lower level of confidence and readiness.

IDG24: All takeover request modalities should be customisable from a performance viewpoint.

Supporting advice and evidence: Both the on-road study data (Chapter 17) and the JLR simulator study (Chapter 12) suggested that customisation has no negative effects on performance post-takeover. Indeed, it may improve performance through enabling drivers to match HMI parameters to their needs, e.g. by adjusting volumes or haptic amplitudes based on sensitivity, thus reducing workload or startle response. Customised HMIs also appeared to improve readiness, perhaps through the preferred placement of only relevant information. There are a range of additional benefits conferred by customisation relating to more subjective aspects such as aesthetics, which are potentially beneficial with respect to marketing a vehicle. However, these also may convey some performance benefits by increasing confidence – for example, drivers are more likely to be happier and familiar with their HMI configuration.

IDG25: A method of indicating the required level of throttle could enable smoother takeovers.

Supporting advice and evidence: Performance data from all studies (Chapters 7, 12, and 17) revealed that immediately post-takeover, vehicle speeds slightly decreased. This was due to participants not knowing the level of throttle required at takeover. The JLR simulation study (Chapter 12) and the on-road study (Chapter 17) indicated that, when using their customised HMIs, participants allowed the vehicle to decelerate less, perhaps due to enhanced readiness or confidence. While this is a positive finding, these decelerations were still occurring. In order to enable the smoothest takeovers, and further improve safety, a calibration method might be considered, allowing drivers to know the required throttle level, and to apply it, prior to switching to manual control. This may further enhance the situation awareness and confidence of drivers.

IDG26: Additional aspects of customisation could be explored.

Supporting advice and evidence: The HMI elements that provided scope for customisation included the adding and removing of information, changes in levels of protocol questions, and adjustments in volumes and amplitude. In the majority of cases, participants selected a range of different settings. The resultant minor improvements in performance suggest that other aspects of the HMI might also benefit from customisation. These might include HMI brightness levels, finer control over position of elements, and text size adjustments. This increases the complexity of the customisation; however, this might be offset by the use of profiles, each of which is customised by the driver for a given circumstance. For example, scenarios that might benefit from profiles could include night-time driving or driving with children in the vehicle.

IDG27: A suitable default set of parameters for all modalities should be carefully calculated.

Supporting advice and evidence: It is useful to have a default set of parameters that provide a holistic initial HMI with parameters high enough enabling the majority of users to experience them. A balance should be found between parameters being too high and potentially startling or irritating drivers, and parameters being too low that they would not feel/hear them or benefit from them. Ideally, drivers should be able to identify their ideal settings with a minimal number of iterations. It should also be noted that the majority of participants chose to customise their interface when provided with the opportunity; however, for the few who did not, it is vital that the default settings provide the necessary information with parameters that will provide a high level of safety. The default settings used in the on-road study (Chapter 17) represent a good starting point, as, despite the majority of participants embracing customisation, the post-takeover performance was very similar when comparing default with customised. It is likely to be important to consider the demographics of those who are reluctant to customise; if there is a correlation, then it may be beneficial to bias the parameters to suit them, providing the settings are still effective and safe for the majority of drivers.

IDG28: Encouragement to customise.

Supporting advice and evidence: While the majority of participants carried out multiple customisations, it was apparent that there were some who preferred to use the default settings. The minor improvements in performance, due to readiness and confidence, combined with additional usability benefits (Chapters 12 and 17), suggest that it might be advantageous to prompt drivers to customise their interface at relevant times – e.g. prior to driving late at night when the driver may be tired, or prior to rush hour when traffic congestion is likely to be higher. Customisations to increase the level of alert and/or the level of information could result in higher awareness and improved safety through readiness.

IDG29: Make customisation available for the takeover interface.

Supporting advice and evidence: Customisation allows drivers to tailor the interface based on their preferences, wants, and needs. It has been applied in many areas in the car in order to accommodate the wide range of drivers' physical characteristics and mental capabilities. This inter-driver variability was also the main motivation to offer customisation for a range of takeover interface settings. For instance, being able to change the number of questions allowed older participants who took longer for the takeover to get more support. The high number of unique customisation profiles (above 80%), observed in both studies looking into customisable takeover interfaces (Chapters 14 and 19), further demonstrates the individuality of each driver. The questionnaire results in Chapter 14 show that people appreciated the possibility to make interface changes. Some even stated the desire for more customisation. In summary, enabling drivers to customise their takeover interface ensures not only inclusivity but also satisfaction and comfort.

IDG30: Make takeover interactions efficient, effective, and short.

Supporting advice and evidence: The takeover from autonomous to manual mode requires substantial effort from the driver, physically and mentally. Building up situation awareness quickly ensures a safe transfer of control. The response-based concept developed throughout the HI:DAVe project has proved to be able to do so. However, while situation awareness and safety need to be ensured, drivers prefer an efficient, effective, and quick interaction. The conducted customisation studies (Chapters 14 and 19) clearly show that for such a response-based, takeover interface, drivers prefer to have three questions instead of the five pre-selected as default. Since our studies showed that three questions enable takeovers as safe as the default number of five (Chapters 12 and 15), the default setting should be set to three in future. In addition, verbal feedback given by drivers underlined that they clearly see the benefit of a response-based approach but consider three questions to be the more appropriate default setting.

IDG31: Present important takeover information on the instrument cluster.

Supporting advice and evidence: The instrument cluster was introduced in cars at an early stage, and drivers have familiarised themselves with and adapted to it over time. Therefore, it is recommended to keep the instrument cluster an essential part of the takeover interface. The customisation studies presented in Chapters 14 and 19 showed that drivers kept the instrument cluster the densest interface and relied on it the most when looking for specific information. Furthermore, some drivers reported to prefer less information on the HUD to not be diverted from the road. Although the application of HUDs has expanded and will continue to do so, it seems like most drivers have not yet got used to it and potentially only want essential driving information on it. Until then, the instrument cluster remains an essential part of a takeover interface, which conveys the majority of driving-related information.

IDG32: Support and implement context-aware and driver-aware adaptation in the takeover interface.

Supporting advice and evidence: Adaptation, in contrast to customisation, does not require active involvement of the driver and would thereby allow context- and driver-aware changes of the interface. This is particularly helpful for drivers who tend to stick to factory settings and therefore miss out on comfort and safety benefits. In the questionnaire of the simulator customisation study (Chapter 14), several participants mentioned that they would like more questions and a stronger seat vibration for longer automated driving periods due to potential sleepiness, giving one potential factor for adaptation. A further factor could be driver state. For example, if a driver is detected to become drowsy, a takeover is initiated to refocus their attention and keep drivers awake. Particularly for longer periods of automated driving, research has suggested to insert manual driving to keep decent alertness levels. Overall, by taking into account contextual and driver-related information and altering the system in the most suitable way, we can achieve the maximum performance, comfort, and safety.

IDG33: Create driver profiles to support adaptation and customisation.

Supporting advice and evidence: Driver profiles contain driver characteristics and preferences, enabling the utilisation of more tailored adaptation and customisation, along with the possibility to allocate drivers to specific driver clusters. Chapters 14 and 19 calculated clusters for drivers' preferences regarding the response-based takeover interface which can be used to improve predictive algorithms (see Chapter 19) and to contribute to an initial customisation profile. These profiles need to be multidimensional and consequently need to be extended beyond drivers' takeover interface preferences. In future, these profiles can not only be used to adapt drivers' own car but also other cars of the same OEM. This will become particularly important on the way to mobility as a service. In this way, customers can get a pre-set car featuring their preferred settings and not the factory settings in all cars of a specific OEM, giving an impression of individuality, ownership, and personal character.

IDG34: Use haptic feedback and audio sounds for the takeover interface.

Supporting advice and evidence: Haptic feedback and audio sounds have been successfully implemented in many areas; in automotive studies, they have shown to reduce reaction time and cognitive load. In the customisation studies, simulator (Chapter 14) and on-road (Chapter 19), the majority of drivers had the seat vibration and audio sounds activated since they considered it a salient approach, raising their attention and alertness. In the simulator study, several drivers mentioned the desire to have a stronger haptic feedback and louder audio sounds for longer journeys of automated driving. Moreover, the questionnaire (Chapter 14) revealed that drivers relied heavily on audio sounds for the takeover process. Their great benefit is that they do not require the driver to spare visual resources and take the eyes off the road at the point of takeover while conveying a specific message. The possibility to easily adapt their intensity based on contextual and driver-related factors is another advantage.

IDG35: Make use of vocalisations for the takeover interface.

Supporting advice and evidence: Vocalisations are powerful in the sense that they can convey a significant amount of information without requiring the driver to take the eyes off the road. While they have been widely used in SATNAV applications, their use has been limited in many other in-car areas. In the simulator customisation study (Chapter 14), drivers reported to rely on vocalisations on the same level as the instrument cluster, underlining the importance of vocalisations in the developed takeover interface concept. In both customisation studies (Chapters 14 and 19), drivers reported that vocalisations were a key factor in supporting their decision-making process regarding their next activity in the takeover interaction. The good prediction performance of the applied OESDs (Chapters 8, 15, and 20), above 80%, clearly supports drivers' reports in the sense that the intended interaction was performed by drivers.

IDG36: Update driver profiles continuously.

Supporting advice and evidence: Although the application of static driver profiles already adds many benefits to drivers' usage of in-car systems, the results in the customisation studies (Chapters 14 and 19) indicate that a continuous data collection, analysis, and integration is crucial for a long-term success. When comparing drivers' preferences in the simulator and on-road studies, a high intra-driver variability was identified, indicating the challenge of transferring inferences from one context to the other. As a consequence, the characteristics of the takeover interface need to continuously be updated with context-, driver-, and interface-related information to always enable the most suitable layout. Since a similarly high intra-driver variability can be assumed across other dimensions as well, such as adaptive cruise control interface and setting preferences, it can be recommended to continuously update all dimensions of the driver profile. Only an accurate profile can ensure appropriate and successful interface adaptation, overcoming the challenge of intra-driver variability.

21.2.3   USER TRIALS GUIDELINES (UTG)

Five guidelines for user trials were identified, as follows.

UTG1: Conduct studies with large samples of drivers with a wide range of ages and abilities (balanced for age and gender).

Supporting advice and evidence: The studies conducted in this research have used reasonably large samples. The desktop simulator study (Cambridge) had 49 drivers, the full vehicle simulator study (Southampton) had 60 drivers, and the final simulator study (JLR) had 66 drivers. The on-road study had planned to have 66 drivers also (those same drivers from the JLR simulator study). Unfortunately, due to delays and technical difficulties, only 24 drivers were actually used in the on-road study, but a large sample is strongly recommended for any follow-on studies. Large sample sizes are recommended due to the large individual differences in the driving population, in terms of skills, experience, training, and so on. Understanding the range of performance is much more important than understanding the average performance, which is why the data is often represented using box-and-whisker plots (as these show the minimum, interquartile range, median, mean, and maximum responses as well as the outliers). It is also important to balance for gender as the male–female driving license holders is almost equal (currently 54% are male). In all of these studies, three age populations were identified as younger drivers (18–34 years), middle-aged drivers (35–56 years), and older drivers (57–82 years). It is important to sample from all three groups (ideally balancing number of drivers) as there will be differences in driving experience, and cognitive and physical functioning.

UTG2: Measuring drivers' perceived workload provides useful insights into how human–machine interaction in SAE Level 2 autonomous driving system could be developed.

Supporting advice and evidence: The NASA-Task Load Index is a robust tool to assess drivers' perceived workload as a non-intrusive method that does not affect their performance while driving. Moreover, the rated scores can be used for comparison of the workload experienced in the situations that the operators were performing the tasks. In Chapter 10, evidence suggested that drivers' perceived workload was higher in SAE Level 2 automated driving than in manual driving. The discrepancy in workload between the driving modes shows room for improvement of the automated system. It is suggested to enable a clear communication between driver and autonomous vehicle in terms of automation mode status, and the systems' limits in which manual interventions are required. Monitoring the automated system and the intermittent interventions which are not required in manual driving could induce workload. Thus, the interface should be designed to be used easily, and the information needs to be understood intuitively to help relieve workload. This in turn could help facilitate drivers' decision-making process about when, and to which degree, to take back control, or to engage automation safely.

UTG3: User trials with differing capabilities are worthy of investigation to identify how to develop vehicle interfaces that could better support user experience.

Supporting advice and evidence: Drivers' perceived workload and acceptance could be influenced by the differing levels of prior experience of automated driving. In Chapter 10, experienced drivers rated lower workload and higher acceptance than novice drivers. The experienced drivers were able to build sufficient situation awareness more rapidly and have better projections about the system's behaviours in response to traffic conditions. This could lead to timely and effective takeover of control. It was based on the understanding about the automation's capabilities and limits, and knowledge about how to react appropriately. It was enabled by mental model developed through the experience of automated driving. In this sense, the interfaces should be able to support drivers to stimulate mental model construction and situation awareness effectively through clear visualisation automation limitations, availability, and alert information settings. This approach seems necessary because we cannot always expect the drivers to be experienced and knowledgeable about the system. This attempt is ultimately to elicit effective driver–autonomous vehicle interaction that could result in a reduction in workload and an increase in acceptance.

UTG4: Age-related limitations need to be considered when designing driver interaction during takeover situations and the interfaces.

Supporting advice and evidence: In Chapter 18, older drivers' perceived workload was higher than younger drivers' in the both first and second trials. Furthermore, mental and physical demands of older drivers were higher in the second trial, while those of younger drivers were lower in the second trial. This may have resulted from fatigue after a long duration of autonomous and manual driving, whose impact was greater on the older drivers. Also, this may have been contributed to older drivers' age-related degradation in physical and cognitive capacities. Therefore, their limitations need to be considered when designing interfaces to better support their performance and experience in takeover situations. Designing interfaces that can assist older drivers to focus better on main tasks without distractions and information overload seems advisable.

UTG5: Train drivers in the use of vehicle automation.

Supporting advice and evidence: In all of the studies conducted in the course of the research, we trained drivers in the use of the vehicle automation. This training included an explanation and demonstration of the limitations of the automated vehicle functions. Simulators provide an ideal training environment as it is possible to present situations to drivers that would be dangerous on real roads. In the on-road study, drivers had already taken part in the simulator study, so were familiar with the operation of the system. In addition, the drivers had the system demonstrated to them on a test track and drove around the test track themselves with the system operating. They were able to experience the vehicle dropping out of automated control and practice recovering manual control. Only when the drivers were comfortable with the operation of the vehicle did the road trials begin.

ACKNOWLEDGEMENTS

This work was supported by Jaguar Land Rover and the UK-EPSRC Grant EP/N011899/1 as part of the jointly funded 'Towards Autonomy: Smart and Connected Control (TASCC) Programme' HI:DAVe Project.

Author Index

Note: **Bold** page numbers refer to tables; *italic* page numbers refer to figures.

Subject Index

Note: **Bold** page numbers refer to tables and *italic* page numbers refer to figures.

Inclusivity Index

Note: **Bold** page numbers refer to tables and *italic* page numbers refer to figures.

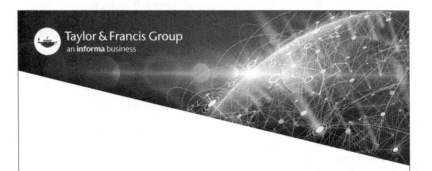